Cyanobacteria

Omics and Manipulation

Edited by

Dmitry A. Los

Institute of Plant Physiology
Russian Academy of Sciences
Moscow
Russia

Caister Academic Press

Caister Academic Press
Norfolk, UK

www.caister.com

British Library Cataloguing-in-Publication Data
A catalogue record for this book is available from the British Library

ISBN: 978-1-910190-55-5 (paperback)
ISBN: 978-1-910190-56-2 (ebook)

Cover design adapted from photograph of *Synechocystis* cells kindly provided by Dr Alexander Voronkov and Dr Maria Sinetova (Institute of Plant Physiology, Russian Academy of Sciences, Moscow, Russia).

Contents

Preface

Cyanobacteria are represented by a diverse group of microorganisms that, being a part of marine and freshwater phytoplankton, significantly contribute to the planetary fixation of atmospheric carbon and evolution of molecular oxygen via photosynthesis. Ancient cyanobacteria (~2.5 billion years of history) participated in the formation of Earth's oil deposits. Modern cyanobacteria grow fast; they do not compete for agricultural lands and resources; they efficiently convert excessive amounts of CO_2 into biomass, thus participating in both carbon fixation and organic chemical production. Many cyanobacterial species are easily transformable and, thus, may be genetically manipulated to produce photosynthetic carbohydrates, fatty acids, or alcohols as renewable sources of fourth-generation biofuels. Genetic modification of strains is a powerful tool to redirect the biosynthetic pathways of cyanobacteria to desirable end-products, including those that have never being produced by these organisms. In addition, cyanobacteria are studied and used as a rich source of bioactive metabolites with unique structural features and biological activities, including antiviral and antibacterial agents, cytotoxins, antioxidants, and other bioactive compounds.

In this volume, the reader will find a collection of chapters devoted to cyanobacterial omics (genomic, transcriptomics, proteomics, etc.) targeted for understanding the basic principles of cyanobacterial metabolism. This fundamental knowledge is then converted into metabolic engineering in order to produce valuable compounds, e.g. pharmaceuticals, biofuels, bioplastics, etc. Such a systemic approach fits well to the concept of the 'Green Planet', which implies the sustainable development on the basis of green (photosynthetic bacteria, plants, algae, cyanobacteria) technologies that produce renewable and clean foods, energy, and materials.

The international team of authors would like to bring the attention of the readers to the latest achievements in biology of cyanobacteria – photosynthetic microorganisms with great academic and industrial potential.

The Cyanobacterial Core Genome: Global and Specific Features with a Focus on Secondary Metabolites

1

Stefan Simm[1], Enrico Schleiff[1,2,3*] and Rafael Pernil[1]

[1]Department of Biosciences, Molecular Cell Biology of Plants, Goethe University, Frankfurt am Main, Germany.
[2]Cluster of Excellence Frankfurt, Frankfurt am Main, Germany.
[3]Buchmann Institute for Molecular Life Sciences (BMLS), Goethe University, Frankfurt am Main, Germany.

*Correspondence: schleiff@bio.uni-frankfurt.de

Abstract

The technological progress done in next generation sequencing (NGS) has enabled to perform genome analyses of environments to understand microbial communities through metagenomics and, by this, has increased the variety of available sequenced genomes. The high number of DNA sequencing genome projects in the last years has created an archive of ~20,000 fully sequenced prokaryotic genomes that has influenced the current research. By this, multi-genome comparison approaches like pangenome analysis become of major importance to boost the understanding of the complexity of biological systems. Such analyses range from the definition of taxa composition in specific environments by metagenomics, investigations of evolutionary relations and assignments of core- and pangenomes to the definition of core gene sets, and even the correlation of specific gene sets for biological functions, phenotypes or traits in cyanobacteria. In this regard, the challenge to handle and use appropriately the increasing amount of information is of major importance. For this reason, data handling, analysis and visualization will be described in this chapter from a bioinformatics point of view. The following sections will explain in detail the general applications of metagenomics and pangenomics and their benefits on the cyanobacteria research field. Metagenomics was used in the past to analyse the taxonomic composition and functional characteristics of cyanobacterial communities in different habitats such as oceans or environments with extreme temperatures or salt concentrations. Furthermore, differences between cultured and wild-living cyanobacteria and their acquisition and loss of genes in relation to changes in the availability of nutrients or metabolites can be investigated. In addition, new genomic elements or specific pathways for secondary metabolites could be also identified and biotechnologically used in the future, like the biosynthesis pathway of the anticancer agent curacin A (Jones *et al.*, 2011).

In the past, pangenome studies have been performed focusing on the identification of intraspecies gene sets. Similarly, pangenomes can be used to obtain evolutionary information by identifying the core-genome of interspecies comparisons in cyanobacteria and get insights into the acquisition and loss of genes within a specific clade or phylum. Furthermore, the core-genome plays a role in the identification of so-called signature genes that only occur in all or most of cyanobacteria. Pangenomics for cyanobacteria has helped to define the origin of photosynthesis and to analyse the diversity of metabolism. Furthermore, pan- and core-genome analyses have been used to identify trait-specific gene sets. In this context, cyanobacteria were clustered regarding their different traits to analyse subfractions of the pan- and core-genome of a specific environmental adaptation. For the correlation of pangenomics analysis and functional annotation of cyanobacteria, discrimination in core-genome and clade-specific functional annotation and determination of specific genes for a subset of cyanobacterial strains with either thermophilic character, growth habitat or the capability to differentiate heterocyst were analysed. In addition, the detection of core- and clade-specific functions in cyanobacteria via multi-genome comparison analyses is explained and exemplified using the gene sets of non-ribosomal peptide synthetases (NRPS) and polyketide synthases (PKS).

Cyanobacteria are a prolific source of natural products and produce a vast array of compounds, including many notorious toxins as well as natural products of huge interest to pharmaceutical and biotechnological industries. Genome mining has enabled the identification and characterization of natural product gene clusters, and mechanisms that are unique to cyanobacteria, or rarely seen in other organisms, have been discovered. Many cyanobacterial secondary metabolites are cyanotoxins, which show a broad range of chemical structures and biological activities, but in addition to toxin production, also several NRPS and PKS gene clusters are devoted to important cellular processes in cyanobacteria such as iron uptake and nitrogen fixation. Most of the biosynthetic clusters identified here have unknown end products, highlighting the power of genome mining for the discovery of new secondary metabolites. These studies show that cyanobacteria encode a huge variety of cryptic gene clusters involved in the production of natural products, and the known chemical diversity to date is likely to be only a fraction of the true biosynthetic capabilities of this fascinating and ancient group of organisms. Furthermore, mechanistic insights obtained from the biochemical studies of cyanobacterial pathways can inspire the development of concepts for the design of bioactive compounds by synthetic-biology approaches in the future. Here, we survey the biosynthetic pathways of the top five most researched cyanobacterial species with the most extensive literature, including *Microcystis aeruginosa* NIES-843, *Synechocystis* sp. PCC 6803, *Anabaena* sp. PCC 7120, *Arthrospira platensis* NIES-39 and *Synechococcus elongatus* PCC 7942. These analyses have enabled the identification of biosynthetic gene clusters for structurally diverse metabolites, including non-ribosomal peptides, polyketides, ribosomal peptides, terpenes and fatty acids. We highlight the unique enzyme mechanisms that were elucidated or can be anticipated for the individual products, but further include different classes of secondary metabolites from cyanobacteria other than NRPS and PKS.

Introduction

Current research is mainly influenced by the accelerated rate of DNA sequencing projects due to the revolution of NGS technologies (Koboldt *et al.*, 2013; Reuter *et al.*, 2015). Today, sequencing of full genomes, especially of prokaryotes, is achievable for modest research institutes and around ~20,000 complete genomes are available to date (http://www.ncbi.nlm.nih.gov/genome/). On the one hand, handling such an increasing amount of information represents a big challenge. On the other hand, the provided information boosts the understanding of the complexity of biological systems and enables a multi-genome comparison (Dai *et al.*, 2011; Hardison, 2003; Miller *et al.*, 2004; Wei *et al.*, 2002). Such analyses range from investigations of evolutionary relations, assignments of core- and pan-genomes of strains or species to define common gene sets, and up to the definition of the taxa composition extracted from a specific environment (Handelsman, 2004; Mira *et al.*, 2010).

The term 'metagenomics' (Table 1.1) was introduced by Handelsman and coworkers to analyse the microflora of soil (Handelsman *et al.*, 1998). It had its beginning by the characterization of the 16S rRNA sequence of microbes from an environmental sample (Ussery *et al.*, 2009). Today, metagenomics describes the analysis of entire communities of microbes without isolation or culturing of individual community members (Riesenfeld *et al.*, 2004) and is divided into two subcategories called 'full shotgun metagenomics' (Xia *et al.*, 2011) and 'marker gene amplification metagenomics' (Handelsman, 2009).

The first metagenomics studies were performed early in the past decade focusing on low-diversity environments like water samples of the Sargasso Sea (Venter *et al.*, 2004), acid mine drainage (Tyson *et al.*, 2004) or the human gut microbiome (Breitbart *et al.*, 2003). Today, this strategy is applied to the description of communities in very diverse environments like oceans (Bohannon, 2007), soils (Daniel, 2005), and the human mouth and gastrointestinal tract (Belda-Ferre *et al.*, 2012; Breitbart *et al.*, 2003; Segata *et al.*, 2012; Wang *et al.*, 2015).

Table 1.1 Terminology of metagenome analysis

Term	Definition
Metagenomics	Metagenomics (from the Greek term 'meta', meaning 'after' or 'beyond') is the culture-independent genomics analysis of microbial communities. Metagenomics describes the functional and sequence-based analysis of the collective microbial genomes contained in an environmental sample. Metagenomics transcends the individual organism to the 'meta level' of the community (Handelsman, 2004, 2009; Handelsman *et al.*, 1998).
Pangenome	The pangenome (from the Greek term 'pan', meaning 'all') is the 'union' of the 'gene sets' – both terms are taken from the set theory, a branch of mathematical logic – of all the selected genomes. The term pangenome was first used to define the gene sets of all strains of a species. It can be divided in a core-genome and a variable genome (Tettelin *et al.*, 2005).
Open pangenome	For open pangenomes an undetermined number of additional genomes is needed to identify the whole gene repertoire of the species within a phylum (Bentley, 2009).
Closed pangenome	Closed pangenomes grant the assumption that the entire gene repertoire is covered by the strains used and newly sequenced genomes will not give additional information (Rouli *et al.*, 2015).
Core genome	The core genome contains gene families shared by all the selected strains or species (intersection of gene families).
Variable genome	The variable genome contains strain-specific genes and gene families shared by two or more organisms, but not by the entirety of all analysed genomes.
Dispensable genome	The dispensable genome is the part of the variable genome that exists in 2^{n-1} organisms of the analysed set (Vernikos *et al.*, 2015).
Unique genes	Unique genes are genes of the variable genome which only exist in one strain (Medini *et al.*, 2005).
CLOG	CLique of Orthologous Genes (CLOG) defines a group of orthologous sequences from at least two different species (Simm *et al.*, 2015).
Signature genes	Signature genes are defined to be specific to a taxonomic rank, occurring exclusively in all or most of the members in a taxon (Dutilh *et al.*, 2008)

Here, analyses of known and unknown organisms by metagenomics lead to a community biodiversity profile (Oulas *et al.*, 2015). In addition, metagenomics has been applied to create fingerprints on specific environments. For instance, the communities in areas of volcanism (Kilias *et al.*, 2013; Urich *et al.*, 2014; Xie *et al.*, 2011), extreme temperature (Bradford *et al.*, 2009; Pearce *et al.*, 2012), alkalinity (Xiong *et al.*, 2012), acidity (Johnson *et al.*, 2015; Mendez-Garcia *et al.*, 2015), low oxygen (Stevens and Ulloa, 2008; Voorhies *et al.*, 2012) and high heavy metal composition (Chodak *et al.*, 2013; Golebiewski *et al.*, 2014) have been studied. Searching in extreme habitats can lead to the discovery of new species, but also of novel enzymes for catalysing reactions of biotechnological commercialization (Segata *et al.*, 2011). At present, technologies and bioinformatics strategies for metagenomics enable the investigation of multitude of existing microbial communities (Fig. 1.1) to reveal solutions for challenges in human health, alternative energy, environmental remediation, biotechnology and environmental stewardship, to name just a few (Bashir *et al.*, 2014; Handelsman, 2004, 2009; Lorenz and Eck, 2005).

In addition to metagenomics, pangenome (Fig. 1.1) analyses were performed to determine the entirety of the gene sets of strains, species or phyla (Vernikos *et al.*, 2015). The term 'pangenome' was introduced by Tettelin and coworkers to describe all sequences shared by genomes of interest (Tettelin *et al.*, 2005). Some recent reports use distinct terms such as supragenome or species-genome, which are all equivalent to the original term pangenome (Broadbent *et al.*, 2012). Pangenome analyses require a set of sequenced genomes with annotated genes and were used to identify the gene set of all strains of a species (Medini *et al.*, 2005; Mongodin *et al.*, 2013; Tettelin *et al.*, 2005; van Schaik *et al.*, 2010) or a genus (Jacobsen *et al.*, 2011; Kettler *et al.*, 2007; Lefebure and Stanhope, 2007), a specific set of species (Collins and Higgs, 2012; Eppinger *et al.*, 2011; Lapidus *et al.*, 2008), a whole phylum (Collingro *et al.*, 2011) and even a super kingdom (Lapierre and Gogarten, 2009). Recently, pangenome analysis was introduced to define gene sets

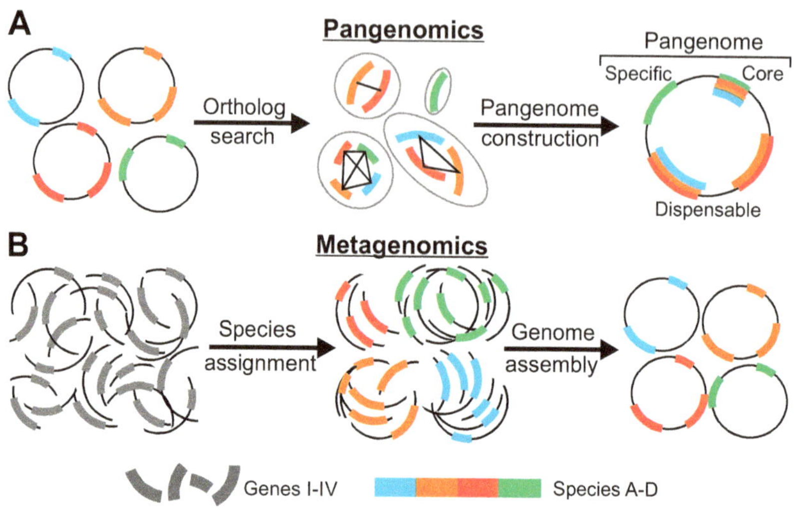

Figure 1.1 Multi-genome analysis. The main steps in pangenome (A) and metagenome (B) analysis are shown as a scheme. Circles stand for genomes of a specific strain/species and grey (unknown species) or coloured (different known species) bars indicate genes of the genome. (A) Pangenomics is defined by orthologue search to determine CLOGs, which are used for the classification in core, dispensable and specific CLOGs assigned to the pangenome. (B) Metagenomics assigns the reads and contigs to the correct species and the contigs are assembled to obtain the full genome.

of organisms based on common phenotypic and environmental features (Simm *et al.*, 2015).

A pangenome includes a core-genome, a dispensable genome as well as unique genes (Table 1.1) (Medini *et al.*, 2005). The core-genome includes the genes that are present in all species analysed (Medini *et al.*, 2005). In general, the majority of these genes contain housekeeping functions (Bentley, 2009). The dispensable genome is defined as the set of genes which occurs in an intersection of at least two genomes but not in all the genomes analysed, while the unique genes are genes which only occur in one genome. The dispensable genome and unique genes can be linked to a certain environmental adaptation based on morphology or physiology. Pangenomes can be open or closed (Table 1.1), which can be correlated to the lifestyle of the studied species or phyla (Beck *et al.*, 2012). Allopatric species, which usually have a small genome and exist in a small niche, more likely tend to have a closed pangenome due to their specialization. Open pangenomes are often found for sympatric species which are included in a community that share genes via horizontal gene transfer and tend to have large genomes.

In addition to the assignment to pan- and core-genomes, the term signature gene for species has been introduced (Table 1.1) (Martin *et al.*, 2003). Signature genes are defined to be specific to a taxonomic rank and are used for phylogenetic classification (Dutilh *et al.*, 2008). For unspecified samples, the occurrence of signature genes in various taxonomic depths gives insights into the composition of the sample (Dutilh *et al.*, 2008). Taxon signature genes of viruses can serve as a useful diagnostic indicator of viral presence but have to be conserved in all or at least most of the members of a particular taxon and further never or only very rarely be observed outside that taxon (Kristensen *et al.*, 2013). Furthermore, identification of signature genes can be used to add a biological function to so far unknown genes and proteins or define feature-dependent gene sets like for endosporulation (Abecasis *et al.*, 2013). In the case of cyanobacteria, core-genome and signature gene definition were used to define a set of genes specific for cyanobacteria against other species containing chloroplasts (Martin *et al.*, 2003) or clades of other bacteria (Gupta and Mathews, 2010). This approach has contributed to expand our knowledge on the origin of photosynthesis (Mulkidjanian *et al.*, 2006) and diversity of metabolism (Beck *et al.*, 2012). Interestingly, signature genes are not immune to horizontal gene transfer (HGT) and in fact may be favoured candidates for HGT among closely related cyanobacterial strains (Yerrapragada *et al.*, 2009).

The following sections focus on the applications of metagenome and pangenome approaches for cyanobacteria. Furthermore, the data handling and visualization from a bioinformatics point of view will be explained. In addition, the detection of core- and clade-specific functions in cyanobacteria via metagenome analysis is explained and exemplified using the gene set of general secondary metabolite pathways but emphasizing NRPSs and PKSs. Secondary metabolites are a huge family of compounds that likely confer an evolutionary advantage to the producer organism (Jenke-Kodama *et al.*, 2008; Metlen *et al.*, 2009). Although primary metabolic processes are found in all species of life, the pathways that lead to secondary metabolites are specialized, with particular pathways found in taxonomically restricted groups of microorganisms (Calteau *et al.*, 2014; Donadio *et al.*, 2007). Secondary metabolites exhibit enormous structural variation and consequently display a wide range of biological activities (O'Connor, 2015). Cyanobacteria are a prominent source of structurally diverse and biologically active secondary metabolites (Kehr *et al.*, 2011; Mejean and Ploux, 2013; Nunnery *et al.*, 2010; Singh *et al.*, 2005; Skulberg, 2000), which are produced by a broad range of enzymes responsible for methylations, oxidations, tailorings, and other alterations (Jones *et al.*, 2010). These enzymes are involved in the synthesis of chemically diverse natural products such as linear peptides (Simmons *et al.*, 2006), cyclic peptides (Sisay *et al.*, 2009), linear lipopeptides (Nogle *et al.*, 2001), depsipeptides (Han *et al.*, 2005), cyclic depsipeptides (Soria-Mercado *et al.*, 2009), fatty acid amides (Chang *et al.*, 2011), swinholides (Andrianasolo *et al.*, 2005), glycomacrolides (Teruya *et al.*, 2009), or macrolactones (Salvador *et al.*, 2010). The known chemical diversity of cyanobacterial natural products includes over 1100 secondary metabolites with intricate and exotic chemical structures reported from the genomes of 37 strains of cyanobacteria (Dittmann *et al.*, 2015), for which the major contributors are the genera *Lyngbya*, *Nostoc*, *Microsystis* and *Hapalosiphon*. Many of these cyanobacterial natural products and their biosynthetic pathways were originally identified and isolated through more traditional bioactivity-guided screening techniques and sequence analysis methods (Burja *et al.*, 2001; Micallef *et al.*, 2015), and have revealed unusual biosynthetic mechanisms that are solely in cyanobacteria or rarely described in other microorganisms (Dittmann *et al.*, 2015; Kehr *et al.*, 2011). Many of these secondary metabolites have a promising therapeutic potential, with anticancer, antibiotic, immunosuppressive, and antiviral bioactivities among others (Burja *et al.*, 2001; Demain and Sanchez, 2009; Fischbach, 2009; Salomon *et al.*, 2004; Welker and Von Döhren, 2006).

Cyanobacterial secondary metabolites are classified according to their biosynthetic origin (O'Connor, 2015) as non-ribosomal peptides, polyketides, ribosomal peptides, alkaloids, terpenes and fatty acids (Jones *et al.*, 2009; Kehr *et al.*, 2011; Micallef *et al.*, 2015; Welker and Von Döhren, 2006). Alternatively, they are defined according to their biological activity including cyanotoxins such as antibiotics, compounds involved in a range of cellular reactions and processes, signalling and stress-response molecules and UV-absorbing pigments (Dittmann *et al.*, 2015; Kehr *et al.*, 2011; Micallef *et al.*, 2015; Pattanaik and Lindberg, 2015). These classification systems, however, have their limitation because some compounds have building blocks from more than one biosynthetic pathway, some others that appear closely related can have completely different biosynthetic origins, and compounds with no related chemical structure can show similar biological activities (Mandal and Rath, 2015).

Peptide metabolites are built from amino acids using either the ribosome with further posttranslational modifications (Patton and van der Donk, 2005) or dedicated enzymes called NRPSs (Wenzel and Muller, 2005). Polyketides are synthesized by PKSs from acyl-CoA monomers, in common with fatty acid biosynthetic pathways, and are also subdivided into several classes (McDaniel *et al.*, 2005). Terpenes are derived from the five-carbon-containing precursors isopentyl pyrophosphate and dimethylallyl pyrophosphate and are synthesized by dedicated terpene synthases (Kitaoka *et al.*, 2015; Micallef *et al.*, 2015; Pattanaik and Lindberg, 2015). Furthermore, some carbohydrates synthesized from nucleoside diphosphate (NDP)-sugars by glycosyltransferases and sugar-tailoring enzymes are also classified as secondary metabolites (McCranie and Bachmann, 2014), and saccharides that are unique to the secondary metabolism often decorate the core structure of secondary metabolites from the classes described above (O'Connor, 2015). In this chapter, we describe the main classes of cyanobacterial secondary metabolites such as non-ribosomal and ribosomal peptides, polyketides, terpenes and lipids, but emphasize non-ribosomal peptides, polyketides or hybrid peptide–polyketide compounds because the majority of cyanobacterial secondary metabolites are synthesized by modular NRPS, PKS and hybrid NRPS-PKS biosynthetic pathways.

Recent advances in genomics are greatly accelerating our understanding of the biosynthetic logic and biochemical mechanisms underpinning the biosynthesis of cyanobacterial natural products (Dittmann *et al.*, 2015). In this context, genome sequence analyses and genome mining techniques give insights into the distribution of biosynthetic pathways and their evolution across species and genera (Micallef *et al.*, 2015), and leads to a better understanding of the physiological function of cyanotoxins and other secondary metabolites in cyanobacteria (Calteau *et al.*, 2014). Here, we search for secondary metabolite clusters in the genomes of the five most extensively studied cyanobacteria (*Microcystis aeruginosa* NIES-843, *Synechocystis* sp. PCC 6803, *Anabaena* sp. PCC 7120, *Arthrospira platensis* NIES-39 and *Synechococcus elongatus* PCC 7942) based on existing publications (www.webofscience.com), compare them to provide evolutionary insights into cyanobacterial secondary metabolite biosynthetic pathways and discuss the most abundant classes of secondary metabolites in this group of photosynthetic organisms.

Metagenome analysis and information in cyanobacterial research

The early metagenomics studies, which contributed to the identification of the steps in data processing, commonly present difficulties and obstacles. Particularly the variety of sampling methods and DNA extraction techniques represent a challenge for the standardization of metagenomics methods (Teeling and Glockner, 2012). In turn, one needs to consider that data quality obtained from metagenomics analysis of any community (Box 1.1) will be directly related to the quality of procedures for DNA extraction. Additionally, the selection of the laboratory strains for analysing the function of genes found in a metagenome is central and affects the value of information (Wooley *et al.*, 2010). Furthermore, procedures to determine the complete genome of individual community members, comparing the community diversity of various environments and assessing changes in the diversity thereof from the pooled sequence data, are still challenging and often not standardized or even not fully developed (Oulas *et al.*, 2015). Thus, improvements of culturing techniques, physical separation methods and bioinformatics tools along with the generation of complete genome sequences for model microorganisms (Box 1.1) will ensure that genome assemblies and interpretations of metagenomics sequence data become reliable (Gabor *et al.*, 2007).

One bioinformatics challenge of metagenomics is the assembly of reads based on more than one reference genome (Handelsman, 2009). Similarly, the binning of reads, contigs and scaffolds for different species or other taxonomic ranks and the annotation of the detected sequences has to be improved to handle more complex and comprehensive data (Thomas *et al.*, 2012). In addition, information about orphans and unknown sequences having perhaps a higher order of similarity like protein fold has to be taken into account of metagenomics analysis (Oulas *et al.*, 2015). Finally, tools for analysis, storage and visualization of metagenomics datasets need to be improved to allow answering fundamental questions like microbial ecology, evolution and diversity (Thomas *et al.*, 2012).

Cyanobacteria are key organisms in many ecosystems and become more and more interesting for biotechnological strategies like secondary metabolite production. Today, more than 100 complete cyanobacterial genomes are available (Simm *et al.*, 2015). In the last 5 years, metagenomics was used to describe cyanobacterial community composition in their natural environment. Here, the taxonomic composition and functional characteristics of marine cyanobacteria in biofilms (Leary *et al.*, 2014; Sanli *et al.*, 2015), in pelagic zones of the sea (Tseng *et al.*, 2015) or in coral reefs (Kwan *et al.*, 2012) were analysed by metagenomics and metaproteomics. Also microbiomes within species like the red seaweed *Laurencia dendroidea* were analysed (de Oliveira *et al.*, 2012)

Box 1.1 From data collection to analysis of metagenomes

DNA is first extracted from all organisms living in a particular environment, irrespectively whether they are assigned to the prokaryotic, archaeal or eukaryotic Domain. In the past, extracted DNA was fragmented, cloned and replicated in laboratory bacterial strains (Handelsman, 2004) but, at present, technologies facilitate the study of a community's DNA directly (Gabor *et al.*, 2007). The DNA is then sequenced by next generation technologies (Thomas *et al.*, 2012). The obtained reads are assembled to contigs, scaffolds or entire genomes based on coverage of the multiple genomes. This can be done with reference-based programs, for example MIRA 4 (Chevreux *et al.*, 2004) or MetaAMOS (Treangen *et al.*, 2013), or *de novo* via programs such as SOAP or Abyss (Oulas *et al.*, 2015) assuming the existence of one genome only. In addition, new tools expecting complex samples containing more than one genome are Meta-Velvet-SL (Afiahayati *et al.*, 2015) or Meta-IDBA-UD (Peng *et al.*, 2011), which are specialized to metagenomics approaches. On the one side, reference-based programs have the advantage of low computational demand but, on the other hand, need available reference genomes to work properly. Binning of reads and contigs for metagenome analysis can be performed by composition-based algorithms (Phylophytia/S+ (Patil *et al.*, 2012); ClaMS (Pati *et al.*, 2011)) or by similarity-based tools [CARMA (Krause *et al.*, 2008); MetaPhyler (Liu *et al.*, 2011)]. Furthermore, hybrid tools like PhymmBL (Brady and Salzberg, 2009) and MetaCluster (Wang *et al.*, 2012b) use similarity- and composition-based approaches in combination which lead to a better accuracy.

In *sequence-based metagenomics*, the sequenced DNA is analysed to identify genes and metabolic pathways by comparing the DNA with genes found in other communities or samples (Thomas *et al.*, 2012). Afterwards, the contigs and reads are binned into individual genomes based on the composition of k-mer sequences or similarity to specific sequences or available databases (Oulas *et al.*, 2015).

In *function-based metagenomics*, the generated DNA libraries are screened for specific functions, such as vitamin production or antibiotic resistance. When a function of interest is detected, the DNA coding for that function is sequenced and compared with DNA from other organisms or communities (Tringe *et al.*, 2005).

Today, metagenomics data are exchanged on web-based portals like the EBI Metagenomics service (Hunter *et al.*, 2014) or CAMERA (Seshadri *et al.*, 2007). For the visualization of reference assembled metagenomics, Artemis (Rutherford *et al.*, 2000) can be used, or the powerful visualization tool called MEGAN 5 (Huson and Weber, 2013) is available to perform taxonomic classification of the complex datasets.

and cyanobacterial strategies to survive in extreme habitats like geothermal (Badhai *et al.*, 2015) or geochemical (Suzuki *et al.*, 2013) springs, high arctic (Varin *et al.*, 2012) and saline (Harris *et al.*, 2013) regions have been analysed by metagenomics. Metagenomics of cyanobacteria is also used to analyse environmental metagenomes and their populations to get insights into the correlation of microbial assembly and its influence on ecosystems (Ansari *et al.*, 2015; Ottesen *et al.*, 2013), which was exemplified by coastal *Synechococcus* populations (Tai *et al.*, 2011). Metagenomics can also provide information about differences between wild-living and cultured cyanobacteria with respect to stress conditions (Kelly *et al.*, 2013) and the gain or loss of trait-specific genes (Berube *et al.*, 2015). In conjunction with this research approach, the response of cyanobacterial communities to changes in nutrient concentrations, like phosphate or iron (Shilova *et al.*, 2014) or dissolved organic matter (Sharma *et al.*, 2014) in marine environments, as well as the influence of abnormal metabolites, like CO_2 in dryland environments (Steven *et al.*, 2012), was also investigated.

Metagenomics so far lead to the identification of a huge amount of gene sets for specific pathways for secondary metabolites alike the non-ribosomal peptide synthetases (NRPS) and polyketide synthases (PKS) in the sphagnum bog, for which a great variety of secondary metabolites was described without a knowledge about the synthesizing enzymes (Muller *et al.*, 2015). Another example for observing new genomic elements was the detection of specific transposable elements of thermophilic cyanobacterial strains, which seem to be less abundant in wild-living cyanobacteria than in cultured ones (Nelson *et al.*, 2012). Application of metagenomics becomes important for agricultural or biotechnological systems as well (Hess, 2011). A prominent example is the study of the impact of cyanobacteria on growth and health of rice (Priya *et al.*, 2015). In turn, culture-independent metagenomics lead for example to the identification of complete biosynthesis gene cluster of the anticancer agent curacin A (Jones *et al.*, 2011). Thus, metagenomics is just at its beginning and will provide in future a major tool for analysis of cyanobacteria.

Recent advances in pangenome assignment

The prerequisites of multi-genomic analyses are a large amount of complete genomic data, good accuracy of the genome assembly and annotation (Vernikos *et al.*, 2015), and databases with detailed information about traits and species- or strain-specific features (Simm *et al.*, 2015). More important for pangenome analysis are the accuracy of genome assembly, annotation and the possibility to use draft genomes, as well as orthologue prediction (Bentley, 2009; Xiao *et al.*, 2015). The orthologue prediction (Box 1.2) can be based on phylogenetic relationships, functional information or sequence similarity, and approaches for multiple closely related strains could improve the accuracy (Xiao *et al.*, 2015). In addition, inclusion of non-coding RNA, pseudogenes, epigenetic elements or genome-organization frameworks (GOFs) can provide new insights into metagenome analysis (Xiao *et al.*, 2015).

Besides quality of input data and orthologue search, an important issue in pangenome analysis is the selection and classification of species and strains. It remains under debate how many strains should be included to yield reliable results and whether there is a risk of overfitting in case there is a huge discrepancy in the available strains for the selected species (Bentley, 2009). Moreover, accuracy of the pangenome could be influenced by the filtering of more diverse strains from less diverse ones.

The application of pangenome analysis is a consequence of the expanding number of sequenced genomes of strains corresponding to microbes (Medini *et al.*, 2005) and bacteria (Tettelin *et al.*,

Box 1.2 Identification of cliques of orthologous genes

Definition of CLOGs (Simm *et al.*, 2015) can be performed by tools for pangenome analysis such as PGAP (Zhao *et al.*, 2012) and Panseq (Laing *et al.*, 2010), or via orthologue and co-orthologue prediction tools for pairwise orthologue prediction (InParanoid (Sonnhammer and Ostlund, 2015)) or multi-organism orthologue prediction (OrthoMCL (Li *et al.*, 2003)). In addition, tools which combine CLOG definition and specific data analysis have been defined. PanCGHweb (Bayjanov *et al.*, 2010) is restricted to Microarray analysis. CAMBer (Wozniak *et al.*, 2011) refines gene function and identifies sequence errors in multi-gene families. Spine and Agent (Ozer *et al.*, 2014) profile pangenomes on draft and fully assembled genomes. The web based PGAT (Brittnacher *et al.*, 2011) is a pangenomic database and is restricted to a limited number of species, but allows SNP (single nucleotide polymorphism) and pathway identification. PanCake (Xiao *et al.*, 2015) identifies singletons (when only one read occurs in the read library) and core regions in genomes. PANNOTATOR (Santos *et al.*, 2013) annotates genomes of closely-related species. Harvest (Treangen *et al.*, 2014) is specialized on variant calling and core-gene alignments. GET_HOMOLOGUES (Contreras-Moreira and Vinuesa, 2013) includes phylogenetic tree construction and genetic feature extraction combined with graphical display. ITEP for exploration of microbial pangenomes (Benedict *et al.*, 2014) integrates different tools, such as orthologue detection and functional domain analysis, and contains a graphical user interface. PanGP (Zhao *et al.*, 2014) includes a graphical interface for analysing pangenome on a larger scale.

The input files should contain the whole proteome of the different strain or species (excluding wrongly annotated proteins), unique identifiers and should avoid isoforms from the same loci (Simm *et al.*, 2015). CLOGs can be clustered in the three categories of the pangenome. CLOGs containing only proteins of the same species or strains are defined as unique genes (Medini *et al.*, 2005). Furthermore, CLOGs containing at least from each selected proteome one representative protein are called core-genome CLOGs, and all CLOGs containing proteins from more than one strain/species proteome ($1 < x \geq n^{-1}$) build the dispensable genome (Tettelin *et al.*, 2005).

2008), as well as of single species like *Prochlorococcus* (Kettler *et al.*, 2007), *Legionella* (D'Auria *et al.*, 2010) or *Streptococcus* (Donati *et al.*, 2010). In the past, pangenome studies were performed with a focus on the identification of intraspecies gene sets. Defining the number of genes needed to fully describe a species using different strains enables to detect marker genes and strain specificities (Tettelin *et al.*, 2005). Today, the application of pangenome analysis is extended to define core-genomes for model organisms such as human (Li *et al.*, 2010b) or yeast (Dunn *et al.*, 2012).

In addition to the identification of intraspecies gene sets, core-genome definition by pangenome analysis is used to obtain evolutionary relations e.g. to core-genome of cyanobacteria (Beck *et al.*, 2012). In addition, the gain and loss of genes within a specific clade or phyla can be detected by this approach. Pangenome analysis of a class of ancient and very important organisms on Earth like cyanobacteria (Garcia-Pichel *et al.*, 2003), that have high variability in live forms and conquered habitats (Carey *et al.*, 2012), provided new insights into common and specific gene sets. Thus, their exploration is important, on the one hand, for their biotechnological use (Wijffels *et al.*, 2013) and, on the other hand, because some of the cyanobacterial species are toxic to their environmental community (Neilan *et al.*, 2013), which might cause significant impact on ecological systems under changing climate conditions. The core-genome of cyanobacteria can also be used to identify signature genes (Box 1.3) (Dutilh *et al.*, 2008; Gupta and Mathews, 2010). For cyanobacteria, the origin of photosynthesis (Mulkidjanian *et al.*, 2006) and the diversity of their metabolism (Beck *et al.*, 2012) were analysed. Moreover, eight cyanobacterial marker genes were defined based on their presence in the cyanobacterial core-genome and their absence in genomes of Viridiplantae (green algae and land plants) or other prokaryotes (Box 1.3). Most of these marker genes are not yet characterized (*alr3411*, *asl1962*) except those whose products are membrane proteins (*alr1215*, *alr2762*, *asr4076*), and only three of them are coding known proteins for transcriptional regulation (*all2080*), pantothenate synthesis (*alr2936*) and nitrogen regulation (*asr0485*).

Core- and pangenomes are also used to define differences between two or more strains or species. Such approaches revealed that uncultivated closely related strains have large divergences between their genomes (Bombar *et al.*, 2014). In combination with transcriptomics, differences in the acclimation of different strains to environmental changes can be detected on the base of common genes (Billis *et al.*, 2014). Thus, it is of major importance to describe core set of proteins in cyanobacteria as well as the connection to species-specific adaptions to the environment. For this reason, core- and pangenome analysis were used to identify trait specific gene sets (Simm *et al.*, 2015), e.g. by clustering bacteria concerning different traits to analyse subfractions of the pan- and core-genome with relation to specific habitats.

Box 1.3 Core genome and signature genes of cyanobacteria

The robustness of the core-genome determined by Simm and co-workers (Simm *et al.*, 2015) was verified by comparison to earlier studies analysing eight (Martin *et al.*, 2003; 179 core genes assigned), 15 (Mulkidjanian *et al.*, 2006; 1044 core genes assigned) or 16 cyanobacterial genomes (Beck *et al.*, 2012; 704 core genes assigned). The overlap between previously assigned core-genomes and the one defined in 2015 by Simm and co-workers (559 core genes) was about 530 sequences for the two largest studies. The overlap to earlier studies suggests that cyanobacteria contain a core-genome of about 500 genes.

For confirmation of feature specific cyanobacterial signature genes, all available genomes for Viridiplantae (containing chloroplasts as cyanobacterial ancestors) and bacteria (except cyanobacteria) available at the NCBI non-redundant (nr) database were selected as reference (Martin *et al.*, 2003; Simm *et al.*, 2015). Using the 559 CLOGs of the core-genome, 57 groups had no BLAST hits to Viridiplantae (80% sequence coverage and e-value ≤ 0.001), and eight groups of them are cyanobacterial specific with no hits in other bacteria, namely *asr0485*, *alr1215*, *asl1962*, *all2080*, *alr2762*, *alr2936*, *alr3411* and *asr4076*. Five of these signature genes encode for uncharacterized proteins (unknown protein: *alr3411*, *asl1962*; unknown membrane protein: *alr1215*, *alr2762*, *asr4076*), but the remaining three are a transcriptional regulator (*all2080*), a pantothenate synthetase/cytidilyte kinase (*alr2936*) and a PII interaction protein (*asr0485*).

Detection of core- and clade-specific functions via metagenome analysis

Pangenomics analysis to define the core-genome of cyanobacteria was initiated by the analysis of eight genomes, leading to a set of 181 signature genes specific for cyanobacteria without any orthologue to eukaryotes containing chloroplasts. Further, 25% of these genes could be associated with some function (Martin *et al.*, 2003). Subsequent metagenomics approaches based on about 15 and 16 cyanobacterial genomes have contributed to the description of the origin of photosynthesis (Mulkidjanian *et al.*, 2006) and diversity of metabolism (Beck *et al.*, 2012). Both studies used the cyanobacterial genomes of *Anabaena* sp. PCC 7120, *Anabaena variabilis* ATCC 29413, *Synechocystis* sp. PCC 6803, *Thermosynechococcus elongatus* BP-1, *Gloeobacter violaceus* PCC 7421 and *Synechococcus elongatus* PCC 7942. Furthermore, several *Synechococcus* and *Prochlorococcus* strains were used to create the pangenome. In contrast to Mulkidjanian and co-workers (2006), which defined the core-genome of cyanobacteria to have a size of 892 cliques of orthologous genes (CLOGs), Beck and co-workers (2012) included some other cyanobacterial species such as *Acaryochloris*, *Cyanothece* and *Trichodesmium* and ended up with a cyanobacterial core-genome size of 660 CLOGs. The most recent cyanobacterial pangenome study included 58 cyanobacterial genomes of strains belonging to 45 different species of six cyanobacterial clades to identify core- and clade-specific genes (Simm *et al.*, 2015). In addition to phylogenetic relations based on the genome analysis, information on 13 different ecological, morphological and physiological features was used to classify cyanobacteria (Simm *et al.*, 2015).

For the selected 58 cyanobacterial strains, a pangenome containing 44,831 CLOGs was built (Simm *et al.*, 2015). 28,520 of all CLOGs are CLOGs of unique genes, whereas 15,752 are CLOGs of the dispensable-genome, but the majority of these CLOGs contain sequences only from up to 10 different strains (Fig. 1.2A and C). The core-genome of the phylum of cyanobacteria comprises 559 CLOGs. The core-genome shows a tendency to converge to ~500 CLOGs, whereas the pangenome increases by a factor α of 0.35 ± 0.07 per strain (Fig. 1.2B). From this value (< 1) it can be concluded that the pangenome is open, meaning that addition of new cyanobacterial strains will increase the number of CLOGs. In addition, the pangenome of a single species of *Prochlorococcus marinus* in an earlier study increased with $\alpha = 0.8$ (Tettelin *et al.*, 2008) and thus, the increase of the pangenome of cyanobacteria is more rapid than the increase of the pangenome of a single species like *P. marinus*.

The trait definition enabled the authors to determine genes specific for a subset of cyanobacterial strains with either thermophilic character, common growth environmental conditions or the capability to differentiate heterocysts (Simm *et al.*, 2015). For the identification of such genes, CLOGs containing exclusively sequences of cyanobacterial strains with a certain trait (like thermophilic character) were chosen. Subsequently, only the CLOGs with sequences of all cyanobacterial strains with this trait were selected. Although the analysis is limited due to an incomplete assignment of features, an initial set of trait-specific genes can be defined.

Three thermophilic cyanobacteria genomes are publicly available, namely *T. elongatus* BP-1 (The1), *Synechococcus* sp. JA-3-3Ab (Syn9), and *Synechococcus* sp. JA-2-3B′a(2–13) (Syn8). With respect to the thermophilic character, four genes in two CLOGs exist for the reference *T. elongatus* BP-1 (Nakamura *et al.*, 2002; Nakamura *et al.*, 2003), which were present in all other thermophilic cyanobacteria exclusively. These CLOGs include *tlr0324*, coding for a protein with a DnaJ domain (heat shock protein binding function), and *tlr0548*, *tlr0547* and *tsr0549*, all coding for proteins of unknown function. Sequences with similarity to *tlr0548* have been identified in bacterial strains with extreme habitats of the genus *Acidithiobacillus* and the species *Haliangium ochraceum*, *Halothiobacillus neapolitanus*, *Sorangium cellulosum*, or *Thiothrix nivea*, but not in plants. In turn, for the other genes (*tlr0324*, *tlr0547*, and *tsr0549*) no orthologues in any other bacterial or plant genome could be detected, and it is the reason why these genes can be used as thermophilic cyanobacteria specific genes.

With respect to the growth habitat for cyanobacterial strains capable of soil-living, five CLOGs were identified, of which one is a possible soil-living cyanobacterial specific gene (*alr7204* in *Anabaena* sp. PCC 7120), because it does not occur in eukaryotic Viridiplantae or other bacterial genomes.

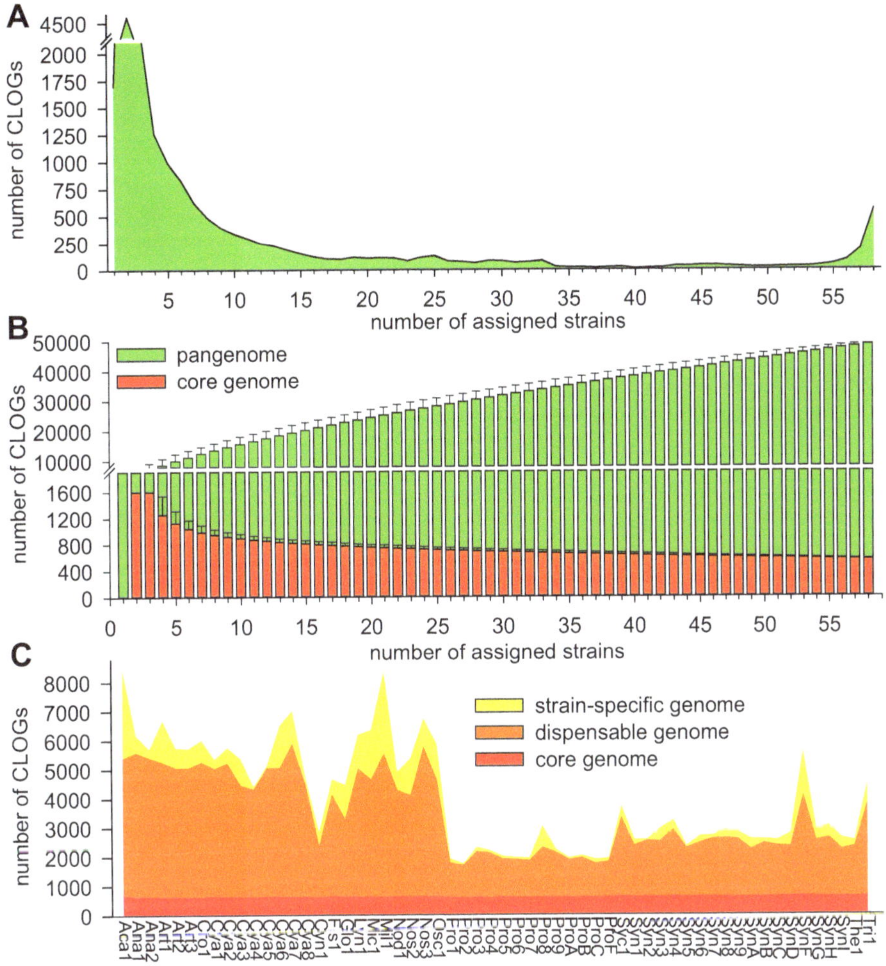

Figure 1.2 Cyanobacterial pangenome of 58 strains. (A) CLOGs assigned to the number of cyanobacteria within the pangenome. (B) Behaviour of the cyanobacterial core- (red) and pangenome (green) concerning the increase of assigned strains. (C) Partitioning in core-genome (red), dispensable genome (orange) and specific genes (yellow) for each cyanobacterial strain.

However, the remaining four specific soil-living gene sets code for an oxidoreductase (*all0827* in *Anabaena* sp. PCC 7120), acetyltransferase (*alr3061*), heterocyst-specific ABC-type transporter (*alr4974*) and a TolB-like protein (*all0351*). All of these genes were identified in many bacterial genomes.

For the six heterocyst-forming cyanobacteria (*Anabaena* sp. PCC 7120, *Anabaena variabilis* ATCC 29413, *Fischerella* sp. JSC-11, *Nodularia spumigena* CCY9414, *Nostoc azollae* 0708, *Nostoc punctiforme* PCC 73102), 93 CLOGs have been identified as trait-specific occurring in at least five (39 CLOGs) or all six (54 CLOGs) strains (Simm *et al.*, 2015). Fourteen of these genes have been experimentally characterized and for 28 additional genes their putative function could be predicted. In combination with expression studies in *Anabaena* sp. PCC 7120, 56 genes in the 93 CLOGs were at least two-fold higher expressed under nitrogen starvation and this strengthens the hypothesis that these genes are involved in nitrogen fixation (Flaherty *et al.*, 2011). In turn, only two of the 93 genes (*asl1933* and *asr1289*) were down-regulated or not expressed under nitrogen starvation (Simm *et al.*, 2015). Remarkably, 66 of these CLOGs contain heterocyst-specific cyanobacterial genes characterized by their absence in other bacterial and plant genomes (Simm *et al.*, 2015). Interestingly, only 14 of 17 heterocyst-forming marker genes in *Anabaena* sp. PCC 7120 described in literature could be detected in all six known heterocyst-forming cyanobacteria. Only *patS* (*asl2301*, *Anabaena* sp. PCC 7120), *hetN* (alr5358), and *pbpB* (*alr5101*) could not be detected in all strains in the related CLOGs, meaning that other putative genes with redundant or similar functions could be involved, or the heterocyst-forming mechanism in cyanobacteria is different in the analysed strains (Simm *et al.*, 2015).

Another important trait of cyanobacteria is their ability to produce notorious toxins, known as cyanotoxins, and a vast variety of other secondary metabolites (Calteau *et al.*, 2014; Dittmann *et al.*, 2013). Cyanotoxins include potent hepatotoxins, neurotoxins and antimicrobials, such as microcystins, anatoxins and lantipeptides, while other secondary metabolites include terpenes, alkaloids and UV-absorbing pigments among others (Dittmann *et al.*, 2015; Kehr *et al.*, 2011; Micallef *et al.*, 2015). These biosynthetic capabilities make the phylum Cyanobacteria a promising source of compounds with relevance to drug development and biotechnological exploitation (Burja *et al.*, 2001; Namikoshi and Rinehart, 1996; Welker and Von Döhren, 2006).

General principles of secondary metabolite biosynthesis in cyanobacteria

The majority of cyanobacterial secondary metabolites are non-ribosomal peptides, polyketides or hybrid peptide-polyketide compounds (Micallef *et al.*, 2015). NRPS and PKS enzymes are both molecular assembly lines that direct the product formation of these natural products using as template a protein sequence within these enzymes (Donadio *et al.*, 2007). Non-ribosomal peptides are biosynthesized by NRPSs, which are multifunctional enzyme complexes that assemble either proteinogenic or non-proteinogenic amino acids into the final peptide structure, while polyketides are biosynthesized by PKSs, which assemble their product from acyl-CoA (Fischbach and Walsh, 2006). Both classes of enzyme complexes synthesize their products by adding building blocks in a sequential manner. NRPS and PKS enzyme complexes are similar in their biochemical logic (Fig. 1.3). An NRPS or PKS assembly line consists of a multifunctional megasynthetase or a multienzyme complex with different modules, being each module responsible for the selection and addition of a single monomer to extend the molecule that is being synthesized (Konz and Marahiel, 1999; Marahiel *et al.*, 1997; Staunton and Weissman, 2001; Welker and Von Döhren, 2006). A single NRPS or PKS polypeptide typically contains multiple modules and, with few exceptions, each biosynthetic pathway incorporates several polypeptides (Weissman, 2015). The order of the modules usually dictates the peptide or polyketide chain sequence since they select the monomers. Thus, NRPS and PKS complexes are used simultaneously as template and biosynthetic machinery, because the amino acid or acyl-CoA, respectively, to be incorporated is determined by the module and it also harbours all necessary catalytic functions (Donadio *et al.*, 2007). Every module contains a series of domains, each responsible for a specific enzymatic reaction. The minimum module in an NRPS contains an adenylation domain (A) responsible for substrate selection and activation, a peptide carrier protein (PCP) responsible for tethering the substrate to the enzymatic complex through a thioester bond, and a condensation domain (C) responsible for catalysing the peptide formation and elongation by C–N bond generation (Fig. 1.3A) (Stachelhaus and Walsh, 2000; Strieker *et al.*, 2010; von Dohren, 2004). There are also auxiliary domains for epimerization (E), which changes L-amino acids into D-amino acids, methylation (MT), and other less frequent reactions (Donadio *et al.*, 2007; Nikolouli and Mossialos, 2012). Additional variations in some domains include the substitution of the C domain for a cyclization domain (Cy) which catalyses both the condensation and the heterocyclization of serine, cysteine or threonine (Challis and Naismith, 2004). Similarly to the NRPS enzyme architecture, a minimum PKS module consists of an acyltransferase domain (AT), which selects the appropriate acyl-CoA monomer (usually malonyl-CoA or methylmalonyl-CoA), an acyl carrier protein (ACP) responsible for binding the acyl-CoA monomer to the enzyme through a thioester bond, and a ketosynthase domain (KS) responsible for the polyketide elongation by C–C bond formation (Fig. 1.3B) (Donadio *et al.*, 2007; Nikolouli and Mossialos, 2012). Additionally, auxiliary or tailoring domains may also be present within a module and act on the activated substrate

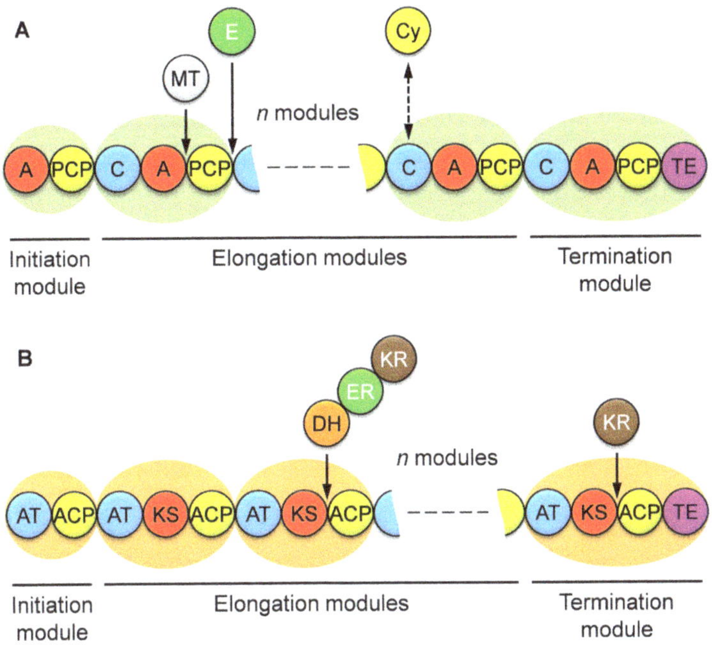

Figure 1.3 NRPS and PKS domain architecture. (A) Schematic of NRPS initiation and chain extension modules. Biosynthesis initiates with the activation of two specific amino acids as their aminoacyl adenylates by the adenylation domain (A) and transfer to the peptidyl carrier protein (PCP) both in the initiation and the first elongation modules. The condensation domain (C) of the first elongation module then catalyses a peptide bond formation using the two aminoacyl-PCPs as substrates. This sequence of reactions is repeated until the final peptide product is released by the thioesterase domain (TE) through cyclization or hydrolysis in the termination module. Chemical modifications by tailoring domains such as N-methyltransferases (MT) and epimerases (E) can occur, and cyclization domains (Cy) can replace the condensation domain in particular modules. (B) Schematic of PKS initiation and chain extension modules. Acyltransferase domains (AT) in the initiation and the first elongation modules initiate the compound building by selection of an appropriate acyl-CoA and transfer to the acyl carrier protein (ACP) of the same module. The ketosynthase domain (KS) of the first elongation module then loads the acyl unit bound to the initiation module and condenses it with the acyl unit bound to the ACP domain of the elongation module. Following the compound extension, the β-keto group can be modified by β-ketoreductase (KR), dehydratase (DH) and enoylreductase (ER) domains. Release of the final product from the multienzyme is catalysed by a TE domain.

to provide structural diversity (Donadio *et al.*, 2007; Liou and Khosla, 2003), such as β-ketoreductase (KR), dehydratase (DH) and enoylreductase (ER) domains.

Reflecting the similarity between these two biosynthetic pathways, NRPS and PKS modules can collaborate effectively and interchange enzymes to generate hybrid assembly lines, in which large multidomain enzymes of NRPS and PKS machineries can be reassembled to generate an almost infinite number of chemical structures (Mares *et al.*, 2014). Some of the end products of these machineries have been found to be important pharmaceuticals, while others are considered important environmental toxins. In cyanobacteria, hepatotoxic microcystins (Dittmann *et al.*, 1997), nodularins (Moffitt and Neilan, 2004), cylindrospermopsin (Mihali *et al.*, 2008), and neurotoxic alkaloids anatoxin (Rantala-Ylinen *et al.*, 2011) and saxitoxins (Kellmann *et al.*, 2008) are produced by these NRPS–PKS hybrid pathways. There are NRPS and PKS enzyme complexes that exhibit this biosynthetic logic in which there is a direct correspondence between the domain content of each module and the structure and composition of the final compound (Weissman, 2015). However, it has been demonstrated that there are exceptions to the product/template co-linearity, and the described biosynthetic logic does not hold for a large proportion of assembly lines, especially for PKS enzyme complexes (Condurso and Bruner, 2012; Gulder *et al.*, 2011; Hill, 2006; McDaniel *et al.*, 2005; Moss *et al.*, 2004; Weissman and Leadlay, 2005). This has led to the classification of PKSs in two distinct classes namely *cis*-AT PKSs, which exhibit a linear mode of synthesis, and *trans*-AT PKSs, when their mode of synthesis is iterative (Weissman, 2015). The first group includes the classical multidomain PKS assembly lines where sets of domains are organized into modules, each being responsible for a complete elongation step in the biosynthesis of the polyketide (Staunton and Weissman, 2001). The second group is an evolutionary distinct type of modular PKS in which the AT domains are present as free enzymes and act iteratively to deliver the same monomer to each module (Piel, 2010). This organizational feature blurs the link between the primary sequence of PKSs and the generated polyketides. While *cis*-AT PKS genes are common among cyanobacteria, *trans*-AT genes have been rarely identified. Functional non-linearity similarly occurs in certain NRPSs (Mootz *et al.*, 2002). These NRPS enzymes also operate in an iteratively way and they are classified as non-collinear NRPS systems (Weissman, 2015). Many NRPS, PKS and hybrid NRPS-PKS biosynthetic pathways are involved in the biosynthesis of relevant cyanotoxins such as microcystin, anatoxin-a, homoanatoxin-a, anabaenopeptilides, anabaenopeptins, aeruginosin and micropeptin among others (Dittmann *et al.*, 2015). However, many of these biosynthetic gene clusters have been predicted by genome-mining tools and their evolutionary and adaptive significance is poorly understood.

Main classes of secondary metabolites in cyanobacteria

Gene clusters predicted via genome mining searches [Box 1.4; here by antiSMASH (Weber *et al.*, 2015)] in five selected cyanobacteria (*Synechocystis* sp. PCC 6803, *Synechococcus elongatus* PCC 7942, *Artrospira platensis* NIES-39, *Microcystis aeruginosa* NIES-843 and *Anabaena* sp. PCC 7120) lead to the identification of putative biosynthetic pathways for four main classes of cyanobacterial secondary metabolites, namely non-ribosomal peptides, polyketides and hybrid compounds; ribosomal peptides; terpenes; and hydrocarbons. *Synechocystis* sp. PCC 6803 (Syc1), *Synechococcus elongatus*

Box 1.4 Genome mining for secondary metabolite gene cluster discovery

Genome mining is a broad term used to describe several processes that exploit information which is genetically encoded within biosynthesis gene clusters in microbial genomes with the ultimate aim of isolating a novel compound or discovering a novel biosynthetic pathway or enzyme (Corre and Challis, 2009; Gross, 2009). However, the identification of target natural products is not limited to the investigation and analysis of genomes, and other approaches such as gene knockout and subsequent comparative metabolic profiling experiments have been used to identify structurally-related secondary metabolites (Corre and Challis, 2009). In genome mining, DNA sequence information is used to assess and evaluate the genetic potential of strains to contain secondary metabolite gene clusters. This approach is possible because, despite the vast diversity and number of compounds, the molecular principles underlying secondary metabolite biosynthesis are highly conserved (Weber and Blin, 2015).

Several bioinformatics tools for genome mining of secondary metabolites have been developed in the last years but most of them are designed to identify NRPS, PKS or hybrid NRPS-PKS gene clusters or C and KS domains, such as NP.searcher (Li *et al.*, 2009), CLUSEAN (Weber *et al.*, 2009), ClustScan (Starcevic *et al.*, 2008), ASMPKS (Tae *et al.*, 2007) and NaPDoS (Ziemert *et al.*, 2012). However, genome mining tools antiSMASH and ClusterFinder have the advantage of screening a wide range of gene cluster classes (Cimermancic *et al.*, 2014; Medema *et al.*, 2011). antiSMASH is the most commonly used tool to analyse the secondary metabolite potential of sequenced genomes and excels in identifying previously characterized clusters and domains, being able to identify up to 24 cluster classes (Blin *et al.*, 2014; Blin *et al.*, 2013; Medema *et al.*, 2011; Weber *et al.*, 2015). It provides protein domain analysis of large multi-domain enzymatic assembly lines, prediction of core chemical structures of their end compounds, and multiple cluster alignments to a database of all sequenced secondary metabolite clusters known to date (Blin *et al.*, 2013; Weber *et al.*, 2015). ClusterFinder, on the other hand, has only been developed recently and has some disadvantages and limitations (Micallef *et al.*, 2015), but can identify more unannotated clusters putatively responsible for small molecule biosynthesis (Cimermancic *et al.*, 2014).

PCC 7942 (Syn7) and *Artrospira platensis* NIES-39 (Art1) contain few (two to three) gene clusters (Syc1: 62 genes; Syn7: 54 genes; Art1: 69 genes), while *Microcystis aeruginosa* NIES-843 (Mic1: 376 genes) and *Anabaena* sp. PCC 7120 (Ana1: 359 genes) have 10 and 14 gene clusters, respectively. Enzymes encoded in these clusters may produce terpenes (Ana1, Art1, Mic1, Syc1, Syn7), non-ribosomal peptides and polyketides (Ana1, Mic1), bacteriocins (Art1, Ana1, Syn7), microviridins (Mic1, Ana1), siderophores (Ana1), ladderanes (Ana1) and cyanobactins (Art1, Mic1). These classes of cyanobacterial secondary metabolites will be described focusing on the heterocyst-forming cyanobacterium *Anabaena* sp. PCC 7120, although non-ribosomal peptides, polyketides and hybrid peptide-polyketide compounds will be emphasized because the majority of cyanobacterial secondary metabolites are synthesized by modular NRPS, PKS and hybrid NRPS-PKS biosynthetic pathways.

The biosynthetic gene cluster for the cyclic heptapeptide toxin microcystin encodes two PKS, three NRPS and a hybrid NRPS-PKS, and has been identified in the genome of nine *Microcystis aeruginosa* strains, eight *Planktothrix* strains, *Anabaena* sp. strain 90 and *Fischerella* sp. PCC 9339 (Fiore *et al.*, 2013; Frangeul *et al.*, 2008; Humbert *et al.*, 2013; Kaneko *et al.*, 2007; Rounge *et al.*, 2009; Shih *et al.*, 2013; Tooming-Klunderud *et al.*, 2013; Wang *et al.*, 2012a; Chapter 3 of this book). Gene clusters for the neurotoxins anatoxin-a and homoanatoxin-a have been identified in the genome of *Oscillatoria* sp. PCC 6505 and *Anabaena* sp. strain 37 (Mejean *et al.*, 2010; Rantala-Ylinen *et al.*, 2011). In *Anabaena* sp. strain 90, a gene cluster has been found to be coding for three PKSs for the synthesis of the cyclic peptide anabaenopeptilide, which is a protease inhibitor (Rouhiainen *et al.*, 2000; Wang *et al.*, 2012a). The five NRPS genes comprising the gene cluster of the protease inhibitor anabaenopeptin, which is a cyclic hexapeptide, have been identified in the genome of eight *Planktothrix* strains, two *M. aeruginosa* strains, *Anabaena* sp. strain 90 and *Nostoc punctiforme* PCC 73102 (Humbert *et al.*, 2013; Rouhiainen *et al.*, 2010; Rounge *et al.*, 2009; Tooming-Klunderud *et al.*, 2013; Voss *et al.*, 2013). Aeruginosins and micropeptins are linear and cyclic peptide toxins, respectively, produced by NRPS (Dittmann *et al.*, 2015; Ishida *et al.*, 2009). They are a vast group of protease inhibitors produced by species of the genera *Microcystis, Oscillatoria, Nodularia* and *Nostoc*. In the genome of *Anabaena* sp. PCC 7120, no gene clusters related to the biosynthesis of these cyanotoxins have been identified based on the antiSMASH predictions. However, a NRPS gene cluster (Fig. 1.4A, cluster 5) similar to that of the biosynthesis of aeruginosins and micropeptins is present in *Anabaena* sp., although no evidences are available regarding its role.

Another group of hybrid NRPS-PKS peptides are the cyclic lipopeptides, which are hybrid peptides that possess a lipid moiety, such as puwainaphycins (Hrouzek *et al.*, 2012; Mares *et al.*, 2014). This group of cyanobacterial metabolites has been described from various cyanobacterial genera and seems to be widely distributed in nature (An *et al.*, 2007; Liu and Rein, 2010; MacMillan *et al.*, 2002; Maru *et al.*, 2010; Moon *et al.*, 1992; Plaza and Bewley, 2006). Puwainaphycins are cyclic decapeptides containing an unusual β-amino fatty acid and some of them cause the influx of calcium ions into the cell, resulting in cell necrosis (Hrouzek *et al.*, 2012). The genome of *Anabaena* sp. PCC 7120 contains two putative clusters that could be involved in the biosynthesis of puwainaphycins according to the antiSMASH predictions. One of them is a candidate hybrid NRPS–PKS cluster (Fig. 1.4A, cluster 9) while the other could be a PKS cluster (Fig. 1.4A, cluster 10).

None of the five NRPS, PKS and hybrid NRPS-PKS gene clusters identified in the genome of *Anabaena* sp. PCC 7120 has more than three genes in the core-genome of cyanobacteria, but two conserved regions were observed for the NRPS gene cluster 5 (Fig. 1.4B, gene regions *all1688–all1693* and *all1699-all1701*). In turn, *Microcystis aeruginosa* NIES-843 genome contains five NRPS, PKS and hybrid NRPS-PKS gene clusters with up to six genes in the core-genome of cyanobacteria, which are partially clustered in conserved regions. However, *Synechococcus elongatus* PCC 7942, *Synechocystis* sp. PCC 6803 and *Arthrospira platensis* NIES-39 contain no NRPS or PKS gene clusters (Table 1.2).

Heterocyst glycolipids are important molecules in heterocyst-forming cyanobacteria which are needed to protect the oxygen-sensitive nitrogenase from inactivation. These compounds are required for the last maturation step of heterocysts rendering them impermeable to the oxygen both present

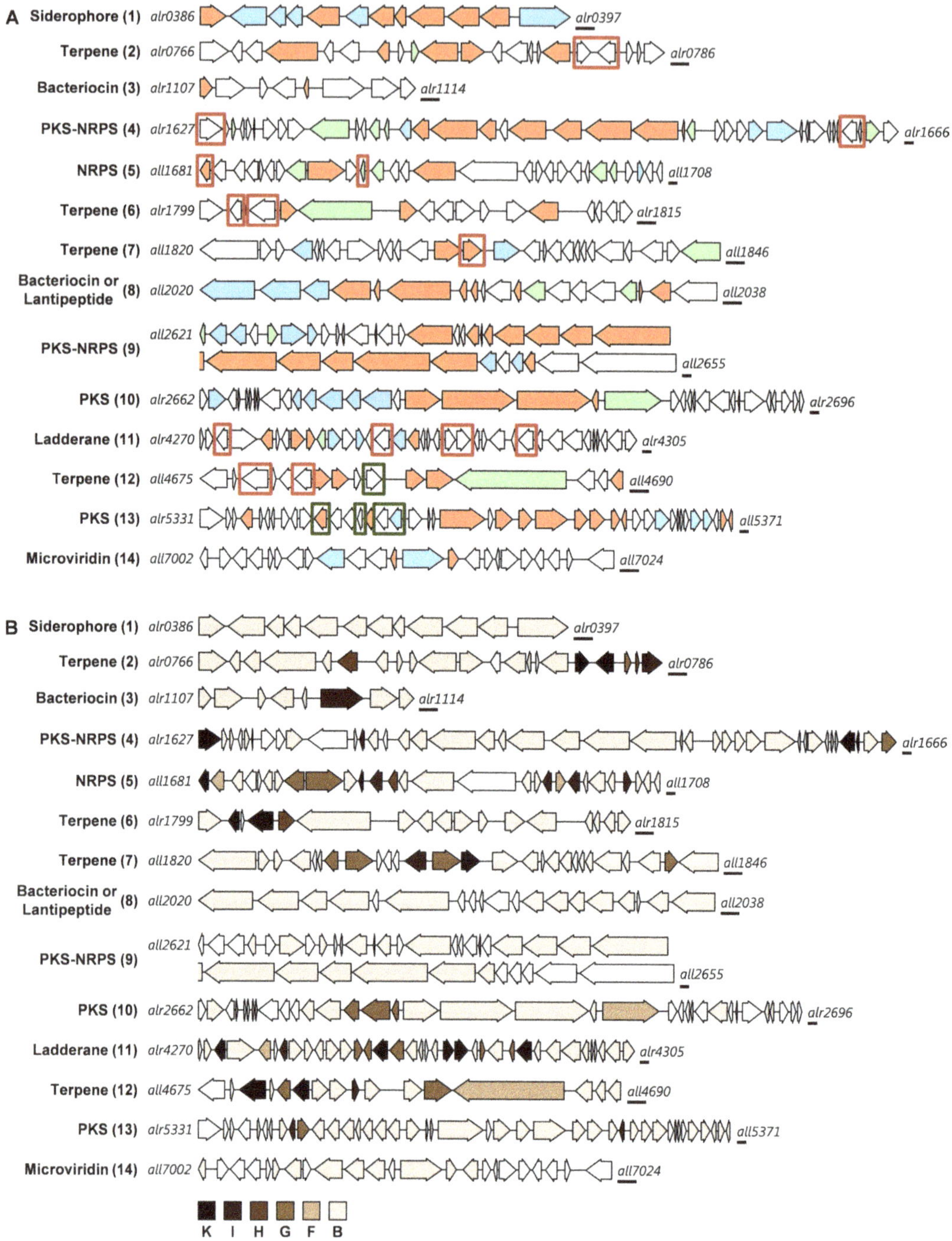

Figure 1.4 Secondary metabolite gene clusters identified in *Anabaena* sp. PCC 7120. The schematic representation to scale of the secondary metabolite gene clusters found in *Anabaena* sp. is shown. The genome of this cyanobacterium was analysed using the genome mining software antiSMASH (Medema *et al.*, 2011), which detects a wide range of secondary metabolite biosynthetic gene clusters, including those encoding the pathways for non-ribosomal peptides, polyketides, ribosomal peptides, terpenoids, aminoglycosides and NRPS-independent siderophores. Clusters 1 to 13 are located in the chromosome while cluster 14 is present in the alpha megaplasmid of *Anabaena* sp. Black bars close to the each gene cluster correspond to 1 kb. (A) Genes coloured according to their putative function. Biosynthetic genes are shown in red, transport-related genes in blue, and regulatory genes in green, while other genes in the cluster with unknown function are shown in white. Red frames indicate core-genome genes of cyanobacteria while green frames indicate genes specific for heterocyst-forming cyanobacteria. (B) Genes coloured according to the deepest evolutionary split (from black to light brown) in the neighbour-joining tree presented in Fig. 1.5 in which they occur.

Table 1.2 Secondary metabolite gene clusters predicted in the selected cyanobacterial species *Synechocystis* sp. PCC 6803 (Syc1), *Synechococcus elongatus* PCC 7942 (Syn7), *Arthrospira platensis* NIES-39 (Art1), *Microcystis aeruginosa* NIES-843 (Mic1) and *Anabaena* sp. PCC 7120 (Ana1)

Species	ID	Metabolite	No. of genes	No. of core genes	No. of genes in CLOGs containing >20 species
Art1	1	Terpene-bacteriocin	23	2	11
	2	Cyanobactin-bacteriocin	46	2	6
Syn7	1	Terpene	22	7	18
	2	Terpene	22	6	21
	3	Bacteriocin	10	2	6
Syc1	1	Terpene	24	2	14
	2	Terpene	21	6	10
	3	Terpene	17	3	12
Mic1	1	Cyanobactin	39	1	8
	2	t1pks-terpene	51	5	21
	3	Other	39	1	9
	4	Microviridin	20	1	7
	5	t1pks-nrps-otherks	40	1	7
	6	Terpene	23	2	11
	7	nrps-t1pks	48	5	25
	8	terpene	22	7	17
	9	nrps	48	6	24
	10	nrps	46	1	13
Ana1	1	Siderophore	12	0	2
	2	Terpene	21	2	14
	3	Bacteriocin	8	0	4
	4	t1pks-otherks nrps	40	3	12
	5	nrps	28	2	15
	6	Terpene	17	2	5
	7	Terpene	27	1	17
	8	Bacteriocin-lantipeptide	19	0	0
	9	t1pks-nrps	35	0	13
	10	t1pks	35	0	9
	11	Ladderane	36	5	21
	12	Terpene	16	2	11
	13	Otherks-t1pks	41	0	4
	14	Microviridin	24	0	1

The number of genes contained in each cluster and the putative secondary metabolites they could produce are indicated, as well as the number of genes in the core-genome and the number of genes in CLOGs containing more than 20 different species.

in the air and produced by adjacent vegetative cells, and thus, contributing to provide an adequate intracellular environment to fix atmospheric nitrogen (Fan *et al.*, 2005; Kumar *et al.*, 2010). The genome of *Anabaena* sp. PCC 7120 shows a PKS cluster (Fig. 1.4A, cluster 13) that is highly related to HglE from *Nostoc punctiforme* ATCC 29133, which has been demonstrated to be required for glycolipid biosynthesis in heterocysts (Campbell *et al.*, 1997), since *hglE* mutants are unable to fix nitrogen under atmospheric conditions. This is in line with the observation that four characterizing genes of heterocyst-forming cyanobacteria are present in this cluster (Simm *et al.*, 2015). In addition, a second hybrid PKS-NRPS gene cluster in the genome of *Anabaena* (Fig. 1.4A, cluster 4) could be also involved in cell wall biogenesis or in the biosynthesis of heterocyst glycolipids (Donadio *et al.*, 2007), although only 28% of the genes in this cluster show similarities to glycolipid biosynthetic genes and no evidence is available regarding its role. In contrast to cluster 13, this cluster does not contain heterocyst-forming cyanobacteria characterizing genes.

Siderophores are small molecules with a high affinity for ferric iron that are secreted by organisms into the environment to scavenge iron (Andrews *et al.*, 2003). Once bound to this metal, iron–siderophore complexes are recognized by specific receptors on the cell surface and actively transported into the cell (Andrews *et al.*, 2003; Wandersman and Delepelaire, 2004). Siderophores are classified according to their chemical structure in three main classes: catecholates, hydroxamates and α-hydroxy-carboxylates (Miethke and Marahiel, 2007). Many catecholate and hydroxamate siderophores are synthesized by NRPS and, in some cases, hybrid NRPS-PKS biosynthetic pathways (Crosa and Walsh, 2002; Haas *et al.*, 2008), and the biosynthetic mechanisms for several structurally diverse siderophores of these groups, such as enterobactin and desferrioxamine B, are well understood (Crosa and Walsh, 2002; Sandy and Butler, 2009). However, α-hydroxy-carboxylate and some hydroxamate siderophores are synthesized by NRPS-independent siderophore (NIS) biosynthesis pathways (Challis, 2005). Very diverse enzymes, such as oxygenases, amino acid ligases, acyltransferases, aldolases, amide synthetases and decarboxylases, are involved in NIS biosynthetic pathways, and have been described for the biosynthesis of siderophores aerobactin, desferrioxamine E and rhizobactin 1021, among others (Challis, 2005). Schizokinen is a siderophore produced by *Anabaena* sp. PCC 7120 whose biosynthetic genes have not been characterized yet. However, the genome of this cyanobacterium contains a putative cluster for a NIS biosynthetic pathway (Fig. 1.4A, cluster 1) which has been speculated to be involved in the biosynthesis of schizokinen (Nicolaisen *et al.*, 2008). Interestingly, this gene cluster seems to be specific for few cyanobacterial strains in addition to *Anabaena* sp. PCC 7120, although only two genes within it (*alr0386*, *all0395*) are conserved in more than 30 cyanobacteria (Table 1.2).

In addition to non-ribosomal peptides, cyanobacteria also produce a group of ribosomally synthesized and post-translationally modified peptides, which are known as RiPPs (Arnison *et al.*, 2013). The biosynthetic pathway of these compounds is referred to as post-ribosomal peptide synthesis (PRPS). RiPPs are usually encoded within a structural gene which produces the precursor peptide, which comprises a signal sequence, a leader peptide, a core peptide and a recognition sequence. Once the precursor peptide is synthesized by the ribosome, the core peptide is post-translationally modified to produce the final RiPP (Arnison *et al.*, 2013). A range of RiPP families have been identified from cyanobacterial genomes, including the cyanotoxins bacteriocin, microviridin, and cyanobactin.

Bacteriocins are characterized by a double glycine motif in the leader sequence, and a major class of bacteriocins is formed by lantipeptides or lantibiotics (Dirix *et al.*, 2004). Lantibiotics are lanthionine/methyllanthionine-containing antimicrobial RiPPs produced by Gram-positive bacteria (Draper *et al.*, 2015). They undergo post-translational modifications which result in the generation of unusual amino acids, such as dehydroalanine, dehydrobutyrine, lanthionine and methyllanthionine residues. The most characterized lantibiotic is nisin A, which is produced by *Lactococcus lactis* and contains five intramolecular ring structures. This compound prevents cell wall biosynthesis and forms pores in the cell membranes of susceptible cells (Wiedemann *et al.*, 2001). Bacteriocin gene clusters have been widely found in cyanobacterial genomes (Begley *et al.*, 2009; Fiore *et al.*, 2013; Lee *et al.*, 2008; Li *et al.*, 2010a; Lima *et al.*, 2014; O'Sullivan *et al.*, 2011; Shih *et al.*, 2013; Wang *et al.*, 2012a), and the genome of *Anabaena* sp. PCC 7120 contains two putative gene clusters for the biosynthesis of a bacteriocin (Fig. 1.4A, cluster 3) and a bacteriocin-lantipeptide (Fig. 1.4A, cluster 8). Microviridins are N-acetylated tri- and tetra-decapeptides that belong to another family of RiPPs and are known for their activity as inhibitor of proteases. Gene clusters for microviridin biosynthesis contain five genes encoding the precursor peptide, two cyclization proteins, an acetyltransferase and an ABC-type transporter gene, and have been identified in the genome of many cyanobacterial strains specifically from the genera *Microcystis, Nodularia, Nostoc, Anabaena* and *Planktothrix* (Fiore *et al.*, 2013; Humbert *et al.*, 2013; Philmus *et al.*, 2008; Rounge *et al.*, 2009; Tooming-Klunderud *et al.*, 2013; Voss *et al.*, 2013). *Anabaena* sp. PCC 7120 contains a putative cluster involved in microviridin biosynthesis in its alpha megaplasmid (Fig. 1.4A, cluster 14), but conditions that activate the expression of this cluster have not been identified to date (Humbert *et al.*, 2013). Cyanobactins are small cyclic RiPPs that are made exclusively of proteinogenic amino acids. Cyanobactin biosynthetic gene clusters comprise a precursor peptide gene and genes encoding proteases. Clusters for the biosynthesis of cyanobactin have been found in a wide range of cyanobacterial strains (Donia *et al.*, 2008; Donia and Schmidt, 2011; Fiore *et al.*, 2013; Humbert *et al.*, 2013; Leikoski *et al.*, 2010; Schmidt *et al.*, 2005; Shih *et al.*, 2013; Sudek *et al.*, 2006; Voss *et al.*, 2013), but not in *Anabaena* sp. PCC 7120.

Only one core-genome CLOG per gene cluster was identified in the microviridin and cyanobactin clusters of *Microcystis aeruginosa* NIES-843. In contrast, the bacteriocin cluster in *Synechococcus elongatus* PCC 7942 has two core-genome CLOGs, and six of the 10 genes within this cluster are conserved in more than 45 different species of cyanobacteria. Both predicted gene clusters in *Arthrospira platensis* NIES-39 could be involved in bacteriocin synthesis and both contain two CLOGs per cluster in the core-genome, which do not overlap each other. However, one of the clusters in *Arthrospira platensis* NIES-39 is catalogued as a hybrid gene cluster involved in terpene-bacteriocin biosynthesis and is hard to determine its putative function with the information that is available. In contrast, the microviridin and the two bacteriocin gene clusters of *Anabaena* sp. PCC 7120 have no core-genome CLOGs (Table 1.2). Regions of gene clusters for bacteriocin, bacteriocin-lantipeptide and microviridin biosynthesis (Fig. 1.4B; clusters 3, 8 and 14, respectively) are specific for *Anabaena* sp. PCC 7120 and closely related species, with the exception of the gene *alr1112* (cluster 3).

Terpenoids are a large family of secondary metabolites, including carotenoids, tocopherol, phytol, sterols and hormones (Pattanaik and Lindberg, 2015, Chapter 9 of this book). They are commonly known for their importance as flavours, pigments, fragrances, insecticides and drugs (Agger *et al.*, 2008; Yamada *et al.*, 2015). Terpenes play a role in respiration chain electron transport and in cell wall and membrane biosynthesis and stability, and are vital for growth and survival of photosynthetic organisms, since they are essential in conversion of light into chemical energy and for assembly and function of photosynthetic reaction centres (Pattanaik and Lindberg, 2015). They also have important roles in stress response and defence mechanisms (Mazid *et al.*, 2011; Tholl, 2006). Terpenoids are made from the five-carbon building blocks isopentenyl pyrophosphate and dimethylallyl pyrophosphate, and there is a large diversity of them in cyanobacteria, notably pigments with different functions and properties but also other compounds with a variety of structures and roles (Pattanaik and Lindberg, 2015). They are classified according to the number of five-carbon monomers (C_5) present in their structures and are synthesized by dedicated terpene synthases (Kitaoka *et al.*, 2015; Micallef *et al.*, 2015; Pattanaik and Lindberg, 2015), but only a few studies have addressed the basis of terpenoid biosynthesis and their role in cyanobacteria and much remain to be elucidated. antiSMASH analysis of the genome of *Anabaena* sp. PCC 7120 has shown the existence of four terpene gene clusters (Fig. 1.4, clusters 2, 6, 7 and 12). Such a variety of gene clusters related to the biosynthesis of terpenes suggests that they may play an important role in cellular and physiological processes in this cyanobacterium. Moreover, terpene biosynthetic gene clusters have been widely found in cyanobacterial genomes (Englund *et al.*, 2014; Fiore *et al.*, 2013; Lima *et al.*, 2014; Shih *et al.*, 2013), and may represent an important source for the discovery of new natural products (Yamada *et al.*, 2015).

At least one gene cluster was detected in the genomes of four cyanobacteria (Syn7, Syc1, Mic1, Ana1) for the biosynthesis of terpenes. However, the genome of *Arthrospira platensis* NIES-39 contains a kind of hybrid terpene–bacteriocin cluster, as mentioned above, which cannot be classified with the information available to date and further analysis will be required to decipher its function and classify it properly. Two terpene clusters have been identified in *Microcystis aeruginosa* NIES-843, being one of them more conserved (seven core-genome

CLOGs) than the other (two core-genome CLOGs). Two and three terpene gene clusters have been found in *Synechococcus elongatus* PCC 7942 and *Synechocystis* sp. PCC 6803, respectively, which contain up to six core-genome CLOGs. Interestingly, all the secondary metabolite gene clusters identified in *Synechocystis* sp. PCC 6803 are related to terpene biosynthesis pathways, while in *Synechococcus elongatus* PCC 7942 two of the three identified clusters are involved in this role. Similarly to the ribosomal peptide gene clusters, the four identified terpene gene clusters in *Anabaena* sp. PCC 7120 showed a lower conservation in contrast to the ones found in the other four cyanobacteria, represented by a maximum of two CLOGs in the core-genome per cluster (Table 1.2).

Cyanobacteria exhibit two separate pathways for hydrocarbon biosynthesis from fatty acids. The first pathway requires two enzymes for the conversion of fatty acids into alkanes, a fatty acyl-ACP reductase and an aldehyde deformylating oxygenase. This biosynthetic pathway has been identified in the genome of many cyanobacterial strains, including *Anabaena* sp. PCC 7120 (Coates *et al.*, 2014; Kothari *et al.*, 2013; Schirmer *et al.*, 2010). The second biosynthetic pathway in cyanobacteria for hydrocarbon biosynthesis is the olefin-producing pathway. This pathway requires a single PKS gene to convert fatty acids into alkenes. This pathway, however, has not been found in the genome of *Anabaena* sp. PCC 7120 (Coates *et al.*, 2014).

Ladderanes are hydrocarbon chains with three or five linearly concatenated cyclobutane rings that are uniquely produced as membrane lipid components by anaerobic ammonia-oxidising (anammox) bacteria (Jetten *et al.*, 2009; Kartal *et al.*, 2012; van Niftrik and Jetten, 2012). The biochemistry and genetics underlying the ladderane biosynthetic pathway are unknown and candidate genes for the biosynthesis of ladderanes have only been found in the anammox bacterium *Kuenenia stuttgartiensis* (Javidpour *et al.*, 2016; Rattray *et al.*, 2009). The structure of the ladderane lipids is unique in nature and they are hypothesized to make membranes highly impermeable (Kuypers *et al.*, 2003; Sinninghe Damste *et al.*, 2005; Sinninghe Damste *et al.*, 2002) to reduce the loss of ions and metabolites and to provide structural integrity to the cell due to a high lipid-packing density (Boumann *et al.*, 2009). Interestingly, and according to the analysis provided by antiSMASH, the genome of *Anabaena* sp. PCC 7120 contains a putative gene cluster involved in ladderane biosynthesis (Fig. 1.4, cluster 11).

The pangenome analysis has also allowed analysing the hypothetical ladderane gene cluster of *Anabaena* sp. PCC 7120 in more detail. Five genes (*all4272, all4284, alr4290, alr4291, all4296*) are included in CLOGs of the core-genome, meaning that they occur in all cyanobacteria. However, these genes are not necessarily involved in the production of ladderanes and could be related to the biosynthesis of other lipids rather than ladderanes. Overall, ~61% of the 36 genes within this gene cluster are present in more than 20 different cyanobacteria and, by this, are not only representatives of the order Nostocales (Table 1.2).

Secondary metabolite gene clusters in the cyanobacterial pangenome

Pangenomics can give insights into the evolutionary conservation of the function and occurrence of secondary metabolite biosynthesis gene clusters in cyanobacteria. The conservation of these clusters can be evaluated by calculating a pairwise CLOG distance neighbour-joining tree (Simm *et al.*, 2015). Orthologue searches for the identified secondary metabolite biosynthesis gene clusters resulted in 46 CLOGs for *Arthrospira platensis* NIES-39 (Art1), 54 CLOGs for *Synechococcus elongatus* PCC 7942 (Syn7), 61 CLOGs for *Synechocystis* sp. PCC 6803 (Syc1), 247 CLOGs for *Microcystis aeruginosa* NIES-843 (Mic1) and 292 CLOGs for *Anabaena* sp. PCC 7120 (Ana1) (Fig. 1.5, Table 1.2). However, only three (Art1) to 30 (Mic1) of the CLOGs for secondary metabolite gene clusters of the different species are present in the cyanobacterial core-genome (Fig. 1.5, split K). An accumulation of core genes is only found for clusters predicted to synthesize terpenes and ladderanes/lipids as shown for *Anabaena* sp. PCC 7120 (Fig. 1.4B, clusters 2, 6, 7, 11 and 12). This is in line with the observation that terpene clusters exist in all the species analysed. Remarkably, only considering these five species, the analysed secondary metabolite gene clusters contain ~13% of the 559 core CLOGs. In turn, only one of these 76 core-genome CLOGs are found in terpene biosynthesis clusters of all

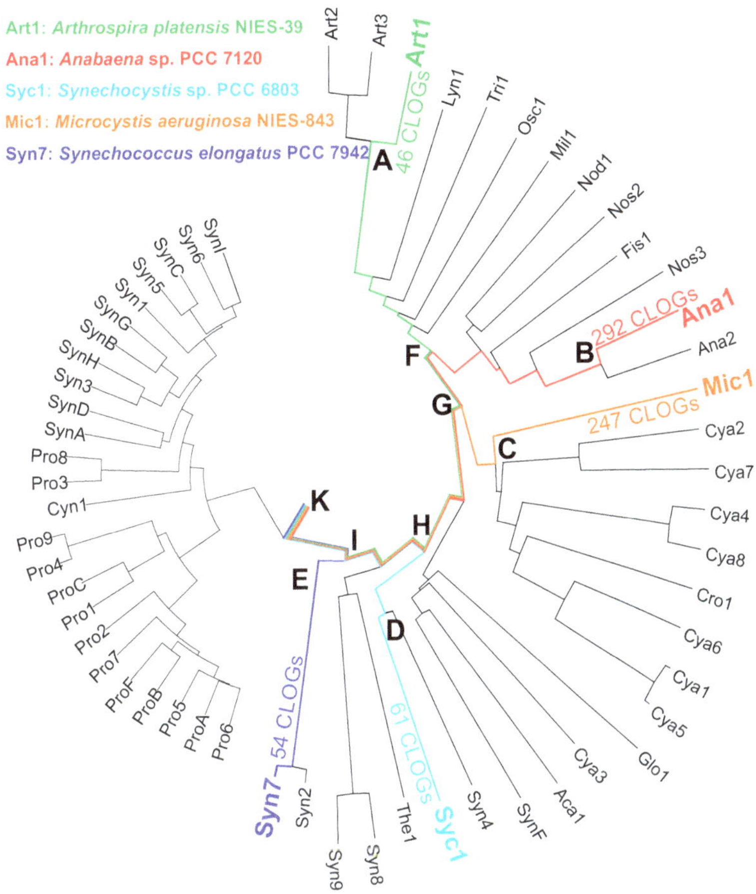

strain	split / cluster	A	B	C	D	E	F	G	H	I	K
Art1	2 cluster	25					8	5	5	3	3
Ana1	14 cluster		235				59	55	33	29	17
Mic1	10 cluster			108				62	41	38	30
Syc1	3 cluster				15				17	16	11
Syn7	3 cluster					19				20	15

Figure 1.5 Secondary metabolite gene cluster conservation in cyanobacteria. CLOG distance neighbour-joining tree based on the CLOG distances between two cyanobacterial species/strains (Simm *et al.*, 2015). The occurrence of genes for secondary metabolite synthesis within CLOGs in the various genomes was analysed starting from the species-specific genes in the branch and following the conservation to core-genome CLOGs in the root. Conservation and occurrence of predicted secondary metabolite CLOGs from cyanobacterial strains Art1 (46 CLOGs; green), Ana1 (292 CLOGs; red), Mic1 (247 CLOGs; orange), Syc1 (61 CLOGs; blue) and Syn7 (54 CLOGs; violet) are represented by coloured branches. The number of secondary metabolite CLOGs for specific splits (A–K) is indicated in the table. The species abbreviations are defined in (Simm *et al.*, 2015).

analysed species containing genes coding for a phytoene synthase. This documents that core CLOGs are not signatures for secondary metabolite biosynthesis gene clusters in general. Moreover, this also suggests that these clusters evolved by functional recycling of existing modules largely represented in the core-genome.

The low content of core-genome CLOGs in secondary metabolite gene clusters documents that these clusters are not globally conserved. In addition, the majority of CLOGs for secondary metabolite gene clusters contain unique genes or common genes in only few different cyanobacteria (Fig. 1.5). This shows that these gene clusters were developed *de novo* during the evolution of different species. However, it is still possible that closely related species share similar biosynthesis gene clusters depending on the time of the evolutionary split of the two species. Supporting this notion, most of the secondary metabolite biosynthesis CLOGs are common to the root between *Anabaena* sp. PCC 7120 and *Anabaena variabilis* (Fig. 1.4B; Fig. 1.5, split B). CLOGs containing genes coding for four of these secondary metabolite gene clusters (Fig. 1.4B, clusters 1, 8, 9, 14) only occur in *Anabaena* species.

Pangenomics approaches provide detailed information on the evolutionary development of organisms. In this respect, pangenomics shows that secondary metabolite gene clusters evolved late during evolution. Nevertheless, the analysis of functional equivalences for these gene clusters based on antiSMASH (Weber *et al.*, 2015) shows that gene clusters that may produce similar secondary metabolites such as terpenes exist in distantly related cyanobacterial species. This also shows that secondary metabolite biosynthesis clusters occur by convergent evolution on the base of pre-existing modules partially present in the core-genome.

Conclusions

Metagenome analysis like pangenomics and metagenomics define common sets of genes between different species and give insights into the composition and diversity of microbial communities in a specific environment (Teeling and Glockner, 2012). In addition, metagenomics provides information on the influence of cyanobacteria on a vast range of environments and microbial communities (Leary *et al.*, 2014; Sanli *et al.*, 2015) and on the formation of cyanobacterial communities in extreme environments (Badhai *et al.*, 2015). In contrast, pangenomics allows defining the conservation of genomes from a set of strains or species, a whole phylum or a super kingdom (Vernikos *et al.*, 2015).

The determination of the core-genome has led to the identification of eight signature genes as unique markers in cyanobacteria (Dutilh *et al.*, 2008). These genes allow the detection of cyanobacterial species in communities. Furthermore, the cyanobacterial pangenome is open, which means that new genes are expected from yet-unknown cyanobacterial genomes. The cyanobacterial core-genome has between 500 and 600 genes and mostly contains housekeeping functions. This allows the formulation of three notions:

1. Genes of the core-genome are of exceptional importance for viability of cyanobacteria as they have not been replaced during the course of evolution.
2. All habitat-specific functions either evolved after the last common ancestor of cyanobacteria or have been drastically modified during the billion years of evolution.
3. Core-genome genes can be used as quality control of newly assembled cyanobacterial genomes.

In addition, clade- and trait-specific gene sets determined by pangenomics (Simm *et al.*, 2015) allow the classification of cyanobacteria. For example, 66 genes specific for heterocyst-forming cyanobacteria were discovered with this approach. Similarly to the eight cyanobacterial unique markers, these genes can be used as signature markers for the classification of newly discovered heterocyst-forming cyanobacteria. Following this example, the classification of cyanobacteria with respect to other traits can be approached. Information on the conservation of metabolic pathways can be extracted from pangenomics as well. We exemplify this capacity on the example of the conservation of secondary metabolite biosynthetic gene clusters. As expected from the high diversity of metabolites in general, gene clusters entirely conserved in all analysed cyanobacteria do not exist. However, NRPS/PKS and ribosomal peptide gene clusters show a lower conservation when compared to gene clusters involved in the synthesis of ladderanes/lipids (*Anabaena* sp.) or terpenes as documented by the existence of genes of the cyanobacterial core-genome in the latter two kind of secondary metabolite gene clusters. Gene clusters for ladderane/lipids and terpenes occur in most of the analysed cyanobacteria, but they do not likely descend from a common ancestor and have evolved independently to provide adaptations to different needs for living in specific environments. The combination of functional predictions and pangenomics uncovers that in general secondary metabolite synthesis clusters evolved convergently by recycling modules existing in the last common ancestor of cyanobacteria. In addition, many of the genes in such clusters are present in unique CLOGs. Together with the observation that the pangenome of cyanobacteria is open, this suggests that new genes for secondary metabolite biosynthesis are yet to be discovered in uncharacterized species.

Future trends and challenges

Specialized, peer-reviewed, and continuously maintained databases for metagenomics and pangenomics data are of major importance. Further, intuitive and easy handling and visualization of such datasets would contribute to the common and frequent usage of these datasets for the daily laboratory work. Metagenome databases should store the sequences generated by NGS associated with the information about the sampling, DNA extraction and sequencing methods employed. Such databases are of higher importance than the storage of the analysed data, since analysis tools are constantly improved. It is also of major importance not to underestimate the second level of information, namely the information about the composition of a microbial community at the time point of the sample collection, which cannot be reproduced in a subsequent collection. However, this is valuable information which will enable the analysis of the long-term community development in the future. In addition, the inclusion of analysing and visualization tools of such data could help to create new hypotheses and answer fundamental questions.

With the generation of detailed metagenome and pangenome databases and the development of novel bioinformatics tools, the biosynthetic logic underpinning the chemical diversity of secondary metabolites in cyanobacteria is being slowly deciphered, establishing links between the known chemical diversity of secondary metabolites and genomes. These efforts in metagenome comparisons are allowing the identification of novel types of compounds and biosynthetic pathways, and are helping to understand the evolution of functional pathways in correlation to their ecological and physiological functions. In this sense, there is controversy about the physiological and ecological function of cyanotoxins and other secondary metabolites, and why cyanobacteria are prolific microorganisms for these compounds. They could be interpreted as the result of multiple pressures in nature to help cyanobacteria to adapt and compete under various environmental conditions, which could explain why these gene clusters are not conserved in all cyanobacteria and why the conservation of secondary metabolite gene clusters seems to be more influenced by habitat-specific traits.

Genome sequencing projects of cyanobacteria have revealed a far greater potential of these organisms to produce secondary metabolites than anticipated. Thus, it is expected that genome mining techniques that start with the theoretical prediction of gene clusters or orphan genes from genomic data will be of increasing importance to discover novel secondary metabolites in the near future. Metagenomics is also opening a new age in cyanobacterial bioactive compound discovery since it provides powerful tools to discover and exploit the vast repertoire of secondary metabolites produced by non-cultivable strains of cyanobacteria, including many marine strains, which have been reluctant to cultivation in laboratories and represent and immense source of novel secondary metabolites that has been kept inaccessible to our knowledge. With the continued development of bioinformatics tools to assemble and analyse genomes, it is feasible that in the near future sequencing and genome mining for biosynthesis genes clusters could routinely become the first step in isolating a novel compound. This will also lead to a better understanding of secondary metabolite functions in cyanobacteria and unveil new biosynthetic mechanisms that can inspire novel medical and biotechnological applications.

References

Abecasis, A.B., Serrano, M., Alves, R., Quintais, L., Pereira-Leal, J.B., and Henriques, A.O. (2013). A genomic signature and the identification of new sporulation genes. J. Bacteriol. *195*, 2101–2115. http://dx.doi.org/10.1128/JB.02110-12.

Afiahayati, Sato, K., and Sakakibara, Y. (2015). MetaVelvet-SL: an extension of the Velvet assembler to a de novo metagenomic assembler utilizing supervised learning. DNA Res. *22*, 69–77. http://dx.doi.org/10.1093/dnares/dsu041.

Agger, S.A., Lopez-Gallego, F., Hoye, T.R., and Schmidt-Dannert, C. (2008). Identification of sesquiterpene synthases from *Nostoc punctiforme* PCC 73102 and *Nostoc* sp. strain PCC 7120. J. Bacteriol. *190*, 6084–6096. http://dx.doi.org/10.1128/JB.00759-08.

An, T., Kumar, T.K., Wang, M., Liu, L., Lay, J.O., Liyanage, R., Berry, J., Gantar, M., Marks, V., Gawley, R.E., *et al.*

(2007). Structures of pahayokolides A and B, cyclic peptides from a *Lyngbya* sp. J. Nat. Prod. *70*, 730–735.

Andrews, S.C., Robinson, A.K., and Rodríguez-Quiñones, F. (2003). Bacterial iron homeostasis. FEMS Microbiol. Rev. *27*, 215–237.

Andrianasolo, E.H., Gross, H., Goeger, D., Musafija-Girt, M., McPhail, K., Leal, R.M., Mooberry, S.L., and Gerwick, W.H. (2005). Isolation of swinholide A and related glycosylated derivatives from two field collections of marine cyanobacteria. Org. Lett. *7*, 1375–1378.

Ansari, M.I., Harb, M., Jones, B., and Hong, P.Y. (2015). Molecular-based approaches to characterize coastal microbial community and their potential relation to the trophic state of Red Sea. Sci. Rep. *5*, 9001. http://dx.doi.org/10.1038/srep09001.

Arnison, P.G., Bibb, M.J., Bierbaum, G., Bowers, A.A., Bugni, T.S., Bulaj, G., Camarero, J.A., Campopiano, D.J., Challis, G.L., Clardy, J., *et al.* (2013). Ribosomally synthesized and post-translationally modified peptide natural products: overview and recommendations for a universal nomenclature. Natural Product Reports *30*, 108–160. http://dx.doi.org/10.1039/c2np20085f.

Badhai, J., Ghosh, T.S., and Das, S.K. (2015). Taxonomic and functional characteristics of microbial communities and their correlation with physicochemical properties of four geothermal springs in Odisha, India. Front. Microbiol. *6*, 1166. http://dx.doi.org/10.3389/fmicb.2015.01166.

Bashir, Y., Singh, S.P., and Konwar, K.B. (2014). Metagenomics: an application based perspective. Chin. J. Biol. *2014*, 7. http://dx.doi.org/10.1155/2014/146030

Bayjanov, J.R., Siezen, R.J., and van Hijum, S.A. (2010). PanCGHweb: a web tool for genotype calling in pangenome CGH data. Bioinformatics *26*, 1256–1257. http://dx.doi.org/10.1093/bioinformatics/btq103.

Beck, C., Knoop, H., Axmann, I.M., and Steuer, R. (2012). The diversity of cyanobacterial metabolism: genome analysis of multiple phototrophic microorganisms. BMC Genomics *13*, 56. http://dx.doi.org/10.1186/1471-2164-13-56.

Begley, M., Cotter, P.D., Hill, C., and Ross, R.P. (2009). Identification of a novel two-peptide lantibiotic, lichenicidin, following rational genome mining for LanM proteins. Appl. Environ. Microbiol. *75*, 5451–5460. http://dx.doi.org/10.1128/AEM.00730-09.

Belda-Ferre, P., Alcaraz, L.D., Cabrera-Rubio, R., Romero, H., Simón-Soro, A., Pignatelli, M., and Mira, A. (2012). The oral metagenome in health and disease. ISME J. *6*, 46–56. http://dx.doi.org/10.1038/ismej.2011.85.

Benedict, M.N., Henriksen, J.R., Metcalf, W.W., Whitaker, R.J., and Price, N.D. (2014). ITEP: an integrated toolkit for exploration of microbial pan-genomes. BMC Genomics *15*, 8. http://dx.doi.org/10.1186/1471-2164-15-8.

Bentley, S. (2009). Sequencing the species pan-genome. Nat. Rev. Microbiol. *7*, 258–259. http://dx.doi.org/10.1038/nrmicro2123.

Berube, P.M., Biller, S.J., Kent, A.G., Berta-Thompson, J.W., Roggensack, S.E., Roache-Johnson, K.H., Ackerman, M., Moore, L.R., Meisel, J.D., Sher, D., *et al.* (2015). Physiology and evolution of nitrate acquisition in *Prochlorococcus*. ISME J. *9*, 1195–1207. http://dx.doi.org/10.1038/ismej.2014.211.

Billis, K., Billini, M., Tripp, H.J., Kyrpides, N.C., and Mavromatis, K. (2014). Comparative transcriptomics between *Synechococcus* PCC 7942 and *Synechocystis* PCC 6803 provide insights into mechanisms of stress acclimation. PLoS One *9*, e109738. http://dx.doi.org/10.1371/journal.pone.0109738.

Blin, K., Kazempour, D., Wohlleben, W., and Weber, T. (2014). Improved lanthipeptide detection and prediction for antiSMASH. PLoS One *9*, e89420. http://dx.doi.org/10.1371/journal.pone.0089420.

Blin, K., Medema, M.H., Kazempour, D., Fischbach, M.A., Breitling, R., Takano, E., and Weber, T. (2013). antiSMASH 2.0 – a versatile platform for genome mining of secondary metabolite producers. Nucleic Acids Res. *41*, W204–12. http://dx.doi.org/10.1093/nar/gkt449.

Bohannon, J. (2007). Metagenomics. Ocean study yields a tidal wave of microbial DNA. Science *315*, 1486–1487.

Bombar, D., Heller, P., Sanchez-Baracaldo, P., Carter, B.J., and Zehr, J.P. (2014). Comparative genomics reveals surprising divergence of two closely related strains of uncultivated UCYN-A cyanobacteria. ISME J. *8*, 2530–2542. http://dx.doi.org/10.1038/ismej.2014.167.

Boumann, H.A., Longo, M.L., Stroeve, P., Poolman, B., Hopmans, E.C., Stuart, M.C., Sinninghe Damste, J.S., and Schouten, S. (2009). Biophysical properties of membrane lipids of anammox bacteria: I. Ladderane phospholipids form highly organized fluid membranes. Biochim. Biophys. Acta *1788*, 1444-1451. http://dx.doi.org/10.1016/j.bbamem.2009.04.008.

Bradford, M.A., Wallenstein, M.D., Allison, S.D., Treseder, K.K., Frey, S.D., Watts, B.W., Davies, C.A., Maddox, T.R., Melillo, J.M., Mohan, J.E., *et al.* (2009). Decreased mass specific respiration under experimental warming is robust to the microbial biomass method employed. Ecol. Lett. *12*, E15–E18. http://dx.doi.org/10.1111/j.1461-0248.2009.01332.x.

Brady, A., and Salzberg, S.L. (2009). Phymm and PhymmBL: metagenomic phylogenetic classification with interpolated Markov models. Nat. Methods *6*, 673–676. http://dx.doi.org/10.1038/nmeth.1358.

Breitbart, M., Hewson, I., Felts, B., Mahaffy, J.M., Nulton, J., Salamon, P., and Rohwer, F. (2003). Metagenomic analyses of an uncultured viral community from human feces. J. Bacteriol. *185*, 6220–6223.

Brittnacher, M.J., Fong, C., Hayden, H.S., Jacobs, M.A., Radey, M., and Rohmer, L. (2011). PGAT: a multistrain analysis resource for microbial genomes. Bioinformatics *27*, 2429–2430. http://dx.doi.org/10.1093/bioinformatics/btr418.

Broadbent, J.R., Neeno-Eckwall, E.C., Stahl, B., Tandee, K., Cai, H., Morovic, W., Horvath, P., Heidenreich, J., Perna, N.T., Barrangou, R., *et al.* (2012). Analysis of the *Lactobacillus casei* supragenome and its influence in species evolution and lifestyle adaptation. BMC Genomics *13*, 533. http://dx.doi.org/10.1186/1471-2164-13-533.

Burja, A.M., Banaigs, B., Abou-Mansour, E., Burgess, J.G., and Wright, P.C. (2001). Marine cyanobacteria – a prolific source of natural products. Tetrahedron 57, 9347–9377. http://dx.doi.org/10.1016/S0040-4020(01)00931-0.

Calteau, A., Fewer, D.P., Latifi, A., Coursin, T., Laurent, T., Jokela, J., Kerfeld, C.A., Sivonen, K., Piel, J., and Gugger, M. (2014). Phylum-wide comparative genomics unravel the diversity of secondary metabolism in Cyanobacteria. BMC Genomics 15, 977. http://dx.doi.org/10.1186/1471-2164-15-977.

Campbell, E.L., Cohen, M.F., and Meeks, J.C. (1997). A polyketide-synthase-like gene is involved in the synthesis of heterocyst glycolipids in *Nostoc punctiforme* strain ATCC 29133. Arch. Microbiol. 167, 251–258.

Carey, C.C., Ibelings, B.W., Hoffmann, E.P., Hamilton, D.P., and Brookes, J.D. (2012). Eco-physiological adaptations that favour freshwater cyanobacteria in a changing climate. Water Res. 46, 1394–1407. http://dx.doi.org/10.1016/j.watres.2011.12.016.

Cimermancic, P., Medema, M.H., Claesen, J., Kurita, K., Wieland Brown, L.C., Mavrommatis, K., Pati, A., Godfrey, P.A., Koehrsen, M., Clardy, J., *et al.* (2014). Insights into secondary metabolism from a global analysis of prokaryotic biosynthetic gene clusters. Cell 158, 412–421. http://dx.doi.org/10.1016/j.cell.2014.06.034.

Coates, R.C., Podell, S., Korobeynikov, A., Lapidus, A., Pevzner, P., Sherman, D.H., Allen, E.E., Gerwick, L., and Gerwick, W.H. (2014). Characterization of cyanobacterial hydrocarbon composition and distribution of biosynthetic pathways. PLoS One 9, e85140. http://dx.doi.org/10.1371/journal.pone.0085140.

Collingro, A., Tischler, P., Weinmaier, T., Penz, T., Heinz, E., Brunham, R.C., Read, T.D., Bavoil, P.M., Sachse, K., Kahane, S., *et al.* (2011). Unity in variety – the pan-genome of the Chlamydiae. Mol. Biol. Evol. 28, 3253–3270. http://dx.doi.org/10.1093/molbev/msr161.

Collins, R.E., and Higgs, P.G. (2012). Testing the infinitely many genes model for the evolution of the bacterial core genome and pangenome. Mol. Biol. Evol. 29, 3413–3425. http://dx.doi.org/10.1093/molbev/mss163.

Condurso, H.L., and Bruner, S.D. (2012). Structure and noncanonical chemistry of nonribosomal peptide biosynthetic machinery. Nat. Prod. Rep. 29, 1099–1110. http://dx.doi.org/10.1039/c2np20023f.

Contreras-Moreira, B., and Vinuesa, P. (2013). GET_HOMOLOGUES, a versatile software package for scalable and robust microbial pangenome analysis. Appl. Environ. Microbiol. 79, 7696–7701. http://dx.doi.org/10.1128/AEM.02411-13.

Corre, C., and Challis, G.L. (2009). New natural product biosynthetic chemistry discovered by genome mining. Nat. Prod. Rep. 26, 977–986. http://dx.doi.org/10.1039/b713024b.

Crosa, J.H., and Walsh, C.T. (2002). Genetics and assembly line enzymology of siderophore biosynthesis in bacteria. Microbiol. Mol. Biol. Rev. 66, 223–249.

Challis, G.L. (2005). A widely distributed bacterial pathway for siderophore biosynthesis independent of nonribosomal peptide synthetases. Chembiochem 6, 601–611.

Challis, G.L., and Naismith, J.H. (2004). Structural aspects of non-ribosomal peptide biosynthesis. Curr. Opin. Struct. Biol. 14, 748–756.

Chang, T.T., More, S.V., Lu, I.H., Hsu, J.C., Chen, T.J., Jen, Y.C., Lu, C.K., and Li, W.S. (2011). Isomalyngamide A, A-1 and their analogs suppress cancer cell migration in vitro. Eur. J. Med. Chem. 46, 3810–3819. http://dx.doi.org/10.1016/j.ejmech.2011.05.049.

Chevreux, B., Pfisterer, T., Drescher, B., Driesel, A.J., Müller, W.E., Wetter, T., and Suhai, S. (2004). Using the miraEST assembler for reliable and automated mRNA transcript assembly and SNP detection in sequenced ESTs. Genome Res. 14, 1147–1159.

Chodak, M., Golebiewski, M., Morawska-Ploskonka, J., Kuduk, K., and Niklinska, M. (2013). Diversity of microorganisms from forest soils differently polluted with heavy metals. Applied soil ecology 64, 7–14. http://dx.doi.org/10.1016/j.apsoil.2012.11.004.

D'Auria, G., Jiménez-Hernández, N., Peris-Bondia, F., Moya, A., and Latorre, A. (2010). *Legionella pneumophila* pangenome reveals strain-specific virulence factors. BMC Genomics 11, 181. http://dx.doi.org/10.1186/1471-2164-11-181.

Dai, J., Chen, Y., Dean, S., Morris, J.G., Salfinger, M., and Johnson, J.A. (2011). Multiple-genome comparison reveals new loci for *Mycobacterium* species identification. J. Clin. Microbiol. 49, 144–153. http://dx.doi.org/10.1128/JCM.00957-10.

Daniel, R. (2005). The metagenomics of soil. Nat. Rev. Microbiol. 3, 470–478.

de Oliveira, L.S., Gregoracci, G.B., Silva, G.G., Salgado, L.T., Filho, G.A., Alves-Ferreira, M., Pereira, R.C., and Thompson, F.L. (2012). Transcriptomic analysis of the red seaweed *Laurencia dendroidea* (Florideophyceae, Rhodophyta) and its microbiome. BMC Genomics 13, 487. http://dx.doi.org/10.1186/1471-2164-13-487.

Demain, A.L., and Sanchez, S. (2009). Microbial drug discovery: 80 years of progress. J. Antibiot. 62, 5–16. http://dx.doi.org/10.1038/ja.2008.16.

Dirix, G., Monsieurs, P., Dombrecht, B., Daniels, R., Marchal, K., Vanderleyden, J., and Michiels, J. (2004). Peptide signal molecules and bacteriocins in Gram-negative bacteria: a genome-wide in silico screening for peptides containing a double-glycine leader sequence and their cognate transporters. Peptides 25, 1425–1440. http://dx.doi.org/10.1016/j.peptides.2003.10.028.

Dittmann, E., Fewer, D.P., and Neilan, B.A. (2013). Cyanobacterial toxins: biosynthetic routes and evolutionary roots. FEMS Microbiol. Rev. 37, 23–43. http://dx.doi.org/10.1111/j.1574-6976.2012.12000.x.

Dittmann, E., Gugger, M., Sivonen, K., and Fewer, D.P. (2015). Natural product biosynthetic diversity and comparative genomics of the cyanobacteria. Trends Microbiol. 23, 642–652. http://dx.doi.org/10.1016/j.tim.2015.07.008.

Dittmann, E., Neilan, B.A., Erhard, M., von Döhren, H., and Börner, T. (1997). Insertional mutagenesis of a peptide synthetase gene that is responsible for hepatotoxin production in the cyanobacterium *Microcystis aeruginosa* PCC 7806. Mol. Microbiol. 26, 779–787.

Donadio, S., Monciardini, P., and Sosio, M. (2007). Polyketide synthases and nonribosomal peptide

synthetases: the emerging view from bacterial genomics. Nat. Prod. Rep. *24*, 1073–1109.
Donati, C., Hiller, N.L., Tettelin, H., Muzzi, A., Croucher, N.J., Angiuoli, S.V., Oggioni, M., Dunning Hotopp, J.C., Hu, F.Z., Riley, D.R., *et al.* (2010). Structure and dynamics of the pan-genome of *Streptococcus pneumoniae* and closely related species. Genome Biol. *11*, R107. http://dx.doi.org/10.1186/gb-2010-11-10-r107.
Donia, M.S., Ravel, J., and Schmidt, E.W. (2008). A global assembly line for cyanobactins. Nat. Chem. Biol. *4*, 341–343. http://dx.doi.org/10.1038/nchembio.84.
Donia, M.S., and Schmidt, E.W. (2011). Linking chemistry and genetics in the growing cyanobactin natural products family. Chem. Biol. *18*, 508–519. http://dx.doi.org/10.1016/j.chembiol.2011.01.019.
Draper, L.A., Cotter, P.D., Hill, C., and Ross, R.P. (2015). Lantibiotic resistance. Microbiol. Mol. Biol. Rev. *79*, 171–191. http://dx.doi.org/10.1128/MMBR.00051-14.
Dunn, B., Richter, C., Kvitek, D.J., Pugh, T., and Sherlock, G. (2012). Analysis of the *Saccharomyces cerevisiae* pan-genome reveals a pool of copy number variants distributed in diverse yeast strains from differing industrial environments. Genome research *22*, 908-924. http://dx.doi.org/10.1101/gr.130310.111.
Dutilh, B.E., Snel, B., Ettema, T.J., and Huynen, M.A. (2008). Signature genes as a phylogenomic tool. Mol. Biol. Evol. *25*, 1659–1667. http://dx.doi.org/10.1093/molbev/msn115.
Englund, E., Pattanaik, B., Ubhayasekera, S.J., Stensjö, K., Bergquist, J., and Lindberg, P. (2014). Production of squalene in *Synechocystis* sp. PCC 6803. PLoS One *9*, e90270. http://dx.doi.org/10.1371/journal.pone.0090270.
Eppinger, M., Bunk, B., Johns, M.A., Edirisinghe, J.N., Kutumbaka, K.K., Koenig, S.S., Creasy, H.H., Rosovitz, M.J., Riley, D.R., Daugherty, S., *et al.* (2011). Genome sequences of the biotechnologically important *Bacillus megaterium* strains QM B1551 and DSM319. J. Bacteriol. *193*, 4199–4213. http://dx.doi.org/10.1128/JB.00449-11.
Fan, Q., Huang, G., Lechno-Yossef, S., Wolk, C.P., Kaneko, T., and Tabata, S. (2005). Clustered genes required for synthesis and deposition of envelope glycolipids in *Anabaena* sp. strain PCC 7120. Mol. Microbiol. *58*, 227–243.
Fiore, M.F., Alvarenga, D.O., Varani, A.M., Hoff-Risseti, C., Crespim, E., Ramos, R.T., Silva, A., Schaker, P.D., Heck, K., Rigonato, J., *et al.* (2013). Draft genome sequence of the Brazilian toxic bloom-forming cyanobacterium *Microcystis aeruginosa* strain SPC777. Genome announcements *1*, e00547–13. http://dx.doi.org/10.1128/genomeA.00547-13.
Fischbach, M.A. (2009). Antibiotics from microbes: converging to kill. Curr. Opin. Microbiol. *12*, 520–527. http://dx.doi.org/10.1016/j.mib.2009.07.002.
Fischbach, M.A., and Walsh, C.T. (2006). Assembly-line enzymology for polyketide and nonribosomal Peptide antibiotics: logic, machinery, and mechanisms. Chem. Rev. *106*, 3468–3496.
Flaherty, B.L., Van Nieuwerburgh, F., Head, S.R., and Golden, J.W. (2011). Directional RNA deep sequencing sheds new light on the transcriptional response of *Anabaena* sp. strain PCC 7120 to combined-nitrogen deprivation. BMC Genomics *12*, 332. http://dx.doi.org/10.1186/1471-2164-12-332.
Frangeul, L., Quillardet, P., Castets, A.M., Humbert, J.F., Matthijs, H.C., Cortez, D., Tolonen, A., Zhang, C.C., Gribaldo, S., Kehr, J.C., *et al.* (2008). Highly plastic genome of *Microcystis aeruginosa* PCC 7806, a ubiquitous toxic freshwater cyanobacterium. BMC Genomics *9*, 274. http://dx.doi.org/10.1186/1471-2164-9-274.
Gabor, E., Liebeton, K., Niehaus, F., Eck, J., and Lorenz, P. (2007). Updating the metagenomics toolbox. Biotechnol. J. *2*, 201–206.
Garcia-Pichel, F., Johnson, S.L., Youngkin, D., and Belnap, J. (2003). Small-scale vertical distribution of bacterial biomass and diversity in biological soil crusts from arid lands in the Colorado plateau. Microb. Ecol. *46*, 312–321.
Gołębiewski, M., Deja-Sikora, E., Cichosz, M., Tretyn, A., and Wróbel, B. (2014). 16S rDNA pyrosequencing analysis of bacterial community in heavy metals polluted soils. Microb. Ecol. *67*, 635–647. http://dx.doi.org/10.1007/s00248-013-0344-7.
Gross, H. (2009). Genomic mining – a concept for the discovery of new bioactive natural products. Curr. Opin. Drug Discov. Devel. *12*, 207-219.
Gulder, T.A., Freeman, M.F., and Piel, J. (2011). The catalytic diversity of multimodular polyketide synthases: Natural product biosynthesis beyond textbook assembly rules. Topics in current chemistry, 1–53. http://dx.doi.org/10.1007/128_2010_113.
Gupta, R.S., and Mathews, D.W. (2010). Signature proteins for the major clades of Cyanobacteria. BMC Evol. Biol. *10*, 24. http://dx.doi.org/10.1186/1471-2148-10-24.
Haas, H., Eisendle, M., and Turgeon, B.G. (2008). Siderophores in fungal physiology and virulence. Annu. Rev. Phytopathol. *46*, 149–187. http://dx.doi.org/10.1146/annurev.phyto.45.062806.094338.
Han, B., Goeger, D., Maier, C.S., and Gerwick, W.H. (2005). The wewakpeptins, cyclic depsipeptides from a Papua New Guinea collection of the marine cyanobacterium *Lyngbya semiplena*. J. Org. Chem. *70*, 3133–3139.
Handelsman, J. (2004). Metagenomics: application of genomics to uncultured microorganisms. Microbiol. Mol. Biol. Rev. *68*, 669–685.
Handelsman, J. (2009). Metagenetics: spending our inheritance on the future. Microb. Biotechnol. *2*, 138–139. http://dx.doi.org/10.1111/j.1751-7915.2009.00090_8.x.
Handelsman, J., Rondon, M.R., Brady, S.F., Clardy, J., and Goodman, R.M. (1998). Molecular biological access to the chemistry of unknown soil microbes: a new frontier for natural products. Chem. Biol. *5*, R245–9.
Hardison, R.C. (2003). Comparative genomics. PLOS Biol. *1*, E58.
Harris, J.K., Caporaso, J.G., Walker, J.J., Spear, J.R., Gold, N.J., Robertson, C.E., Hugenholtz, P., Goodrich, J., McDonald, D., Knights, D., *et al.* (2013). Phylogenetic stratigraphy in the Guerrero Negro hypersaline microbial mat. ISME J. *7*, 50–60. http://dx.doi.org/10.1038/ismej.2012.79.

Hess, W.R. (2011). Cyanobacterial genomics for ecology and biotechnology. Curr. Opin. Microbiol. *14*, 608–614. http://dx.doi.org/10.1016/j.mib.2011.07.024.

Hill, A.M. (2006). The biosynthesis, molecular genetics and enzymology of the polyketide-derived metabolites. Nat. Prod. Rep. 23, 256–320.

Hrouzek, P., Kuzma, M., Cerny, J., Novak, P., Fiser, R., Simek, P., Lukesova, A., and Kopecky, J. (2012). The cyanobacterial cyclic lipopeptides puwainaphycins F/G are inducing necrosis via cell membrane permeabilization and subsequent unusual actin relocalization. Chemical research in toxicology 25, 1203-1211. http://dx.doi.org/10.1021/tx300044t.

Humbert, J.F., Barbe, V., Latifi, A., Gugger, M., Calteau, A., Coursin, T., Lajus, A., Castelli, V., Oztas, S., Samson, G., *et al.* (2013). A tribute to disorder in the genome of the bloom-forming freshwater cyanobacterium *Microcystis aeruginosa*. PLoS One *8*, e70747. http://dx.doi.org/10.1371/journal.pone.0070747.

Hunter, S., Corbett, M., Denise, H., Fraser, M., Gonzalez-Beltran, A., Hunter, C., Jones, P., Leinonen, R., McAnulla, C., Maguire, E., *et al.* (2014). EBI metagenomics – a new resource for the analysis and archiving of metagenomic data. Nucleic Acids Res. *42*, D600–6. http://dx.doi.org/10.1093/nar/gkt961.

Huson, D.H., and Weber, N. (2013). Microbial community analysis using MEGAN. Methods Enzymol. *531*, 465–485. http://dx.doi.org/10.1016/B978-0-12-407863-5.00021-6.

Ishida, K., Welker, M., Christiansen, G., Cadel-Six, S., Bouchier, C., Dittmann, E., Hertweck, C., and Tandeau de Marsac, N. (2009). Plasticity and evolution of aeruginosin biosynthesis in cyanobacteria. Appl. Environ. Microbiol. *75*, 2017–2026. http://dx.doi.org/10.1128/AEM.02258-08.

Jacobsen, A., Hendriksen, R.S., Aaresturp, F.M., Ussery, D.W., and Friis, C. (2011). The *Salmonella enterica* pan-genome. Microb. Ecol. *62*, 487–504. http://dx.doi.org/10.1007/s00248-011-9880-1.

Javidpour, P., Deutsch, S., Mutalik, V.K., Hillson, N.J., Petzold, C.J., Keasling, J.D., and Beller, H.R. (2016). Investigation of proposed ladderane biosynthetic genes from anammox bacteria by heterologous expression in *E. coli*. PLoS One *11*, e0151087. http://dx.doi.org/10.1371/journal.pone.0151087.

Jenke-Kodama, H., Müller, R., and Dittmann, E. (2008). Evolutionary mechanisms underlying secondary metabolite diversity. Prog. Drug Res. *65*, 119, 121–140.

Jetten, M.S., Niftrik, L.v., Strous, M., Kartal, B., Keltjens, J.T., and Op den Camp, H.J. (2009). Biochemistry and molecular biology of anammox bacteria. Crit. Rev. Biochem. Mol. Biol. *44*, 65–84. http://dx.doi.org/10.1080/10409230902722783.

Johnson, S.S., Chevrette, M.G., Ehlmann, B.L., and Benison, K.C. (2015). Insights from the metagenome of an acid salt lake: the role of biology in an extreme depositional environment. PLoS One *10*, e0122869. http://dx.doi.org/10.1371/journal.pone.0122869.

Jones, A.C., Gu, L., Sorrels, C.M., Sherman, D.H., and Gerwick, W.H. (2009). New tricks from ancient algae: natural products biosynthesis in marine cyanobacteria. Curr. Opin. Chem. Biol. *13*, 216–223. http://dx.doi.org/10.1016/j.cbpa.2009.02.019.

Jones, A.C., Monroe, E.A., Eisman, E.B., Gerwick, L., Sherman, D.H., and Gerwick, W.H. (2010). The unique mechanistic transformations involved in the biosynthesis of modular natural products from marine cyanobacteria. Nat. Prod. Rep. 27, 1048–1065. http://dx.doi.org/10.1039/c000535e.

Jones, A.C., Monroe, E.A., Podell, S., Hess, W.R., Klages, S., Esquenazi, E., Niessen, S., Hoover, H., Rothmann, M., Lasken, R.S., *et al.* (2011). Genomic insights into the physiology and ecology of the marine filamentous cyanobacterium *Lyngbya majuscula*. Proc. Natl. Acad. Sci. U.S.A. *108*, 8815–8820. http://dx.doi.org/10.1073/pnas.1101137108.

Kaneko, T., Nakajima, N., Okamoto, S., Suzuki, I., Tanabe, Y., Tamaoki, M., Nakamura, Y., Kasai, F., Watanabe, A., Kawashima, K., *et al.* (2007). Complete genomic structure of the bloom-forming toxic cyanobacterium *Microcystis aeruginosa* NIES-843. DNA Res. *14*, 247–256. http://dx.doi.org/10.1093/dnares/dsm026.

Kartal, B., van Niftrik, L., Keltjens, J.T., Op den Camp, H.J., and Jetten, M.S. (2012). Anammox – growth physiology, cell biology, and metabolism. Adv. Microb. Physiol. *60*, 211–262. http://dx.doi.org/10.1016/B978-0-12-398264-3.00003-6.

Kehr, J.C., Gatte Picchi, D., and Dittmann, E. (2011). Natural product biosyntheses in cyanobacteria: A treasure trove of unique enzymes. Beilstein J. Org. Chem. 7, 1622–1635. http://dx.doi.org/10.3762/bjoc.7.191.

Kellmann, R., Mihali, T.K., Jeon, Y.J., Pickford, R., Pomati, F., and Neilan, B.A. (2008). Biosynthetic intermediate analysis and functional homology reveal a saxitoxin gene cluster in cyanobacteria. Appl. Environ. Microbiol. 74, 4044–4053. http://dx.doi.org/10.1128/AEM.00353-08.

Kelly, L., Ding, H., Huang, K.H., Osburne, M.S., and Chisholm, S.W. (2013). Genetic diversity in cultured and wild marine cyanomyoviruses reveals phosphorus stress as a strong selective agent. ISME J. 7, 1827–1841. http://dx.doi.org/10.1038/ismej.2013.58.

Kettler, G.C., Martiny, A.C., Huang, K., Zucker, J., Coleman, M.L., Rodrigue, S., Chen, F., Lapidus, A., Ferriera, S., Johnson, J., *et al.* (2007). Patterns and implications of gene gain and loss in the evolution of *Prochlorococcus*. PLOS Genet. 3, e231.

Kilias, S.P., Nomikou, P., Papanikolaou, D., Polymenakou, P.N., Godelitsas, A., Argyraki, A., Carey, S., Gamaletsos, P., Mertzimekis, T.J., Stathopoulou, E., *et al.* (2013). New insights into hydrothermal vent processes in the unique shallow-submarine arc-volcano, Kolumbo (Santorini), Greece. Sci. Rep. 3, 2421. http://dx.doi.org/10.1038/srep02421.

Kitaoka, N., Lu, X., Yang, B., and Peters, R.J. (2015). The application of synthetic biology to elucidation of plant mono-, sesqui-, and diterpenoid metabolism. Mol. Plant. *8*, 6–16. http://dx.doi.org/10.1016/j.molp.2014.12.002.

Koboldt, D.C., Steinberg, K.M., Larson, D.E., Wilson, R.K., and Mardis, E.R. (2013). The next-generation sequencing revolution and its impact on genomics.

Cell 155, 27–38. http://dx.doi.org/10.1016/j.cell.2013.09.006.

Konz, D., and Marahiel, M.A. (1999). How do peptide synthetases generate structural diversity? Chem. Biol. 6, R39–48.

Kothari, A., Vaughn, M., and Garcia-Pichel, F. (2013). Comparative genomic analyses of the cyanobacterium, *Lyngbya aestuarii* BL J, a powerful hydrogen producer. Front. Microbiol. 4, 363. http://dx.doi.org/10.3389/fmicb.2013.00363.

Krause, L., Diaz, N.N., Goesmann, A., Kelley, S., Nattkemper, T.W., Rohwer, F., Edwards, R.A., and Stoye, J. (2008). Phylogenetic classification of short environmental DNA fragments. Nucleic Acids Res. 36, 2230–2239. http://dx.doi.org/10.1093/nar/gkn038.

Kristensen, D.M., Waller, A.S., Yamada, T., Bork, P., Mushegian, A.R., and Koonin, E.V. (2013). Orthologous gene clusters and taxon signature genes for viruses of prokaryotes. J. Bacteriol. 195, 941–950. http://dx.doi.org/10.1128/JB.01801-12.

Kumar, K., Mella-Herrera, R.A., and Golden, J.W. (2010). Cyanobacterial heterocysts. Cold Spring Harbor Perspect. Biol. 2, a000315. http://dx.doi.org/10.1101/cshperspect.a000315.

Kuypers, M.M., Sliekers, A.O., Lavik, G., Schmid, M., Jorgensen, B.B., Kuenen, J.G., Sinninghe Damste, J.S., Strous, M., and Jetten, M.S. (2003). Anaerobic ammonium oxidation by anammox bacteria in the Black Sea. Nature 422, 608–611. http://dx.doi.org/10.1038/nature01472.

Kwan, J.C., Donia, M.S., Han, A.W., Hirose, E., Haygood, M.G., and Schmidt, E.W. (2012). Genome streamlining and chemical defense in a coral reef symbiosis. Proc. Natl. Acad. Sci. U.S.A. 109, 20655–20660. http://dx.doi.org/10.1073/pnas.1213820109.

Laing, C., Buchanan, C., Taboada, E.N., Zhang, Y., Kropinski, A., Villegas, A., Thomas, J.E., and Gannon, V.P. (2010). Pan-genome sequence analysis using Panseq: an online tool for the rapid analysis of core and accessory genomic regions. BMC Bioinf. 11, 461. http://dx.doi.org/10.1186/1471-2105-11-461.

Lapidus, A., Goltsman, E., Auger, S., Galleron, N., Ségurens, B., Dossat, C., Land, M.L., Broussolle, V., Brillard, J., Guinebretiere, M.H., *et al.* (2008). Extending the Bacillus cereus group genomics to putative food-borne pathogens of different toxicity. Chem. Biol. Interact. 171, 236–249.

Lapierre, P., and Gogarten, J.P. (2009). Estimating the size of the bacterial pan-genome. Trends Genet. 25, 107–110. http://dx.doi.org/10.1016/j.tig.2008.12.004.

Leary, D.H., Li, R.W., Hamdan, L.J., Hervey, W.J., Lebedev, N., Wang, Z., Deschamps, J.R., Kusterbeck, A.W., and Vora, G.J. (2014). Integrated metagenomic and metaproteomic analyses of marine biofilm communities. Biofouling 30, 1211–1223. http://dx.doi.org/10.1080/08927014.2014.977267.

Lee, S.W., Mitchell, D.A., Markley, A.L., Hensler, M.E., Gonzalez, D., Wohlrab, A., Dorrestein, P.C., Nizet, V., and Dixon, J.E. (2008). Discovery of a widely distributed toxin biosynthetic gene cluster. Proc. Natl. Acad. Sci. U.S.A. 105, 5879–5884. http://dx.doi.org/10.1073/pnas.0801338105.

Lefébure, T., and Stanhope, M.J. (2007). Evolution of the core and pan-genome of *Streptococcus*: positive selection, recombination, and genome composition. Genome Biol. 8, R71.

Leikoski, N., Fewer, D.P., Jokela, J., Wahlsten, M., Rouhiainen, L., and Sivonen, K. (2010). Highly diverse cyanobactins in strains of the genus *Anabaena*. Appl. Environ. Microbiol. 76, 701–709. http://dx.doi.org/10.1128/AEM.01061-09.

Li, B., Sher, D., Kelly, L., Shi, Y., Huang, K., Knerr, P.J., Joewono, I., Rusch, D., Chisholm, S.W., and van der Donk, W.A. (2010a). Catalytic promiscuity in the biosynthesis of cyclic peptide secondary metabolites in planktonic marine cyanobacteria. Proc. Natl. Acad. Sci. U.S.A. 107, 10430–10435. http://dx.doi.org/10.1073/pnas.0913677107.

Li, L., Stoeckert, C.J., and Roos, D.S. (2003). OrthoMCL: identification of ortholog groups for eukaryotic genomes. Genome Res. 13, 2178–2189.

Li, M.H., Ung, P.M., Zajkowski, J., Garneau-Tsodikova, S., and Sherman, D.H. (2009). Automated genome mining for natural products. BMC Bioinf. 10, 185. http://dx.doi.org/10.1186/1471-2105-10-185.

Li, R., Li, Y., Zheng, H., Luo, R., Zhu, H., Li, Q., Qian, W., Ren, Y., Tian, G., Li, J., *et al.* (2010b). Building the sequence map of the human pan-genome. Nat. Biotechnol. 28, 57–63. http://dx.doi.org/10.1038/nbt.1596.

Lima, A.R., Siqueira, A.S., Dos Santos, B.G., da Silva, F.D., Lima, C.P., Cardoso, J.F., Vianez Júnior, J.L., Dall'Agnol, L.T., McCulloch, J.A., Nunes, M.R., *et al.* (2014). Draft genome sequence of the Brazilian *Cyanobium* sp. strain CACIAM 14. Genome Announc. 2, e00669–14. http://dx.doi.org/10.1128/genomeA.00669-14.

Liou, G.F., and Khosla, C. (2003). Building-block selectivity of polyketide synthases. Curr. Opin. Chem. Biol. 7, 279–284.

Liu, B., Gibbons, T., Ghodsi, M., Treangen, T., and Pop, M. (2011). Accurate and fast estimation of taxonomic profiles from metagenomic shotgun sequences. BMC Genomics 12 (Suppl. 2), S4. http://dx.doi.org/10.1186/1471-2164-12-S2-S4.

Liu, L., and Rein, K.S. (2010). New peptides isolated from *Lyngbya* species: a review. Mar. Drugs 8, 1817–1837. http://dx.doi.org/10.3390/md8061817.

Lorenz, P., and Eck, J. (2005). Metagenomics and industrial applications. Nat. Rev. Microbiol. 3, 510–516.

MacMillan, J.B., Ernst-Russell, M.A., de Ropp, J.S., and Molinski, T.F. (2002). Lobocyclamides A-C, lipopeptides from a cryptic cyanobacterial mat containing *Lyngbya confervoides*. J. Org. Chem. 67, 8210–8215.

Mandal, S., and Rath, J. (2015). Extremophilic cyanobacteria for novel drug development. Springer International Publishing. http://dx.doi.org/10.1007/978-3-319-12009-6.

Marahiel, M.A., Stachelhaus, T., and Mootz, H.D. (1997). Modular peptide synthetases involved in nonribosomal peptide synthesis. Chem. Rev. 97, 2651–2674.

Mares, J., Hajek, J., Urajova, P., Kopecky, J., and Hrouzek, P. (2014). A hybrid non-ribosomal peptide/polyketide synthetase containing fatty-acyl ligase (FAAL) synthesizes the beta-amino fatty acid lipopeptides puwainaphycins in the cyanobacterium *Cylindrospermum alatosporum*. PloS ONE 9, e111904. http://dx.doi.org/10.1371/journal.pone.0111904.

Martin, K.A., Siefert, J.L., Yerrapragada, S., Lu, Y., McNeill, T.Z., Moreno, P.A., Weinstock, G.M., Widger, W.R., and Fox, G.E. (2003). Cyanobacterial signature genes. Photosynth. Res. 75, 211–221.

Maru, N., Ohno, O., and Uemura, D. (2010). Lyngbyacyclamides A and B, novel cytotoxic peptides from marine cyanobacteria *Lyngbya* sp. Tetrahedron letters *51*, 6384-6387. http://dx.doi.org/10.1016/j.tetlet.2010.06.105.

Mazid, M., Khan, T.A., and Mohammad, F. (2011). Role of secondary metabolites in defense mechanisms of plants. Biology and medicine 3, 232-249.

McCranie, E.K., and Bachmann, B.O. (2014). Bioactive oligosaccharide natural products. Nat. Prod. Rep. *31*, 1026–1042. http://dx.doi.org/10.1039/c3np70128j.

McDaniel, R., Welch, M., and Hutchinson, C.R. (2005). Genetic approaches to polyketide antibiotics. 1. Chem. Rev. *105*, 543–558.

Medema, M.H., Blin, K., Cimermancic, P., de Jager, V., Zakrzewski, P., Fischbach, M.A., Weber, T., Takano, E., and Breitling, R. (2011). antiSMASH: rapid identification, annotation and analysis of secondary metabolite biosynthesis gene clusters in bacterial and fungal genome sequences. Nucleic Acids Res. 39, W339-346. http://dx.doi.org/10.1093/nar/gkr466.

Medini, D., Donati, C., Tettelin, H., Masignani, V., and Rappuoli, R. (2005). The microbial pan-genome. Curr. Opin. Genet. Dev. *15*, 589–594.

Méjean, A., Mazmouz, R., Mann, S., Calteau, A., Médigue, C., and Ploux, O. (2010). The genome sequence of the cyanobacterium *Oscillatoria* sp. PCC 6506 reveals several gene clusters responsible for the biosynthesis of toxins and secondary metabolites. J. Bacteriol. *192*, 5264–5265. http://dx.doi.org/10.1128/JB.00704-10.

Méjean, A., and Ploux, O. (2013). A genomic view of secondary metabolite production in cyanobacteria. Adv. Bot. Res. *65*, 189–234. http://dx.doi.org/10.1016/B978-0-12-394313-2.00006-8.

Méndez-García, C., Peláez, A.I., Mesa, V., Sánchez, J., Golyshina, O.V., and Ferrer, M. (2015). Microbial diversity and metabolic networks in acid mine drainage habitats. Front. Microbiol. *6*, 475. http://dx.doi.org/10.3389/fmicb.2015.00475.

Metlen, K.L., Aschehoug, E.T., and Callaway, R.M. (2009). Plant behavioural ecology: dynamic plasticity in secondary metabolites. Plant. Cell. Environ. *32*, 641–653. http://dx.doi.org/10.1111/j.1365-3040.2008.01910.x.

Micallef, M.L., D'Agostino, P.M., Al-Sinawi, B., Neilan, B.A., and Moffitt, M.C. (2015). Exploring cyanobacterial genomes for natural product biosynthesis pathways. Mar. Genomics *21*, 1–12. http://dx.doi.org/10.1016/j.margen.2014.11.009.

Miethke, M., and Marahiel, M.A. (2007). Siderophore-based iron acquisition and pathogen control. Microbiol. Mol. Biol. Rev. *71*, 413–451.

Mihali, T.K., Kellmann, R., Muenchhoff, J., Barrow, K.D., and Neilan, B.A. (2008). Characterization of the gene cluster responsible for cylindrospermopsin biosynthesis. Appl. Environ. Microbiol. *74*, 716–722.

Miller, W., Makova, K.D., Nekrutenko, A., and Hardison, R.C. (2004). Comparative genomics. Annu. Rev. Genomics. Hum. Genet. *5*, 15–56.

Mira, A., Martín-Cuadrado, A.B., D'Auria, G., and Rodríguez-Valera, F. (2010). The bacterial pan-genome: a new paradigm in microbiology. Int. Microbiol. *13*, 45–57.

Moffitt, M.C., and Neilan, B.A. (2004). Characterization of the nodularin synthetase gene cluster and proposed theory of the evolution of cyanobacterial hepatotoxins. Appl. Environ. Microbiol. *70*, 6353-6362.

Mongodin, E.F., Casjens, S.R., Bruno, J.F., Xu, Y., Drabek, E.F., Riley, D.R., Cantarel, B.L., Pagan, P.E., Hernandez, Y.A., Vargas, L.C., *et al.* (2013). Inter- and intra-specific pan-genomes of *Borrelia burgdorferi* sensu lato: genome stability and adaptive radiation. BMC Genomics *14*, 693. http://dx.doi.org/10.1186/1471-2164-14-693.

Moon, S.S., Chen, J.L., Moore, R.E., and Patterson, G.M.L. (1992). Calophycin, a fungicidal cyclic decapeptide from the terrestrial blue-green alga *Calothrix fusca*. The journal of organic chemistry 57, 1097–1103. http://dx.doi.org/10.1021/jo00030a013.

Mootz, H.D., Schwarzer, D., and Marahiel, M.A. (2002). Ways of assembling complex natural products on modular nonribosomal peptide synthetases. Chembiochem 3, 490–504.

Moss, S.J., Martin, C.J., and Wilkinson, B. (2004). Loss of co-linearity by modular polyketide synthases: a mechanism for the evolution of chemical diversity. Nat. Prod. Rep. *21*, 575–593.

Mulkidjanian, A.Y., Koonin, E.V., Makarova, K.S., Mekhedov, S.L., Sorokin, A., Wolf, Y.I., Dufresne, A., Partensky, F., Burd, H., Kaznadzey, D., *et al.* (2006). The cyanobacterial genome core and the origin of photosynthesis. Proc. Natl. Acad. Sci. U.S.A. *103*, 13126–13131.

Müller, C.A., Oberauner-Wappis, L., Peyman, A., Amos, G.C., Wellington, E.M., and Berg, G. (2015). Mining for nonribosomal peptide synthetase and polyketide synthase genes revealed a high level of diversity in the *Sphagnum* Bog Metagenome. Appl. Environ. Microbiol. *81*, 5064–5072. http://dx.doi.org/10.1128/AEM.00631-15.

Nakamura, Y., Kaneko, T., Sato, S., Ikeuchi, M., Katoh, H., Sasamoto, S., Watanabe, A., Iriguchi, M., Kawashima, K., Kimura, T., *et al.* (2002). Complete genome structure of the thermophilic cyanobacterium *Thermosynechococcus elongatus* BP-1 (supplement). DNA Res. *9*, 135–148. http://dx.doi.org/10.1093/dnares/9.4.135

Nakamura, Y., Kaneko, T., Sato, S., Mimuro, M., Miyashita, H., Tsuchiya, T., Sasamoto, S., Watanabe, A., Kawashima, K., Kishida, Y., *et al.* (2003). Complete genome structure of *Gloeobacter violaceus* PCC 7421, a cyanobacterium that lacks thylakoids (supplement). DNA Res. *10*, 181–201. http://dx.doi.org/10.1093/dnares/10.4.181

Namikoshi, M., and Rinehart, K.L. (1996). Bioactive compounds produced by cyanobacteria. J. Ind. Microbiol. Biotechnol. *17*, 373–384. http://dx.doi.org/10.1007/Bf01574768.

Neilan, B.A., Pearson, L.A., Muenchhoff, J., Moffitt, M.C., and Dittmann, E. (2013). Environmental conditions that influence toxin biosynthesis in cyanobacteria. Environ. Microbiol. *15*, 1239–1253. http://dx.doi.org/10.1111/j.1462-2920.2012.02729.x.

Nelson, W.C., Bhaya, D., and Heidelberg, J.F. (2012). Novel miniature transposable elements in thermophilic *Synechococcus* strains and their impact on an environmental population. J. Bacteriol. *194*, 3636–3642. http://dx.doi.org/10.1128/JB.00333-12.

Nicolaisen, K., Moslavac, S., Samborski, A., Valdebenito, M., Hantke, K., Maldener, I., Muro-Pastor, A.M., Flores, E., and Schleiff, E. (2008). Alr0397 is an outer membrane transporter for the siderophore schizokinen in *Anabaena* sp. strain PCC 7120. J. Bacteriol. *190*, 7500–7507. http://dx.doi.org/10.1128/JB.01062-08.

Nikolouli, K., and Mossialos, D. (2012). Bioactive compounds synthesized by non-ribosomal peptide synthetases and type-I polyketide synthases discovered through genome-mining and metagenomics. Biotechnology letters *34*, 1393-1403. http://dx.doi.org/10.1007/s10529-012-0919-2.

Nogle, L.M., Okino, T., and Gerwick, W.H. (2001). Antillatoxin B, a neurotoxic lipopeptide from the marine cyanobacterium *Lyngbya majuscula*. J. Nat. Prod. *64*, 983–985.

Nunnery, J.K., Mevers, E., and Gerwick, W.H. (2010). Biologically active secondary metabolites from marine cyanobacteria. Curr. Opin. Biotechnol. *21*, 787–793. http://dx.doi.org/10.1016/j.copbio.2010.09.019.

O'Connor, S.E. (2015). Engineering of Secondary Metabolism. Annu. Rev. Genet. *49*, 71–94. http://dx.doi.org/10.1146/annurev-genet-120213-092053.

O'Sullivan, O., Begley, M., Ross, R.P., Cotter, P.D., and Hill, C. (2011). Further Identification of Novel Lantibiotic Operons Using LanM-Based Genome Mining. Probiotics. Antimicrob. Proteins. *3*, 27–40. http://dx.doi.org/10.1007/s12602-011-9062-y.

Ottesen, E.A., Young, C.R., Eppley, J.M., Ryan, J.P., Chavez, F.P., Scholin, C.A., and DeLong, E.F. (2013). Pattern and synchrony of gene expression among sympatric marine microbial populations. Proc. Natl. Acad. Sci. U.S.A. *110*, E488–97. http://dx.doi.org/10.1073/pnas.1222099110.

Oulas, A., Pavloudi, C., Polymenakou, P., Pavlopoulos, G.A., Papanikolaou, N., Kotoulas, G., Arvanitidis, C., and Iliopoulos, I. (2015). Metagenomics: tools and insights for analyzing next-generation sequencing data derived from biodiversity studies. Bioinf. Biol. Insights 9, 75–88. http://dx.doi.org/10.4137/BBI.S12462.

Ozer, E.A., Allen, J.P., and Hauser, A.R. (2014). Characterization of the core and accessory genomes of Pseudomonas aeruginosa using bioinformatic tools Spine and AGEnt. BMC Genomics *15*, 737. http://dx.doi.org/10.1186/1471-2164-15-737.

Pati, A., Heath, L.S., Kyrpides, N.C., and Ivanova, N. (2011). ClaMS: A Classifier for Metagenomic Sequences. Stand. Genomic. Sci. 5, 248–253. http://dx.doi.org/10.4056/sigs.2075298.

Patil, K.R., Roune, L., and McHardy, A.C. (2012). The PhyloPythiaS web server for taxonomic assignment of metagenome sequences. PLoS One 7, e38581. http://dx.doi.org/10.1371/journal.pone.0038581.

Pattanaik, B., and Lindberg, P. (2015). Terpenoids and their biosynthesis in cyanobacteria. Life 5, 269–293. http://dx.doi.org/10.3390/life5010269.

Patton, G.C., and van der Donk, W.A. (2005). New developments in lantibiotic biosynthesis and mode of action. Curr. Opin. Microbiol. *8*, 543–551.

Pearce, D.A., Newsham, K.K., Thorne, M.A., Calvo-Bado, L., Krsek, M., Laskaris, P., Hodson, A., and Wellington, E.M. (2012). Metagenomic analysis of a southern maritime antarctic soil. Front. Microbiol. 3, 403. http://dx.doi.org/10.3389/fmicb.2012.00403.

Peng, Y., Leung, H.C., Yiu, S.M., and Chin, F.Y. (2011). Meta-IDBA: a *de* Novo assembler for metagenomic data. Bioinformatics *27*, i94–101. http://dx.doi.org/10.1093/bioinformatics/btr216.

Philmus, B., Christiansen, G., Yoshida, W.Y., and Hemscheidt, T.K. (2008). Post-translational modification in microviridin biosynthesis. Chembiochem *9*, 3066–3073. http://dx.doi.org/10.1002/cbic.200800560.

Piel, J. (2010). Biosynthesis of polyketides by trans-AT polyketide synthases. Nat. Prod. Rep. *27*, 996–1047. http://dx.doi.org/10.1039/b816430b.

Plaza, A., and Bewley, C.A. (2006). Largamides A-H, unusual cyclic peptides from the marine cyanobacterium *Oscillatoria* sp. J. Org. Chem. *71*, 6898–6907.

Priya, H., Prasanna, R., Ramakrishnan, B., Bidyarani, N., Babu, S., Thapa, S., and Renuka, N. (2015). Influence of cyanobacterial inoculation on the culturable microbiome and growth of rice. Microbiol. Res. *171*, 78–89. http://dx.doi.org/10.1016/j.micres.2014.12.011.

Rantala-Ylinen, A., Känä, S., Wang, H., Rouhiainen, L., Wahlsten, M., Rizzi, E., Berg, K., Gugger, M., and Sivonen, K. (2011). Anatoxin-a synthetase gene cluster of the cyanobacterium *Anabaena* sp. strain 37 and molecular methods to detect potential producers. Appl. Environ. Microbiol. *77*, 7271–7278. http://dx.doi.org/10.1128/AEM.06022-11.

Rattray, J.E., Strous, M., Op den Camp, H.J., Schouten, S., Jetten, M.S., and Damsté, J.S. (2009). A comparative genomics study of genetic products potentially encoding ladderane lipid biosynthesis. Biol. Direct. *4*, 8. http://dx.doi.org/10.1186/1745-6150-4-8.

Reuter, J.A., Spacek, D.V., and Snyder, M.P. (2015). High-throughput sequencing technologies. Mol. Cell. *58*, 586–597. http://dx.doi.org/10.1016/j.molcel.2015.05.004.

Riesenfeld, C.S., Schloss, P.D., and Handelsman, J. (2004). Metagenomics: genomic analysis of microbial communities. Annu. Rev. Genet. *38*, 525–552.

Rouhiainen, L., Jokela, J., Fewer, D.P., Urmann, M., and Sivonen, K. (2010). Two alternative starter modules for the non-ribosomal biosynthesis of specific anabaenopeptin variants in *Anabaena* (Cyanobacteria). Chem. Biol. *17*, 265–273. http://dx.doi.org/10.1016/j.chembiol.2010.01.017.

Rouhiainen, L., Paulin, L., Suomalainen, S., Hyytiäinen, H., Buikema, W., Haselkorn, R., and Sivonen, K. (2000). Genes encoding synthetases of cyclic depsipeptides, anabaenopeptilides, in *Anabaena* strain 90. Mol. Microbiol. *37*, 156–167.

Rouli, L., Merhej, V., Fournier, P.E., and Raoult, D. (2015). The bacterial pangenome as a new tool for analysing pathogenic bacteria. New. Microbes. New. Infect. 7, 72–85. http://dx.doi.org/10.1016/j.nmni.2015.06.005.

Rounge, T.B., Rohrlack, T., Nederbragt, A.J., Kristensen, T., and Jakobsen, K.S. (2009). A genome-wide analysis of nonribosomal peptide synthetase gene clusters and their peptides in a *Planktothrix rubescens* strain. BMC Genomics *10*, 396. http://dx.doi.org/10.1186/1471-2164-10-396.

Rutherford, K., Parkhill, J., Crook, J., Horsnell, T., Rice, P., Rajandream, M.A., and Barrell, B. (2000). Artemis: sequence visualization and annotation. Bioinformatics *16*, 944–945.

Salomon, C.E., Magarvey, N.A., and Sherman, D.H. (2004). Merging the potential of microbial genetics with biological and chemical diversity: an even brighter future for marine natural product drug discovery. Nat. Prod. Rep. *21*, 105–121.

Salvador, L.A., Paul, V.J., and Luesch, H. (2010). Caylobolide B, a macrolactone from symplostatin 1-producing marine cyanobacteria *Phormidium* spp. from Florida. J. Nat. Prod. 73, 1606-1609. http://dx.doi.org/10.1021/np100467d.

Sandy, M., and Butler, A. (2009). Microbial iron acquisition: marine and terrestrial siderophores. Chem. Rev. *109*, 4580–4595. http://dx.doi.org/10.1021/cr9002787.

Sanli, K., Bengtsson-Palme, J., Nilsson, R.H., Kristiansson, E., Alm Rosenblad, M., Blanck, H., and Eriksson, K.M. (2015). Metagenomic sequencing of marine periphyton: taxonomic and functional insights into biofilm communities. Front. Microbiol. 6, 1192. http://dx.doi.org/10.3389/fmicb.2015.01192.

Santos, A.R., Barbosa, E., Fiaux, K., Zurita-Turk, M., Chaitankar, V., Kamapantula, B., Abdelzaher, A., Ghosh, P., Tiwari, S., Barve, N., *et al.* (2013). PANNOTATOR: an automated tool for annotation of pan-genomes. Genet. Mol. Res. *12*, 2982–2989. http://dx.doi.org/10.4238/2013.August.16.2.

Schirmer, A., Rude, M.A., Li, X., Popova, E., and del Cardayre, S.B. (2010). Microbial biosynthesis of alkanes. Science *329*, 559–562. http://dx.doi.org/10.1126/science.1187936.

Schmidt, E.W., Nelson, J.T., Rasko, D.A., Sudek, S., Eisen, J.A., Haygood, M.G., and Ravel, J. (2005). Patellamide A and C biosynthesis by a microcin-like pathway in *Prochloron didemni*, the cyanobacterial symbiont of *Lissoclinum patella*. Proc. Natl. Acad. Sci. U.S.A. *102*, 7315–7320.

Segata, N., Haake, S.K., Mannon, P., Lemon, K.P., Waldron, L., Gevers, D., Huttenhower, C., and Izard, J. (2012). Composition of the adult digestive tract bacterial microbiome based on seven mouth surfaces, tonsils, throat and stool samples. Genome Biol. *13*, R42. http://dx.doi.org/10.1186/gb-2012-13-6-r42.

Segata, N., Izard, J., Waldron, L., Gevers, D., Miropolsky, L., Garrett, W.S., and Huttenhower, C. (2011). Metagenomic biomarker discovery and explanation. Genome Biol. *12*, R60. http://dx.doi.org/10.1186/gb-2011-12-6-r60.

Seshadri, R., Kravitz, S.A., Smarr, L., Gilna, P., and Frazier, M. (2007). CAMERA: a community resource for metagenomics. PLOS Biol. 5, e75.

Sharma, A.K., Becker, J.W., Ottesen, E.A., Bryant, J.A., Duhamel, S., Karl, D.M., Cordero, O.X., Repeta, D.J., and DeLong, E.F. (2014). Distinct dissolved organic matter sources induce rapid transcriptional responses in coexisting populations of *Prochlorococcus, Pelagibacter* and the OM60 clade. Environ. Microbiol. *16*, 2815–2830. http://dx.doi.org/10.1111/1462-2920.12254.

Shih, P.M., Wu, D., Latifi, A., Axen, S.D., Fewer, D.P., Talla, E., Calteau, A., Cai, F., Tandeau de Marsac, N., Rippka, R., *et al.* (2013). Improving the coverage of the cyanobacterial phylum using diversity-driven genome sequencing. Proc. Natl. Acad. Sci. U.S.A. *110*, 1053–1058. http://dx.doi.org/10.1073/pnas.1217107110.

Shilova, I.N., Robidart, J.C., James Tripp, H., Turk-Kubo, K., Wawrik, B., Post, A.F., Thompson, A.W., Ward, B., Hollibaugh, J.T., Millard, A., *et al.* (2014). A microarray for assessing transcription from pelagic marine microbial taxa. ISME J. *8*, 1476–1491. http://dx.doi.org/10.1038/ismej.2014.1.

Simm, S., Keller, M., Selymesi, M., and Schleiff, E. (2015). The composition of the global and feature specific cyanobacterial core-genomes. Front. Microbiol. *6*, 219. http://dx.doi.org/10.3389/fmicb.2015.00219.

Simmons, T.L., McPhail, K.L., Ortega-Barría, E., Mooberry, S.L., and Gerwick, W.H. (2006). Belamide A, a new antimitotic tetrapeptide from a Panamanian marine cyanobacterium. Tetrahedron Lett. *47*, 3387-3390. http://dx.doi.org/10.1016/j.tetlet.2006.03.082.

Singh, S., Kate, B.N., and Banerjee, U.C. (2005). Bioactive compounds from cyanobacteria and microalgae: an overview. Crit. Rev. Biotechnol. *25*, 73–95.

Sinninghe Damsté, J.S., Rijpstra, W.I., Geenevasen, J.A., Strous, M., and Jetten, M.S. (2005). Structural identification of ladderane and other membrane lipids of planctomycetes capable of anaerobic ammonium oxidation (anammox). FEBS. J. 272, 4270–4283.

Sinninghe Damsté, J.S., Strous, M., Rijpstra, W.I., Hopmans, E.C., Geenevasen, J.A., van Duin, A.C., van Niftrik, L.A., and Jetten, M.S. (2002). Linearly concatenated cyclobutane lipids form a dense bacterial membrane. Nature *419*, 708–712.

Sisay, M.T., Hautmann, S., Mehner, C., König, G.M., Bajorath, J., and Gütschow, M. (2009). Inhibition of human leukocyte elastase by brunsvicamides a-C: cyanobacterial cyclic peptides. ChemMedChem *4*, 1425–1429. http://dx.doi.org/10.1002/cmdc.200900139.

Skulberg, O.M. (2000). Microalgae as a source of bioactive molecules – experience from cyanophyte research. J. Appl. Phycol. *12*, 341–348. http://dx.doi.org/10.1023/A:1008140403621.

Sonnhammer, E.L., and Östlund, G. (2015). InParanoid 8: orthology analysis between 273 proteomes, mostly eukaryotic. Nucleic Acids Res. *43*, D234–9. http://dx.doi.org/10.1093/nar/gku1203.

Soria-Mercado, I.E., Pereira, A., Cao, Z., Murray, T.F., and Gerwick, W.H. (2009). Alotamide A, a novel neuropharmacological agent from the marine cyanobacterium *Lyngbya bouillonii*. Org. Lett. *11*, 4704–4707. http://dx.doi.org/10.1021/ol901438b.

Stachelhaus, T., and Walsh, C.T. (2000). Mutational analysis of the epimerization domain in the initiation module

PheATE of gramicidin S synthetase. Biochemistry *39*, 5775–5787.
Starcevic, A., Zucko, J., Simunkovic, J., Long, P.F., Cullum, J., and Hranueli, D. (2008). ClustScan: an integrated program package for the semi-automatic annotation of modular biosynthetic gene clusters and in silico prediction of novel chemical structures. Nucleic Acids Res. *36*, 6882–6892. http://dx.doi.org/10.1093/nar/gkn685.
Staunton, J., and Weissman, K.J. (2001). Polyketide biosynthesis: a millennium review. Nat. Prod. Rep. *18*, 380–416.
Steven, B., Gallegos-Graves, L.V., Yeager, C.M., Belnap, J., Evans, R.D., and Kuske, C.R. (2012). Dryland biological soil crust cyanobacteria show unexpected decreases in abundance under long-term elevated CO_2. Environ. Microbiol. *14*, 3247–3258. http://dx.doi.org/10.1111/1462-2920.12011.
Stevens, H., and Ulloa, O. (2008). Bacterial diversity in the oxygen minimum zone of the eastern tropical South Pacific. Environ. Microbiol. *10*, 1244–1259. http://dx.doi.org/10.1111/j.1462-2920.2007.01539.x.
Strieker, M., Tanović, A., and Marahiel, M.A. (2010). Nonribosomal peptide synthetases: structures and dynamics. Curr. Opin. Struct. Biol. *20*, 234–240. http://dx.doi.org/10.1016/j.sbi.2010.01.009.
Sudek, S., Haygood, M.G., Youssef, D.T., and Schmidt, E.W. (2006). Structure of trichamide, a cyclic peptide from the bloom-forming cyanobacterium *Trichodesmium erythraeum*, predicted from the genome sequence. Appl. Environ. Microbiol. *72*, 4382-4387. http://dx.doi.org/10.1128/AEM.00380-06.
Suzuki, S., Ishii, S., Wu, A., Cheung, A., Tenney, A., Wanger, G., Kuenen, J.G., and Nealson, K.H. (2013). Microbial diversity in The Cedars, an ultrabasic, ultrareducing, and low salinity serpentinizing ecosystem. Proc. Natl. Acad. Sci. U.S.A. *110*, 15336–15341. http://dx.doi.org/10.1073/pnas.1302426110.
Tae, H., Kong, E.B., and Park, K. (2007). ASMPKS: an analysis system for modular polyketide synthases. BMC Bioinf. *8*, 327.
Tai, V., Poon, A.F., Paulsen, I.T., and Palenik, B. (2011). Selection in coastal *Synechococcus* (cyanobacteria) populations evaluated from environmental metagenomes. PLoS One *6*, e24249. http://dx.doi.org/10.1371/journal.pone.0024249.
Teeling, H., and Glöckner, F.O. (2012). Current opportunities and challenges in microbial metagenome analysis – a bioinformatic perspective. Briefings Bioinf. *13*, 728–742. http://dx.doi.org/10.1093/bib/bbs039.
Teruya, T., Sasaki, H., Kitamura, K., Nakayama, T., and Suenaga, K. (2009). Biselyngbyaside, a macrolide glycoside from the marine cyanobacterium *Lyngbya* sp. Org. Lett. *11*, 2421–2424. http://dx.doi.org/10.1021/ol900579k.
Tettelin, H., Masignani, V., Cieslewicz, M.J., Donati, C., Medini, D., Ward, N.L., Angiuoli, S.V., Crabtree, J., Jones, A.L., Durkin, A.S., *et al.* (2005). Genome analysis of multiple pathogenic isolates of *Streptococcus agalactiae*: implications for the microbial "pan-genome". Proc. Natl. Acad. Sci. U.S.A. *102*, 13950–13955.
Tettelin, H., Riley, D., Cattuto, C., and Medini, D. (2008). Comparative genomics: the bacterial pan-genome. Curr. Opin. Microbiol. *11*, 472–477.
Tholl, D. (2006). Terpene synthases and the regulation, diversity and biological roles of terpene metabolism. Curr. Opin. Plant Biol. *9*, 297–304.
Thomas, T., Gilbert, J., and Meyer, F. (2012). Metagenomics – a guide from sampling to data analysis. Microb. Inf. Exp. *2*, 3. http://dx.doi.org/10.1186/2042-5783-2-3.
Tooming-Klunderud, A., Sogge, H., Rounge, T.B., Nederbragt, A.J., Lagesen, K., Glöckner, G., Hayes, P.K., Rohrlack, T., and Jakobsen, K.S. (2013). From green to red: horizontal gene transfer of the phycoerythrin gene cluster between *Planktothrix* strains. Appl. Environ. Microbiol. *79*, 6803–6812. http://dx.doi.org/10.1128/AEM.01455-13.
Treangen, T.J., Koren, S., Sommer, D.D., Liu, B., Astrovskaya, I., Ondov, B., Darling, A.E., Phillippy, A.M., and Pop, M. (2013). MetAMOS: a modular and open source metagenomic assembly and analysis pipeline. Genome Biol. *14*, R2. http://dx.doi.org/10.1186/gb-2013-14-1-r2.
Treangen, T.J., Ondov, B.D., Koren, S., and Phillippy, A.M. (2014). The Harvest suite for rapid core-genome alignment and visualization of thousands of intraspecific microbial genomes. Genome Biol. *15*, 524.
Tringe, S.G., von Mering, C., Kobayashi, A., Salamov, A.A., Chen, K., Chang, H.W., Podar, M., Short, J.M., Mathur, E.J., Detter, J.C., *et al.* (2005). Comparative metagenomics of microbial communities. Science *308*, 554–557.
Tseng, C.H., Chiang, P.W., Lai, H.C., Shiah, F.K., Hsu, T.C., Chen, Y.L., Wen, L.S., Tseng, C.M., Shieh, W.Y., Saeed, I., *et al.* (2015). Prokaryotic assemblages and metagenomes in pelagic zones of the South China Sea. BMC Genomics *16*, 219. http://dx.doi.org/10.1186/s12864-015-1434-3.
Tyson, G.W., Chapman, J., Hugenholtz, P., Allen, E.E., Ram, R.J., Richardson, P.M., Solovyev, V.V., Rubin, E.M., Rokhsar, D.S., and Banfield, J.F. (2004). Community structure and metabolism through reconstruction of microbial genomes from the environment. Nature *428*, 37–43.
Urich, T., Lanzen, A., Stokke, R., Pedersen, R.B., Bayer, C., Thorseth, I.H., Schleper, C., Steen, I.H., and Ovreas, L. (2014). Microbial community structure and functioning in marine sediments associated with diffuse hydrothermal venting assessed by integrated meta-omics. Environ. Microbiol. *16*, 2699–2710. http://dx.doi.org/10.1111/1462-2920.12283.
Ussery, D.W., Wassenaar, T.M., and Borini, S. (2009). Computing for comparative microbial genomics: bioinformatics for microbiologists (London: Springer).
van Niftrik, L., and Jetten, M.S. (2012). Anaerobic ammonium-oxidizing bacteria: unique microorganisms with exceptional properties. Microbiol. Mol. Biol. Rev. *76*, 585–596. http://dx.doi.org/10.1128/MMBR.05025-11.
van Schaik, W., Top, J., Riley, D.R., Boekhorst, J., Vrijenhoek, J.E., Schapendonk, C.M., Hendrickx, A.P., Nijman, I.J., Bonten, M.J., Tettelin, H., *et al.* (2010). Pyrosequencing-based comparative genome analysis

of the nosocomial pathogen *Enterococcus faecium* and identification of a large transferable pathogenicity island. BMC Genomics *11*, 239. http://dx.doi.org/10.1186/1471-2164-11-239.

Varin, T., Lovejoy, C., Jungblut, A.D., Vincent, W.F., and Corbeil, J. (2012). Metagenomic analysis of stress genes in microbial mat communities from Antarctica and the High Arctic. Appl. Environ. Microbiol. *78*, 549–559. http://dx.doi.org/10.1128/AEM.06354-11.

Venter, J.C., Remington, K., Heidelberg, J.F., Halpern, A.L., Rusch, D., Eisen, J.A., Wu, D., Paulsen, I., Nelson, K.E., Nelson, W., *et al.* (2004). Environmental genome shotgun sequencing of the Sargasso Sea. Science *304*, 66–74.

Vernikos, G., Medini, D., Riley, D.R., and Tettelin, H. (2015). Ten years of pan-genome analyses. Curr. Opin. Microbiol. *23*, 148–154. http://dx.doi.org/10.1016/j.mib.2014.11.016.

von Döhren, H. (2004). Biochemistry and general genetics of nonribosomal peptide synthetases in fungi. Adv. Biochem. Eng. Biotechnol. *88*, 217–264.

Voorhies, A.A., Biddanda, B.A., Kendall, S.T., Jain, S., Marcus, D.N., Nold, S.C., Sheldon, N.D., and Dick, G.J. (2012). Cyanobacterial life at low O(2): community genomics and function reveal metabolic versatility and extremely low diversity in a Great Lakes sinkhole mat. Geobiology *10*, 250–267. http://dx.doi.org/10.1111/j.1472-4669.2012.00322.x.

Voss, B., Bolhuis, H., Fewer, D.P., Kopf, M., Möke, F., Haas, F., El-Shehawy, R., Hayes, P., Bergman, B., Sivonen, K., *et al.* (2013). Insights into the physiology and ecology of the brackish-water-adapted Cyanobacterium *Nodularia spumigena* CCY9414 based on a genome-transcriptome analysis. PLoS One *8*, e60224. http://dx.doi.org/10.1371/journal.pone.0060224.

Wandersman, C., and Delepelaire, P. (2004). Bacterial iron sources: from siderophores to hemophores. Annu. Rev. Microbiol. *58*, 611–647.

Wang, H., Sivonen, K., Rouhiainen, L., Fewer, D.P., Lyra, C., Rantala-Ylinen, A., Vestola, J., Jokela, J., Rantasärkkä, K., Li, Z., *et al.* (2012a). Genome derived insights into the biology of the hepatotoxic bloom-forming cyanobacterium *Anabaena* sp. strain 90. BMC Genomics *13*, 613. http://dx.doi.org/10.1186/1471-2164-13-613.

Wang, W.L., Xu, S.Y., Ren, Z.G., Tao, L., Jiang, J.W., and Zheng, S.S. (2015). Application of metagenomics in the human gut microbiome. World. J. Gastroenterol. *21*, 803–814. http://dx.doi.org/10.3748/wjg.v21.i3.803.

Wang, Y., Leung, H.C., Yiu, S.M., and Chin, F.Y. (2012b). MetaCluster 5.0: a two-round binning approach for metagenomic data for low-abundance species in a noisy sample. Bioinformatics *28*, i356–i362. http://dx.doi.org/10.1093/bioinformatics/bts397.

Weber, T., and Blin, K. (2015). CLUSEAN, overview. In Encyclopedia of Metagenomics: Genes, Genomes and Metagenomes: Basics, Methods, Databases and Tools, K.E. Nelson, ed. (Boston, MA: Springer US), pp. 93–95. http://dx.doi.org/10.1007/978-1-4899-7478-5_706

Weber, T., Blin, K., Duddela, S., Krug, D., Kim, H.U., Bruccoleri, R., Lee, S.Y., Fischbach, M.A., Müller, R., Wohlleben, W., *et al.* (2015). antiSMASH 3.0-a comprehensive resource for the genome mining of biosynthetic gene clusters. Nucleic Acids Res. *43*, W237–43. http://dx.doi.org/10.1093/nar/gkv437.

Weber, T., Rausch, C., Lopez, P., Hoof, I., Gaykova, V., Huson, D.H., and Wohlleben, W. (2009). CLUSEAN: a computer-based framework for the automated analysis of bacterial secondary metabolite biosynthetic gene clusters. J. Biotechnol. *140*, 13–17.

Wei, L., Liu, Y., Dubchak, I., Shon, J., and Park, J. (2002). Comparative genomics approaches to study organism similarities and differences. J. Biomed. Inform. *35*, 142–150.

Weissman, K.J. (2015). The structural biology of biosynthetic megaenzymes. Nat. Chem. Biol. *11*, 660–670. http://dx.doi.org/10.1038/nchembio.1883.

Weissman, K.J., and Leadlay, P.F. (2005). Combinatorial biosynthesis of reduced polyketides. Nat. Rev. Microbiol. *3*, 925–936.

Welker, M., and von Döhren, H. (2006). Cyanobacterial peptides– nature's own combinatorial biosynthesis. FEMS Microbiol. Rev. *30*, 530–563.

Wenzel, S.C., and Müller, R. (2005). Formation of novel secondary metabolites by bacterial multimodular assembly lines: deviations from textbook biosynthetic logic. Curr. Opin. Chem. Biol. *9*, 447–458.

Wiedemann, I., Breukink, E., van Kraaij, C., Kuipers, O.P., Bierbaum, G., de Kruijff, B., and Sahl, H.G. (2001). Specific binding of nisin to the peptidoglycan precursor lipid II combines pore formation and inhibition of cell wall biosynthesis for potent antibiotic activity. J. Biol. Chem. *276*, 1772–1779.

Wijffels, R.H., Kruse, O., and Hellingwerf, K.J. (2013). Potential of industrial biotechnology with cyanobacteria and eukaryotic microalgae. Curr. Opin. Biotechnol. *24*, 405–413. http://dx.doi.org/10.1016/j.copbio.2013.04.004.

Wooley, J.C., Godzik, A., and Friedberg, I. (2010). A primer on metagenomics. PLOS Comput. Biol. *6*, e1000667. http://dx.doi.org/10.1371/journal.pcbi.1000667.

Wozniak, M., Wong, L., and Tiuryn, J. (2011). CAMBer: an approach to support comparative analysis of multiple bacterial strains. BMC Genomics *12* (Suppl. 2), S6. http://dx.doi.org/10.1186/1471-2164-12-S2-S6.

Xia, L.C., Cram, J.A., Chen, T., Fuhrman, J.A., and Sun, F. (2011). Accurate genome relative abundance estimation based on shotgun metagenomic reads. PLoS One *6*, e27992. http://dx.doi.org/10.1371/journal.pone.0027992.

Xiao, J., Zhang, Z., Wu, J., and Yu, J. (2015). A brief review of software tools for pangenomics. Genomics. Proteomics. Bioinformatics. *13*, 73–76. http://dx.doi.org/10.1016/j.gpb.2015.01.007.

Xie, S.C., Pancost, R.D., Wang, Y.B., Yang, H., Wignall, P.B., Luo, G.M., Jia, C.L., and Chen, L. (2011). Cyanobacterial blooms tied to volcanism during the 5 m.y. Permo-Triassic biotic crisis: Reply. Geology 39, E249–E249. http://dx.doi.org/10.1130/G32293y.1.

Xiong, J., Liu, Y., Lin, X., Zhang, H., Zeng, J., Hou, J., Yang, Y., Yao, T., Knight, R., and Chu, H. (2012). Geographic distance and pH drive bacterial distribution in alkaline lake sediments across Tibetan Plateau. Environ. Microbiol. *14*, 2457–2466. http://dx.doi.org/10.1111/j.1462-2920.2012.02799.x.

Yamada, Y., Kuzuyama, T., Komatsu, M., Shin-Ya, K., Omura, S., Cane, D.E., and Ikeda, H. (2015). Terpene synthases are widely distributed in bacteria. Proc. Natl. Acad. Sci. U.S.A. *112*, 857–862. http://dx.doi.org/10.1073/pnas.1422108112.

Yerrapragada, S., Siefert, J.L., and Fox, G.E. (2009). Horizontal gene transfer in cyanobacterial signature genes. Methods Mol. Biol. *532*, 339–366. http://dx.doi.org/10.1007/978-1-60327-853-9_20.

Zhao, Y., Wu, J., Yang, J., Sun, S., Xiao, J., and Yu, J. (2012). PGAP: pan-genomes analysis pipeline. Bioinformatics *28*, 416–418. http://dx.doi.org/10.1093/bioinformatics/btr655.

Zhao, Y., Jia, X., Yang, J., Ling, Y., Zhang, Z., Yu, J., Wu, J., and Xiao, J. (2014). PanGP: a tool for quickly analyzing bacterial pan-genome profile. Bioinformatics *30*, 1297–1299. http://dx.doi.org/10.1093/bioinformatics/btu017.

Ziemert, N., Podell, S., Penn, K., Badger, J.H., Allen, E., and Jensen, P.R. (2012). The natural product domain seeker NaPDoS: a phylogeny based bioinformatic tool to classify secondary metabolite gene diversity. PLoS One *7*, e34064. http://dx.doi.org/10.1371/journal.pone.0034064.

Genome-wide Analysis of Cyanobacterial Evolution: The Example of *Synechococcus*

2

Petr Dvořák

Department of Botany, Faculty of Science, Palacký University Olomouc, Olomouc, Czech Republic.

Correspondence: p.dvorak@upol.cz

Abstract

Cyanobacteria are an ancient group of prokaryotes with an immense impact on the evolution of other organisms and the global ecosystem. In this chapter, I will focus on a genetically diverse yet morphologically similar group of cyanobacteria merged under the name *Synechococcus*. Together with *Prochlorococcus*, *Synechococcus* may be the most abundant cyanobacterium and amongst the most ancient organisms on Earth. However, the evolutionary trajectories of *Synechococcus* are entangled, resulting in polyphyletic relationships based on whole-genome phylogenetic reconstructions. Moreover, there is a high variability of the genome composition, size, and complexity within *Synechococcus*, potentially accounting for and being an example of speciation patterns within cyanobacteria, which may be largely driven by genome alterations (e.g. horizontal gene transfer). In conclusion, I will show that the cyanobacterial evolution is a dynamic process, characterized by drastic genome alterations and gene transfer (both homologous and non-homologous), often resulting in convergent evolutionary patterns as typified by *Synechococcus*.

Introduction

Synechococcus represents one of the simplest cyanobacteria in terms of morphology: unicellular, solitary living cyanobacteria with minute dimensions (up to 4 μm). Some species are capable of pseudofilament formation and involution cells (unusually prolonged cells, reviewed in Komárek and Anagnostidis, 1998). Within this phenotypically simple and coherent group, there is hidden a vast genetic and physiological variability, putatively allowing *Synechococcus* to be among the most evolutionary and ecologically successful cyanobacteria in terms of biodiversity, ecological range, and abundance. *Synechococcus* may have originated three billion years ago in the hot spring (Dvořák *et al.*, 2014). Hot spring is a habitat originated by a venting of geothermally heated groundwater with a temperature higher than 36.7°C (human body temperature; Pentecost *et al.*, 2003). From an ecological perspective, marine picoplanktic *Synechococcus*, together with *Prochlorococcus*, are the most abundant primary producers in the ocean with the global abundances reaching $2.9 \pm 0.1 \times 10^{27}$ and $7.0 \pm 0.3 \times 10^{26}$ cells, respectively (Flombaum *et al.*, 2013).

Cyanobacteria exhibit relatively high morphological variability for bacteria, and thus their evolutionary relationships were originally constructed based on their phenotypic features. Initially, they were classified under the botanical code due to their ecological and morphological similarity to the algae. Subsequently, polymerase chain reaction (PCR) and advances in sequencing allowed to use simple and cheap techniques to get insights into the evolution of cyanobacterial genomes. As the 'gold standard' for evolutionary relationships inference, the gene coding small ribosomal subunit has been introduced (16S rRNA; e.g. Stackebrandt and Goebel, 1994). However, an evolutionary resolution of 16S rRNA is limited because the 16S represents a very conservative gene, while the rest of the genome may have experienced different rates of change. Further, the conserved nature of the 16S rRNA makes its resolution at the species level spurious. Thus, additional molecular markers have been proposed for phylogenetic reconstructions, such as sequencing the 16S–23S internal transcribed spacer region (16S–23S ITS) (e.g. Boyer *et al.*, 2001) and multilocus sequence typing (MLST; Maiden *et al.*, 1998), which is characterized by a use of a combination of the protein-coding loci to allow more sensitive resolution. The most recently, some authors argue that only the whole genome sequencing approach is capable of such a fine resolution that it would be able to recognize small evolutionary steps, which are undertaken by a diversification of ecotypes (Rocap *et al.*, 2003; Kopac *et al.*, 2014; Olsen *et al.*, 2015). However, in terms of systematics, the 16S rRNA is still the defined species level identifier (Gevers *et al.*, 2006). GenBank (https://www.ncbi.nlm.nih.gov/) currently holds 180 genome sequencing BioProjects of cyanobacteria (24 March 2016). The taxon sampling of these cyanobacteria is strongly biased with 51 genomes belonging to *Synechococcus* (28) and *Prochlorococcus* (23). It should be noted that the majority of these genomes belongs the marine picoplanktic group (see Fig. 2.1 for its phylogenetic position) and the other existing lineages are still undersampled.

In this chapter, I will review the most recent advances in the evolutionary genomics of the cyanobacteria, concentrating on the genus *Synechococcus*. This is an ideal model group for a research involved in both microevolution and macroevolution of cyanobacteria and by extension of other prokaryotes. I will not disregard *Prochlorococcus*, as this cyanobacterium is tightly connected to *Synechococcus* by their common evolutionary history and ecological importance. To be more specific, I will concentrate on (1) speciation and comparative genomics of *Synechococcus*, and (2) a reconstruction of evolutionary relationships using whole-genome data (phylogenomics).

The origin of *Synechococcus*

I begin with a brief excursion to the first half of the nineteenth century in Switzerland, where the professor of the University of Zürich, Carl Nägeli, investigated the cyanobacterial flora of the Zürich surroundings. There, in the Katzensee near Zürich, Nägeli found a minute unicellular cyanobacterium on the mud, which he later described as *Synechococcus* (Nägeli 1849). The species was called *S. elongatus* and it has been later designed as a type species for the genus (Komárek and Anagnostidis, 1998). Nägeli did not fully appreciate an importance of his discovery, naming one of the most abundant genera of phototrophs on earth.

If we search for the first evidence of the *Synechococcus*-like cyanobacterium, we cannot rely much on the fossil record, where more morphologically complex cyanobacteria may be identified with more confidence. Nevertheless, the oldest fossil of the *Synechococcus*-like cyanobacterium was identified in the 1.9 billion years old stromatolitic dolostone, subsequently named *Eosynechococcus* (Hofmann, 1976). Molecular data indicates a much earlier origin of *Synechococcus*, which itself is tightly connected with the origin of the oxygenic phototrophs. It should be noted here that we cannot reliably decide which evolutionary lineage of *Synechococcus* might be affiliated with *Eosynechococcus* due to the morphological convergence within *Synechococcus*, explained in further detail below (Dvořák *et al.*, 2014).

Phylogenomic dating based on molecular clocks puts the origin of *Synechococcus* to the Archean (ca. 2.5–4 BYA). The oldest group of *Synechococcus* is composed of cyanobacteria inhabiting hot springs in Yellowstone National Park, putatively derived three billion years ago (Dvořák *et al.*, 2014). Since cyanobacteria were likely the first oxygenic phototrophs, it seems likely that this ability first evolved in the hot springs (reviewed in Butterfield, 2015), contrary to Blank and Sanchez-Baracaldo (2010), who proposed a freshwater origin of oxygenic phototrophs. Based on the most comprehensive phylogeny of *Synechococcus* using single gene and whole genome dataset (Dvořák *et al.*, 2014), I postulate that *Synechococcus* has subsequently radiated into the marine habitats and finally to the freshwater habitats.

The genome and pan-genome of cyanobacteria

Cyanobacteria inhabit a variety of environments from hot springs to polar areas, from freshwaters to marine systems, and from aquatic to terrestrial environments. To conquer these habitats, cyanobacteria possess a great variability of physiological

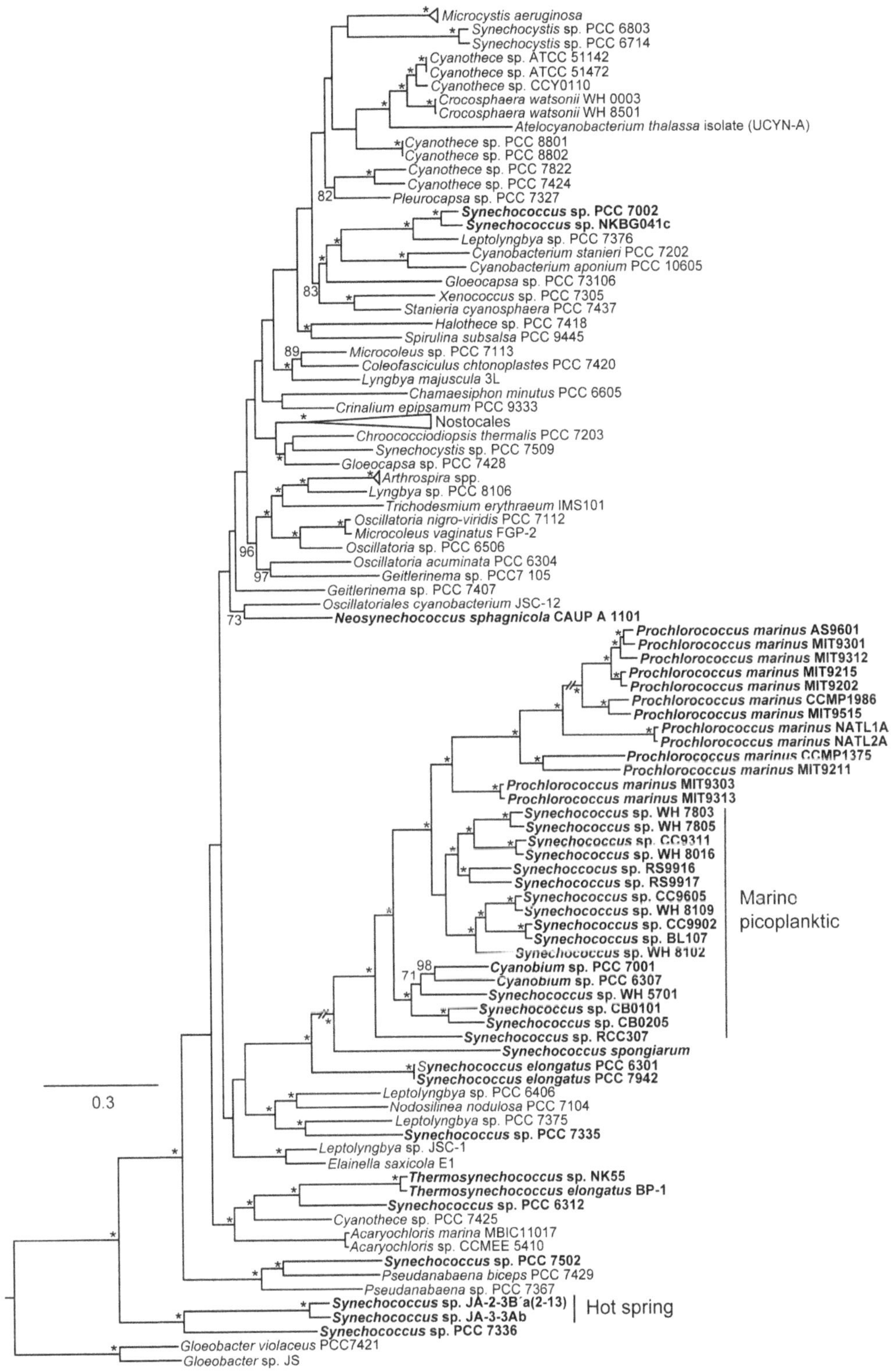

Figure 2.1 Phylogenomic reconstruction of cyanobacterial evolution based on 69 concatenated, orthologous protein coding loci. The alignment has been obtained from Jahodářová *et al.* (submitted). The tree was reconstructed using maximum likelihood optimality criterion and the CAT + LG model in RaxML 8.0.2 (Stamatakis 2006). The topology was tested by 500 rapid bootstrap replicates. Bootstrap support is located at the nodes with an asterisk representing 100% bootstrap support. *Synechococcus* strains are in bold.

adaptations reflected in their genome (Larsson *et al.*, 2011). As seen in the other prokaryotes, the gene composition of cyanobacteria may vary greatly, but an elementary genome architecture is stable: the DNA is concentrated into a single chromosome. However, there may be only one (monoploidy), several (oligoploidy) or even hundreds of copies (polyploidy) of the chromosome in each cell. The highest polyploidy has been identified in *Synechocystis* sp. PCC 6803 by Griese *et al.* (2011), who found 258 copies of the chromosome during the exponential growth phase. *Synechococcus* is mostly monoploid or oligoploid (Binder and Chisholm, 1995; Griese *et al.*, 2011).

Cyanobacterial genome may vary significantly in size. The smallest genome has been identified in the symbiotic marine cyanobacterium UCYN-A (*Candidatus Atelocyanobacterium thalassa*), which possesses only 1.44 Mb. On the other hand, the largest complete genome sequence belongs to the heterocystous cyanobacterium *Nostoc punctiforme* PCC 73102, which has 9.06 Mb. Interestingly, the largest genome from a unicellular cyanobacterium is the chlorophyll-*d* containing *Acaryochloris marina* at 8.36 Mb (Larsson *et al.*, 2011). It has been suggested that prokaryotic genome size distribution has one peak around 2 Mb and second around 5 Mb (Koonin and Wolf, 2008). *Synechococcus* fits within this range as well. The marine picoplanktic *Synechococcus* (Fig. 2.1) belongs to the group with genome sizes around 2, while the hot spring *Synechococcus* has a slightly larger genome with 3 Mb. *Prochlorococcus* is an atypical cyanobacterium considering the genome size since it has undergone extensive genome reduction, ranging from 1.66 to 2.68 Mb (Table 2.1). Thus, *Prochlorococcus* possesses a very small genome considering fact that it is free-living cyanobacterium (I will discuss further details below in section Genome streamlining). Larger genome sizes are present in the other clades of *Synechococcus*. For example, *Synechococcus* PCC 7335 and PCC 7336 belong to the group with genome size larger than 5 Mb (Table 2.1).

Genome size is correlated with the number of protein-coding genes in prokaryotes with an average of 0.8–1.2 gene to 1 kb genome size (Koonin and Wolf, 2008). We may observe similar relationship within cyanobacteria and in *Synechococcus* (Larsson *et al.*, 2011). This contrasts sharply to the eukaryotic genome, in which genome size does not correlate with the number of genes (as one of possible measure of genome complexity), leading to the C-value paradox, possibly a result of a majority of non-coding regions, whose size varies over several orders of magnitude in comparison with coding regions (Doolittle, 2013).

The G + C content varies among *Synechococcus* strains between 48.2% and 65.4%. On the other hand, the G + C content of *Prochlorococcus* may be only 30.08%. Although the G + C content of cyanobacterial genomes vary widely, this does not seem to be correlated with environmental factors, e.g. temperature (Herdman *et al.*, 1979; Galtier and Lobry, 1997), and it seems to be more or less haphazard. However, bacteria with streamlined genomes tend to have lower G + C content (Moran *et al.*, 2009). A variability of G + C content is also observed within the genome and may be related to horizontal gene transfer (McLean *et al.*, 1998; see details further in this chapter).

Koonin and Wolf (2008) analysed clusters of orthologous genes (COGs) stored in the eggNOG database (Jensen *et al.*, 2008). They found that there exist three groups of COGs among bacteria and archea, with sometimes uncertain boundaries: (1) the smallest set of genes is conserved and it represents a core part of all genomes (~70); (2) moderately conserved genes (a group two orders of magnitude larger; ~5700); and (3) unconserved genes, the largest group (~24,000). An alternative, two-group division based on the conservativeness of the genes in cyanobacteria, has been proposed by Shi and Falkowski (2008). They identified 682 conservative gene families, which were shared by all 13 genomes of cyanobacteria analysed. Shi and Falkowski (2008) called these conservative genes the 'core genome' and they showed that a relative size of the core genome largely differs in cyanobacteria. For instance, the core genome of *Nostoc punctiforme* PCC 73102 is only 8.9% of its total number of genes, while it is 39.7% of *Prochlorococcus marinus* CCMP1986.

Coutinho *et al.* (2016) suggested that the *Synechococcus/Prochlorococcus* pan-genome (all protein-coding genes present in their genomes) contains 82,703 genes, of which only 744 represent the core genome. Thus, although the methods of Shi and Falkowski (2008) are not entirely identical to the Coutinho *et al.* (2016), the core genome of *Synechococcus* and *Prochlorococcus* is strikingly similar

Table 2.1 Basic features of genomes of *Synechococcus*-like cyanobacteria

Taxon	CRISPR/ spacers	Genome size (Mb)	GC%	No. of protein coding genes	Habitat	GenBank No.
Cyanobium sp. PCC 6307	0/0	3.34	68.7	3094	Freshwater	GCA_000316515
Cyanobium sp. PCC 7001	0/0	2.83	68.7	2764	Marine soil	GCA_000155635.1
Neosynechococcus sphagnicola CAUP A 1101	2/9	4.33	51.6	3269	Sphagnum bog	JJML00000000
Prochlorococcus marinus AS9601	0/0	1.67	31.3	1769	Marine	GCA_000015645.1
Prochlorococcus marinus CCMP1375	0/0	1.75	36.4	1882	Marine	GCA_000007925.1
Prochlorococcus marinus MIT 9202	0/0	1.69	31.1	1795	Marine	GCA_000158595.1
Prochlorococcus marinus MIT9211	0/0	1.69	38	1748	Marine	CP000878
Prochlorococcus marinus MIT9215	0/0	1.74	31.1	1840	Marine	CP000825
Prochlorococcus marinus MIT9301	0/0	1.64	31.3	1774	Marine	GCA_000015965.1
Prochlorococcus marinus MIT9303	1/7	2.68	50	2997	Marine	GCA_000015705.1
Prochlorococcus marinus MIT9312	0/0	1.7	31.2	1815	Marine	GCA_000012645.1
Prochlorococcus marinus MIT9313	0/0	2.41	50.7	2339	Marine	GCA_000011485.1
Prochlorococcus marinus MIT9515	0/0	1.7	30.8	1784	Marine	GCA_000015665.1
Prochlorococcus marinus NATL1A	0/0	1.87	35	1954	Marine	GCA_000015685.1
Prochlorococcus marinus NATL2A	0/0	1.85	35.1	1930	Marine	GCA_000012465.1
Prochlorococcus marinus pastoris CCMP1986	0/0	1.66	30.8	1777	Marine	GCA_000011465.1
Synechococcus sp. BL107	0/0	2.29	54.2	2322	Marine	GCA_000153805.1
Synechococcus sp. CB0101	0/0	2.69	64.2	2757	Marine	GCA_000179235.1
Synechococcus sp. CB0205	0/0	2.43	63	2588	Marine	GCA_000179255.1
Synechococcus sp. CC9311	0/0	2.61	52.4	2627	Marine	CP000435
Synechococcus sp. CC9605	0/0	2.51	59.2	2583	Marine	CP000110
Synechococcus sp. CC9902	0/0	2.23	54.2	2289	Marine	CP000097
Synechococcus elongatus PCC 6301	1/6	2.7	55.5	2576	Freshwater	GCA_000010065.1
Synechococcus elongatus PCC 7942	1/3	2.74	55.5	2655	Freshwater	CP000100
Synechococcus sp. JA-2–3B'a(2–13)	6/119	3.5	58.5	2705	Thermal	CP000240
Synechococcus sp. JA-3–3Ab	6/86	2.93	60.2	2590	Thermal	CP000239
Synechococcus sp. PCC 7335	1/4	5.97	48.2	4870	Snail shell marine	GCA_000155595
Synechococcus sp. PCC 7336	8/285	5.1	53.7	4249	Sea water tank	GCA_000332275.1
Synechococcus sp. PCC 6312	3/149	3.72	48.5	3484	Freshwater	CP003558
Synechococcus sp. PCC 7002	3/88	3.41	49.2	3121	Marine	CP000951
Synechococcus sp. PCC 7502	1/59	3.58	40.6	3442	Sphagnum bog	CP003594
Synechococcus sp. RCC307	0/0	2.22	60.8	2348	Freshwater	CT978603
Synechococcus sp. RS9916	0/0	2.99	59.8	2603	Marine	GCA_000153825
Synechococcus sp. RS9917	0/0	2.43	59.4	2575	Marine	BX548020
Synechococcus sp. NKBG041c	3/11	2.11	60.1	2848	Marine	CP006882
Candidatus Synechococcus spongiarum	3/16	2.17	60	1386	Symbiotic	GCA_000586015.1
Synechococcus sp. WH 5701	0/0	3.28	65.4	2917	Marine	GCA_000153045
Synechococcus sp. WH 7803	0/0	2.37	60.2	2439	Marine	NC_009481
Synechococcus sp. WH 7805	0/0	2.37	60.2	2595	Marine	CT971583

Table 2.1 Continued

Taxon	CRISPR/ spacers	Genome size (Mb)	GC%	No. of protein coding genes	Habitat	GenBank No.
Synechococcus sp. WH 8016	1/5	2.69	541	2762	Marine	GCA_000230675
Synechococcus sp. WH 8102	0/0	2.43	59.4	2461	Marine	GCA_000195975.1
Synechococcus sp. WH 8109	0/0	2.11	60.1	2167	Marine	GCA_000161795.2
Thermosynechococcus elongatus BF-1	0/0	2.59	53.9	2476	Thermal	BA000039
Thermosynechococcus sp. NK55	1/24	2.52	53.8	2302	Thermal	GCA_000505665.1

to the core genome of all cyanobacteria inferred by Shi and Falkowski (2008). However, Coutinho *et al.* (2016) did not take into account to the polyphyletic origin of *Synechococcus*, which biased their results. They selected strains of *Synechococcus* based on names of deposited sequences and not based on evolutionary relationships, although evolutionary trajectories of the marine picoplanktic *Synechococcus* and other *Synechococcus* lineages has diverged billions years ago (Dvořák *et al.*, 2014). Thus, many of their compared genomes do not represent members of the same genus, but rather separate genera. I will illustrate the impact of the biased taxon sampling on an example of the marine picoplanktic *Synechococcus* and *Prochlorococcus* lineages. Dufresne *et al.* (2008) estimated that the total pangenome of marine picoplanktic *Synechococcus* and *Prochlorococcus* is composed of 35,946 genes, thus the difference to Coutinho *et al.* (2016) estimates is striking. Furthermore, Dufresne *et al.* (2008) suggested that a size of the core genome within the marine picoplanktic *Synechococcus* is 1572 and within both *Synechococcus* and *Prochlorococcus* 1228, which is almost twice more that estimates of Coutinho *et al.* (2016).

Speciation in cyanobacteria

Prokaryotic diversity is immense, with total numbers of species likely greatly exceeding the number of eukaryotic species (Dykhuizen, 1998) by several orders of magnitude (Locey and Lennon, 2016). It is difficult to estimate the true diversity of this lineage as we are unable to cultivated most of them with the most pessimistic estimates declare that we may cultivate only 1% of existing prokaryotic species (Amann *et al.*, 1995). In any case, in terms of species numbers prokaryotes are the most successful organisms on Earth. Of course, these estimates also largely depend on the species concept and species definition. Understanding of the speciation within prokaryotes holds key evidence for uncovering this remarkable evolutionary success. In this section, I will attempt to synthesize advances in a research of factors forming cyanobacterial species. I will use examples of hot spring and marine picoplanktic *Synechococcus* (see their phylogenetic definition in Fig. 2.1), because the evolutionary patterns within these cyanobacteria have been explored for a long time and in case of marine *Synechococcus* also on a large geographical scale.

Prokaryotes do not possess any form of sexual reproduction, have large populations, small size, and their propagules are able to travel over the considerable distances (for comprehensive reviews see Cohan 2001, 2002; Cohan and Perry, 2007; Achtman and Wagner, 2008; Polz *et al.*, 2013; Shapiro and Polz, 2015). Thus, it was supposed that the evolutionary forces forming prokaryotic species largely differ from the macrobes, and that prokaryotes have only clonal population with an occasional adaptive mutation. However, Avery *et al.* (1944) showed that bacteria have the capability for transformation, or the uptake of DNA from the environment. Later, Lederberg and Tatum (1946) performed experiments showing that bacteria possess mobile elements, which may be transferred and that allow the recipients to receive a new metabolic feature (cf. resistance to antibiotics). These elements are now called plasmids. This indicates that transfer of generic information impacts the prokaryotic evolution and that prokaryotic populations (species) are not entirely clonal. In following paragraphs, I will investigate some aspects of genome dynamics during the speciation of cyanobacteria via mechanisms such as homologous recombination (HR), non-homologous recombination–horizontal (lateral) gene transfer (HGT). During HR, short fragments of the genome are replaced by the corresponding fragments from another organisms (Didelot and Maiden, 2010). On the other hand, through the HGT, a new fragment of the genome from another organisms is acquired (Ochman *et al.*, 2000).

Mechanisms of the DNA uptake

As seen in other bacteria, foreign fragments of DNA may be acquired via transformation, conjugation, and transduction. During transformation, exogenous DNA is acquired from the surrounding environment and incorporated in to the chromosome via homologous recombination. In Gram-negative bacteria (such as cyanobacteria), DNA uptake is mediated by type IV pili, with subsequent passage through cytoplasmic membrane mediated by the membrane-anchored dsDNA binding proteins *comE* or *comA* (in case of *Synechocystis* sp. PCC 6803; Yoshihara *et al.*, 2001) and *comEC* membrane protein. The transferred DNA may be incorporated to the recipient genome by *recA*-dependent homologous recombination (reviewed in Johnsborg *et al.*, 2007). Transformation seems to be a domain of unicellular cyanobacteria, in which non-induced (natural) transformation has been frequently observed under the laboratory conditions in strains of *Synechococcus* (e.g. PCC 7002, PCC 7942) and *Synechocystis* (Shestakov and Khyen, 1970; Stevens and Porter, 1980; Grigorieva and Shestakov, 1982; Lightfoot *et al.*, 1988). Filamentous cyanobacteria appear to largely resist to the natural transformation (reviewed in Flores *et al.*, 2008), but in some strains of the heterocystous cyanobacterium *Anabaena* transformation has been induced by an electroporation (Thiel and Poo, 1989).

Conjugation may be induced in a wide range of both unicellular and filamentous cyanobacteria. Conjugation is an intercellular transfer of mobile elements such as plasmids. It may proceed among phylogenetically distant organisms, for example between *Escherichia coli* and *Anabaena* (Wolk *et al.*, 1984) or different proteobacteria and *Synechococcus elongatus* PCC 7942 (Delaye *et al.*, 2011). In general, the conjugation system is based on the type IV protein secretion system, relaxase (a site- and strand-specific endonuclease), and a protein mediating a connection of relaxosome and secretion system (reviewed in Llosa and de la Cruz, 2005).

Viral (bacteriophages) mediated HGT, transduction, has not yet been directly observed in cyanobacteria. However, comparative genomics of some cyanobacteria and their phages exhibited a striking similarity of some orthologous genes contained in a phage and cyanobacterial genome. For instance, Mann *et al.* (2003) recognized *psbA* and *psbD* genes (photosystem II proteins) encoded in the bacteriophages of marine picoplantic *Synechococcus.* Recent evidence for frequent cyanobacterium/bacteriophage interaction also provides CRISPR-Cas systems (Clustered regularly interspaced short palindromic repeats-CRISPR associated proteins). CRISPR-Cas systems seem to play a pivotal role in bacterial reactions to phage infection, and thus an important part of the bacterial cell immune system. They also represent a good model for research involved in the bacteria/phage coevolution (reviewed in Bhaya *et al.*, 2011). Interestingly, CRISPRs are largely missing in the marine picoplantic *Synechococcus* and *Prochlorococcus* (Table 2.1).

The marine picoplanktic *Synechococcus* and *Prochlorococcus* produce 'lipid vesicles', which contain DNA, RNA and proteins (Biller *et al.*, 2014). There are abundant throughout the oceans and are continuously produced by *Synechococcus* and *Prochlorococcus.* These lipid vesicles seem to be involved in the transfer of DNA (Biller *et al.*, 2014). Whether these special features are restricted to marine picoplanktic *Synechococcus* and *Prochlorococcus,* or if they may be bound in the other environments and other cyanobacteria, has to be further tested. In any case, with regard to the ecological and evolutionary importance of marine picoplanktic *Synechococcus* and *Prochlorococcus,* it is clear that the vesicle transport have significantly affected the global ecosystem.

In silico HGT detection

HGT may be difficult to discern as it does not typically lead to the phenotype modification in cyanobacteria. Thus, we largely rely on a whole genome sequence data when pursuing a detection of HGT. There are two distinct approaches in *in silico* HGT detection – parametric and phylogenetic methods (Eisen, 2000; Koonin *et al.*, 2001; Ravenhall *et al.*, 2015 for review).

Parametric methods seek an alteration of a genomic signature within a particular portion of the genome since the deviation is likely a result of a HGT. The simplest genomic signature is an observation of G + C content variation within a genome. However, G + C content tends to vary significantly among prokaryotes and within the genome, especially at the third codon position (cf. McLean *et al.*, 1998). G + C content variation may be also illustrated in *Synechococcus* (Table 2.1). Further genomic signatures commonly used signatures are nucleotide composition, oligonucleotide frequencies (Lawrence and Ochman, 1998), and structural features of genomes (Worning *et al.*, 2000).

Phylogenetic methods may help to compare the evolutionary history between phylogenetic species and a particular gene phylogeny. In the other words, deviations between traditional phylogenies (e.g. based on the 16S rDNA gene sequence) and relationships within a gene tree may likely be a result of HGT. These methods are called explicit. Furthermore, there are implicit phylogenetic methods, which utilize evolutionary distances and sequence similarity. If the distance is unexpectedly short or long, it might suggest HGT (reviewed by Ravenhall *et al.*, 2015).

A genome dynamics during the speciation of cyanobacteria

Regardless of the mechanism, horizontal gene transfer has played an important role in the speciation of cyanobacteria as it allowed rapid innovations of the genome through the acquirement of the new genes with variable functions (Ochman *et al.*, 2000; Boucher *et al.*, 2003; Achtman and Wagner, 2008; Koonin and Wolf, 2008; Wiedenbeck and Cohan, 2011; Polz *et al.*, 2013; Shapiro and Polz, 2015). HGT and gene loss seem to be two dominant forces shaping bacterial species (Puigbò *et al.*, 2014) and thus far has only been intensively studied in *Synechococcus*-like cyanobacteria.

HGT seems to affect a broad range of functional genes (Beiko *et al.*, 2005; Zhaxybayeva *et al.*, 2006; Dvořák *et al.*, 2014) and about 50% of gene families seem to have a history of HGT (Zhaxybayeva *et al.*, 2006). However, Popa *et al.* (2011) found that HGT is more likely to take place within gene families with unknown function and the non-conserved (shell) part of the genome seems to be more affected by HGT. For example, there at least 250 HGT events were identified in marine *Synechococcus* and *Prochlorococcus* (Beiko *et al.*, 2005; Zhaxybayeva *et al.*, 2006; Zhaxybayeva *et al.*, 2009). The most striking feature of HGT is that it is prevalent among more closely related strains or species while transfers on an interphylum level or above is less likely (Beiko *et al.*, 2005; Popa *et al.*, 2011). Thus, HGT seems to decrease with the genetic distance. Furthermore, HGT is not randomly dispersed throughout the genome, but it is rather concentrated in the variable genome regions where rapid innovation occurs, called 'genomic islands' (reviewed in Dobrindt *et al.*, 2004). For example, genomic islands represents 11–31% of the marine picoplanktic *Synechococcus* and *Prochlorococcus* genome (Dufresne *et al.*, 2008). Thus, features of HGT may be summarized as follows: it is unselective (to functional gene category), local (short geographical distance), potent, and it is more likely among closely related organisms.

DNA may also be exchanged among prokaryotes also via a homologous recombination. In cyanobacteria, *in situ* HR has been reported in the cosmopolitan, benthic, marine taxon *Microcoleus* (*Coleofasciculus*) *chthonoplastes* (Lodders *et al.*, 2005). Furthermore, it has been proposed that HR in marine picoplanktic *Synechococcus* and *Prochlorococcus* is mediated via marine phages. Since the phages contain genes crucial for the photosynthesis, the HR seems to be a crucial factor of a photosynthesis evolution in the marine picoplankton (e.g. Zeidner *et al.*, 2005; Sullivan *et al.*, 2006). Furthermore, freshwater planktic cyanobacterium *Microcystis aeruginosa* also exhibit high HR rates (Vos and Didelot, 2009). From the analysis of the human microbiome, it has been suggested that an intensity of HR decreases with the genetic distance (Smillie *et al.*, 2011).

HGT and HR are reflected in phylogenetic reconstructions by conflicting tree topologies of different genes (identification of the HGT using explicit phylogenetic method is based on this fact; see details above). Thus, it has become debatable if bacterial species are actually coherent units through the time and space. The evidence rising from the HGT suggested that species boundaries within prokaryotes may be 'fuzzy' (Lawrence,

2002; Hanage *et al.*, 2005). However, there are multiple lines of evidence that suggest that despite the dynamic nature of the prokaryotic genomes, there are possibly coherent identifiable units, which may be called species in the animal and plant sense. First, from the genomic point of view, since the frequency of HGT and HR is negatively correlated with the genetic distance, it may indicate that newly formed genotypes are stabilizing through the time. The diverging lineages may appear to have conflicting phylogenies at the beginning of the speciation before becoming clearly separated after some time (Polz *et al.*, 2013). Beiko *et al.*, (2005) suggested that the shell, less conserved portion of the genome, undergoes more HGT events and therefore phylogenetic inferences based on the core genes may exhibit less entangled relationships. Second, from the ecological point of view, we may seek evidence for coherence within the stable ecotypes (Cohan 2001; 2002; Cohan and Perry, 2007), such as occurs in the hot spring *Synechococcus* and *Prochlorococcus*. It has been shown that these organisms have two putative ecotypes – high light adapted and low light adapted. Although the ecotypes were almost identical in 16S rRNA, they differed in the genome composition (Olsen *et al.*, 2015), and they may be recognized based on the *psaA* gene sequence (Becraft *et al.*, 2015).

There are four distinct ecotypes recognized in *Prochlorococcus*. Two of them are more abundant in high light conditions (abbreviated HL). HL ecotype is further separated based on geographical distribution to HLll, which is more abundant in lower latitudes, and HLl in higher latitudes. On the other hand, the low light ecotypes prefer low light conditions (LL) and they are also further separated into two groups based on the vertical distribution in the ocean and geographical distribution. Ecotype LLl is more abundant in higher latitudes in upper layers of oceans and in lower latitudes in deeper layers around the thermocline. The last ecotype, LLlV, lives only in low latitudes and in the deepest areas of occurrence under the thermocline (Moore *et al.*, 1998; West and Scanlan, 1999; Biller *et al.*, 2015). Thus, it indicates that there exist stable ecotypes, which may be considered separate species. Similar results were obtained from the whole-genome sequences of strains of the soil dwelling *Bacillus subtilis* (Kopac *et al.*, 2014).

Genome streamlining in cyanobacteria

HGT and other processes, such as duplication, are responsible for expansion of recipient genomes. However, a genome size has its limits and while the genome size of prokaryotes most likely fluctuates around 2 and 5 Mb (Koonin and Wolf, 2008). Thus, the prokaryotic genome is not of an infinite complexity and gene gain is balanced by gene loss, which may lead to the genome streamlining – a minimization of the genome favoured by selection. In general, streamlined genomes are characterized by small size, highly conserved core genes, and low numbers of pseudo-genes and intergenic spacers (reviewed, for example, by Batut *et al.*, 2014; Giovannoni *et al.*, 2014). Genome streamlining has been observed in many prokaryotic lineages including cyanobacteria such as *Prochlorococcus* (for example Partensky and Garczarek, 2010).

Prochlorococcus is one of the most abundant organisms on Earth: Flombaum *et al.* (2013) estimate $7.0 \pm 0.3 \times 10^{26}$ cells in oceans. It is likely that the genome streamlining is one of the factors responsible for such evolutionary success (Batut *et al.*, 2014). It has been hypothesized that although the tropical subtropical parts of the ocean (where *Prochlorococcus* has a centre of distribution) are nutrient poor, they are also stable. Genome streamlining leads to the adaptation to the stable environment, where a reduced number of metabolic pathways may provide more of an advantage than in variable environments (Rocap *et al.*, 2003). However, this does not necessarily account for the diversification of *Prochlorococcus* about 150 million year ago from the marine picoplanktic *Synechococcus* (Dufresne *et al.*, 2005). Morris *et al.* (2012) proposed an alternative hypothesis to explain *Prochlorococcus* genome streamlining: the Black Queen hypothesis. The central idea of this hypothesis is that coexisting bacterial species within the populations may share products of metabolism with each other and they may reduced their genome as they do not need to have a complete set of all metabolic pathways. Therefore, they may live in an energy efficient mode of life similar to 'symbiosis'. Morris *et al.* (2012) provides an example: production of catalase peroxidase (*katG* gene; a defence against external peroxide oxide) is lacking in *Prochlorococcus* and present in marine picoplanktic *Synechococcus*. Thus, *Prochlorococcus* benefits by saving energy although it still possesses a defence against external peroxide oxide by the catalase peroxidase present in the environment, which is produced by marine picoplanktic *Synechococcus* cell machinery.

Is everything everywhere in cyanobacteria?

Allopatric speciation (speciation by a geographical isolation of gene flow among populations) is a major speciation force in macroorganisms. However, barriers to gene flow have not been evidenced in microbes. The first hypothesis describing a geographical distribution of free-living microorganisms has been postulated by Baas Becking (1934) in his famous tenet: '*everything is everywhere: but the environment selects*'. Thus, local species composition is shaped by local environmental factors and it is not restricted by a geographical position. This hypothesis is appealing considering the immense dispersal capabilities of microbes and coupled with their morphological similarity (which was the main criterion for identification in that time). Baas Becking's idea was subsequently taken as canon with myriad evidence supporting it – mainly morphological data (Finlay, 2002). However, an introduction of molecular markers provided evidence that a cohesive morphology might mask changes at the genome level. This occurs very frequently in *Synechococcus* (Erwin and Thacker, 2008; Dvořák *et al.*, 2014). It should be kept in mind that obligatory parasites and symbionts have their distribution largely driven by the distribution of their host. I will focus on the free-living cyanobacteria and on some of recent molecular evidences, which will provide an insight into an influence of dispersal barriers in the speciation of cyanobacteria and thus a possible role of allopatric speciation.

Traditionally, most cyanobacteria reportedly have cosmopolitan geographical distribution. For example, a 16S rRNA phylogeny *Microcoleus* spp. (soil species) shows cosmopolitan distribution (Garcia-Pichel *et al.*, 2001). Very similar results revealed an analysis of the phylogeography of freshwater bloom-forming cyanobacterium *Microcystis aeruginosa* (van Gremberghe *et al.*, 2011). In this instance, authors use less conservative 16S–23S ITS (internal transcribed spacer) locus. On the other hand, hot spring dwelling *Synechococcus* exhibit geographical barriers in distribution (Papke *et al.*, 2003). Furthermore, marine picoplanktic *Synechococcus* and *Prochlorococcus* exhibit a geographical patterning on the global scale (Zwirglmaier *et al.*, 2008), which concurs with our observation (Dvořák *et al.*, 2012), which pointed to geographical structuring on the continental scale within the soil dwelling cyanobacterium *Microcoleus vaginatus* using 16S rRNA and 16S–23S ITS phylogenies. Moreover, since some lineages seem to have repeatedly radiated to different continents, suggesting only temporary geographic barriers of *M. vaginatus* distribution. More evidence for periodic geographical patterning has been proposed with the soil dwelling genus *Chroococcidiopsis* based on 16S rRNA, 16S–23S ITS and 23S rRNA phylogeny (Bahl *et al.*, 2011).

At this state, I may safely conclude that geographical isolation and gene flow limits (allopatric speciation) may impact the evolution and diversification of cyanobacteria and other prokaryotes. However, we are unable to precisely describe its extent. If we look at the other prokaryotes such as hot spring archeon *Sulfolobus islandicus*, in which Reno *et al.* (2009) found a clear geographical patterning based on comparative genome analysis of seven *S. islandicus* genomes, we may further conclude that genomic data especially within closely related strains or species (other than marine *Synechococcus* and *Prochlorococcus*) will provide important evidence for the biogeography of cyanobacteria.

Genome-scale reconstruction of evolutionary relationships: phylogenomics

The number of whole-genome sequences of cyanobacteria (and other prokaryotes) is steadily growing and the price of sequencing is decreasing. This progress has allowed hundreds genomic loci for phylogenetic reconstructions, a process known as phylogenomics (Eisen and Fraser, 2003). It is worth noting that phylogenomic reconstructions may be affected by HGT (this topic is extensively discussed above).

The data preparation for the phylogenomic analysis involves several steps. To adequately describe this complex task is beyond the scope of this chapter and therefore I provide some useful and comprehensive reviews (Philippe *et al.*, 2005;

Swingley *et al.*, 2008; Schreiber *et al.*, 2009; Chan and Ragan, 2013).

The convoluted evolutionary relationships within *Synechococcus*

Synechococcus has been long considered as a coherent group of species with a simple morphology. However, the first attempts to reconstruct its phylogenetic relationships based on a single gene suggested that the morphological coherence may mask immense genetic variability (Honda *et al.*, 1999; Turner *et al.*, 1999; Robertson *et al.* 2001). The most comprehensive study revealed 12 lineages based on 16S rRNA phylogeny (Dvořák *et al.*, 2014) and six of these lineages were present in the phylogenetic reconstruction based on the whole-genome data (Shih *et al.*, 2013; Dvořák *et al.*, 2014). In the following lines, I will focus on evolutionary history of some of these lineages.

The deepest branches of *Synechococcus* are those isolated from hot springs (Ferris and Ward, 1997). These taxa form a separate clade from the other *Synechococcus* strains, but have not been formally named yet. Another hot spring dwelling *Synechococcus* diverged later (it is unrelated to previous lineage). It has been called *Thermosynechococcus* (Katoh *et al*, 2001). It should be noted that albeit this name is not valid under any code of nomenclature, it is widely used. *Thermosynechococcus* have been isolated from hot springs in both Japan and the Untied States. Both clades exhibits large internal genetic variety, thus especially the hot spring *Synechococcus* is a perfect model for an investigation of the microevolutionary patterns, speciation in (cyano)bacteria, and adaptation for a life in the extreme temperatures. *Synechococcus* has possibly radiated to the marine environment as many as six times (Dvořák *et al.* 2014). While the majority of the strains from these clades inhabit marine sediments, those of to the marine picoplanktic *Synechococcus* (Fig. 2.1), together with *Prochlorococcus*, are the most abundant oxygenic phototrophs. The genetic variability within this clade is immense. The marine picoplanktic *Synechococcus* shares a monophyletic clade with *Candidatus* taxon *S. spongiarum*, a symbiont of the marine sponges; Erwin and Thacker, 2008), and some '*Prochlorococcus*' strains. Moreover, we may recognize several ecological, geographical, and physiological groups within this clade. Finally, based on Dvořák *et al.* (2014) 16S rRNA phylogeny, *Synechococcus* has also radiated twice to the *Sphagnum*-bogs, another lineage is solely freshwater, and the last lineage isolated from aerophytic habitats (the hair of the polar bear in San Diego Zoo; Lewin and Robinson, 1979). Most of the *Synechococcus* lineages have evolved as predecessors to the more morphologically complex cyanobacteria from the orders Oscillatorialles, Pleurocapsales, and Nostocales (sensu Rippka *et al.*, 1979). However, one marine lineage (containing *Synechococcus* sp. PCC 7002 and NKBG041c; Fig. 2.1) has derived after the emergence of mentioned more morphologically complex cyanobacteria.

Thus, I may conclude that, while *Synechococcus* is morphologically coherent, it has a polyphyletic evolutionary origin based on genetics (major clades of *Synechococcus* have no common ancestor). Moreover, since *Synechococcus*-like cyanobacteria are spread over the evolutionary tree of cyanobacteria, they provide an ideal model for a study of patterns in the macroevolution of cyanobacteria.

Macroevolution of cyanobacteria: the serial convergence

Before the introduction of the molecular methods, the systematics of cyanobacteria was rather simple and based on a handful of morphological characters. Even so, several taxonomic systems have been proposed, the best known of which was the division into five groups by Rippka *et al.* (1979). Group I contained unicellular cyanobacteria with simple binary fission reproduction (including *Synechococcus*), group II unicellular with division in multiple planes, group III filamentous cyanobacteria without specialized cells, group IV filamentous, heterocystous cyanobacteria without true branching, and group V filamentous heterocystous cyanobacteria with true branching.

It has soon become clear that this systematic view was overly simplified and did not reflect evolutionary relationships (e.g. Turner *et al.*, 1999; Robertson *et al.* 2001; Shih *et al.*, 2013; Dvořák *et al.* 2014; Komárek *et al.* 2014). In cladistics terms, very few lineages are monophyletic, with most genera currently described as polyphyletic (Komárek *et al.* 2014). One of the most important evolutionary innovations was the origin of cell differentiation into the heterocytes and akinetes, which may have happened 2.1–2.45 billion years ago (Tomitani *et al.*, 2006). Within the group of heterocystous cyanobacteria, phylogenetic reconstructions have also revealed polyphyletic origin, which may be illustrated on the polyphyly within the genus *Cylindrospermum* (Johansen *et al.*, 2014) or multiple origin of the true branching (Komárek *et al.*, 2014). Furthermore, a peculiar cyanobacterium, *Starria*, has been found in the granite rock in Zimbabwe and possesses a unique filament formation composed of star-like cells (Lang, 1977), a trait that has not been observed in other cyanobacteria.

These taxonomic examples, together with complicated phylogenetic relationships, were used to illustrate an immense number of convergent evolutionary events through the history of cyanobacteria. Moreover, similar trends may be detected in the other group of bacteria, e.g. the giant Sulphur bacteria (gammaproteobacteria, Thiotrichaceae; Salman *et al.*, 2011), which similarly possess a complex morphology and relatively large cells like cyanobacteria. In the following lines, I will elucidate potential evolutionary forces responsible for the entangled relationships among cyanobacteria exhibited by the convergence phenotype.

We have suggested in Dvořák *et al.* (2014) that the frequent convergence of phenotypic traits in cyanobacteria has its nature in the speciation patterns of cyanobacteria. We have proposed the 'model of serial convergence' (Fig. 2.2). As I have already shown in this chapter, cyanobacterial genomes are largely modified during the time by HGT and HR. The genome dynamics seem to be also key factor for the macroevolutionary patterns in cyanobacteria. All cyanobacteria have a common gene pool because the gene flow among the populations has only periodic obstacles (e.g. Bahl *et al.*, 2011; Dvořák *et al.*, 2012; see details above). Although there may be exogenous input of genes through the HGT from the other bacteria, the gene pool of cyanobacteria is more or less restricted (Koonin and Wolf, 2008). Finally, genetic exchange takes places locally via local gene pools and it largely has adaptive function (Polz *et al.* 2013, see details above). Therefore, after speciation events, the subsequent phenotype (or morphotype) may be more complex (filamentous) or less complex (unicellular). It depends on the environmental factor and on the genes, which has been acquired via HGT.

The convergence of morphological traits may also suggest an explanation for the putative 'evolutionary stasis' proposed by Schopf (1994). He showed that cyanobacteria found in the fossil record (up to 2 BY old) and recent cyanobacteria are morphologically indistinguishable and exist in similar environments. Thus, he proposed they are

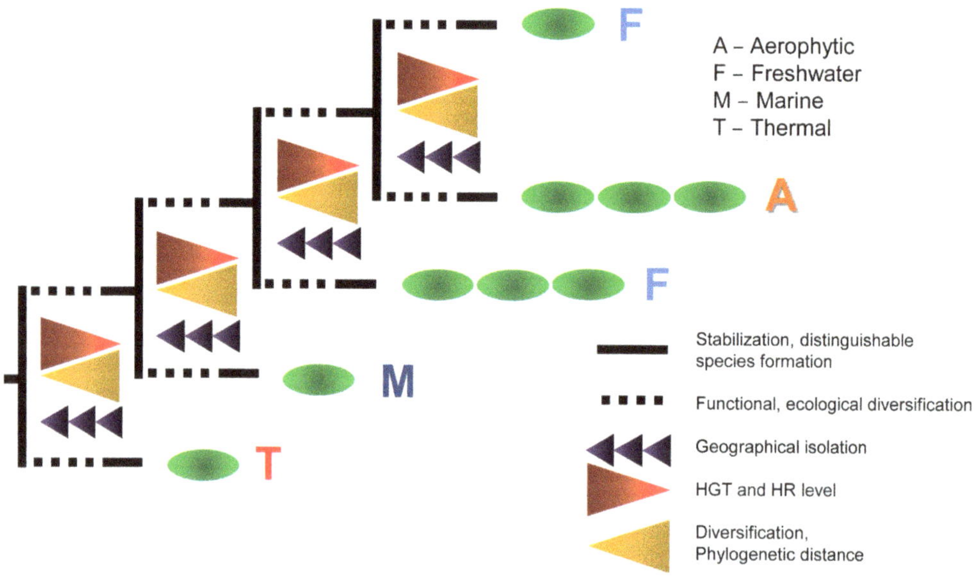

Figure 2.2 A model of cyanobacterial evolution, which summarizes speciation factors and macroevolutionary patterns characterized by convergence. Adopted from Dvořák *et al.* (2014). One green oval represents unicellular and three ovals – filamentous cyanobacteria.

in evolutionary stasis, because they do not need to evolve within the stable environment, a process termed 'hypobradytelly'. A similar pattern has been proposed for the 2-BY-old giant sulfur bacteria (Schopf *et al.*, 2015) and advocated by Rudi *et al.* (1998). They analysed phylogenies of 16S rRNA and *rbcL* genes, finding the phylogenies incongruent, which suggest an influence of recombination at these loci. Furthermore, they suggested that since there is greater genetic distance within the clade than among clades, the lineages are stable during the time. Thus, Rudi *et al.* (1998) proposed that cyanobacteria undergo mostly neutral changes during evolution without any change in morphology. I would like to stress here, that Rudi *et al.* (1998) did use only restricted dataset, which was unable to exhibit any convergence of phenotype.

However, since the bacterial world is so dynamic and the bacterial genome undergoes drastic changes, it is unlikely that it would have been stable during such a long time. Moreover, bacteria are able to disperse over a significant distances, therefore a long lasting barriers of the gene flow among populations are improbable. Thus, we proposed that the hypobradytelly is only elusive and it is masks actual genome dynamics and phenotype convergence, because the major macroevolutionary pattern of cyanobacterial evolution seems to be the 'serial convergence' (Dvořák *et al.*, 2014, 2015a,b).

The systematics of *Synechococcus* in the genomic era

Synechococcus has baffled taxonomists since the introduction of molecular methods for the identification of species. Early molecular methods, such as DNA–DNA hybridization or RFLP (restriction fragment length polymorphism), clearly demonstrated that the genus *Synechococcus* is composed of several genetically distant lineages (e.g. Douglas and Carr, 1988; Garrity *et al.* 2001). 16S rRNA sequencing confirmed this notion. Honda *et al.* (1999) and Robertson *et al.* (2001) suggested that *Synechococcus* is composed of at least five lineages, which may be considered as genera. We reviewed all *Synechococcus* 16S rRNA sequences stored in GenBank (Dvořák *et al.*, 2014). The phylogenetic reconstruction revealed at least 12 separate lineages, which may be ranked genera. Some authors suggested that *Synechococcus* has unique evolutionary trajectories, which are exhibited by these enigmatic relationships. This hypothesis was supported by the fact that *Synechococcus* possesses very simple morphological features (Rippka *et al.*, 1979; Honda *et al.*, 1999; Robertson *et al.*, 2001). However, some *Synechococcus* strains (PCC 7335, PCC 7336) have larger and more complex genomes than, for example, *Cylindrospermum raciborskii*, which is diazotrophic, heterocystous, and it complex filament formation (Stucken *et al.*, 2010). Moreover, morphological convergence appears to be spread over the whole evolutionary history of cyanobacteria. Therefore, I suppose that *Synechococcus* provided the impetus to recognize this general pattern. From the taxonomic point of view, *Synechococcus* should be divided into the at least 12 separate genera, because all *Synechococcus* lineages have different evolutionary histories and confusion may arise from a using term *Synechococcus* without any further specification (Dvořák *et al.*, 2014).

Conclusion

Synechococcus is undoubtedly an extraordinarily successful group of cyanobacteria from both ecological and evolutionary points of view. I have shown that *Synechococcus* is the oldest known cyanobacterium and its evolution may be dated back to the Archean, where the first lineage of thermophilic *Synechococcus* diverged. Since its origin, *Synechococcus* radiated to freshwater, marine habitats (both benthic and planktic), and soil. Finally, it spread over the vast pelagic portion of the global ocean and become one of the most abundant organisms inhabiting Earth. However, *Synechococcus* does not represent a coherent group showed by extreme polyphyletic origin. It has appeared at least 12 times. Thus, all lineages of *Synechococcus* have coherent morphological characteristics, which masks an immense genetic variability measured by size, composition, and complexity of their genome. The evolutionary success of *Synechococcus* seems to be largely connected with a nature of prokaryotic speciation, which is driven by factors such as horizontal gene transfer and homologous recombination. Moreover, it is apparent that convergent evolutionary events are very frequent among cyanobacteria and almost all known cyanobacterial lineages, including more morphologically complex cyanobacteria, exhibit this pattern. The purpose of this chapter was to not to give a review of all factors and events shaping an evolution of *Synechococcus* and other cyanobacteria, but rather illustrate that the knowledge of evolutionary biology of cyanobacteria is now rapidly expanding and the *Synechococcus* provides a great example for study of both microevolution and macroevolution of prokaryotic microbes.

Future trends

From a very pessimistic point of view, we know only a tiny portion of the potentially trillions of prokaryotic species (Locey and Lennon, 2016), which may inhabit Earth, and therefore lack a considerable amount of information about an evolution within prokaryotes. Moreover, only 1% of this diversity may be cultured by recent methods. At this state, we are unsure what the true biodiversity of prokaryotic lineages may entail. It is questionable how much will the data obtained from the uncultured species alter our understanding of evolutionary processes. Fortunately, recent rapid expansion of the methods allowing an investigation of environmental samples (e.g. metagenomics and single-cell genomics) promise rapid advancement in the evolutionary biology of prokaryotes in general. Moreover, relatively intensively studied groups, such as cyanobacteria, may provide a significant asset to the knowledge of evolutionary processes of all prokaryotes.

Acknowledgements

I would like to express my gratitude and admiration to Dale A. Casamatta (University of North Florida) for a valuable discussion and English editing of this chapter. This work has been supported by Internal Grant Agency of Palacký University Prf-2016-001.

Web resources

World Wide Web resources are rather focused on cyanobacteria as a whole group than only on *Synechococcus*. For example, JGI (Joined Genome Institute) offers quite extensive overview of ongoing sequencing projects (http://genome.jgi.doe.gov/cyanobacteria/cyanobacteria.info.html). Further, most of the genomes available in JGI are stored in GenBank with an addition of the other sequencing projects, which are not listed in JGI (http://www.ncbi.nlm.nih.gov). An extensive recent asset to the omics of cyanobacteria represents CyanoOmics (Cyanobacterial Omics Anlysis Database). It contains available data from genomics, proteomics, and transcriptomics *Synechococcus* sp. PCC 7002 (Yang *et al.*, 2015; http://lag.ihb.ac.cn/cyanomics).

References

Achtman, M., and Wagner, M. (2008). Microbial diversity and the genetic nature of microbial species. Nat. Rev. Microbiol. *6*, 431–440. http://dx.doi.org/10.1038/nrmicro1872.

Amann, R.I., Ludwig, W., and Schleifer, K.H. (1995). Phylogenetic identification and in situ detection of individual microbial cells without cultivation. Microbiol. Rev. *59*, 143–169.

Avery, O.T., MacLeod, C.M., and McCarty, M. (1944). Studies on the chemical nature of the substance inducing transformation of pneumococcal types. Introduction of transformation by a desoxyribonicleic acid fraction isolate from *Pneumococcus* type III. J. Exp. Med. *79*, 137–158.

Bahl, J., Lau, M.C., Smith, G.J., Vijaykrishna, D., Cary, S.C., Lacap, D.C., Lee, C.K., Papke, R.T., Warren-Rhodes, K.A., Wong, F.K., *et al.* (2011). Ancient origins determine global biogeography of hot and cold desert cyanobacteria. Nat. Commun. *2*, 163. http://dx.doi.org/10.1038/ncomms1167.

Baas Becking, L.G.M. (1934). Geobiologie of inleiding tot de miliekunde (Hague, Netherlands: W. P. van Stockum).

Batut, B., Knibbe, C., Marais, G., and Daubin, V. (2014). Reductive genome evolution at both ends of the bacterial population size spectrum. Nat. Rev. Microbiol. *12*, 841–850. http://dx.doi.org/10.1038/nrmicro3331.

Becraft, E.D., Wood, J.M., Rusch, D.B., Kühl, M., Jensen, S.I., Bryant, D.A., Roberts, D.W., Cohan, F.M., and Ward, D.M. (2015). The molecular dimension of microbial species: 1. Ecological distinctions among, and homogeneity within, putative ecotypes of *Synechococcus* inhabiting the cyanobacterial mat of Mushroom Spring, Yellowstone National Park. Front. Microbiol. *6*, 590. http://dx.doi.org/10.3389/fmicb.2015.00590.

Beiko, R.G., Harlow, T.J., and Ragan, M.A. (2005). Highways of gene sharing in prokaryotes. Proc. Natl. Acad. Sci. U.S.A. *102*, 14332–14337. http://dx.doi.org/10.1073/pnas.0504068102.

Bhaya, D., Davison, M., and Barrangou, R. (2011). CRISPR-Cas systems in bacteria and archaea: versatile small RNAs for adaptive defense and regulation. Annu. Rev. Genet. *45*, 273–297. http://dx.doi.org/10.1146/annurev-genet-110410-132430.

Biller, S.J., Berube, P.M., Lindell, D., and Chisholm, S.W. (2015). *Prochlorococcus*: the structure and function of collective diversity. Nat. Rev. Microbiol. *13*, 13–27. http://dx.doi.org/10.1038/nrmicro3378.

Biller, S.J., Schubotz, F., Roggensack, S.E., Thompson, A.W., Summons, R.E., and Chisholm, S.W. (2014). Bacterial vesicles in marine ecosystems. Science *343*, 183–186. http://dx.doi.org/10.1126/science.1243457.

Binder, B.J., and Chisholm, S.W. (1995). Cell cycle regulation in marine *Synechococcus* sp. strains. Appl. Environ. Microbiol. *61*, 708–717.

Blank, C.E., and Sánchez-Baracaldo, P. (2010). Timing of morphological and ecological innovations in the cyanobacteria – a key to understanding the rise in atmospheric oxygen. Geobiology *8*, 1–23. http://dx.doi.org/10.1111/j.1472-4669.2009.00220.x.

Boucher, Y., Douady, C.J., Papke, R.T., Walsh, D.A., Boudreau, M.E., Nesbø, C.L., Case, R.J., and Doolittle, W.F. (2003). Lateral gene transfer and the origins of prokaryotic groups. Annu. Rev. Genet. *37*, 283–328. http://dx.doi.org/10.1146/annurev.genet.37.050503.084247.

Boyer, S.L., Flechtner, V.R., and Johansen, J.R. (2001). Is the 16S–23S rRNA internal transcribed spacer region a good tool for use in molecular systematics and population genetics? A case study in cyanobacteria. Mol. Biol. Evol. *18*, 1057–1069. http://dx.doi.org/10.1093/oxfordjournals.molbev.a003877.

Butterfield, N.J. (2015). Proterozoic photosynthesis – a critical review. Paleontology *58*, 953–972. http://dx.doi.org/10.1111/pala.12211.

Chan, C.X., and Ragan, M.A. (2013). Next-generation phylogenomics. Biol. Direct *8*, 3. http://dx.doi.org/10.1186/1745-6150-8-3.

Coutinho, F., Tschoeke, D.A., Thompson, F., and Thompson, C. (2016). Comparative genomics of *Synechococcus* and proposal of the new genus *Parasynechococcus*. PeerJ *4*, e1522. http://dx.doi.org/10.7717/peerj.1522.

Cohan, F.M. (2001). Bacterial species and speciation. Syst. Biol. *50*, 513–524. http://dx.doi.org/10.1080/10635150118398.

Cohan, F.M. (2002). What are bacterial species? Annu. Rev. Microbiol. *56*, 457–487. http://dx.doi.org/10.1146/annurev.micro.56.012302.160634.

Cohan, F.M., and Perry, E.B. (2007). A systematics for discovering the fundamental units of bacterial diversity. Curr. Biol. *17*, R373–86. http://dx.doi.org/10.1016/j.cub.2007.03.032.

Delaye, L., González-Domenech, C.M., Garcillán-Barcia, M.P., Peretó, J., de la Cruz, F., and Moya, A. (2011). Blueprint for a minimal photoautotrophic cell: conserved and variable genes in *Synechococcus elongatus* PCC 7942. BMC Genomics *12*, 25. http://dx.doi.org/10.1186/1471-2164-12-25.

Didelot, X., and Maiden, M.C. (2010). Impact of recombination on bacterial evolution. Trends Microbiol. *18*, 315–322. http://dx.doi.org/10.1016/j.tim.2010.04.002.

Dobrindt, U., Hochhut, B., Hentschel, U., and Hacker, J. (2004). Genomic islands in pathogenic and environmental microorganisms. Nat. Rev. Microbiol. *2*, 414–424. http://dx.doi.org/10.1038/nrmicro884.

Doolittle, W.F. (2013). Is junk DNA bunk? A critique of ENCODE. Proc. Natl. Acad. Sci. U.S.A. *110*, 5294–5300. http://dx.doi.org/10.1073/pnas.1221376110.

Douglas, S.E., and Carr, N. (1988). Examination of genetic relatedness of Marine *Synechococcus* spp. by using restriction fragment length polymorphisms. Appl. Environ. Microbiol. *54*, 3071–3078.

Dvořák, P., Casamatta, D.A., Poulíčková, A., Hašler, P., Ondřej, V., and Sanges, R. (2014). *Synechococcus*: 3 billion years of global dominance. Mol. Ecol. *23*, 5538–5551. http://dx.doi.org/10.1111/mec.12948.

Dvořák, P., Casamatta, D.A., Hašler, P., and Poulíčková, A. (2015a). Putative extremely long evolutionary stasis in bacteria might be explained by serial convergence. Proc. Natl. Acad. Sci. U.S.A. *112*, E2559. http://dx.doi.org/10.1073/pnas.1502924112.

Dvořák, P., Hašler, P., and Poulíčková, A. (2012). Phylogeography of the *Microcoleus vaginatus* (Cyanobacteria) from three continents – a spatial and temporal characterization. PLoS One *7*, e40153. http://dx.doi.org/10.1371/journal.pone.0040153.

Dvořák, P., Poulíčková, A., Hašler, P., Belli, M., Casamatta, D.A., and Papini, A. (2015b). Species concepts and speciation factors in cyanobacteria, with connection to the problems of diversity and classification. Biodivers. Conserv. *24*, 739–757. http://dx.doi.org/10.1007/s10531-015-0888-6.

Dufresne, A., Garczarek, L., and Partensky, F. (2005). Accelerated evolution associated with genome reduction in a free-living prokaryote. Genome Biol. *6*, R14. http://dx.doi.org/10.1186/gb-2005-6-2-r14.

Dufresne, A., Ostrowski, M., Scanlan, D.J., Garczarek, L., Mazard, S., Palenik, B.P., Paulsen, I.T., de Marsac, N.T., Wincker, P., Dossat, C., *et al.* (2008). Unraveling the genomic mosaic of a ubiquitous genus of marine cyanobacteria. Genome Biol. *9*, R90. http://dx.doi.org/10.1186/gb-2008-9-5-r90.

Dykhuizen, D.E. (1998). Santa Rosalia revisited: why are there so many species of bacteria? Antonie Van Leeuwenhoek 73, 25–33.

Eisen, J.A. (2000). Horizontal gene transfer among microbial genomes: new insights from complete genome analysis. Curr. Opin. Genet. Dev. *10*, 606–611. http://dx.doi.org/10.1016/s0959-437x(00)00143-x.

Eisen, J.A., and Fraser, C.M. (2003). Phylogenomics: intersection of evolution and genomics. Science *300*, 1706–1707. http://dx.doi.org/10.1126/science.1086292.

Erwin, P.M., and Thacker, R.W. (2008). Cryptic diversity of the symbiotic cyanobacterium *Synechococcus spongiarum* among sponge hosts. Mol. Ecol. *17*, 2937–2947. http://dx.doi.org/10.1111/j.1365-294X.2008.03808.x.

Ferris, M.J., and Ward, D.M. (1997). Seasonal distributions of dominant 16S rRNA-defined populations in a hot spring microbial mat examined by denaturing gradient gel electrophoresis. Appl. Environ. Microbiol. *63*, 1375–1381.

Finlay, B.J. (2002). Global dispersal of free-living microbial eukaryote species. Science *296*, 1061–1063. http://dx.doi.org/10.1126/science.1070710.

Flombaum, P., Gallegos, J.L., Gordillo, R.A., Rincón, J., Zabala, L.L., Jiao, N., Karl, D.M., Li, W.K., Lomas, M.W., Veneziano, D., *et al.* (2013). Present and future global distributions of the marine Cyanobacteria *Prochlorococcus* and *Synechococcus*. Proc. Natl. Acad. Sci. U.S.A. *110*, 9824–9829. http://dx.doi.org/10.1073/pnas.1307701110.

Flores, E., Muro-Pastor, A.M., and Meeks, J.C. (2008). Gene transfer to cyanobacteria in the laboratory and in nature. In The cyanobacteria: molecular biology, genomics and evolution, A. Herrero and E. Flores, eds. (Norfolk, UK: Caister Academic Press), pp. 45–57.

Galtier, N., and Lobry, J.R. (1997). Relationships between genomic G+C content, RNA secondary structures, and optimal growth temperature in prokaryotes. J. Mol. Evol. *44*, 632–636. http://dx.doi.org/10.1007/PL00006186.

Garcia-Pichel, F., López-Cortés, A., and Nübel, U. (2001). Phylogenetic and morphological diversity of cyanobacteria in soil desert crusts from the Colorado plateau. Appl. Environ. Microbiol. *67*, 1902–1910.

Garrity, G.M., Boone, D.R., and Castenholz, R.W. (2001). Bergey's Manual of Systematic Bacteriology Volume 1: The *Archaea* and the Deeply Branching and Phototrophic *Bacteria* (New York, USA: Springer).

Gevers, D., Dawyndt, P., Vandamme, P., Willems, A., Vancanneyt, M., Swings, J., and De Vos, P. (2006). Stepping stones towards a new prokaryotic taxonomy. Philos. Trans. R. Soc. Lond. B Biol. Sci. *361*, 1911–1916. http://dx.doi.org/10.1098/rstb.2006.1915.

Giovannoni, S.J., Cameron Thrash, J., and Temperton, B. (2014). Implications of streamlining theory for microbial ecology. ISME J. *8*, 1553–1565. http://dx.doi.org/10.1038/ismej.2014.60.

Griese, M., Lange, C., and Soppa, J. (2011). Ploidy in cyanobacteria. FEMS Microbiol. Lett. *323*, 124–131. http://dx.doi.org/10.1111/j.1574-6968.2011.02368.x.

Grigorieva, G.A., and Shestakov, S.V. (1982). Transformation in the cyanobacterium *Synechocystis* PCC 6803. FEMS Microbiol. Lett. *13*, 367–370.

Hanage, W.P., Fraser, C., and Spratt, B.G. (2005). Fuzzy species among recombinogenic bacteria. BMC Biol. *3*, 6. http://dx.doi.org/10.1186/1741-7007-3-6.

Herdman, M., Janvier, M., Waterbury, J.B., Rippka, R., Stanier, R.Y., and Mandel, M. (1979). Deoxyribonucleic acid base composition of cyanobacteria. J. Gen. Microbiol. *111*, 63–71. http://dx.doi.org/10.1099/00221287-111-1-63.

Hofmann, H.J. (1976). Precambrian microflora, Belcher Islands, Canada: significance and systematics. J. Paleo. *50*, 1040–1073. http://dx.doi.org/10.1038/249087a0.

Honda, D., Yokota, A., and Sugiyama, J. (1999). Detection of seven major evolutionary lineages in cyanobacteria based on the 16S rRNA gene sequence analysis with new sequences of five marine *Synechococcus* strains. J. Mol. Evol. *48*, 723–739. http://dx.doi.org/10.1007/pl00006517.

Jahodářová, E., Dvořák, P., Hašler, P., Holušová, K., and Poulíčková, A. (2016). *Elainella* gen. nov.: a new tropical cyanobacterium characterized using a complex genomic approach. PLoS One. (submitted).

Jensen, L.J., Julien, P., Kuhn, M., von Mering, C., Muller, J., Doerks, T., and Bork, P. (2008). eggNOG: automated construction and annotation of orthologous groups of genes. Nucleic Acids Res. *36*, D250–4. http://dx.doi.org/10.1093/nar/gkm796.

Johnsborg, O., Eldholm, V., and Håvarstein, L.S. (2007). Natural genetic transformation: prevalence, mechanisms and function. Res. Microbiol. *158*, 767–778. http://dx.doi.org/10.1016/j.resmic.2007.09.004.

Johansen, J.R., Bohunická, M., Lukešová, A., Hrčková, K., Vaccarino, M.A., and Chesarino, N.M. (2014). Morphological and molecular characterization within 26 strains of the genus *Cylindrospermum* (Nostocaceae, Cyanobacteria), with descriptions of three new species. J. Phycol. *50*, 187–202. http://dx.doi.org/10.1111/jpy.12150.

Katoh, H., Itoh, S., and Shen, J.-R. (2001). Functional analysis of *psbV* and a novel *c*-type cytochrome gene *psbV2* of the thermophilic cyanobacterium *Thermosynechococcus elongatus* strain BP-1. Plant Cell Physiol. *42*, 599–607. http://dx.doi.org/10.1093/pcp/pce074

Komárek, J., and Anagnostidis, K. (1998). Cyanoprokaryota 1. Teil: Chroococcales (Jena-Stuttgart-Lübeck-Ulm, Germany: Gustav Fischer).

Komárek, J., Kaštovský, J., Mareš, J., and Johansen, J.R. (2014). Taxonomic classification of cyanoprokaryotes (cyanobacterial genera) 2014 using a polyphasic approach. Preslia *86*, 295–335.

Koonin, E.V., Makarova, K.S., and Aravind, L. (2001). Horizontal gene transfer in prokaryotes: quantification and classification. Annu. Rev. Microbiol. *55*, 709–742. http://dx.doi.org/10.1146/annurev.micro.55.1.709.

Koonin, E.V., and Wolf, Y.I. (2008). Genomics of bacteria and archaea: the emerging dynamic view of the prokaryotic world. Nucleic Acids Res. *36*, 6688–6719. http://dx.doi.org/10.1093/nar/gkn668.

Kopac, S., Wang, Z., Wiedenbeck, J., Sherry, J., Wu, M., and Cohan, F.M. (2014). Genomic heterogeneity and ecological speciation within one subspecies of *Bacillus subtilis*. Appl. Environ. Microbiol. *80*, 4842–4853. http://dx.doi.org/10.1128/AEM.00576-14.

Lang, N.J. (1977). *Starria zimbabweësnis* (Cyanophyceae) gen. nov. et sp. nov.: a filament triradiate in transverse section. J. Phycol. *13*, 288–296.

Larsson, J., Nylander, J.A., and Bergman, B. (2011). Genome fluctuations in cyanobacteria reflect evolutionary, developmental and adaptive traits. BMC Evol. Biol. *11*, 187. http://dx.doi.org/10.1186/1471-2148-11-187.

Lawrence, J.G. (2002). Gene transfer in bacteria: speciation without species? Theor. Popul. Biol. *61*, 449–460. http://dx.doi.org/10.1006/tpbi.2002.1587.

Lawrence, J.G., and Ochman, H. (1998). Molecular archaeology of the *Escherichia coli* genome. Proc. Natl. Acad. Sci. U.S.A. *95*, 9413–9417. http://dx.doi.org/10.1073/pnas.95.16.9413.

Lederberg, J., and Tatum, E.L. (1946). Novel genotypes in mixed cultured of biochemical mutants of bacterial. Cold Spring Harbor Symp. Quant. Biol. *11*, 113–114.

Lewin, R.A., and Robinson, P.T. (1979). The greening of polar bears in zoos. Nature *278*, 445–447. http://dx.doi.org/10.1038/278445a0.

Lightfoot, D.A., Walters, D.E., 3 and Wootton, J.C., (1988). Transformation of the cyanobacterium *Synechococcus* PCC 6301 using cloned DNA. J. Gen. Microbiol. *134*, 1509–1514. http://dx.doi.org/10.1099/00221287-134-6-1509.

Llosa, M., and de la Cruz, F. (2005). Bacterial conjugation: a potential tool for genomic engineering. Res. Microbiol. *156*, 1–6. http://dx.doi.org/10.1016/j.resmic.2004.07.008.

Locey, K.J., and Lennon, J.T. (2016). Scaling laws predict global microbial diversity. Proc. Natl. Acad. Sci. U.S.A. *113*, 5970–5975. http://dx.doi.org/10.1073/pnas.1521291113.

Lodders, N., Stackebrandt, E., and Nubel, U. (2005). Frequent genetic recombination in natural populations of the marine cyanobacterium *Microcoleus chthonoplastes*. Environ. Microbiol. 7, 434–442. http://dx.doi.org/10.1111/j.1462-2920.2005.00730.x.

Maiden, M.C., Bygraves, J.A., Feil, E., Morelli, G., Russell, J.E., Urwin, R., Zhang, Q., Zhou, J., Zurth, K., Caugant, D.A., *et al.* (1998). Multilocus sequence typing: a portable approach to the identification of clones within populations of pathogenic microorganisms. Proc. Natl. Acad. Sci. U.S.A. *95*, 3140–3145. http://dx.doi.org/10.1073/pnas.95.6.3140.

Mann, N.H., Cook, A., Millard, A., Bailey, S., and Clokie, M. (2003). Marine ecosystems: bacterial photosynthesis genes in a virus. Nature *424*, 741. http://dx.doi.org/10.1038/424741a.

McLean, M.J., Wolfe, K.H., and Devine, K.M. (1998). Base composition skews, replication orientation, and gene orientation in 12 prokaryote genomes. J. Mol. Evol. *47*, 691–696. http://dx.doi.org/10.1007/pl00006428.

Moore, L.R., Rocap, G., and Chisholm, S.W. (1998). Physiology and molecular phylogeny of coexisting *Prochlorococcus* ecotypes. Nature *393*, 464–467. http://dx.doi.org/10.1038/nature01947.

Moran, N.A., McLaughlin, H.J., and Sorek, R. (2009). The dynamics and time scale of ongoing genomic erosion in symbiotic bacteria. Science *323*, 379–382. http://dx.doi.org/10.1126/science.1167140.

Morris, J.J., Lenski, R.E., and Zinser, E.R. (2012). The Black Queen Hypothesis: evolution of dependencies through adaptive gene loss. MBio *3*, e00036-12. http://dx.doi.org/10.1128/mBio.00036-12.

Nägeli, C. (1849) Gattungen einzellinger Algen. Neue Denkschriften der Allgemeinen Schweizerischen Gesellschaft für die Gesammten Naturwissenschaften *10*, 1–139.

Ochman, H., Lawrence, J.G., and Groisman, E.A. (2000). Lateral gene transfer and the nature of bacterial innovation. Nature *405*, 299–304. http://dx.doi.org/10.1038/35012500.

Olsen, M.T., Nowack, S., Wood, J.M., Becraft, E.D., LaButti, K., Lipzen, A., Martin, J., Schackwitz, W.S., Rusch, D.B., Cohan, F.M. *et al.* (2015). The molecular dimension of microbial species: 3. Comparative genomics of *Synechococcus* strains with different light responses and in situ diel transcription patterns of associated putative ecotypes in the Mushroom Spring microbial mat. Front. Microbiol. *6*, 604. http://dx.doi.org/10.3389/fmicb.2015.00604.

Papke, R.T., Ramsing, N.B., Bateson, M.M., and Ward, D.M. (2003). Geographical isolation in hot spring cyanobacteria. Environ. Microbiol. *5*, 650–659. http://dx.doi.org/10.1046/j.1462-2920.2003.00460.x.

Partensky, F., and Garczarek, L. (2010). *Prochlorococcus*: advantages and limits of minimalism. Ann. Rev. Mar. Sci. *2*, 305–331. http://dx.doi.org/10.1146/annurev-marine-120308-081034.

Pentecost, A., Jones, B., and Renaut, R.W. (2003). What is a hot spring? Can. J. Earth Sci. *40*, 1447–1457. http://dx.doi.org/10.1139/e03-08.

Philippe, H., Delcus, F., Brinkmann, H., and Lartillot, N. (2005). Phylogenomics. Annu. Rev. Ecol. Evol. Syst. *36*, 541–562. http://dx.doi.org/10.1146/annurev.ecolsys.35.112202.130205.

Polz, M.F., Alm, E.J., and Hanage, W.P. (2013). Horizontal gene transfer and the evolution of bacterial and archaeal population structure. Trends Genet. *29*, 170–175. http://dx.doi.org/10.1016/j.tig.2012.12.006.

Popa, O., Hazkani-Covo, E., Landan, G., Martin, W., and Dagan, T. (2011). Directed networks reveal genomic barriers and DNA repair bypasses to lateral gene transfer among prokaryotes. Genome Res. *21*, 599–609. http://dx.doi.org/10.1101/gr.115592.110.

Puigbò, P., Lobkovsky, A.E., Kristensen, D.M., Wolf, Y.I., and Koonin, E.V. (2014). Genomes in turmoil: quantification of genome dynamics in prokaryote supergenomes. BMC Biol. *12*, 66. http://dx.doi.org/10.1186/s12915-014-0066-4.

Ravenhall, M., Škunca, N., Lassalle, F., and Dessimoz, C. (2015). Inferring horizontal gene transfer. PLOS Comput. Biol. *11*, e1004095. http://dx.doi.org/10.1371/journal.pcbi.1004095.

Reno, M.L., Held, N.L., Fields, C.J., Burke, P.V., and Whitaker, R.J. (2009). Biogeography of the *Sulfolobus islandicus* pan-genome. Proc. Natl. Acad. Sci. U.S.A. *106*, 8605–8610. http://dx.doi.org/10.1073/pnas.0808945106.

Rippka, R., Deruelles, J., Waterbury, J.B., Herdman, M., and Stanier, R.Y. (1979). Generic assignments, strain histories and properties of pure cultures of cyanobacteria. Microbiology *111*, 1–61. http://dx.doi.org/10.1099/00221287-111-1-1.

Robertson, B.R., Tezuka, N., and Watanabe, M. (2001). Phylogenetic analyses of *Synechococcus* strains (cyanobacteria) using sequences of 16S rDNA and part of the phycocyanin operon reveal multiple evolutionary lines and reflect phycobilin content. Int. J. Syst. Evol. Microbiol. *51*:861–871. http://dx.doi.org/10.1099/00207713-51-3-861.

Rocap, G., Larimer, F.W., Lamerdin, J., Malfatti, S., Chain, P., Ahlgren, N.A., Arellano, A., Coleman, M., Hauser, L., Hess, W.R., *et al.* (2003). Genome divergence in two *Prochlorococcus* ecotypes reflects oceanic niche differentiation. Nature *424*, 1042–1047. http://dx.doi.org/10.1038/nature01947.

Rudi, K., Skulberg, O.M., and Jakobsen, K.S. (1998). Evolution of cyanobacteria by exchange of genetic material among phyletically related strains. J. Bacteriol. *180*, 3453–3461.

Salman, V., Amann, R., Girnth, A.C., Polerecky, L., Bailey, J.V., Høgslund, S., Jessen, G., Pantoja, S., and Schulz-Vogt, H.N. (2011). A single-cell sequencing approach to the classification of large, vacuolated sulfur bacteria. Syst. Appl. Microbiol. *34*, 243–259. http://dx.doi.org/10.1016/j.syapm.2011.02.001.

Schopf, J.W. (1994). Disparate rates, differing fates: tempo and mode of evolution changed from the Precambrian to the Phanerozoic. Proc. Natl. Acad. Sci. U.S.A. *91*, 6735–6742. http://dx.doi.org/10.1073/pnas.91.15.6735.

Schopf, J.W., Kudryavtsev, A.B., Walter, M.R., Van Kranendonk, M.J., Williford, K.H., Kozdon, R., Valley, J.W., Gallardo, V.A., Espinoza, C., and Flannery, D.T. (2015). Sulfur-cycling fossil bacteria from the 1.8-Ga Duck Creek Formation provide promising evidence of evolution's null hypothesis. Proc. Natl. Acad. Sci. U.S.A. *112*, 2087–2092. http://dx.doi.org/10.1073/pnas.1419241112.

Schreiber, F., Pick, K., Erpenbeck, D., Wörheide, G., and Morgenstern, B. (2009). OrthoSelect: a protocol for selecting orthologous groups in phylogenomics. BMC Bioinf. *10*, 219. http://dx.doi.org/10.1186/1471-2105-10-219.

Shapiro, B.J., and Polz, M.F. (2015). Microbial Speciation. Cold Spring Harbor Perspect. Biol. *7*, a018143. http://dx.doi.org/10.1101/cshperspect.a018143.

Shestakov, S.V., and Khyen, N.T. (1970). Evidence for genetic transformation in blue-green alga *Anacystis nidulans*. Mol. Gen. Genet. *107*, 372–375. http://dx.doi.org/10.1007/bf00441199.

Shi, T., and Falkowski, P.G. (2008). Genome evolution in cyanobacteria: the stable core and the variable shell. Proc. Natl. Acad. Sci. U.S.A. *105*, 2510–2515. http://dx.doi.org/10.1073/pnas.0711165105.

Shih, P.M., Wu, D., Latifi, A., Axen, S.D., Fewer, D.P., Talla, E., Calteau, A., Cai, F., Tandeau de Marsac, N., Rippka, R., *et al.* (2013). Improving the coverage of the cyanobacterial phylum using diversity-driven genome sequencing. Proc. Natl. Acad. Sci. U.S.A. *110*, 1053–1058. http://dx.doi.org/10.1073/pnas.1217107110.

Smillie, C.S., Smith, M.B., Friedman, J., Cordero, O.X., David, L.A., and Alm, E.J. (2011). Ecology drives a global network of gene exchange connecting the human microbiome. Nature *480*, 241–244. http://dx.doi.org/10.1038/nature10571.

Stackebrandt, E., and Goebel, B.M. (1994). Taxonomic note: A place for DNA–DNA reassociation and 16S rRNAsequenceanalysisinthepresentspeciesdefinition in bacteriology. Int. J. Syst. Bacteriol. *44*, 846–849. http://dx.doi.org/10.1099/00207713-44-4-846.

Stamatakis, A. (2006). RAxML-VI-HPC: maximum likelihood-based phylogenetic analyses with thousands of taxa and mixed models. Bioinformatics *22*, 2688–2690. http://dx.doi.org/10.1093/bioinformatics/btl446.

Stevens, S.E., and Porter, R.D. (1980). Transformation in *Agmenellum quadruplicatum*. Proc. Natl. Acad. Sci. U.S.A. *77*, 6052–6056. http://dx.doi.org/10.1073/pnas.77.10.6052.

Stucken, K., John, U., Cembella, A., Murillo, A.A., Soto-Liebe, K., Fuentes-Valdés, J.J., Friedel, M., Plominsky, A.M., Vásquez, M., and Glöckner, G. (2010). The smallest known genomes of multicellular and toxic cyanobacteria: comparison, minimal gene sets for linked traits and the evolutionary implications. PLoS One *5*, e9235. http://dx.doi.org/10.1371/journal.pone.0009235

Sullivan, M.B., Lindell, D., Lee, J.A., Thompson, L.R., Bielawski, J.P., and Chisholm, S.W. (2006). Prevalence and evolution of core photosystem II genes in marine cyanobacterial viruses and their hosts. PLOS Biol. *4*, e234. http://dx.doi.org/10.1371/journal.pbio.0040234.

Swingley, W.D., Blankenship, R.E., and Raymond, J. (2008). Insights into cyanobacterial evolution from comparative genomics. In The cyanobacteria: molecular biology, genomics and evolution, A. Herrero and E. Flores, eds. (Norfolk, UK: Caister Academic Press), pp. 21–43.

Thiel, T., and Poo, H. (1989). Transformation of a filamentous cyanobacterium by electroporation. J. Bacteriol. *171*, 5743–5746.

Tomitani, A., Knoll, A.H., Cavanaugh, C.M., and Ohno, T. (2006). The evolutionary diversification of cyanobacteria: molecular-phylogenetic and paleontological perspectives. Proc. Natl. Acad. Sci. U.S.A. *103*, 5442–5447. http://dx.doi.org/10.1073/pnas.0600999103.

Turner, S., Pryer, K.M., Miao, V.P., and Palmer, J.D. (1999). Investigating deep phylogenetic relationships among cyanobacteria and plastids by small subunit rRNA sequence analysis. J. Eukaryot. Microbiol. *46*, 327–338. http://dx.doi.org/10.1111/j.1550-7408.1999.tb04612.x.

Vos, M., and Didelot, X. (2009). A comparison of homologous recombination rates in bacteria and archaea. ISME J. *3*, 199–208. http://dx.doi.org/10.1038/ismej.2008.93.

van Gremberghe, I., Leliaert, F., Mergeay, J., Vanormelingen, P., Van der Gucht, K., Debeer, A.E., Lacerot, G., De Meester, L., and Vyverman, W. (2011). Lack of phylogeographic structure in the freshwater cyanobacterium *Microcystis aeruginosa* suggests global dispersal. PLoS One *6*, e19561. http://dx.doi.org/10.1371/journal.pone.0019561.

West, N.J., and Scanlan, D.J. (1999). Niche-partitioning of *Prochlorococcus* populations in a stratified water column in the eastern North Atlantic Ocean. Appl. Environ. Microbiol. *65*, 2585–2591.

Wiedenbeck, J., and Cohan, F.M. (2011). Origins of bacterial diversity through horizontal genetic transfer and adaptation to new ecological niches. FEMS Microbiol. Rev. *35*, 957–976. http://dx.doi.org/10.1111/j.1574-6976.2011.00292.x.

Wolk, C.P., Vonshak, A., Kehoe, P., and Elhai, J. (1984). Construction of shuttle vectors capable of conjugative transfer from *Escherichia coli* to nitrogen-fixing filamentous cyanobacteria. Proc. Natl. Acad. Sci. U.S.A. *81*, 1561–1565. http://dx.doi.org/10.1073/pnas.81.5.1561.

Worning, P., Jensen, L.J., Nelson, K.E., Brunak, S., and Ussery, D.W. (2000). Structural analysis of DNA sequence: evidence for lateral gene transfer in *Thermotoga maritima*. Nucleic Acids Res. *28*, 706–709. http://dx.doi.org/10.1093/nar/28.3.706.

Yang, Y., Feng, J., Li, T., Ge, F., and Zhao, J. (2015). CyanOmics: an integrated database of omics for the model cyanobacterium *Synechococcus* sp. PCC 7002. Database *2015*, bau127. http://dx.doi.org/10.1093/database/bau127.

Yoshihara, S., Geng, X.X., Okamoto, S., Yura, K., Murata, T., Go, M., Ohmori, M., and Ikeuchi, M. (2001). Mutational analysis of gene involved in pilus structure, motility and transformation competency in the unicellular motile cyanobacterium *Synechocystis* sp. PCC 6803. Plant Cell Physiol. *42*, 63–73. http://dx.doi.org/10.1093/pcp/pce007

Zeidner, G., Bielawski, J.P., Shmoish, M., Scanlan, D.J., Sabehi, G., and Béjà, O. (2005). Potential photosynthesis gene recombination between *Prochlorococcus* and *Synechococcus* via viral intermediates. Environ.

Microbiol. *7*, 1505–1513. http://dx.doi.org/10.1111/j.1462-2920.2005.00833.x.

Zhaxybayeva, O., Doolittle, W.F., Papke, R.T., and Gogarten, J.P. (2009). Intertwined evolutionary histories of marine *Synechococcus* and *Prochlorococcus marinus*. Genome Biol. Evol. *1*, 325–339. http://dx.doi.org/10.1093/gbe/evp032.

Zhaxybayeva, O., Gogarten, J.P., Charlebois, R.L., Doolittle, W.F., and Papke, R.T. (2006). Phylogenetic analyses of cyanobacterial genomes: quantification of horizontal gene transfer events. Genome Res. *16*, 1099–1108. http://dx.doi.org/10.1101/gr.5322306.

Zwirglmaier, K., Jardillier, L., Ostrowski, M., Mazard, S., Garczarek, L., Vaulot, D., Not, F., Massana, R., Ulloa, O., and Scanlan, D.J. (2008). Global phylogeography of marine *Synechococcus* and *Prochlorococcus* reveals a distinct partitioning of lineages among oceanic biomes. Environ. Microbiol. *10*, 147–161. http://dx.doi.org/10.1111/j.1462-2920.2007.01440.x

Genomics of NRPS/PKS Biosynthetic Gene Clusters in Cyanobacteria

3

Claire Pancrace[1,2], Muriel Gugger[1] and Alexandra Calteau[3*]

[1]Institut Pasteur, Collection of Cyanobacteria, Paris, France.
[2]Sorbonne Universités, UPMC Univ Paris 06, UPEC, UDD, CNRS, INRA, IRD, IEES-Paris, Paris, France.
[3]French Alternative Energies and Atomic Energy Commission (CEA), Genoscope & CNRS, UMR 8030, Laboratory of Bioinformatics Analyses for Genomics and Metabolism, Evry, France.

*Correspondence: acalteau@genoscope.cns.fr

Abstract

Cyanobacteria are well-known producers of a broad range of natural products of diverse chemical structures. Many of these compounds are the products of non-ribosomal peptide synthetase (NRPS) or polyketide synthase (PKS) pathways. In the last 15 years, genome mining of biosynthetic gene clusters has become a key methodology for studying NRPS/PKS biosynthetic gene clusters in cyanobacteria. This particularly rich phylum revealed an unexpected diversity for these clusters with only 20% of them that could be linked to a known natural product. The present chapter reviews recent literature on genomics of NRPS/PKS biosynthetic gene clusters in cyanobacteria from strategies of genome mining to comparative genomics. We then tackle the gap from prediction to characterization of natural products before discussing the benefits of combined genomic and metagenomic investigations. Finally we propose future trends for the study of these natural products in cyanobacteria to overcome the current gap between genome mining and natural products discovery.

Introduction

Cyanobacteria are well-known producers of a broad range of natural products or so called secondary metabolites of diverse chemical structures (Welker and von Dohren, 2006; Kehr *et al.*, 2011). Among these compounds, the cyanotoxins are the most studied natural products of cyanobacteria, notably the well-known hepatotoxin microcystin which concerns more than 50% of the cyanotoxin's literature, followed by the paralytic shellfish toxin, saxitoxin (Merel *et al.*, 2013). These toxins, especially in the context of cyanobacterial blooms, cause ecosystem disturbance and are a major threat for animal and human health through recreational and drinking water (Dittmann *et al.*, 2013; van Apeldoorn *et al.*, 2007). Although the cellular function of the natural products remains unknown in most cases (Holland and Kinnear, 2013), the great diversity of chemical structures and bioactivities of these compounds may play an important potential for pharmaceutical or biotechnological interest (Niedermeyer, 2015).

Historically, the toxicity associated with bloom events attracted scientists to uncover the cyanobacterial compounds, with the first reported cattle intoxication (Francis, 1878) followed by several testimonies, then with a real interest on the pathologies associated to these events (Ashworth and Mason, 1946). Consequently, the toxic natural products were isolated following bioactivity-guided investigations. This strategy was carried on and implemented to discover most of the currently known cyanobacterial natural products (Burja *et al.*, 2001). Later, the genetic bases of the biosynthesis of the most emblematic cyanotoxin, microcystin, were established in comparison with other natural product pathways already described for yersiniabactin and the antibiotic TA (Gehring *et al.*, 1998; Paitan *et al.*, 1999; Tillett *et al.*, 2000). Several cyanobacterial toxin pathways, notably anatoxin-a, cylindrospermopsin, and saxitoxin, were discovered in 2000–2010, which revealed the genetic bases of their peptidic and polyketidic structures. As more than 1000 cyanobacterial compounds, involved or not in known toxicities, are of the same peptidic and polyketidic nature, genome sequencing greatly helps to uncover the genes behind chemical compounds (Dittmann *et al.*, 2015).

The peptidic natural products can be synthesized through two types of pathways: the classic ribosomal synthesis with post-translational modification of the peptide, or via non-ribosomal pathways assembling the final compound through modular mega-enzyme NRPS (non-ribosomal peptide synthetase). While ribosomes can only use proteinogenic amino acids, NRPS enable the integration in the final molecule of substrates ranging from L- and D-amino acids to sugars and fatty acid chains (Welker and von Dohren, 2006). Similarly the polyketides are produced via modular mega-enzyme, PKS (polyketide synthase), which will assemble mostly acyl or malonyl. NRPS and PKS can also be assembled into a hybrid modular mega-enzyme, such as in the case of the microcystin biosynthetic pathway. Genes encoding the biosynthetic pathway responsible for the production of a given natural product are spatially clustered together on the chromosome as a biosynthesis gene cluster (see also Chapter 1). These clusters, spanning from 4 to 99 kb in cyanobacteria, contain NRPS and PKS encoding genes and also include tailoring genes, coding enzymes involved in post-synthesis modifications of the metabolite. The diversity of the tailoring enzymes is mainly responsible for the apparent singularity of the cyanobacterial compounds produced by the modular NRPS and PKS biosynthesis (Jones *et al.*, 2010).

Until recently, attempts to link the unknown gene clusters to their encoded natural products remained based on individual bottom-up approach, it was time-consuming and often unsuccessful. In a few cases though, new molecules and their biosynthetic pathways were uncovered. The mining for natural product biosynthetic gene clusters was initially performed in cyanobacterial strains by molecular screening, specific sequence pattern of the modular enzymes being targeted by PCR (Christiansen *et al.*, 2001). But these strategies were limited and biased towards already described biosynthetic gene clusters. Recent advances in sequencing technology and data handling enabled the release of a subsequent number of genomes of cyanobacterial strains (Bandyopadhyay *et al.*, 2011; Dagan *et al.*, 2013; Humbert *et al.*, 2013; Shih *et al.*, 2013). Genome-mining applied on these genomes confirmed cyanobacteria as a major host of NRPS and PKS pathways involved in peptides and polyketides types of natural products (Calteau *et al.*, 2014). It also evidenced an extended and unexpected potential, as most of the gene clusters were yet unknowns and orphans (Calteau *et al.*, 2014; Wang *et al.*, 2014). More and more environmental metagenomes are also released and provide insights into the diversity of studied sample and its functional potential, including the potential to produce natural products (Crits-Christoph *et al.*, 2016). But even with both the biosynthetic gene cluster and its corresponding natural product characterized, the links between the two remain mostly based on *in silico* prediction with no experimental data supporting it (Dittmann *et al.*, 2015).

The present chapter reviews recent literature on genomics of NRPS/PKS biosynthetic gene clusters in cyanobacteria from strategies of genome mining to comparative genomics. We then tackle the gap from prediction to characterization of natural products before discussing the benefits of combined genomic and metagenomic investigations. Finally, we propose future trends for the study of these natural products in cyanobacteria to overcome the current gap between genome mining and natural products discovery.

Genome mining for NRPS/PKS biosynthetic gene clusters

The recent rise in whole genome sequencing projects has led to a wealth of genetic information allowing the exploration of full genomic content of biosynthetic gene clusters in prokaryotes (Cimermancic *et al.*, 2014; Wang *et al.*, 2014). Genome mining has become a key-methodology. Recently, software and databases have been developed to predict and identify NRPS/PKS type biosynthetic gene clusters from genomic data (see web resources section for more details). *In silico* prediction of NRPS/PKS type biosynthetic gene clusters in genomic sequences combines established knowledge on secondary metabolism and Hidden Markov Model-based algorithms. More particularly, these tools exploit the characteristic domain composition of NRPS and PKS giant multi-domain enzymes. Indeed NRPS type mega-enzymes are composed at minima of a condensation (C) domain for catalysing peptide bond formation, an adenylation domain (A) for selection of the substrate, and a peptide carrier protein (PCP) domain. This minimal core NRPS can accept about 300 proteinogenic and nonproteinogenic substrates and may contain further domains introducing tailoring modifications or epimerizing the amino acid substrates (Donia *et al.*, 2011). Similarly minimum PKS type mega-enzymes are composed of a ketosynthase (KS) domain, an acetyltransferase (AT) domain and an acyl carrier protein (ACP) domain. Additional auxiliary domains can be present like ketoreductase (KR), dehydrogenase (DH) or enoyl-reductase (ER). PKS mega-enzymes assemble polyketides from acyl-CoA in a linear manner and are classified into different classes (Weissman, 2009). Hybrids NRPS/PKS will be composed by a combination of domains from both types. Thus, bioinformatics tools will exploit the typical domain composition to identify genomic regions containing NRPS/PKS type biosynthetic gene clusters in assembled genomes using profile hidden Markov models generated from multiple sequence alignments. They can also predict structural elements of the products based upon substrate specificities of their conserved biosynthetic domains. Substrate specificity prediction modules are based on the active sites of the acyltransferase (AT) or the ketosynthase (KS) for the PKS, and of the adenylation (A) for NRPS. The amino acid selected and incorporated by the A domain can be predicted through the ten critical amino acids comprising the A-domain binding pocket (Rausch *et al.*, 2005). Similarly for most PKS, AT domain specificities are predicted using a 24 amino acid signature sequence of the active site (Yadav *et al.*, 2003). The AT will incorporate acyl, malonyl or their 2-substituted derivatives. In the case of *trans*-AT PKS, the building block incorporated is a malonyl, while the KS are substrate specific to the nascent polyketide from the upstream module (Helfrich and Piel, 2016).

Bioinformatic predictors will easily identify NRPS/PKS type genes. However, they struggle at defining correct biosynthetic gene cluster borders since these clusters are also composed of tailoring enzymes and eventually transporters and regulators. Only comparisons with already characterized biosynthetic gene clusters allow setting up the correct cluster boundaries. For this purpose, the recently published reference database MIBiG (Minimal Information about a Biosynthetic Gene cluster), aims at developing a genomic standard for experimentally characterized biosynthetic gene clusters in bacteria, fungi and plants (Medema *et al.*, 2015). However to date, the MIBiG database gathers 748 NRPS/PKS type biosynthetic gene cluster descriptions among which 275 full clusters from diverse organisms (24 December 2015), which is few comparatively to the huge diversity of these types of biosynthetic gene clusters. It also has to be stressed that cyanobacteria typically produce multiple variants of the same natural products, e.g. anabaenopeptin or microcystin with respectively 38 and 63 variants described so far (Dittmann *et al.*, 2015). Modifications in the tailoring regions of homologous gene clusters can also lead to diverse congeners.

More than 384 genomes of cyanobacteria were referenced at NCBI in April 2016 (http://www.ncbi.nlm.nih.gov/genome/browse/), with only 91 complete sequences. Unfinished genomic sequences render the identification of biosynthetic gene clusters more complex. Indeed, due to the repeated nature of their domains, long genes coding for mega-enzymes might be difficult to assemble (Klassen and Currie, 2012). The predictors will thus miss clusters or identify partial clusters with missing and/or partial gene information rendering impossible the prediction of core peptide. However, the development of third generation sequencing technologies (such as those from Pacific Biosciences and Oxford Nanopore) with long-read sequencing data will overcome issues related to repeated sequences and render possible to generate complete genome assemblies for prokaryotes at low cost in a near future (Koren and Phillippy, 2015; Laver *et al.*, 2015).

Genome mining for natural product biosynthetic gene cluster in 126 cyanobacterial genomes showed an extended potential as only 20% of the clusters could be associated to known products (Calteau *et al.*, 2014). To date, 33 NRPS/PKS type biosynthetic gene clusters have been associated to a family of compounds in cyanobacteria (Table 3.1). Among these, experimental demonstrations

Table 3.1 Biosynthetic gene clusters encoding known natural products and biosynthetic pathways characterization status

Status of the biosynthetic pathway	Biosynthetic gene cluster name	Size (kb)	Biosynthetic pathway type	Biosynthetic gene cluster reference	Producer strain
Demonstrated	Anabaenopeptilide	29	NRPS	Rouhiainen *et al.* (2000)	*Anabaena* sp. 90
Demonstrated	Anatoxin	29	PKS	Mejean *et al.* (2009)	*Oscillatoria* sp. PCC 6506
Demonstrated	Apratoxin	57.4	NRPS-PKS	Grindberg *et al.* (2011)	*Lyngbya bouillonii*
Demonstrated	Barbamide	26	NRPS-PKS	Chang *et al.* (2002)	*Lyngbya majuscula* 19L
Demonstrated	Cryptophycin	40.3	NRPS-PKS	Magarvey *et al.* (2006)	*Nostoc* sp. ATCC 53789
Demonstrated	Hassallidin	59	NRPS	Vestola *et al.* (2014)	*Anabaena* sp. SYKE 748A
Demonstrated	Lyngbyatoxin	11.3	NRPS	Edwards and Gerwick (2004)	*Lyngbya majuscula*
Demonstrated	Microcystin	55	NRPS-PKS	Tillett *et al.* (2000)	*Microcystis aeruginosa* PCC 7806
Demonstrated	Mycosporine-2-glycine	4	NRPS-like	Waditee-Sirisattha *et al.* (2014)	*Aphanothece halophytica*
Demonstrated	Saxitoxin	35	PKS	Tsuchiya *et al.* (2016)	*Anabaena circinalis* TA04
Partially demonstrated	Anabaenopeptin	32	NRPS	Rouhiainen *et al.* (2010)	*Anabaena* sp. 90
Partially demonstrated	Curacin	64	NRPS-PKS	Chang *et al.* (2004)	*Lyngbya majuscula* 19L
Partially demonstrated	Heterocyst glycolipid	29	PKS	Fan *et al.* (2005)	*Nostoc punctiforme* PCC 7120
Partially demonstrated	Jamaicamide	58	NRPS-PKS	Edwards *et al.* (2004)	*Lyngbya majuscula* JHB
Partially demonstrated	Nostocyclopeptide	33	NRPS-PKS	Becker *et al.* (2004)	*Nostoc* sp. ATCC53789
Partially demonstrated	Nostopeptolide	40	NRPS	Hoffmann *et al.* (2003)	*Nostoc* sp. GSV224
Partially demonstrated	Nostophycin	45.1	NRPS-PKS	Fewer *et al.* (2011)	*Nostoc* sp. 152
Predicted	Aeruginosin	27	NRPS	Ishida *et al.* (2009)	*Microcystis aeruginosa* PCC 7806
Predicted	Anachelin	34	NRPS-PKS	Calteau *et al.* (2014)	*Anabaena cylindrica* PCC 7122
Predicted	Columbamide	8.5	NRPS-PKS	Kleigrewe *et al.* (2015)	*Moorea producens* 3L
Predicted	Cylindrocyclophane	26.2	PKS	Nakamura *et al.* (2012)	*Cylindrospermum licheniforme* UTEX 'B 2014'
Predicted	Cylindrospermopsin	43	NRPS-PKS	Mihali *et al.* (2008)	*Cylindrospermopsis raciborskii* AWT205
Predicted	Hapalosin	25	NRPS-PKS	Micallef *et al.* (2015)	*Hapalosiphon*
Predicted	Hectochlorin	38	NRPS-PKS	Ramaswamy *et al.* (2007)	*Lyngbya majuscula* JHB
Predicted	Luminaolide B	99	PKS	Ueoka *et al.* (2015)	*Planktothrix paucivesiculata* PCC 9631
Predicted	Microginin	20.5	NRPS	Rounge *et al.* (2009)	*Planktothrix agardhii* NIVA-CYA 126/8

Table 3.1 Continued

Status of the biosynthetic pathway	Biosynthetic gene cluster name	Size (kb)	Biosynthetic pathway type	Biosynthetic gene cluster reference	Producer strain
Predicted	Nodularin	48	NRPS-PKS	Moffit and Neilan (2004)	*Nodularia spumigena* NSOR10
Predicted	Nosperin	22.4	PKS	Kampa *et al.* (2013)	*Nostoc* sp. 'Peltigera membranacea cyanobiont'
Predicted	Phormidolide	98.3	PKS	Bertin *et al.* (2016)	*Phormidium* sp.
Predicted	Puwainaphycins	56.7	NRPS-PKS	Mares *et al.* (2014)	*Cylindrospermum alatosporum* CCALA 988
Predicted	Shinorine	6.8	NRPS-like	Balskus and Walsh (2010)	*Anabaena variabilis* ATCC 29413
Predicted	Spumigin	21	NRPS	Fewer *et al.* (2009)	*Nodularia spumigena* CCY 9414
Predicted	Tolytoxin	90	PKS	Ueoka *et al.* (2015)	*Scytonema* sp. PCC 10023

The given reference and producer strain refer to the first reference of the biosynthetic gene cluster in the literature. Demonstrated: knockout mutant, heterologous expression or full biosynthetic pathway characterization has evidenced the link between biosynthetic gene cluster and natural product. Partially characterized: some enzymes of the pathway have been characterized or biochemical assays have been performed to assess substrate specificity. Predicted: link between the biosynthetic gene cluster and the natural product is based solely on *in silico* prediction.

(knockout mutants, heterologous expression or full biosynthetic pathway characterization) linked the biosynthetic gene clusters to ten natural products (e.g. microcystin, lyngbyatoxin and anabaenopeptilide). Seven biosynthetic gene clusters have been partially characterized with some enzymes of the pathway characterized or biochemical assays performed to assess substrate specificity (e.g. anabaenopeptin, curacin and nostopeptolide). Finally, 16 biosynthetic gene clusters have been linked to natural products solely based on *in silico* prediction (e.g. anachelin, luminaolide B and shinorine). In this latter case, the predictions have to be taken with caution due to the complexity of NRPS/PKS biosynthetic pathways. Indeed, it is extremely difficult to be sure of the prediction because the bioinformatics software cannot predict the mechanistic processes (linear, non-linear, iterative, or both), nor identify all tailoring enzymes. Other modifications such as adding sugar or fatty acid cannot be anticipated by bioinformatic either. Also, a chemically complex natural product can be produced by ribosomally synthesized and post-translationally modified peptides (RiPP) type biosynthetic gene clusters (e.g. microvidirin or natural products from the cyanobactin family). The nicest example of chemically complex peptidic natural product was recently described in marine sponges with the giant cytotoxin polytheonamides, which result indeed from a RiPP biosynthesis and not from an expected NRPS pathway (Inoue *et al.*, 2010; Freeman *et al.*, 2012).

Comparative genomics of biosynthetic gene clusters

The development of whole genome sequencing has enabled the rapid identification of entire biosynthetic gene clusters, which can be easily compared between multiple cyanobacterial strains to identify common and unique genes (Humbert *et al.*, 2013; Shih *et al.*, 2013; Calteau *et al.*, 2014; Micallef *et al.*, 2015). 70% of cyanobacteria sequenced to date encode NRPS/PKS type biosynthetic gene clusters with an average of five biosynthetic gene clusters per genome (Shih *et al.*, 2013).

Genome mining for natural product biosynthetic gene clusters in 126 cyanobacterial genomes identified 452 biosynthetic gene clusters (190 NRPS, 162 PKS and 100 hybrid biosynthetic gene clusters) from 89 cyanobacterial genomes, revealing a clear burst in late-branching lineages (Calteau *et al.*, 2014). The distribution of biosynthetic gene clusters is widespread in the cyanobacterial phylum (Fig. 3.1). NRPS/PKS biosynthetic gene cluster content varies greatly from one cyanobacterium to another one, with for example no gene cluster in the genomes of *Oscillatoria* sp. PCC

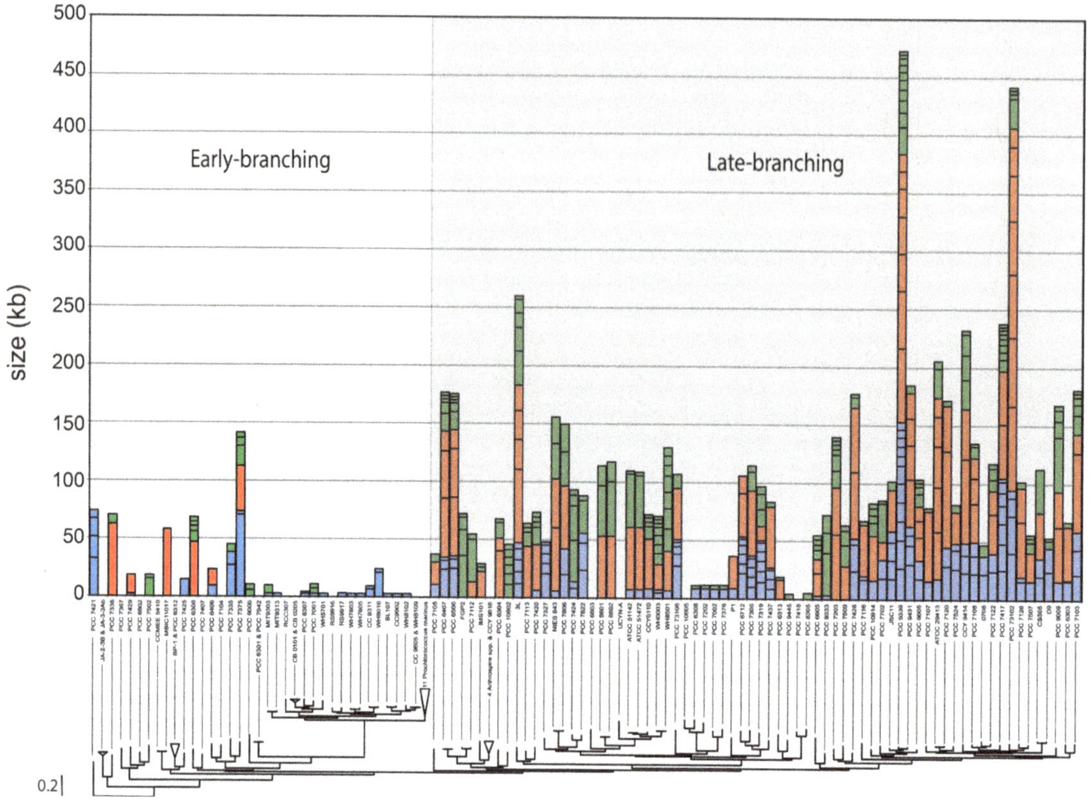

Figure 3.1 Biosynthetic gene clusters found in 126 cyanobacterial genomes. The biosynthetic gene clusters found in each genome are represented by their size (in kb) and their nature, with PKS indicated in blue, NRPS in green and the hybrids of thereof in red. As illustrated, the early branched cyanobacteria are less doted in biosynthetic gene clusters than the late-branched one. Data originated from Calteau *et al.* (2014).

6802, *Halothece* sp. PCC 7418 and all *Arthrospira* strains, but with more than 400 kb dedicated to these biosynthetic gene clusters in *Fischerella* sp. PCC 9339 (Fig. 3.1). It is also noteworthy to highlight that very close species possess similar biosynthetic gene cluster content profiles such as fresh-water species *Kamptonema* spp. PCC 6506 and PCC 6407 or marine *Cyanothece* sp. ATCC 51472 and ATCC 51142. Finally, no link could be established so far between the ecology or the geographic origin of the strains and their biosynthetic gene cluster contents.

Comparison of 452 biosynthetic gene clusters allowed classifying them into 286 biosynthetic gene cluster families (Calteau *et al.*, 2014). Only 6.6% of the biosynthetic gene cluster families gathered 91 biosynthetic pathways corresponding to 19 described natural products. Up to 14.7% of the biosynthetic gene cluster families gathered 136 biosynthetic gene clusters without associated products. More importantly, 78.7% of the biosynthetic gene cluster families corresponded to orphan biosynthetic gene clusters, each forming a putative family by its own (Fig. 3.2). These results suggest that more diversity will be brought with more genome data and some orphan biosynthetic gene cluster families will not remain orphans. Indeed, an orphan biosynthetic gene cluster in *Fischerella* sp. PCC 9431 (Calteau *et al.*, 2014) has been identified as the candidate biosynthetic gene cluster for hapalosin in this strain and two others (*Westiella intricata* UH strain HT-29-1 and *Hapalosiphon welwitschii* UH strain IC-52-3; Micallef *et al.*, 2015). Finally, the construction of biosynthetic gene cluster families can serve as a tool to set up correct cluster borders based on the conservation of gene architecture across several species. It will thus ease the identification of potential tailoring enzymes realizing post synthesis peptide modifications.

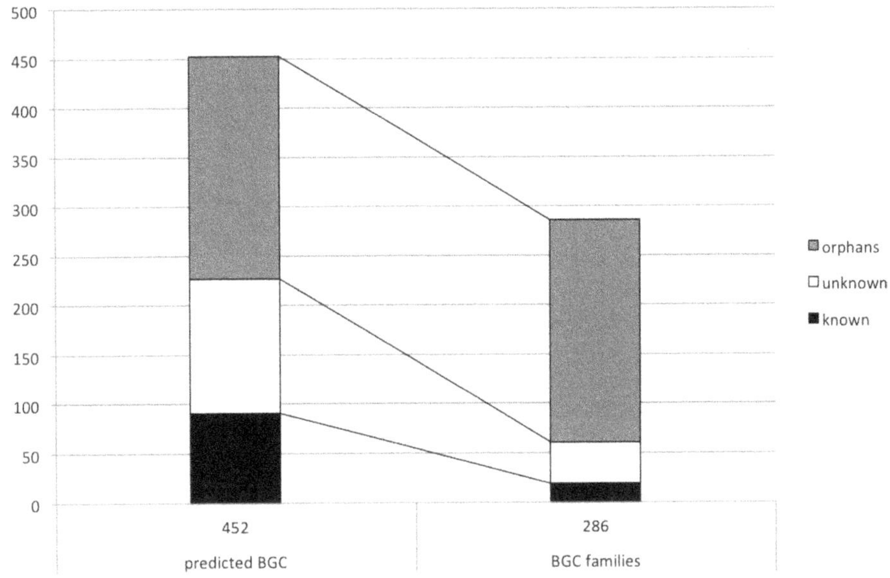

Figure 3.2 Distribution of predicted biosynthetic gene clusters according to their family classification in Cyanobacteria. 452 gene clusters encoding NRPS, PKS and hybrid NRPS/PKS were found in 89 out of 126 cyanobacterial genomes (Calteau *et al.*, 2014). Several gene clusters were shared among Cyanobacteria, forming biosynthetic gene cluster families (BGCFs). Some BGCFs gather biosynthetic pathways of known products, whereas others without associated products were recovered using BLASTP and a transitive link criterion to build families. 225 biosynthetic gene clusters were orphans, each forming a putative family by its own.

Biosynthetic gene clusters have been shown to evolve through different evolutionary scenarios such as horizontal gene transfer, gene shuffling, or recombination (Jenke-Kodama *et al.*, 2005; Calteau *et al.*, 2014; Ziemert *et al.*, 2014). Recently, Medema and collaborators (Medema *et al.*, 2014, 2016) demonstrated that large gene clusters appear to evolve through the merger of subclusters, which function independently, and that a subset of gene clusters evolve by concerted evolution, which generates sets of interoperable domains that may enable predictable domain swapping. They showed that many biosynthetic gene clusters evolve in family specific modes that differ greatly from each other. Indeed, small modifications in the genomic sequence have an immense impact on the resulting chemical structures and their associated biological activities. Recently, several examples of the genetic modifications of clusters were shown. In marine ecosystems, curacin A, a potent inhibitor of tubulin polymerization, jamaicamide A, a sodium channel blocker, and apratoxin A, a highly cytotoxic compound causing G1 cell cycle arrest and apoptosis, were revealed to be encoded by closely related biosynthetic gene clusters despite their diverse functions (Kleigrewe *et al.*, 2015). Genomic analyses of their clusters showed a parallel interplay of evolutionary events in multi-enzyme systems leading to functional group diversity in secondary metabolites (Gu *et al.*, 2009; Grindberg *et al.*, 2011).

Global similarity comparisons of cyanobacterial biosynthetic gene clusters against public databases revealed that only tailoring enzymes seem to be specific to cyanobacteria (Calteau *et al.*, 2014). Indeed, several pathways have shown to be shared with other phyla. For instance, saxitoxin and its derivatives are produced by a small number of marine dinoflagellates and several freshwater filamentous cyanobacteria (Plumley, 2001). These carbamate alkaloid neurotoxins can cause paralytic shellfish poisoning by blocking voltage-gated Na+ channels. More recently, Piel and collaborators have studied actin-targeting polyketides biosynthetic gene clusters (Ueoka *et al.*, 2015). Two clusters encoding for such compounds were uncovered in two genomes of cyanobacteria: tolytoxin in *Scytonema* sp. PCC 10023 and luminaolide B in *Planktothrix paucivesiculata* PCC 9631. These *trans*-AT PKS biosynthetic gene clusters are linked to the misakinolide, which is supposed to be produced by an uncultivated sponge symbiont belonging to the candidate phylum Tectomicrobia (Wilson *et al.*, 2014a, b). Phylogenetic analyses of these biosynthetic gene clusters demonstrated that misakinolide, tolytoxin and luminaolide B pathways arose from a common ancestor and diversified by deletion or acquisition of DNA regions coding parts of the PKS (Ueoka *et al.*, 2015).

Biosynthetic gene clusters conservation

The information linked to each biosynthetic gene cluster differs greatly as some are highly conserved through several representatives while other went through large recombination events. The study of 24 heterocyst glycolipid pathways of different heterocystous strain genomes indicated the high conservation of six genes *hglA-E,* with 12 cases where *hglC* and *hglD* were separated and 11 cases where the two genes were merged, *hglCD* (Calteau *et al.*, 2014). Similar high conservation was observed in anatoxin-a biosynthetic gene cluster in which the genes *anaB-G* are in the same order with a high sequence identity (Rantala-Ylinen *et al.*, 2011). Both examples are mainly constituted of modular NRPS or PKS genes, and proportionally contained only few tailoring genes. In the same way, the biosynthetic gene clusters corresponding to structurally different actin-binding macrolides, (tolytoxin/scytophycin and luminaolide B), originate from same ancestral *trans*-AT PKS pathways, one resulting after final lactonization, oxygenation and *O*-methylation steps to the synthesis of variants of tolytoxin and scytophycin, while the other one with five modules shorter at the genetic level, conducted to a dimer molecule of luminaolide B (Ueoka *et al.*, 2015).

In Cyanobacteria, diverse biosynthetic gene clusters are composed of one or few NRPS or PKS genes and several tailoring enzyme genes. The tailoring enzymes can be part of the 'core cluster', essential to the primary synthesis of a stable metabolite, or be part of an accessory set of modifying enzymes. In those, the genetic variation observed within the biosynthetic gene cluster is wider. A nice example of the specificity and diversity of the tailoring enzymes can be observed in the saxitoxin biosynthetic gene cluster. This highly potent neurotoxin is produced by a biosynthetic gene cluster composed of one PKS gene, *stxA*, and up to 23 tailoring enzymes in *Cylindrospermopsis raciborskii* T3, 22 in *Aphanizomenon* sp. NH-5 and 21 in *Anabaena circinalis* AWQC131C (Mihali *et al.*, 2009). While 20 *stx* genes are shared among the three described biosynthetic gene clusters, seven other *stx* genes (*stxF, O, W, X, Y, Z* and *PER*) are present only in one or two saxitoxin biosynthetic gene clusters. Another example of the requirement of tailoring enzymes in the production of the final molecule is illustrated by the aeruginosin biosynthetic gene cluster found in *Microcystis* (Fig. 3.3). The gene *aerK-F* involved in the Choi motif of the final molecule, some ORFs such as orf1 and orf2, and the transporter appeared always present. In contrary, the presence and size of the four NRPS genes (*aerA, aerB, aerG1* and *aerM*), as well as the halogenases, *aerJ* and *aerJ'*, vary from one cluster to another. Thus, the size of aeruginosin biosynthetic gene cluster varies from 13 kb in *Microcystis aeruginosa* PCC 9432 to 25 kb in *Microcystis aeruginosa* PCC 7806 to produce aeruginosin 574 and aeruginosin 686 respectively (Cadel-Six *et al.*, 2008; Humbert *et al.*, 2013). Such variations in different but related biosynthetic gene clusters encoding the same compound family should be carefully studied prior any development of genetic tools intended to detect toxin potential or presence of the potential producers in environmental samples. Notably, it is demonstrated that the gene order can be rearranged (e.g. microcystin biosynthetic gene cluster; Dittmann *et al.*, 2013). It is noteworthy to notice that genetic events, such as single mutation and gene deletion, occur also in these pathways inactivating them (Christiansen *et al.*, 2008).

From prediction to characterization

The last decade, several compounds and their biosynthetic pathways were discovered with the help of genomic investigations. When the first NRPS and PKS gene clusters were described from cyanobacteria to be associated to toxic compounds production, the genetic information was obtained through library construction, cloning

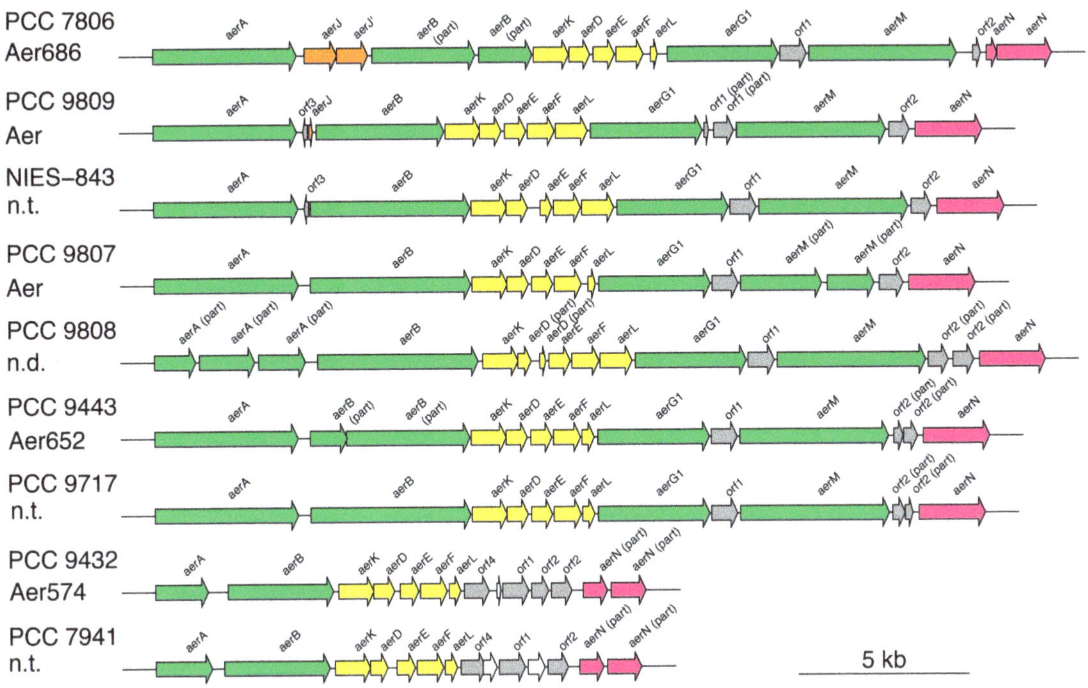

Figure 3.3 Aeruginosin biosynthetic gene cluster in nine *Microcystis*. NRPS genes are indicated in green, halogenases in orange, genes encoding the Choi motif in yellow, transporter genes in pink, the genes potentially involved and present in several clusters in grey, and the other genes are in white. The gene clusters encoding aeruginosin and chemical detection originate from Cadel-Six *et al.* (2008) and Humbert *et al.* (2013); Aer, aeruginosin; n.t., not tested; n.d., not detected.

and sequencing, e.g. microcystin and curacin gene clusters (Tillett *et al.*, 2000; Chang *et al.*, 2004). Then, with the sequencing technology evolving, it became much simpler to sequence the genome of a strain producing the natural product of interest to find the dedicated cluster that encodes the mega-enzymatic complex needed for its biosynthesis. Nowadays, single genome and genome mining of several genomes assist the search for pathways and novel products as exemplified by biosynthetic gene clusters and peptides of *Planktothrix rubescens* (Rounge *et al.*, 2009) or by gene clusters needed to produce tolytoxins and related compounds (Ueoka *et al.*, 2015).

Indeed, in several cases, prediction of the function of each enzyme of the biosynthetic gene clusters helped to find the compounds, but some others failed to find it (Nishizawa *et al.*, 2007). Such examples of failure to find the desired compound despite the information on the biosynthetic gene cluster are rarely published, but encountered in many laboratories. Investigations mainly based on microbiology, chemistry, mass spectrometry, together with genome sequencing have made a step forward in assisting the discovery of natural products, but the biochemical approach to validate the function of the enzymes is still not straight forward. Thus, most of the time, the NRPS and PKS gene clusters are well predicted, but the tailoring enzymes are still poorly characterized. Their functions are predicted by comparison with other genetic data; however, the biochemical proof is rarely established. For NRPS/PKS as for ribosomal pathways, these predictions are iteratively used. Recently, a new family of radical S-adenosyl methionime epimerase (rSAM) was characterized, demonstrating that biochemical studies are essential to reveal the true functionalities of the predicted enzymes. While rSAM were already described as a superfamily, a novel family of architecturally and functionally distinct enzyme rSAM was then evidenced (Morinaka *et al.*, 2014).

Therefore, for natural products four cases can be encountered regarding the links with their pathways, biosynthetic gene clusters and molecules (Table 3.2):

Table 3.2 Description of the four categories of information linking the biosynthetic gene cluster, the pathways and the produced molecule for a natural product

	Biosynthetic gene cluster	Pathway	Molecule
Characterized biosynthetic gene cluster	+	+	+
Predicted biosynthetic gene cluster	+	–	+
Orphan biosynthetic gene cluster	+	–	–
Orphan molecule	–	–	+

1 The natural products, for which a characterized biosynthetic gene cluster, a described pathway, and a molecule produced are known. Although the way of producing the variants of the dedicated product are not always understood (e.g. variants of microcystins and hassallidins);
2 The natural products, for which no pathway is described, despite the knowledge of structurally characterized molecule and the identification of the biosynthetic gene cluster coding different parts of the known molecule (e.g. several siderophores);
3 With the plethora of new genomes sequenced, orphan biosynthetic gene clusters represent up to 80% of the recent genome mining of 126 cyanobacteria (Calteau *et al.*, 2014), indicating the huge potential of the phylum, but not much is known about the product, the gene cluster boundaries and the biosynthetic pathway;
4 Finally, as mirror to the orphan biosynthetic gene clusters, there are 1000 orphan molecules described for the cyanobacteria not linked to genetic data or to biosynthetic pathways (Dittmann *et al.*, 2015).

Genomics from isolated strains versus metagenomics

Isolating new compounds via laboratory culture can be more practical, but it implies a selection bias of cultivable free-living organism and under-represents the wide potential for natural products in the ecosystems (Cuadrat *et al.*, 2015; Müller *et al.*, 2015). In this context, another type of data is to be considered: the metagenomic studies of environmental samples. Just as new sequencing technologies enable more and more single genomes to be released, application to environmental samples induces a burst of metagenomic datasets in databases. Metagenomes reveal qualitative and quantitative information on the functional potential of the studied microbial communities, including their potential for bioactive compounds of pharmaceutical of biotechnological interest (Trindade *et al.*, 2015).

Unfortunately, few bioinformatics resources are specifically dedicated to biosynthetic gene cluster genome mining in metagenomes. Only eSNAP (Environmental Surveyor of Natural Product Diversity; see Web resources) is meant to re-cluster fragmented biosynthetic gene cluster in metagenomic dataset (Reddy *et al.*, 2014). The usual tools dedicated to genome mining can also be used though for metagenomic dataset, considering most genomes are not closed and biosynthetic gene cluster can also span several contigs in these cases. As an example of such application, very recently, microbial communities of hyper-arid environment were investigated through metagenome sequencing (Crits-Christoph *et al.*, 2016). The microbial community on the rock substrates from the Atacama Desert are often dominated by primary producers, such as cyanobacteria in this study case. In the metagenomic dataset, cyanobacterial natural products biosynthetic gene clusters were uncovered using the usual genome mining software antiSMASH and the NORINE database. Several NRPS and PKS gene clusters were predicted, with a majority encoding a wide range of siderophores, natural products for iron acquisition. One difficulty to overcome with metagenomic data can be to assess the true host organism of a biosynthetic gene cluster. Indeed, NRPS and PKS biosynthetic gene clusters can be similar to the ones found in other organisms and involved in synthesis of the same family of natural products. Tailoring enzymes can then help the identification of the host organism, as they are more phylum-specific (Calteau *et al.*, 2014).

Symbionts are also potent sources of natural products, even though they are unfit to traditional culture isolation and investigation. Furthermore, the natural products produced by the symbiont can even be involved in the establishment and stability of the symbiotic interaction (Donia *et al.*, 2011; Liaimer *et al.*, 2015). The natural product nostopeptolide and its NRPS type biosynthetic gene cluster had been described in a free-living *Nostoc* strain (Hoffmann *et al.*, 2003), but like in most cases, its biological function remained unknown. More recently, nostopeptolide has been shown to play multiple roles in *Nostoc punctiforme* in free-living state versus symbiotic interaction, suggesting a role of this natural product and of its regulation in the interaction with its plant host (Liaimer *et al.*, 2015). The natural products detected in cases of symbiotic interactions are often not the ones encountered in free-living organism and represent an under-explored source of new natural products (Crawford and Clardy, 2011). Cyanobacteria are no exception when found in symbiotic association with plants or lichen. To date only one link between a biosynthetic gene cluster and a natural product was established following metagenomic data: the nosperin compound in the *Nostoc* sp. photosymbiont of the lichen *Peltidera membranacea* (Kampa *et al.*, 2013). This 59-kb-long biosynthetic gene cluster was not only the first *trans*-AT PKS discovered in cyanobacteria, but was also a unique mosaic of *trans*-AT PKS, *cis*-AT PKS and NRPS modules. On one hand, mapping on single genomes can reduce problems in assembly of metagenomes, and on the other hand, metagenomics is also a necessary mean to assess the relevance of laboratory culture findings in natural ecosystems and to study usually inaccessible or uncultivable resources. Altogether, the huge amount of orphan biosynthetic gene clusters will decrease with input of more genomic and metagenomic data alike.

One aspect though to consider in culture collections is the change of selective pressure on the isolated cyanobacterial cells. This change of environment might not favour the conservation of functional gene clusters with high metabolic cost. Cases of loss of function of the biosynthetic gene cluster over time in culture collections have been reported. For instance, the hassallidin gene cluster has been found inactivated by a deletion mutation in present-day cultured strain *Anabaena* sp. 90, while freeze-dried material sampled prior to the deletion event showed indeed that the strain used to be a hassallidin producer (Vestola *et al.*, 2014). Compared to environmental samples, isolated strains offer most of the time higher genome quality obtained from sufficient biological material. Working with pure bacterial culture and single species genome is the best way to ensure the natural host of the biosynthetic gene cluster and the direct link between the biosynthetic gene cluster and the molecule. Working on cultured isolated strains is also non-destructive of sometimes rare, non-accessible or protected ecosystems. In any case, contribution of other research fields will be necessary to fill in the gap between biosynthetic gene clusters and natural products.

Conclusion

In the last 15 years, genome mining of biosynthetic gene clusters has become a key methodology for studying NRPS/PKS type biosynthetic gene clusters in Cyanobacteria. As expected from chemical data, Cyanobacteria happened to be very rich in natural product biosynthetic gene clusters. Although thousands of natural products have been described in the phylum, only 33 have been linked to a biosynthetic gene cluster sequence, and even less have their pathways totally characterized. Comparative genomics of biosynthetic gene cluster revealed that some pathways were shared and conserved in the phylum while other had been subject to recombination, gene shuffling and/or horizontal gene transfer leading to the huge diversity of pathways observed at the genomics level. New genomic data and comparative genomics studies give insights into the evolutionary mechanisms underlying natural products biosynthetic gene cluster diversity. We evidenced that genomics of NRPS and PKS gene clusters in cyanobacteria will gain from current technological advances. Third generation sequencing technologies as well as growing databases dedicated to natural products biosynthetic pathways will ease *in silico* analyses. With more and more genomic data on one side, and chemical characterization on the other, the main difficulty to overcome will be the gap between the biosynthetic gene cluster and the natural product, notably to better characterize the function of the enzymes of these pathways. For this purpose, a joint effort with molecular biologists, chemists and biochemists will be necessary. Finally, the comparative genomics gain to be performed at intra and inter-phylum, as similar molecules often arise from similar pathways.

Future trends

Recent advances in high throughput mass spectrometry analysed by molecular networking are also very promising for microbial natural products discovery (Moss *et al.*, 2016). A better comprehension of the regulation of expression of the biosynthetic gene cluster and natural product can benefit from a combination of genomics, (meta-) transcriptomics (Hilton *et al.*, 2015) and metabolomics (Kleigrewe *et al.*, 2015). In some cases, thorough investigations did not enable natural product characterization and led only to the conclusion that the considered gene cluster was silent in all conditions tested or cryptic, or a possible remnant of a previously functional pathway. In other organisms, strategies have been implemented to access to the potential for natural products silent in standard laboratory condition. Indeed, it is impossible to reproduce artificially the unique and dynamic properties of environments ranging from physiochemical parameters (pH, pressure, temperature, osmolarity, light exposure, etc.) to communities' interaction. To overcome this issue, silent gene clusters have been activated in *Streptomyces* (Ochi and Hosaka, 2013) or in the filamentous fungus *Aspergillus* (Brakhage and Schroeckh, 2011). Successful strategies in these cases range from molecular biology (gene knock-outs, promoter exchange, overexpression of transcription factors) to microbiology (co-cultivation). But in cyanobacteria, with no knowledge of the regulatory mechanisms underlying natural product synthesis, biosynthetic gene cluster can be found in genomes and never be linked to the corresponding natural product. Not only are Cyanobacteria a difficult model to work with when it comes to mutagenesis and use of genetic tools, but the size and nature of the NRPS and PKS biosynthetic gene clusters are also a challenge. Indeed, using DNA fragment long of tens of kilobases in prokaryote heterologous expression systems is still uncommon. Moreover, the weakly expressed, silent and cryptic biosynthetic gene cluster and the current technical limits have evidenced the need for an alternative to genetic screening or strain selection. Efficiency of investigating new natural products could benefit from a strain prioritization strategy as shown for marine microbial natural products with LC/MS-based secondary metabolomics (Hou *et al.*, 2012).

With a wider and long-term perspective, the wide potential for natural products in cyanobacteria is also promising for pharmaceutical and biotechnological purposes. Indeed, all natural products exhibit bioactive properties depending on their structural characteristics. Reported functions include antibacterial, antifungal, antitumoral, inhibitors of eukaryote protein phosphatases (Niedermeyer, 2015), and make them valuable products to consider. Though nowadays, the natural products documented functions need to be carefully considered. The compounds are mostly screened against a narrow range of target for a few potential expected properties. A more systematic and wider bioactivity screening will be necessary to assess the true potential for compounds of interest. In the future, improvement of genetic tools adapted to more cyanobacteria like mutagenesis, heterologous expression, and genome engineering (Ramey *et al.*, 2015), will enable more efficient investigation on each and every biosynthetic gene cluster. Translational, multidisciplinary approaches need to be developed to overcome the current gap between genome mining and natural products discovery.

Web resources

Several programs and databases have been developed during these last years to predict and describe NRPS/PKS type biosynthetic gene clusters. Table 3.3 provides a list of relevant items with a short description for each one. *In silico* prediction of NRPS/PKS type biosynthetic gene clusters in genomic sequences combines established knowledge on secondary metabolism and Hidden Markov Model-based algorithms. Some programs are specifically dedicated to NRPS/PKS clusters prediction such as NP.searcher (Li *et al.*, 2009), Clustscan (Starcevic *et al.*, 2008), NaPDoS (Ziemert *et al.*, 2012), NRPS-PKS/SBPKS (Anand *et al.*, 2010) and 2metdb (Bachmann and Ravel, 2009). Other programs are more generalist and can predict several other types of biosynthetic gene clusters, such as antiSMASH (Weber *et al.*, 2015) or ClusterFinder (Cimermancic *et al.*, 2014). More recent tools allow to link bioinformatics predictions to chemical and or LC-MS/MS data, like PRISM (Skinnider *et al.*, 2015) and GNP (Johnston *et al.*, 2015). The only tool developed for prediction of biosynthetic gene clusters in metagenome data is eSNaPD for environmental surveyor of natural product diversity (Reddy *et al.*, 2014). Using PCR-generated sequence tags, the eSNaPD data-analysis pipeline profiles

Table 3.3 Overview of computational tools and databases of interest for NRPS/PKS biosynthetic gene cluster analyses

Tool or database	Web server URL	Reference	Web server	Available for download	Short description
Biosynthetic gene cluster databases					
MIBiG repository	http://mibig.secondarymetabolites.org	Medema *et al.* (2015)		+	The Minimum Information about a Biosynthetic Gene cluster MIBiG specification provides a robust community standard for annotations and metadata on biosynthetic gene clusters and their molecular products
IMG-ABC	https://img.jgi.doe.gov/cgi-bin/abc/main.cgi	Hadjithomas *et al.* (2015)	+		IMG-ABC is an atlas of biosynthetic gene clusters within the Integrated Microbial Genomes IMG system
DoBISCUIT	http://www.bio.nite.go.jp/pks/top	Ichikawa *et al.* (2013)	+	+	Database Of BIoSynthesis clusters CUrated and InTegrated is a literature-based, manually curated database of gene clusters for secondary metabolite biosynthesis in actinomycetes
ClusterMine360	http://www.clustermine360.ca	Conway and Boddy (2013)	+	+	ClusterMine360 is a database of microbial polyketide and non-ribosomal peptide gene clusters. The database is organized around compound families. Allows data crowd-sourcing
CSDB	http://csdb.bioserv.pbf.hr/csdb/ClustScanWeb.html	Diminic *et al.* (2013); Starcevic *et al.* (2008)	+		The ClustScan DataBase (CSDB) is a bioinformatics database containing genetic and biochemical information on natural products synthetized by thiotemplate modular systems
Prediction and analysis of NRPS/PKS of Biosynthetic gene clusters					
antiSMASH 3.0	http://antismash.secondarymetabolites.org	Weber *et al.* (2015)	+	+	antiSMASH allows the rapid genome-wide identification, annotation and analysis of secondary metabolite biosynthesis gene clusters in bacterial and fungal genomes
ClusterFinder	https://github.com/petercim/ClusterFinder	Cimermancic *et al.* (2014)	+	+	ClusterFinder detects BGCs of both known and unknown classes using a hidden Markov model-based probabilistic algorithm
NP.searcher	http://dna.sherman.lsi.umich.edu/	Li *et al.* (2009)	+	+	NP.searcher identifies type I PKS, NRPS and mixed PKS/NRPS in genomic data and use the information on the identified catalytic domains to predict structures of the metabolites encoded by the BGCs
ClustScan-Professional	http://bioserv.pbf.hr/cms/index.php?page=clustscan	Starcevic *et al.* (2008)		+	ClustScan-Professional the Cluster Scanner is a commercial Java-based tool to identify PKS and NRPS gene clusters in genomic data
NaPDoS	http://napdos.ucsd.edu/	Ziemert *et al.* (2012)	+		The Natural Products Domain Seeker uses a phylogenetic approach to identify and classify C domains of NRPS and KS domains of PKS enzymes from DNA or amino acid sequence data

Table 3.3 Continued

Tool or database	Web server URL	Reference	Web server	Available for download	Short description
NRPS-PKS/ SBSPKS	http://www.nii.ac.in/ sbspks.html	Anand *et al.* (2010)	+		SBSPKS is a webserver for carrying out sequence and structure-based analysis of NRPS and PKS BGC
eSNaPD	http://esnapd2. rockefeller.edu/	Reddy *et al.* (2014)	+		Environmental Surveyor of Natural Product Diversity eSNaPD is a web-based bioinformatics and data aggregation platform. It predicts secondary metabolite gene cluster diversity in metagenomic DNA sequences, also taking into account metadata of the data, i.e. geographic sampling location
PKS/NRPS Analysis Web-site					
2metDB	http://nrps.igs. umaryland.edu	Bachmann and Ravel (2009)	+	+	2metDB implements methods for prediction of structural elements for NRPS/PKS by performing *in silico* predictions of microbial secondary metabolic pathways from DNA sequence data
PRISM	http://magarveylab.ca/ prism/	Skinnider *et al.* (2015)	+		PRISM is a computational resource for the identification of biosynthetic gene clusters, prediction of genetically encoded nonribosomal peptides and type I and II polyketides, and bio- and cheminformatic dereplication of known natural products
GNP	http://magarveylab. ca/gnp/	Johnston *et al.* (2015)	+		GNP Genes to Natural Products is an integrated online genomics and chemoinformatics platform for natural product discovery linking gene cluster data for PKS/NRPS clusters to LC-MS/MS data
Substrate specificity prediction for NRPS and PKS enzymes					
NRPS predictor	http://nrps.informatik. uni-tuebingen.de/	Röttig *et al.* (2011)	+	+	NRPS predictor uses a Support Vector Machine based approach to classify NRPS A domain specificities
LSI predictor	http://bioserv7.bioinfo. pbf.hr/LSIpredictor/	Baranasic *et al.* (2014)	+		LSI predictor uses Latent Semantic Indexing LSI to predict NRPS A domains specificities
NRPSsp	http://www.nrpssp. com/	Prieto *et al.* (2012)	+		NRPSsp uses a HMM-based approach to predict specificities of NRPS A domains
NRPS/PKS substrate predictor	http://www.cmbi.ru.nl/ NRPS-PKS-substrate-predictor/	Khayatt *et al.* (2013)	+		NRPS/PKS substrate predictor uses a HMM-based approach to predict specificities of NRPS A domains and PKS AT domains

biosynthetic diversity hidden within (meta-) genomes by comparing sequence tags to a reference dataset of characterized gene clusters. Among the programs dedicated to biosynthetic gene cluster prediction, the most successful until now is antiSMASH (antibiotics and Secondary Metabolite Analysis Shell; Weber *et al.*, 2015), which is unanimously used in the scientific community and has been cited in more than 900 times since its first publication in 2011 (source Google Scholar). This tool integrates or cross-links other secondary-metabolite specific gene analysis methods in one interactive view. Results are cross-linked with the MIBiG database (Medema *et al.*, 2015) and with the database of non-ribosomal pepides, NORINE (Caboche *et al.*, 2008).

Different databases dedicated to NRPS/PKS biosynthetic gene clusters have been published. ClusterMine360 (Conway and Boddy, 2013) is a database containing more than 290 curated entries of biosynthetic gene clusters including classification of the produced compound(s), taxonomic information of the producers, and links to sequence data and antiSMASH results. The ClustScan database (Diminic *et al.*, 2013; Starcevic *et al.*, 2008) is a manually curated database containing about 170 PKS/NRPS clusters. The information in the database was partially retrieved from the ClustScan software developed in the same laboratory. The database of biosynthesis clusters curated and integrated, DoBISCUIT (Ichikawa *et al.*, 2013) contains a literature-based collection of PKS and NRPS biosynthetic gene clusters from actinomycetes. This database offers extended search capabilities on the different features of the clusters and contains entries for about 100 biosynthetic gene clusters. More recently, two generalist biosynthetic gene cluster databases have been released. The Integrated Microbial Genomes – Atlas of Biosynthetic gene Clusters (IMG-ABC; Hadjithomas *et al.*, 2015) is a comprehensive resource of biosynthetic gene clusters identified in microbial genome sequences. This database is fully integrated into JGI's IMG resource (Chen *et al.*, 2016) covering more than 36,000 microbial genomes and 5000 metagenomes. IMG-ABC gathers automatically identified gene clusters (using the ClusterFinder algorithm), clusters with known biosynthesis products and metabolites. The recently published reference database MIBiG (Medema *et al.*, 2015) aims at developing a genomic standard for experimentally characterized biosynthetic gene clusters of all types (NRPS, PKS or RIPP) and from all phyla (bacteria, fungi or plants). It provides a comprehensive and standardized specification for biosynthetic gene cluster annotations and gene cluster-associated metadata. This database is open to crowdsourcing and researchers are encouraged to submit new biosynthetic gene clusters, i.e. genomic sequences, their annotations and associated experimental data. Furthermore, data can be cross linked with external chemical resources like PubChem (Kim *et al.*, 2016), ChEMBL (Bento *et al.*, 2014) or ChEBI (Hastings *et al.*, 2016). The most recent release of MiBIG contains 1294 entries, among which 748 are of NRPS/PKS type (24 December 2015).

References

Anand, S., Prasad, M.V., Yadav, G., Kumar, N., Shehara, J., Ansari, M.Z., and Mohanty, D. (2010). SBSPKS: Structure based sequence analysis of polyketide synthases. Nucleic Acids Res. *38*, W487–496. http://dx.doi.org/10.1093/nar/gkq340.

Ashworth, C.T., and Mason, M.F. (1946). Observations on the pathological changes produced by a toxic substance present in blue-green algae (*Microcystis aeruginosa*). Am. J. Pathol. *22*, 369–383.

Bachmann, B.O., and Ravel, J. (2009). Methods for *in silico* prediction of microbial polyketide and nonribosomal peptide biosynthetic pathways from DNA sequence data. Methods Enzymol. *458*, 181–217. http://dx.doi.org/10.1016/S0076-6879(09)04808-3.

Balskus, E.P., and Walsh, C.T. (2010). The genetic and molecular basis for sunscreen biosynthesis in cyanobacteria. Science *329*, 1653–1656. http://dx.doi.org/10.1126/science.1193637.

Bandyopadhyay, A., Elvitigala, T., Welsh, E., Stöckel, J., Liberton, M., Min, H., Sherman, L.A., and Pakrasi, H.B. (2011). Novel metabolic attributes of the genus *Cyanothece*, comprising a group of unicellular nitrogen-fixing *Cyanothece*. MBio *2*, e00214–11. http://dx.doi.org/10.1128/mBio.00214-11.

Baranasic, D., Zucko, J., Diminic, J., Gacesa, R., Long, P.F., Cullum, J., Hranueli, D., and Starcevic, A. (2014). Predicting substrate specificity of adenylation domains of nonribosomal peptide synthetases and other protein properties by latent semantic indexing. J. Ind. Microbiol. Biotechnol. *41*, 461–467. http://dx.doi.org/10.1007/s10295-013-1322-2.

Becker, J.E., Moore, R.E., and Moore, B.S. (2004). Cloning, sequencing, and biochemical characterization of the nostocyclopeptide biosynthetic gene cluster: Molecular basis for imine macrocyclization. Gene *325*, 35–42. http://dx.doi.org/10.1016/j.gene.2003.09.034.

Bento, A.P., Gaulton, A., Hersey, A., Bellis, L.J., Chambers, J., Davies, M., Krüger, F.A., Light, Y., Mak, L., McGlinchey, S., *et al.* (2014). The ChEMBL bioactivity database: An update. Nucleic Acids Res. *42*, D1083–90. http://dx.doi.org/10.1093/nar/gkt1031.

Bertin, M.J., Vulpanovici, A., Monroe, E.A., Korobeynikov, A., Sherman, D.H., Gerwick, L., and Gerwick, W.H. (2016). The phormidolide biosynthetic gene cluster: A *trans*-AT PKS pathway encoding a toxic macrocyclic polyketide. Chembiochem *17*, 164–173. http://dx.doi.org/10.1002/cbic.201500467.

Brakhage, A.A., and Schroeckh, V. (2011). Fungal secondary metabolites – strategies to activate silent gene clusters. Fungal. Genet. Biol. *48*, 15–22. http://dx.doi.org/10.1016/j.fgb.2010.04.004.

Burja, A.M., Banaigs, B., Abou-Mansour, E., Grant Burgess, J., and Wright, P.C. (2001). Marine cyanobacteria – a prolific source of natural products. Tetrahedron *57*, 9347–9377. http://dx.doi.org/10.1016/S0040-4020(01)00931-0.

Caboche, S., Pupin, M., Leclère, V., Fontaine, A., Jacques, P., and Kucherov, G. (2008). NORINE: A database of nonribosomal peptides. Nucleic Acids Res. *36*, D326–31. http://dx.doi.org/10.1093/nar/gkm792.

Cadel-Six, S., Dauga, C., Castets, A.M., Rippka, R., Bouchier, C., Tandeau de Marsac, N., and Welker, M. (2008). Halogenase genes in nonribosomal peptide synthetase gene clusters of *Microcystis* (cyanobacteria): Sporadic distribution and evolution. Mol. Biol. Evol. *25*, 2031–2041. http://dx.doi.org/10.1093/molbev/msn150.

Calteau, A., Fewer, D.P., Latifi, A., Coursin, T., Laurent, T., Jokela, J., Kerfeld, C.A., Sivonen, K., Piel, J., and Gugger, M. (2014). Phylum-wide comparative genomics unravel the diversity of secondary metabolism in Cyanobacteria. BMC Genomics *15*, 977. http://dx.doi.org/10.1186/1471-2164-15-977.

Chang, Z., Flatt, P., Gerwick, W.H., Nguyen, V.A., Willis, C.L., and Sherman, D.H. (2002). The barbamide biosynthetic gene cluster: A novel marine cyanobacterial system of mixed polyketide synthase (PKS)-non-ribosomal peptide synthetase (NRPS) origin involving an unusual trichloroleucyl starter unit. Gene *296*, 235–247. http://dx.doi.org/10.1016/S0378-1119(02)00860-0.

Chang, Z., Sitachitta, N., Rossi, J.V., Roberts, M.A., Flatt, P.M., Jia, J., Sherman, D.H., and Gerwick, W.H. (2004). Biosynthetic pathway and gene cluster analysis of curacin A, an antitubulin natural product from the tropical marine cyanobacterium *Lyngbya majuscula*. J. Nat. Prod. *67*, 1356–1367. http://dx.doi.org/10.1021/np0499261.

Chen, I.M., Markowitz, V.M., Palaniappan, K., Szeto, E., Chu, K., Huang, J., Ratner, A., Pillay, M., Hadjithomas, M., Huntemann, M., *et al.* (2016). Supporting community annotation and user collaboration in the integrated microbial genomes (IMG) system. BMC Genomics *17*, 307. http://dx.doi.org/10.1186/s12864-016-2629-y.

Christiansen, G., Dittmann, E., Via Ordorika, L., Rippka, R., Herdman, M., and Börner, T. (2001). Nonribosomal peptide synthetase genes occur in most cyanobacterial genera as evidenced by their distribution in axenic strains of the PCC. Arch. Microbiol. *176*, 452–458. http://dx.doi.org/10.1007/s002030100349.

Christiansen, G., Molitor, C., Philmus, B., and Kurmayer, R. (2008). Nontoxic strains of cyanobacteria are the result of major gene deletion events induced by a transposable element. Mol. Biol. Evol. 25, 1695–1704. http://dx.doi.org/10.1093/molbev/msn120.

Cimermancic, P., Medema, M.H., Claesen, J., Kurita, K., Wieland Brown, L.C., Mavrommatis, K., Pati, A., Godfrey, P.A., Koehrsen, M., Clardy, J., *et al.* (2014). Insights into secondary metabolism from a global analysis of prokaryotic biosynthetic gene clusters. Cell *158*, 412–421. http://dx.doi.org/10.1016/j.cell.2014.06.034.

Conway, K.R., and Boddy, C.N. (2013). ClusterMine360: a database of microbial PKS/NRPS biosynthesis. Nucleic Acids Res. *41*, D402–7. http://dx.doi.org/10.1093/nar/gks993.

Crawford, J.M., and Clardy, J. (2011). Bacterial symbionts and natural products. Chem. Commun. *47*, 7559–7566. http://dx.doi.org/10.1039/c1cc11574j.

Crits-Christoph, A., Robinson, C.K., Ma, B., Ravel, J., Wierzchos, J., Ascaso, C., Artieda, O., Souza-Egipsy, V., Casero, M.C., and DiRuggiero, J. (2016). Phylogenetic and functional substrate specificity for endolithic microbial communities in hyper-arid environments. Front. Microbiol. *7*, 301. http://dx.doi.org/10.3389/fmicb.2016.00301.

Cuadrat, R.R., Cury, J.C., and Dávila, A.M. (2015). Metagenomic analysis of upwelling-affected Brazilian coastal seawater reveals sequence domains of type I PKS and modular NRPS. Int. J. Mol. Sci. *16*, 28285–28295. http://dx.doi.org/10.3390/ijms161226101.

Dagan, T., Roettger, M., Stucken, K., Landan, G., Koch, R., Major, P., Gould, S.B., Goremykin, V.V., Rippka, R., Tandeau de Marsac, N., *et al.* (2013). Genomes of Stigonematalean cyanobacteria (subsection V) and the evolution of oxygenic photosynthesis from prokaryotes to plastids. Genome. Biol. Evol. 5, 31–44. http://dx.doi.org/10.1093/gbe/evs117.

Diminic, J., Zucko, J., Ruzic, I.T., Gacesa, R., Hranueli, D., Long, P.F., Cullum, J., and Starcevic, A. (2013). Databases of the thiotemplate modular systems (CSDB) and their *in silico* recombinants (r-CSDB). J. Ind. Microbiol. Biotechnol. *40*, 653–659. http://dx.doi.org/10.1007/s10295-013-1252-z.

Dittmann, E., Fewer, D.P., and Neilan, B.A. (2013). Cyanobacterial toxins: Biosynthetic routes and evolutionary roots. FEMS Microbiol. Rev. *37*, 23–43. http://dx.doi.org/10.1111/j.1574-6976.2012.12000.x.

Dittmann, E., Gugger, M., Sivonen, K., and Fewer, D.P. (2015). Natural product biosynthetic diversity and comparative genomics of the Cyanobacteria. Trends Microbiol. *23*, 642–652. http://dx.doi.org/10.1016/j.tim.2015.07.008.

Donia, M.S., Fricke, W.F., Partensky, F., Cox, J., Elshahawi, S.I., White, J.R., Phillippy, A.M., Schatz, M.C., Piel, J., Haygood, M.G., *et al.* (2011). Complex microbiome underlying secondary and primary metabolism in the tunicate-*Prochloron* symbiosis. Proc. Natl. Acad. Sci. U.S.A. *108*, E1423–32. http://dx.doi.org/10.1073/pnas.1111712108.

Edwards, D.J., and Gerwick, W.H. (2004). Lyngbyatoxin biosynthesis: sequence of biosynthetic gene cluster and identification of a novel aromatic prenyltransferase. J. Am. Chem. Soc. *126*, 11432–11433. http://dx.doi.org/10.1021/ja047876g.

Edwards, D.J., Marquez, B.L., Nogle, L.M., McPhail, K., Goeger, D.E., Roberts, M.A., and Gerwick, W.H. (2004). Structure and biosynthesis of the jamaicamides, new mixed polyketide-peptide neurotoxins from the marine cyanobacterium *Lyngbya majuscula*. Chem. Biol. *11*, 817–833. http://dx.doi.org/10.1016/j.chembiol.2004.03.030.

Fan, Q., Huang, G., Lechno-Yossef, S., Wolk, C.P., Kaneko, T., and Tabata, S. (2005). Clustered genes required for synthesis and deposition of envelope glycolipids in *Anabaena* sp. strain PCC 7120. Mol. Microbiol. *58*, 227–243. http://dx.doi.org/10.1111/j.1365-2958.2005.04818.x.

Fewer, D.P., Jokela, J., Rouhiainen, L., Wahlsten, M., Koskenniemi, K., Stal, L.J., and Sivonen, K. (2009). The non-ribosomal assembly and frequent occurrence of the protease inhibitors spumigins in the bloom-forming cyanobacterium *Nodularia spumigena*. Mol. Microbiol. *73*, 924–937. http://dx.doi.org/10.1111/j.1365-2958.2009.06816.x.

Fewer, D.P., Osterholm, J., Rouhiainen, L., Jokela, J., Wahlsten, M., and Sivonen, K. (2011). Nostophycin biosynthesis is directed by a hybrid polyketide synthase-nonribosomal peptide synthetase in the toxic cyanobacterium *Nostoc* sp. strain 152. Appl. Environ. Microbiol. *77*, 8034–8040. http://dx.doi.org/10.1128/AEM.05993-11.

Francis, G. (1878). Poisonous Australian lakes. Nature *18*, 11–12. http://dx.doi.org/10.1038/018011d0.

Freeman, M.F., Gurgui, C., Helf, M.J., Morinaka, B.I., Uria, A.R., Oldham, N.J., Sahl, H.G., Matsunaga, S., and Piel, J. (2012). Metagenome mining reveals polytheonamides as posttranslationally modified ribosomal peptides. Science *338*, 387–390. http://dx.doi.org/10.1126/science.1226121.

Gehring, A.M., DeMoll, E., Fetherston, J.D., Mori, I., Mayhew, G.F., Blattner, F.R., Walsh, C.T., and Perry, R.D. (1998). Iron acquisition in plague: Modular logic in enzymatic biogenesis of yersiniabactin by *Yersinia pestis*.

Chem. Biol. 5, 573–586. http://dx.doi.org/10.1016/S1074-5521(98)90115-6.

Grindberg, R.V., Ishoey, T., Brinza, D., Esquenazi, E., Coates, R.C., Liu, W.T., Gerwick, L., Dorrestein, P.C., Pevzner, P., Lasken, R., et al. (2011). Single cell genome amplification accelerates identification of the apratoxin biosynthetic pathway from a complex microbial assemblage. PLoS One *6*, e18565. http://dx.doi.org/10.1371/journal.pone.0018565.

Gu, L., Wang, B., Kulkarni, A., Geders, T.W., Grindberg, R.V., Gerwick, L., Håkansson, K., Wipf, P., Smith, J.L., Gerwick, W.H., *et al.* (2009). Metamorphic enzyme assembly in polyketide diversification. Nature *459*, 731–735. http://dx.doi.org/10.1038/nature07870.

Hadjithomas, M., Chen, I.M., Chu, K., Ratner, A., Palaniappan, K., Szeto, E., Huang, J., Reddy, T.B., Cimermančič, P., Fischbach, M.A., *et al.* (2015). IMG-ABC: A knowledge base to fuel discovery of biosynthetic gene clusters and novel secondary metabolites. MBio *6*, e00932. http://dx.doi.org/10.1128/mBio.00932-15.

Hastings, J., Owen, G., Dekker, A., Ennis, M., Kale, N., Muthukrishnan, V., Turner, S., Swainston, N., Mendes, P., and Steinbeck, C. (2016). ChEBI in 2016: Improved services and an expanding collection of metabolites. Nucleic Acids Res. *44*, D1214–9. http://dx.doi.org/10.1093/nar/gkv1031.

Helfrich, E.J., and Piel, J. (2016). Biosynthesis of polyketides by *trans*-AT polyketide synthases. Nat. Prod. Rep. *33*, 231–316. http://dx.doi.org/10.1039/c5np00125k.

Hilton, J.A., Satinsky, B.M., Doherty, M., Zielinski, B., and Zehr, J.P. (2015). Metatranscriptomics of N2-fixing cyanobacteria in the Amazon River plume. ISME J. *9*, 1557–1569. http://dx.doi.org/10.1038/ismej.2014.240.

Hoffmann, D., Hevel, J.M., Moore, R.E., and Moore, B.S. (2003). Sequence analysis and biochemical characterization of the nostopeptolide A biosynthetic gene cluster from *Nostoc* sp. GSV224. Gene *311*, 171–180. http://dx.doi.org/10.1016/S0378-1119(03)00587-0.

Holland, A., and Kinnear, S. (2013). Interpreting the possible ecological role(s) of cyanotoxins: Compounds for competitive advantage and/or physiological aide? Mar. Drugs *11*, 2239–2258. http://dx.doi.org/10.3390/md11072239.

Hou, Y., Braun, D.R., Michel, C.R., Klassen, J.L., Adnani, N., Wyche, T.P., and Bugni, T.S. (2012). Microbial strain prioritization using metabolomics tools for the discovery of natural products. Anal. Chem. *84*, 4277–4283. http://dx.doi.org/10.1021/ac202623g.

Humbert, J.F., Barbe, V., Latifi, A., Gugger, M., Calteau, A., Coursin, T., Lajus, A., Castelli, V., Oztas, S., Samson, G., *et al.* (2013). A tribute to disorder in the genome of the bloom-forming freshwater cyanobacterium *Microcystis aeruginosa*. PLoS One *8*, e70747. http://dx.doi.org/10.1371/journal.pone.0070747.

Ichikawa, N., Sasagawa, M., Yamamoto, M., Komaki, H., Yoshida, Y., Yamazaki, S., and Fujita, N. (2013). DoBISCUIT: A database of secondary metabolite biosynthetic gene clusters. Nucleic Acids Res. *41*, D408–14. http://dx.doi.org/10.1093/nar/gks1177.

Inoue, M., Shinohara, N., Tanabe, S., Takahashi, T., Okura, K., Itoh, H., Mizoguchi, Y., Iida, M., Lee, N., and Matsuoka, S. (2010). Total synthesis of the large non-ribosomal peptide polytheonamide B. Nat. Chem. *2*, 280–285. http://dx.doi.org/10.1038/nchem.554.

Ishida, K., Welker, M., Christiansen, G., Cadel-Six, S., Bouchier, C., Dittmann, E., Hertweck, C., and Tandeau de Marsac, N. (2009). Plasticity and evolution of aeruginosin biosynthesis in cyanobacteria. Appl. Environ. Microbiol. *75*, 2017–2026. http://dx.doi.org/10.1128/AEM.02258-08.

Jenke-Kodama, H., Sandmann, A., Müller, R., and Dittmann, E. (2005). Evolutionary implications of bacterial polyketide synthases. Mol. Biol. Evol. *22*, 2027–2039. http://dx.doi.org/10.1093/molbev/msi193.

Johnston, C.W., Skinnider, M.A., Wyatt, M.A., Li, X., Ranieri, M.R., Yang, L., Zechel, D.L., Ma, B., and Magarvey, N.A. (2015). An automated Genomes-to-Natural Products platform (GNP) for the discovery of modular natural products. Nat. Commun. *6*, 8421. http://dx.doi.org/10.1038/ncomms9421.

Jones, A.C., Monroe, E.A., Eisman, E.B., Gerwick, L., Sherman, D.H., and Gerwick, W.H. (2010). The unique mechanistic transformations involved in the biosynthesis of modular natural products from marine cyanobacteria. Nat. Prod. Rep. *27*, 1048–1065. http://dx.doi.org/10.1039/c000535e.

Kampa, A., Gagunashvili, A.N., Gulder, T.A., Morinaka, B.I., Daolio, C., Godejohann, M., Miao, V.P., Piel, J., and Andrésson, Ó. (2013). Metagenomic natural product discovery in lichen provides evidence for a family of biosynthetic pathways in diverse symbioses. Proc. Natl. Acad. Sci. U.S.A. *110*, E3129–37. http://dx.doi.org/10.1073/pnas.1305867110.

Kehr, J.C., Gatte Picchi, D., and Dittmann, E. (2011). Natural product biosyntheses in cyanobacteria: A treasure trove of unique enzymes. Beilstein J. Org. Chem. *7*, 1622–1635. http://dx.doi.org/10.3762/bjoc.7.191.

Khayatt, B.I., Overmars, L., Siezen, R.J., and Francke, C. (2013). Classification of the adenylation and acyl-transferase activity of NRPS and PKS systems using ensembles of substrate specific hidden Markov models. PLoS One *8*, e62136. http://dx.doi.org/10.1371/journal.pone.0062136.

Kim, S., Thiessen, P.A., Bolton, E.E., Chen, J., Fu, G., Gindulyte, A., Han, L., He, J., He, S., Shoemaker, B.A., *et al.* (2016). PubChem Substance and Compound databases. Nucleic Acids Res. *44*, D1202–13. http://dx.doi.org/10.1093/nar/gkv951.

Klassen, J.L., and Currie, C.R. (2012). Gene fragmentation in bacterial draft genomes: Extent, consequences and mitigation. BMC Genomics *13*, 14. http://dx.doi.org/10.1186/1471-2164-13-14.

Kleigrewe, K., Almaliti, J., Tian, I.Y., Kinnel, R.B., Korobeynikov, A., Monroe, E.A., Duggan, B.M., Di Marzo, V., Sherman, D.H., Dorrestein, P.C., *et al.* (2015). Combining mass spectrometric metabolic profiling with genomic analysis: A powerful approach for discovering natural products from Cyanobacteria. J. Nat. Prod. *78*, 1671–1682. http://dx.doi.org/10.1021/acs.jnatprod.5b00301.

Klementz, D., Döring, K., Lucas, X., Telukunta, K.K., Erxleben, A., Deubel, D., Erber, A., Santillana, I., Thomas, O.S., Bechthold, A., *et al.* (2016). StreptomeDB 2.0 – an extended resource of natural products produced by streptomycetes. Nucleic Acids Res. *44*, D509–14. http://dx.doi.org/10.1093/nar/gkv1319.

Koren, S., and Phillippy, A.M. (2015). One chromosome, one contig: Complete microbial genomes from long-read sequencing and assembly. Curr. Opin. Microbiol. *23*, 110–120. http://dx.doi.org/10.1016/j.mib.2014.11.014.

Laver, T., Harrison, J., O'Neill, P.A., Moore, K., Farbos, A., Paszkiewicz, K., and Studholme, D.J. (2015). Assessing the performance of the Oxford Nanopore Technologies MinION. Biomol. Detect. Quantif. *3*, 1–8. http://dx.doi.org/10.1016/j.bdq.2015.02.001.

Li, M.H., Ung, P.M., Zajkowski, J., Garneau-Tsodikova, S., and Sherman, D.H. (2009). Automated genome mining for natural products. BMC Bioinf. *10*, 185. http://dx.doi.org/10.1186/1471-2105-10-185.

Liaimer, A., Helfrich, E.J., Hinrichs, K., Guljamow, A., Ishida, K., Hertweck, C., and Dittmann, E. (2015). Nostopeptolide plays a governing role during cellular differentiation of the symbiotic cyanobacterium *Nostoc punctiforme*. Proc. Natl. Acad. Sci. U.S.A. *112*, 1862–1867. http://dx.doi.org/10.1073/pnas.1419543112.

Magarvey, N.A., Beck, Z.Q., Golakoti, T., Ding, Y., Huber, U., Hemscheidt, T.K., Abelson, D., Moore, R.E., and Sherman, D.H. (2006). Biosynthetic characterization and chemoenzymatic assembly of the cryptophycins. Potent anticancer agents from cyanobionts. ACS Chem. Biol. *1*, 766–779. http://dx.doi.org/10.1021/cb6004307.

Mares, J., Hajek, J., Urajova, P., Kopecky, J., and Hrouzek, P. (2014). A hybrid non-ribosomal peptide/polyketide synthetase containing fatty-acyl ligase (FAAL) synthesizes the beta-amino fatty acid lipopeptides puwainaphycins in the cyanobacterium *Cylindrospermum alatosporum*. PLoS One *9*, e111904. http://dx.doi.org/10.1371/journal.pone.0111904.

Medema, M.H., Cimermancic, P., Sali, A., Takano, E., and Fischbach, M.A. (2014). A systematic computational analysis of biosynthetic gene cluster evolution: Lessons for engineering biosynthesis. PLOS Comput. Biol. *10*, e1004016. http://dx.doi.org/10.1371/journal.pcbi.1004016.

Medema, M.H., Cimermancic, P., Sali, A., Takano, E., and Fischbach, M.A. (2016). A systematic computational analysis of biosynthetic gene cluster evolution: Lessons for engineering biosynthesis. Erratum. PLOS Comput. Biol. *12*, e1004767. http://dx.doi.org/10.1371/journal.pcbi.1004016.

Medema, M.H., Kottmann, R., Yilmaz, P., Cummings, M., Biggins, J.B., Blin, K., de Bruijn, I., Chooi, Y.H., Claesen, J., Coates, R.C., *et al.* (2015). Minimum information about a biosynthetic gene cluster. Nat. Chem. Biol. *11*, 625–631. http://dx.doi.org/10.1038/nchembio.1890.

Méjean, A., Mann, S., Maldiney, T., Vassiliadis, G., Lequin, O., and Ploux, O. (2009). Evidence that biosynthesis of the neurotoxic alkaloids anatoxin-a and homoanatoxin-a in the cyanobacterium *Oscillatoria* PCC 6506 occurs on a modular polyketide synthase initiated by L-proline. J. Am. Chem. Soc. *131*, 7512–7513. http://dx.doi.org/10.1021/ja9024353.

Merel, S., Villarín, M.C., Chung, K., and Snyder, S. (2013). Spatial and thematic distribution of research on cyanotoxins. Toxicon *76*, 118–131. http://dx.doi.org/10.1016/j.toxicon.2013.09.008.

Micallef, M.L., D'Agostino, P.M., Sharma, D., Viswanathan, R., and Moffitt, M.C. (2015). Genome mining for natural product biosynthetic gene clusters in the Subsection V cyanobacteria. BMC Genomics *16*, 669. http://dx.doi.org/10.1186/s12864-015-1855-z.

Mihali, T.K., Kellmann, R., Muenchhoff, J., Barrow, K.D., and Neilan, B.A. (2008). Characterization of the gene cluster responsible for cylindrospermopsin biosynthesis. Appl. Environ. Microbiol. *74*, 716–722. http://dx.doi.org/10.1128%2FAEM.01988-07.

Mihali, T.K., Kellmann, R., and Neilan, B.A. (2009). Characterisation of the paralytic shellfish toxin biosynthesis gene clusters in *Anabaena circinalis* AWQC131C and *Aphanizomenon* sp. NH-5. BMC Biochem. *10*, 8. http://dx.doi.org/10.1186/1471-2091-10-8.

Moffitt, M.C., and Neilan, B.A. (2004). Characterization of the nodularin synthetase gene cluster and proposed theory of the evolution of cyanobacterial hepatotoxins. Appl. Environ. Microbiol. *70*, 6353–6362. http://dx.doi.org/10.1128%2FAEM.70.11.6353-6362.2004.

Morinaka, B.I., Vagstad, A.L., Helf, M.J., Gugger, M., Kegler, C., Freeman, M.F., Bode, H.B., and Piel, J. (2014). Radical S-adenosyl methionine epimerases: Regioselective introduction of diverse D-amino acid patterns into peptide natural products. Angew Chem. Int. Ed. Engl. *53*, 8503–8507. http://dx.doi.org/10.1002/anie.201400478.

Moss, N.A., Bertin, M.J., Kleigrewe, K., Leão, T.F., Gerwick, L., and Gerwick, W.H. (2016). Integrating mass spectrometry and genomics for cyanobacterial metabolite discovery. J. Ind. Microbiol. Biotechnol. *43*, 313–324. http://dx.doi.org/10.1007/s10295-015-1705-7.

Müller, C.A., Oberauner-Wappis, L., Peyman, A., Amos, G.C., Wellington, E.M., and Berg, G. (2015). Mining for nonribosomal peptide synthetase and polyketide synthase genes revealed a high level of diversity in the sphagnum bog metagenome. Appl. Environ. Microbiol. *81*, 5064–5072. http://dx.doi.org/10.1128/AEM.00631-15.

Nakamura, H., Hamer, H.A., Sirasani, G., and Balskus, E.P. (2012). Cylindrocyclophane biosynthesis involves functionalization of an unactivated carbon center. JACS *134*, 18518–18521. http://dx.doi.org/10.1021/ja308318p.

Nakamura, Y., Afendi, F.M., Parvin, A.K., Ono, N., Tanaka, K., Hirai Morita, A., Sato, T., Sugiura, T., Altaf-Ul-Amin, M., and Kanaya, S. (2014). KNApSAcK metabolite activity database for retrieving the relationships between metabolites and biological activities. Plant Cell Physiol. *55*, e7. http://dx.doi.org/10.1093/pcp/pct176.

Niedermeyer, T.H. (2015). Anti-infective natural products from Cyanobacteria. Planta Med. *81*, 1309–1325. http://dx.doi.org/10.1055/s-0035-1546055.

Nishizawa, A., Arshad, A.B., Nishizawa, T., Asayama, M., Fujii, K., Nakano, T., Harada, K., and Shirai, M. (2007). Cloning and characterization of a new hetero-gene cluster of nonribosomal peptide synthetase and polyketide synthase from the cyanobacterium

Microcystis aeruginosa K-139. J. Gen. Appl. Microbiol. 53, 17–27. http://dx.doi.org/10.2323/jgam.53.17.

Ochi, K., and Hosaka, T. (2013). New strategies for drug discovery: activation of silent or weakly expressed microbial gene clusters. Appl. Microbiol. Biotechnol. 97, 87–98. http://dx.doi.org/10.1007/s00253-012-4551-9.

Paitan, Y., Alon, G., Orr, E., Ron, E.Z., and Rosenberg, E. (1999). The first gene in the biosynthesis of the polyketide antibiotic TA of *Myxococcus xanthus* codes for a unique PKS module coupled to a peptide synthetase. J. Mol. Biol. *286*, 465–474. http://dx.doi.org/10.1006/jmbi.1998.2478.

Plumley, F.G. (2001). Purification of an enzyme involved in saxitoxin synthesis. J. Phycol. *37*, 926–928. http://dx.doi.org/10.1046/j.1529-8817.2001.37601.x.

Prieto, C., García-Estrada, C., Lorenzana, D., and Martín, J.F. (2012). NRPSsp: Non-ribosomal peptide synthase substrate predictor. Bioinformatics *28*, 426–427. http://dx.doi.org/10.1093/bioinformatics/btr659.

Ramaswamy, A.V., Sorrels, C.M., and Gerwick, W.H. (2007). Cloning and biochemical characterization of the hectochlorin biosynthetic gene cluster from the marine cyanobacterium *Lyngbya majuscula*. J. Nat. Prod. *70*, 1977–1986. http://dx.doi.org/10.1021/np0704250.

Ramey, C.J., Barón-Sola, Á., Aucoin, H.R., and Boyle, N.R. (2015). Genome engineering in Cyanobacteria: Where we are and where we need to go. ACS Synth. Biol. *4*, 1186–1196. http://dx.doi.org/10.1021/acssynbio.5b00043.

Rantala-Ylinen, A., Känä, S., Wang, H., Rouhiainen, L., Wahlsten, M., Rizzi, E., Berg, K., Gugger, M., and Sivonen, K. (2011). Anatoxin-a synthetase gene cluster of the cyanobacterium *Anabaena* sp. strain 37 and molecular methods to detect potential producers. Appl. Environ. Microbiol. *77*, 7271–7278. http://dx.doi.org/10.1128/AEM.06022-11.

Rausch, C., Weber, T., Kohlbacher, O., Wohlleben, W., and Huson, D. (2005). Specificity prediction of adenylation domains in nonribosomal peptide synthetases (NRPS) using transductive support vector machines (TSVMs). Nucleic Acids Res. *33*, 5799–5808. http://dx.doi.org/10.1093/nar/gki885.

Reddy, B.V., Milshteyn, A., Charlop-Powers, Z., and Brady, S.F. (2014). eSNaPD: A versatile, web-based bioinformatics platform for surveying and mining natural product biosynthetic diversity from metagenomes. Chem. Biol. *21*, 1023–1033. http://dx.doi.org/10.1016/j.chembiol.2014.06.007.

Röttig, M., Medema, M.H., Blin, K., Weber, T., Rausch, C., and Kohlbacher, O. (2011). NRPSpredictor2 – a web server for predicting NRPS adenylation domain specificity. Nucleic Acids Res. *39*, W362–7. http://dx.doi.org/10.1093/nar/gkr323.

Rouhiainen, L., Jokela, J., Fewer, D.P., Urmann, M., and Sivonen, K. (2010). Two alternative starter modules for the non-ribosomal biosynthesis of specific anabaenopeptin variants in *Anabaena* (Cyanobacteria). Chem. Biol. *17*, 265–273. http://dx.doi.org/10.1016/j.chembiol.2010.01.017.

Rouhiainen, L., Paulin, L., Suomalainen, S., Hyytiäinen, H., Buikema, W., Haselkorn, R., and Sivonen, K. (2000). Genes encoding synthetases of cyclic depsipeptides, anabaenopeptilides, in *Anabaena* strain 90. Mol. Microbiol. *37*, 156–167. http://dx.doi.org/10.1046/j.1365-2958.2000.01982.x.

Rounge, T.B., Rohrlack, T., Nederbragt, A.J., Kristensen, T., and Jakobsen, K.S. (2009). A genome-wide analysis of nonribosomal peptide synthetase gene clusters and their peptides in a *Planktothrix rubescens* strain. BMC Genomics *10*, 396. http://dx.doi.org/10.1186/1471-2164-10-396.

Shih, P.M., Wu, D., Latifi, A., Axen, S.D., Fewer, D.P., Talla, E., Calteau, A., Cai, F., Tandeau de Marsac, N., Rippka, R., *et al.* (2013). Improving the coverage of the cyanobacterial phylum using diversity-driven genome sequencing. Proc. Natl. Acad. Sci. U.S.A. *110*, 1053–1058. http://dx.doi.org/10.1073/pnas.1217107110.

Skinnider, M.A., Dejong, C.A., Rees, P.N., Johnston, C.W., Li, H., Webster, A.L., Wyatt, M.A., and Magarvey, N.A. (2015). Genomes to natural products PRediction Informatics for Secondary Metabolomes (PRISM). Nucleic Acids Res. *43*, 9645–9662. http://dx.doi.org/10.1093/nar/gkv1012.

Starcevic, A., Zucko, J., Simunkovic, J., Long, P.F., Cullum, J., and Hranueli, D. (2008). ClustScan: An integrated program package for the semi-automatic annotation of modular biosynthetic gene clusters and *in silico* prediction of novel chemical structures. Nucleic Acids Res. *36*, 6882–6892. http://dx.doi.org/10.1093/nar/gkn685.

Tillett, D., Dittmann, E., Erhard, M., von Dohren, H., Borner, T., and Neilan, B.A. (2000). Structural organization of microcystin biosynthesis in *Microcystis aeruginosa* PCC7806: An integrated peptide-polyketide synthetase system. Chem. Biol. *7*, 753–764. http://dx.doi.org/10.1016/S1074-5521(00)00021-1.

Trindade, M., van Zyl, L.J., Navarro-Fernández, J., and Abd Elrazak, A. (2015). Targeted metagenomics as a tool to tap into marine natural product diversity for the discovery and production of drug candidates. Front. Microbiol. *6*, 890. http://dx.doi.org/10.3389/fmicb.2015.00890.

Tsuchiya, S., Cho, Y., Konoki, K., Nagasawa, K., Oshima, Y., and Yotsu-Yamashita, M. (2016). Biosynthetic route towards saxitoxin and shunt pathway. Sci. Rep. *6*, 20340. http://dx.doi.org/10.1038/srep20340.

Ueoka, R., Uria, A.R., Reiter, S., Mori, T., Karbaum, P., Peters, E.E., Helfrich, E.J., Morinaka, B.I., Gugger, M., Takeyama, H., *et al.* (2015). Metabolic and evolutionary origin of actin-binding polyketides from diverse organisms. Nat. Chem. Biol. *11*, 705–712. http://dx.doi.org/10.1038/nchembio.1870.

van Apeldoorn, M.E., van Egmond, H.P., Speijers, G.J., and Bakker, G.J. (2007). Toxins of cyanobacteria. Mol. Nutr. Food. Res. *51*, 7–60. http://dx.doi.org/10.1002/mnfr.200600185.

Vestola, J., Shishido, T.K., Jokela, J., Fewer, D.P., Aitio, O., Permi, P., Wahlsten, M., Wang, H., Rouhiainen, L., and Sivonen, K. (2014). Hassallidins, antifungal glycolipopeptides, are widespread among cyanobacteria and are the end-product of a nonribosomal pathway. Proc. Natl. Acad. Sci. U.S.A. *111*, E1909–17. http://dx.doi.org/10.1073/pnas.1320913111.

Waditee-Sirisattha, R., Kageyama, H., Sopun, W., Tanaka, Y., and Takabe, T. (2014). Identification and upregulation of biosynthetic genes required for accumulation of mycosporine-2-glycine under salt stress conditions in the halotolerant cyanobacterium *Aphanothece halophytica*. Appl. Environ. Microbiol. *80*, 1763–1769. http://dx.doi.org/10.1128/AEM.03729-13.

Wang, H., Fewer, D.P., Holm, L., Rouhiainen, L., and Sivonen, K. (2014). Atlas of nonribosomal peptide and polyketide biosynthetic pathways reveals common occurrence of nonmodular enzymes. Proc. Natl. Acad. Sci. U.S.A. *111*, 9259–9264. http://dx.doi.org/10.1073/pnas.1401734111.

Weber, T., Blin, K., Duddela, S., Krug, D., Kim, H.U., Bruccoleri, R., Lee, S.Y., Fischbach, M.A., Müller, R., Wohlleben, W., *et al.* (2015). antiSMASH 3.0– a comprehensive resource for the genome mining of biosynthetic gene clusters. Nucleic Acids Res. *43*, W237–43. http://dx.doi.org/10.1093/nar/gkv437.

Weissman, K.J. (2009). Introduction to polyketide biosynthesis. Methods Enzymol. *459*, 3–16. http://dx.doi.org/10.1016/S0076-6879(09)04601-1.

Welker, M., and von Döhren, H. (2006). Cyanobacterial peptides– nature's own combinatorial biosynthesis. FEMS Microbiol. Rev. *30*, 530–563. http://dx.doi.org/10.1111/j.1574-6976.2006.00022.x.

Wilson, M.C., Mori, T., Rückert, C., Uria, A.R., Helf, M.J., Takada, K., Gernert, C., Steffens, U.A., Heycke, N., Schmitt, S., *et al.* (2014a). An environmental bacterial taxon with a large and distinct metabolic repertoire. Nature *506*, 58–62. http://dx.doi.org/10.1038/nature12959.

Wilson, M.C., Mori, T., Rückert, C., Uria, A.R., Helf, M.J., Takada, K., Gernert, C., Steffens, U.A., Heycke, N., Schmitt, S., *et al.* (2014b). Corrigendum: An environmental bacterial taxon with a large and distinct metabolic repertoire. Nature *507*, 262. http://dx.doi.org/10.1038/nature13126.

Yadav, G., Gokhale, R.S., and Mohanty, D. (2003). Computational approach for prediction of domain organization and substrate specificity of modular polyketide synthases. J. Mol. Biol. *328*, 335–363. http://dx.doi.org/10.1016/S0022-2836(03)00232-8.

Ziemert, N., Lechner, A., Wietz, M., Millán-Aguiñaga, N., Chavarria, K.L., and Jensen, P.R. (2014). Diversity and evolution of secondary metabolism in the marine actinomycete genus *Salinispora*. Proc. Natl. Acad. Sci. U.S.A. *111*, E1130–9. http://dx.doi.org/10.1073/pnas.1324161111.

Ziemert, N., Podell, S., Penn, K., Badger, J.H., Allen, E., and Jensen, P.R. (2012). The natural product domain seeker NaPDoS: A phylogeny based bioinformatic tool to classify secondary metabolite gene diversity. PLoS One *7*, e34064. http://dx.doi.org/10.1371/journal.pone.0034064.

RNA-seq Based Transcriptomic Analysis of Single Cyanobacterial Cells

4

Zixi Chen[1,2,3], Jiangxin Wang[1,2,3], Lei Chen[1,2,3] and Weiwen Zhang[1,2,3*]

[1]Laboratory of Synthetic Microbiology, School of Chemical Engineering and Technology, Tianjin University, Tianjin, People's Republic of China.
[2]Key Laboratory of Systems Bioengineering, Ministry of Education of China, Tianjin, People's Republic of China.
[3]Collaborative Innovation Center of Chemical Science and Engineering, Tianjin, People's Republic of China.

*Correspondence: wwzhang8@tju.edu.cn

Abstract

Gene-expression heterogeneity among individual cells will eventually determine the fate of a bacterial population. We describe the first bacterial single-cell RNA sequencing (RNA-seq), BaSiC RNA-seq, a method integrating RNA isolation, cDNA synthesis and amplification, and RNA-seq analysis of the whole transcriptome of single cells of the cyanobacterium *Synechocystis* sp. PCC 6803, which typically contain approximately 5–7 fg of total RNA per cell. We applied the method to three *Synechocystis* single cells at 24 h and three single cells at 72 h after a nitrogen starvation stress treatment, as well as their bulk cell controls for the same conditions, to determine the heterogeneity upon the environmental stress. With 82–98% and 31–48% of all putative *Synechocystis* genes identified in single cells of 24 and 72 h, respectively, the results demonstrated the method could achieve good identification of the transcripts in single bacterial cells. In addition, the preliminary results from nitrogen-starved cells also showed a possible increasing gene expression heterogeneity from 24 h to 72 h after nitrogen starvation. Moreover, preliminary analysis of single-cell transcriptomic datasets revealed that genes from the 'Mobile elements' functional category have the most significant increase of gene expression heterogeneity under stress, which was further confirmed by single-cell RT-qPCR analysis of gene expression in 24 randomly selected cells.

Introduction

Microbiologists typically assume that microbial cells that grow under the same conditions are uniform and, thus, can be described by average values of their physiologic, phenotypic, genotypic or other measurements (Brehm-Stecher and Johnson, 2004). While this approach has been applied to generate current knowledge of microbial biology successfully, it was recently found that even isogenic populations of exponentially growing microorganisms showed substantial cell-to-cell heterogeneity at both cellular and molecular levels with an order of magnitude greater than previously thought (Lidstrom and Meldrum, 2003). Heterogeneity in isogenic cell populations could arize from the intrinsically stochastic processes of the expression of individual genes (Becskei *et al.*, 2005; Colman-Lerner *et al.*, 2005), and such stochasticity, once amplified to certain amplitude threshold, will result in heterogeneity at the cellular level, and eventually lead to different fates for a microbial population (Newman *et al.*, 2006; Strovas *et al.*, 2007). Such cell-to-cell heterogeneities could be further magnified in natural environments when complicated chemical, physical and biological factors are involved (Stewart and Franklin, 2008). It is now increasingly recognized that by using conventional averaged molecular or phenotypic measurements of a whole population for describing cell behaviours, conclusions could be biased as the unique patterns related to distinct functional subpopulations cannot be revealed.

Another major drive to pursue single-cell analysis stems from the fact that more than 99% of microbes from environmental samples are 'unculturable' in laboratory and, thus, cannot be studied using the conventional microbiological methods (Cardenas and Tiedje, 2008). Meanwhile, many of these 'unculturable' microbes represent new phylotypes, families and divisions in domain bacteria and archaea, and could be of direct relevance to various valuable biological processes involved in bioremediation, global warming, alternative energy, and the basic sciences of global cycling of carbon, nitrogen and metals (Fenical and Jensen, 2006). Although several 'culture-independent' single-cell genomics technologies have been established and successfully applied in microbial ecology studies (Lloyd *et al.*, 2013; Parkhomchuk *et al.*, 2009; Wang and Bodovitz, 2010; Zhang *et al.*, 2006), they only provide information related to genetic architecture and metabolic potential of cells, and do not reveal metabolic functionality and gene expression dynamics related to various environmental parameters.

Several approaches have been developed in the past to define gene expression heterogeneity at the level of single bacterial cells (e.g. reporter gene, fluorescence *in situ* hybridization (FISH), *in situ* PCR combined with *in situ* reverse transcription (*in situ* RT-PCR), RT-qPCR, and various imaging-based gene expression measurements), leading to an increasingly recognized view that gene expression heterogeneity was common phenomena even in isogenic bacterial population (Stewart and Franklin, 2008). However, most of these previous methods can measure only a small number of genes in each single cell. Transcriptomic analysis of single bacterial cells has obvious advantages in deciphering heterogeneity at a whole-genome level and its underlying biological significance. In recent years, methods to perform transcriptomic profiling of single eukaryotic cells with average ~10 pg of total RNA per cell have been established and applied to the developing pancreas, retina and olfactory systems, stem cells, cancer cells and fungi (Goetz and Trimarchi, 2012; Tang *et al.*, 2009). However, since all these methods for eukaryotic single cell transcriptomics employed oligo dT as the primers for cDNA synthesis, none could be adapted directly for transcriptomic analysis of single bacterial cells whose messenger RNA molecules have no unique poly(A) structures. In the NuGen RNA amplification systems utilized in this study, the use of both oligo-dT and hexamer primers provides the potential of prokaryotic mRNA amplification (Filiatrault, 2011; Head *et al.*, 2011).

Due to the challenging facts that most bacterial cells are small and difficult to lyse efficiently, and contain extremely low amount of short half-lived mRNA (Gao *et al.*, 2011), the first transcriptomic analysis of single bacterial cells was reported only recently (Kang *et al.*, 2011). In the study, the whole transcriptome of single *Burkholderia thailandensis* cells was amplified using the φ29 polymerase multiple displacement amplification (MDA) of circularized cDNA, with estimated starting RNA of 2 pg using random DNA hexamers as primers, and then analysed by a DNA microarray, the results showed lower fold-change bias (less than two-fold difference and *Pearson* correlation coefficient *R* ~0.87–0.89) and drop-outs (4–6% of 2842 detectable genes) as compared with the data obtained from non-amplified RNA samples. However, the estimated total RNA in a single *B. thailandensis* cell is about 2 pg, which is several orders of magnitude higher than the estimated total RNA amount of a typical bacterial cell, such as 3–9 fg in *Escherichia coli* and marine planktons (Lee and Kemp, 1994; Lee *et al.*, 1993; Schmid *et al.*, 2010).

We recently reported a new bacterial single-cell RNA-sequencing method (known as BaSiC RNA-seq) to amplify whole transcriptome of single cyanobacterial cells with total RNA of ~ 5 fg, and to analyse the single-cell transcriptome through a next-generation RNA-seq sequencing technology (Wang *et al.*, 2015). We applied the BaSiC RNA-seq protocol to obtain high-coverage transcriptomic profiles of single cells of *Synechocystis* sp. strain PCC 6803 (hereafter *Synechocystis*) grown under a common environmental stress, nitrogen starvation condition (Richaud *et al.*, 2001), as well as the bulk cell controls for the same treatments (Fig. 4.1). The results showed significant gene expression heterogeneity among single cells, especially for cells after long-term stress treatments. In addition, the results revealed varying heterogeneity among transcripts of different functional categories upon stress.

Quantitation of total RNA in single *Synechocystis* cells

We first performed a bulk cells-based analysis to estimate total RNA in single *Synechocystis* cells.

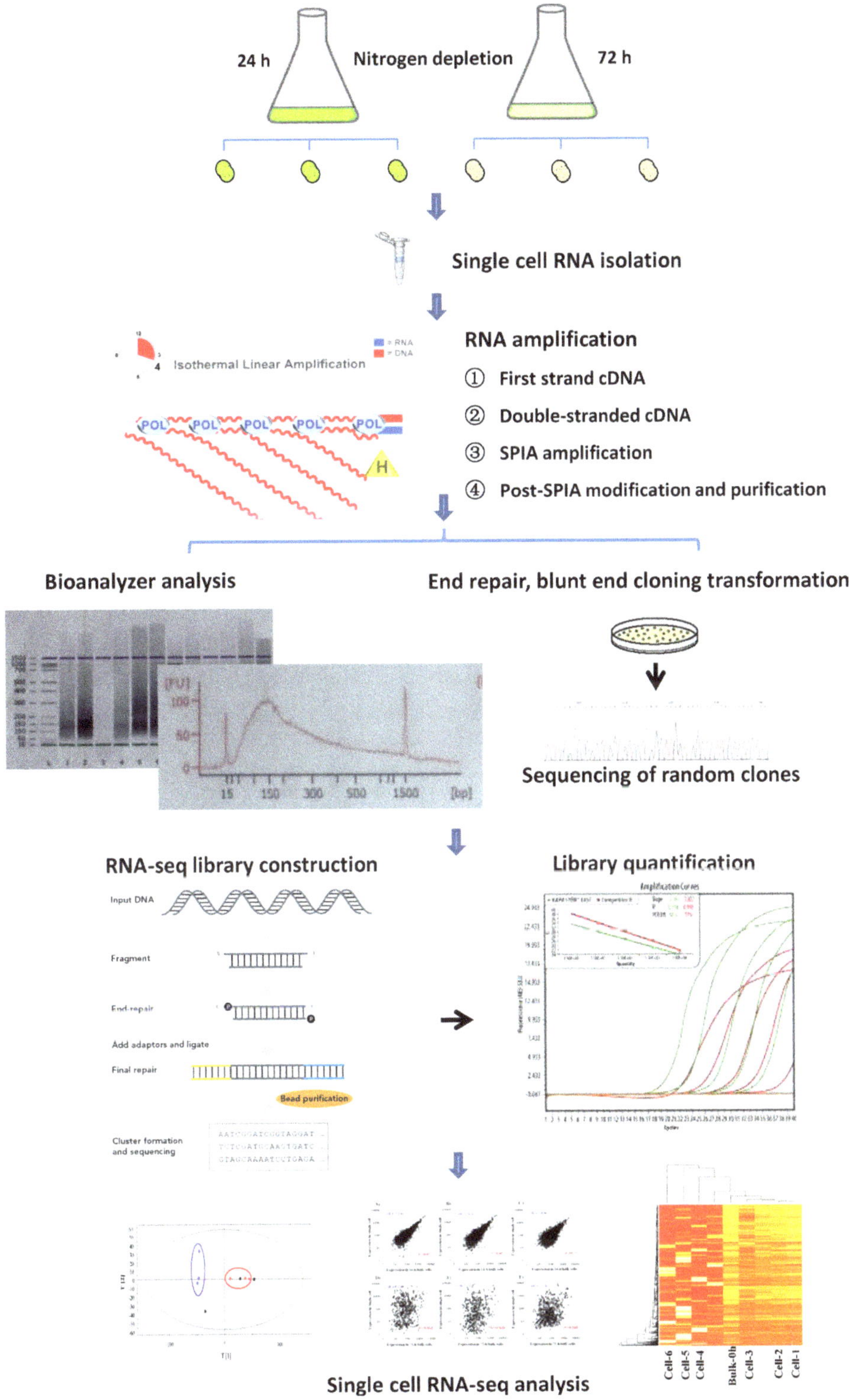

Figure 4.1 Scheme of the bacterial single-cell transcriptomics procedures. Single bacterial cells were isolated and total RNA from each single cell was purified, then RNA amplification was applied to obtain microgram-level cDNAs from single cells. After quality controls (i.e. no-template negative control, gel electrophoresis, cDNA quantification and size distribution analysis, clone library construction and sequencing), RNA-seq library derived from the amplified single-cell cDNA was subjected for RNA-seq analysis.

Synechocystis was grown in BG11 medium (pH 7.5) under a light intensity of approximately 50 µmol photons/m^2/s in a shaking incubator with 130 rpm speed at 30°C (HNY-211B Illuminating Shaker, Honour, China). Cell density was measured with a UV-1750 spectrophotometer (Shimadzu, Japan). For nitrogen starvation treatment, 10 ml of fresh cultures grown in BG11 medium at exponential phase ($OD_{730}=0.8$) were collected by centrifugation (8000 *g*), washed twice with deionized water, and then were inoculated into 25 ml of fresh BG11 liquid medium without nitrogen source. Cultivation continued in 250-ml flasks at condition described above. Growth experiments were repeated, at least, three times to confirm the growth patterns. Cell samples (~10^7 cells) were collected at 0, 24 and 72 h by centrifugation at 12,000 *g* for 2 min. Supernatant was discarded, and cells were resuspended in 5 mL of RNALater™ (Ambion, Austin, TX, USA). RNA isolation was performed with ZR RNA MicroPrep Kit (ZYMO, Irvine, CA, USA). RNA concentration in each sample was determined immediately with Nanodrop 2000 (Thermal Scientific, Waltham, MA) in 3 analytical replicates. The linear regression of the total RNA content to the cell number was conducted, and the slope was designated as the total RNA content in a single *Synechocystis* cell. The cell number was determined by directly cell counting under microscope.

It was estimated that a typical single bacterial cell contains ca 1.5–100 fg of total RNA, compared with 10–30 pg total RNA per eukaryotic cell (Westermann *et al.*, 2012). Our analysis showed that the total RNA content was 7.5 ± 0.8 fg per single *Synechocystis* cell in control; and 7.1 ± 1.2 and 5 ± 0.6 fg per single cells kept under nitrogen starvation conditions for 24 and 72 h, respectively. This estimation was in the similar range as the previous estimation for a single *E. coli* cell (3.4 fg) (Gao *et al.*, 2011; Schmid *et al.*, 2010) and marine bacterial plankton cells (1.5–9.0 fg) (Lee and Kemp, 1994; Lee *et al.*, 1993).

Single-cell RNA isolation

Synechocystis single cells were isolated using an Olympus single cell manipulator (Olympus Inc., Japan) with the Inverted microscope IX71 equipped with a micromanipulator for ICSI/Injection (Narishige Group, Tokyo, Japan) and an ICSI 10-µm-diameter micropipette (Origio, Charlottesville, VA, USA). Briefly, cells were centrifuged at 12,000 *g* for 2 min at room temperature. Supernatant was removed, and cells were immediately resuspended with 5 ml of 1× phosphate buffer (PBS) (pH 7.0) containing 20% RNALater™ (v/v) (Ambion, Austin, TX, USA). The cell suspension was further diluted with PBS-RNALater™ solution if needed, and then pipetted onto a microscope slide. Under light microscope (40× objective), cells were located randomly, aspirated into a micropipette, visually confirmed, and then loaded into a 0.2 ml PCR tube preloaded with 40 µl of 1 × PBS containing 20% RNALater (*v*/*v*).

Whole-transcriptome amplification

Isolation of total RNA from the harvested single cells was performed immediately after harvesting with ZYMO RNA MicroPrep Isolation Kit (ZYMO, Irvine, CA, USA), following the manufacturer's instruction with a minor modification to omit the step of IIIC column due to limited amount of lysed cell residue derived from a single cell (Gao *et al.*, 2011). This method has been verified to yield high-quality total RNA without DNA contamination from both single eukaryotic and bacterial cells in our previous study (Gao *et al.*, 2011). Cell lysis and RNA isolation have been successfully conducted in *Desulfovibrio vulgaris* with *Methanosarcina barkeri* (Qi *et al.*, 2014; 2016), *E. coli* (Gao *et al.*, 2011), diatom (Shi *et al.*, 2013), and human single cells (Wang *et al.*, 2013; Zeng *et al.*, 2011). Then the purified total RNA was used as a template for amplification width a modified One-Direct RNA Amplification System to generate high quality cDNAs.

Single-cell RNA amplification was conducted with NuGen WT-Ovation One-Direct RNA Amplification System (NuGen, San Carlos, CA, USA) according to the manufacturer's instructions, with the following modifications: (i) instead of using 2 µl of lysis buffer provided in the kit to lyse single bacterial cells, we started from the isolated single-cell total RNA using the ZYMO RNA isolation kit; (ii) the input volumes of primers for the first- and second-strand cDNA synthesis were reduced due to the low amount of started RNA from single cells; and (iii) each step (cDNA synthesis, Single Primer

Isothermal Amplification – SPIA, and post-SPIA modification) was performed in a separated PCR workstation to avoid the possible contamination during the synthesis and amplification process. To eliminate the potential exogenous DNA or bacterial contamination, all the related kits and reagents were put in the incubators in advance and treated with ultraviolet light for at least 30 min before manipulation. The no-template negative controls containing only RNase free water were used in parallel with total RNA samples isolated from single cells. Single cell harvesting, RNA isolation and RNA amplification were conducted immediately one after another to avoid possible RNA degradation and cross-contamination. The whole process took approximately 8 h. Amplified cDNA from single cells was purified using Qiagen MinElute Spin Column (Qiagen, Shanghai) according to the manufacturer's instruction.

Quality control

For quality control of the amplified whole transcriptome, concentration, and purity of the amplified cDNAs from both bulk- and single-cells were firstly evaluated with Nanodrop as the ratio of absorbance readings at 260 and 280 nm, and then by a standard 1.5% agarose gel electrophoresis. The ratios of A_{260}:A_{280} of amplified products obtained were around 1.8–1.9. Since the amplified cDNA product from One-Direct RNA Amplification System contained both sense and antisense cDNA strands, it was recommended by the manufacturer to use the single stranded cDNA conversion factors (1 A_{260} unit = 33 ng / µL of cDNA) for quantitation. Based on this calculation, we were able to obtain 6.85~16.78 µg of cDNAs after amplification form single-cell RNA samples. Although no-template control (NTC) also generated around 2.8 µg cDNA, gel analysis showed that they were at significantly lower DNA sizes (i.e. 150–1,000 bp in NTC versus 200–5,000 bp in single-cell samples), suggesting they were probably random non-specific cDNA products. The median length of the amplified cDNA varied from 200 to 295 bp, with a peak length around 150–180 bp measured by Agilent Bioanalyzer using the DNA 100 size markers. Moreover, to confirm that the amplified cDNA products were derived from the target bacterial cells instead of contaminated sources, blunt-end clonal library (End-It DNA end repair Kit, Epicentre Biotechnologies, Madison, WI, USA) and SMART Blunt Cloning Kit (Lucigen Corporation, Middleton, WI, USA) were used to construct a clone library in *Escherichia coli* for the amplified cDNA products. Thirty randomly selected *E. coli* clones were subjected to standard Sanger sequencing analysis with ABI3700 (Life Science Inc., Carlsbad, CA, USA). The BlastN results showed that all inserts in the library matched to the *Synechocystis* genome, indicating the successful amplification of the target transcripts and no RNA contamination from any other organism. Meanwhile, sequences obtained from NTC had no homologues in GenBank database, indicating a random non-specific amplification in NTC.

RNA-seq analysis of single-cell transcriptome

The NuGen Encore Rapid DR Multiplex System 1–8 (NuGen, San Carlos, CA, USA) was used to construct libraries for next-generation sequencing on the Illumina NGS platform. Briefly, 1 µg of purified double-stranded cDNA was subjected to library preparation through a four-step protocol including end repairing, adaptor ligation, final repair and library purification. To determine the quality of the libraries, a KAPPA Library Quantification Kit (KAPPA Biosystems, Woburn, MA, USA) was first used to determine the DNA concentration of the libraries, and then Agilent Technologies 2000 Bioanalyzer (Agilent, Santa Clara, CA, USA) was used to determine the product size of the libraries, with good libraries typically around 300 bp. The products were then used directly for cluster generation using Illumina's Solexa Sequencer (Illumina, San Diego, CA, USA) according to the manufacturer's instructions.

Libraries for nine cDNA samples, including three bulk cells samples (one each from 0, 24 h and 72 h of nitrogen-starved cultures, respectively) and six single cells (biological triplicate cell 1, 2 and 3 at 24 h, and biological triplicate cell-4, 5 and 6 at 72 h after nitrogen starvation treatment, respectively) were prepared and sequenced with Illumina HiSeq with 1 × 50 bp configuration. Multiple cDNA libraries from four single cell samples were loaded onto a single lane of an Illumina flow cell. The image deconvolution and calculation of quality value were performed using Goat module (Firecrest v. 1.4.0 and Bustard v. 1.4.0 programs) of Illumina pipeline v. 1.4. Sequenced reads were generated by base calling using the standard Illumina sequencing pipeline. The analysis produced 332–938 Mb yields and 6.5–18.4 million reads for bulk cells samples, 344–3283 Mb and 6.7–64.3 million reads for single cells. Analysis of the sequence reads showed that the mean Q scores of reads were 38.66–39.15, with 98.08–99.22% of reads having Q scores ≥ 30, and 90.1% of reads within 38–40 Q scores, suggesting overall good quality of sequencing.

For transcriptomic data analysis, reads were pre-processed using FASTX Toolkit (Version: 0.0.13) to remove low-quality bases and reads shorter than 20 bp. Genome sequences (including *nc*RNA sequences) and annotation information of *Synechocystis* were downloaded from NCBI and the Comprehensive Microbial Resource (CMR) of TIGR (http://www.tigr.org/CMR). After successful RNA amplification and labelled with normal RNA-seq preparation kit, the single cell transcriptomic mapping was conducted following standard protocol for RNA-seq data. The sequence reads were pre-processed using the FASTX toolkit to remove low-quality bases and reads shorter than 20 bp, and then mapped to the *Synechocystis* genome using Bowtie (version 2.0.0) with default parameters. The expression level of the genes was normalized using a standard RPKM (reads per kilobase of gene per million mapped reads) method before they were subjected to further analysis. Analysis of cDNAs detected in single *Synechocystis* cells showed that 82.9, 98.6 and 98.6% of the putative transcripts in the *Synechocystis* genome (3,620 genes) were detected in single cell 1, 2 and 3 from the 24 h nitrogen-starved cultures, respectively. Meanwhile, 33.5, 38.4 and 50.6% of the putative transcripts were detected in single cell 4, 5 and 6 from the 72 h nitrogen-starved culture, respectively (Table 4.1). The low transcript coverage at 72 h was probably due to the increasing dormant status caused by the extended nitrogen starvation stress, as similar trends were also observed in the bulk cell samples, 99.4% putative transcripts detected in 0 h sample, while only 92.8% and 63.2% in bulk cell samples of 24 h and 72 h after nitrogen starvation treatment, respectively (Table 4.1).

Hierarchical clustering of both single-cell and bulk cell transcriptomic data were conducted on basis of the *Z*-scores of the log2 signals of the robust multi-array analysis (RMA) using the 3,117 genes with RPKM > 0 in at least one single *Synechocystis* cell (Fig. 4.2A), and 424 genes detected in all six single-cell and three bulk cell transcriptomic profiles (Fig. 4.2C). The clustering results showed the similar trends that profiles of single cells at 24 h or 72 h were grouped together with those of bulk cells of the same time points. Further principal component analysis (PCA) of the same data sets showed that:

1. Changes along the nitrogen starvation time were clearly observed for the transcriptomic profiles of bulk cells at 0, 24 and 72 h, suggesting that treatment time is a important factor of separating the transcriptomic profiles.
2. Although none of the single-cell transcriptomic profiles were identical with the corresponding bulk cells profile, transcriptomic profiles of single cells of 24 and 72 h were clustered together correspondingly with profiles of their bulk cells, while the group of three single cells at 24 h were well separated from the group of 72 h.
3. Compared to the single-cell profiles of 24 h treatment, higher degree of gene expression heterogeneity was observed among three single-cell transcriptomic profile of 72 h treatment (Fig. 4.2B and D).

To further confirm the trends, the correlation between transcriptomic profiles of single and bulk cells of the same time points was plotted (Fig. 4.3). The results showed that the transcriptomic profiles of three single cells shared higher correlation with the bulk cell transcriptomic profile at 24 h ($R^2 = 0.77$, 0.85 and 0.85 for single cell 1, 2 and 3, respectively), compared with very low correlation of the transcriptomic profiles of three single cells of 72 h with their corresponding bulk cell transcriptomic profile ($R^2 = 0.014$, 0.028 and 0.088 for single cell 4, 5 and 6, respectively). While the high correlation of three single-cell transcriptomic profiles with the bulk cell profile at 24 h is indicative of good reliability of the BaSiC method for single-cell bacterial analysis, the low correlation of three single-cell transcriptomic profiles with the bulk cell profile at 72 h suggested that the increased gene expression heterogeneity among single cells may have occurred

Table 4.1 Summary of transcriptome identified

	0 h	24 h				72 h				Genes detected in all samples
	Bulk cells	Bulk cells	Single cell 1	Single cell 2	Single cell 3	Bulk cells	Single cell 4	Single cell 5	Single cell 6	
Total genes identified	3117	2937	3117	3102	2615	1961	1132	1521	982	
Percentage of the genome (%)	98	92	98	98	82	62	36	48	31	
Amino acid metabolism	75	73	75	75	64	57	25	38	18	6
Biosynthesis of cofactors, prosthetic groups, and carriers	82	81	82	81	71	47	23	34	21	6
Cell envelope	69	69	69	69	63	46	31	36	23	12
Cellular process	91	84	91	90	79	64	28	48	31	10
Central intermediary metabolism	57	53	57	57	48	35	20	33	14	6
DNA metabolism	50	49	50	50	44	35	23	38	21	10
Energy metabolism	269	246	269	267	217	159	90	118	80	26
Fatty acid and phospholipid metabolism	35	33	35	35	32	26	15	18	8	1
Hypothetical protein	1275	1193	1275	1272	1027	753	437	582	387	180
Mobile and extrachromosomal element functions	81	70	81	77	65	62	54	51	42	28
Protein fate	62	61	62	62	57	46	29	30	21	12
Protein synthesis	98	84	98	96	67	43	20	34	16	5
Purines, pyrimidines, nucleosides, and nucleotides	42	42	42	41	38	26	13	24	15	6
Regulatory functions	140	135	140	140	126	102	62	82	55	21
Transcription	24	22	24	24	15	11	4	11	8	4
Transport and binding proteins	166	164	166	166	158	118	70	98	66	16
Unclassified	501	478	501	500	444	331	188	246	156	87

for the *Synechocystis* population under the nitrogen starvation condition for an extended period of time. The same results were also obtained when we plotted the gene expression variation among single cells (i.e. standard deviation divided by the median measurement of three single cells) with the median gene expression measurement of all genes or genes with expression level greater than RPKM values of 500.

To further decipher the gene expression heterogeneity, the overrepresentation of gene functional categories was analysed for the 424 transcripts detected in all six single-cell and three bulk cell profiles. Based on Spearman's correlation analysis, the functional category 'Mobile elements' was detected as significantly correlated ($P < 0.001$) among all 3 single cells under both 24 and 72 h nitrogen starvation. ANOVA *F*-test was performed between single cells against each functional category to determine whether a given functional category is more different from the others at each of the time points (i.e. 24 or 72 h). The functional category of 'Mobile elements' showed significant difference from other functional categories among three single cells at 24 h using BaSiC RNA-seq data, and the coefficient of variation (CV) values at both 24 and 72 h time points (standard deviation/average from single cell 1, 2, 3 and single cell 4, 5, 6 for 24 h and 72 h respectively, either with 424 genes detected in all single cells or all genes detected in each single cells), while 'Protein synthesis' functional category was significantly distinguished from others only at 24 h. The results suggested that transcripts from different functional categories could contribute differently to the overall gene expression heterogeneity. Among others, the 'Mobile elements' category was the main contributor to the increased heterogeneity caused by the nitrogen starvation stress.

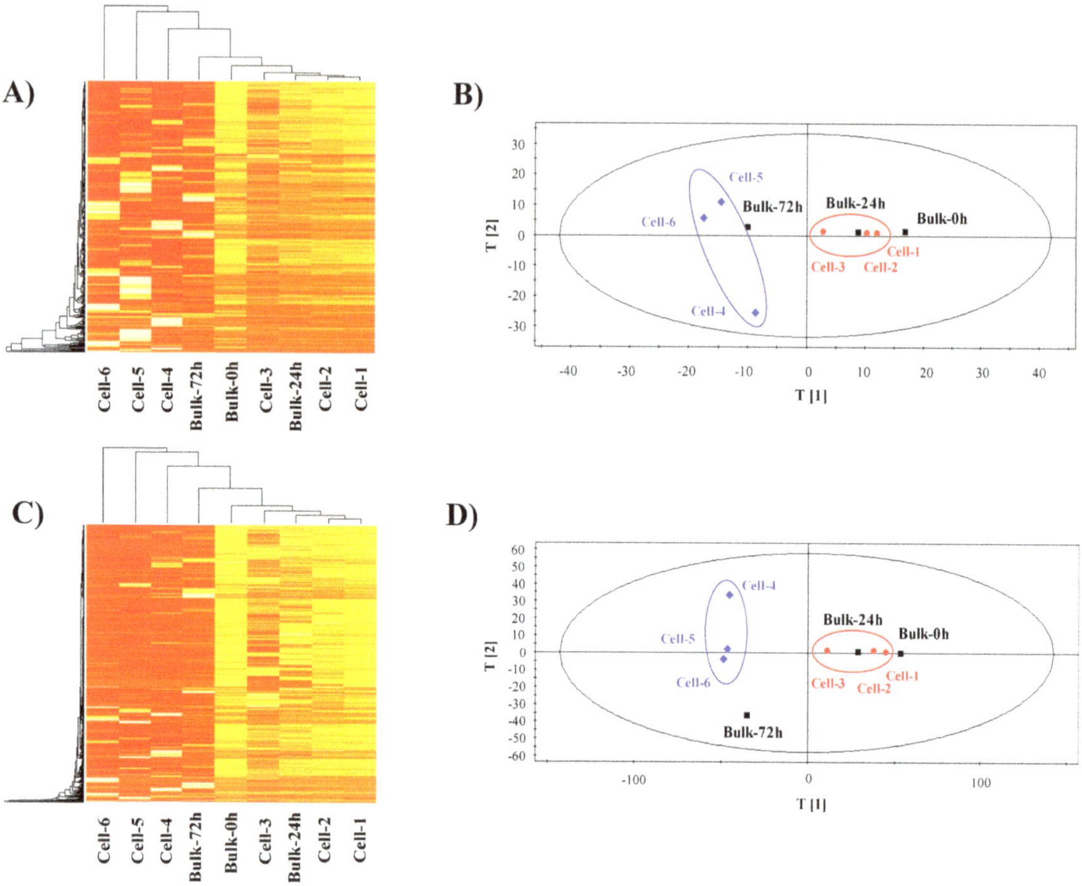

Figure 4.2 Hierarchal clustering and PCA analysis of the single-cell transcriptomic data. Hierarchal clustering of transcriptomic profiles of 3 bulk cell samples (i.e. 0, 24 and 72h) and transcriptomic profiles of 6 single Synechocystis cells (i.e. 3 from 24h and 3 from 72h) using datasets containing 3117 genes which were identified in at least one profile (A) using datasets containing 424 genes that were identified in all profiles (C). PCA analysis of nine transcriptomic profiles using datasets containing 3117 genes which were identified in at least one profile (B) and using datasets containing 424 genes which were identified in all profiles (D).

Verification of RNA-seq results

Considering the technical challenges related to the analysis of single bacterial cells (i.e. low abundance of RNA and difficult lysis), the results derived from the single-cell transcriptomics needed to be verified by other approaches. To this end, some methodologies have been recently introduced, such as pre-addition of unique molecular identifiers (UMIs) (Grun *et al.*, 2014; Islam *et al.*, 2014; Jaitin *et al.*, 2014; Shalek *et al.*, 2013) and spike-in control RNA (Katayama *et al.*, 2013) during RNA amplification or single cell FISH (Shalek *et al.*, 2013), and single cell RT-qPCR (Shi *et al.*, 2013; Wang *et al.*, 2013). Similar to the majority of other single-cell transcriptomics protocols established previously (Goetz and Trimarchi, 2012; Tang *et al.*, 2009), BaSiC RNA-seq produces ds-cDNA, and, therefore it cannot retain strand specificity for non-coding RNAs, which is preferable for antisense RNA detection. However, the issue can be addressed by coupling the BaSiC RNA-seq analysis with single-cell RT-qPCR analysis (Gao *et al.*, 2011), which can be employed to validate results from single-cell bacterial transcriptomic analysis. It is difficult to split the RNA from a single bacterial cell for both RT-qPCR and amplification, which makes it impossible to validate the linear amplification with RT-qPCR at both pre- and post-amplification steps of a few selected genes.

Since the statistical analysis based on only limited single-cell RNA-seq transcriptomic data (in our case three cells for each condition) may be not

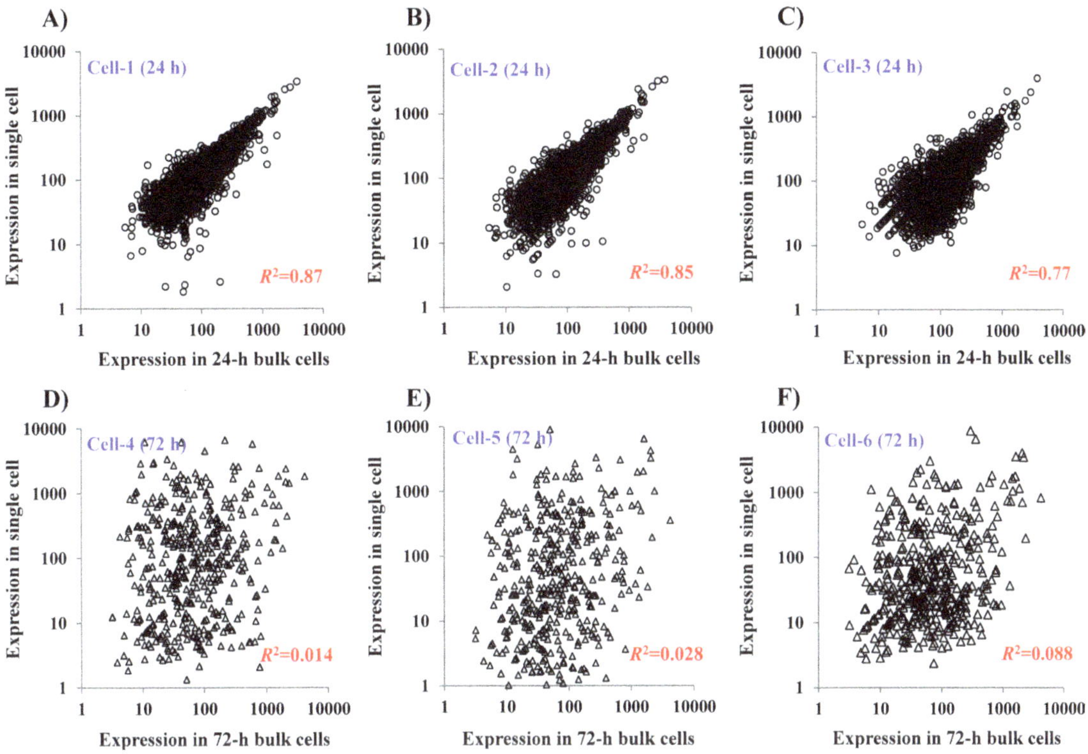

Figure 4.3 Comparison of single-cell and bulk cell transcriptomic profiles. X-axis represents gene expression level from the bulk cells at 24 h (A, B and C), and at 72 h (D, E and F), while Y-axis represents expression level in individual single cell at 24 h (A, B and C) and at 72 h (D, E and F). For 24 h, 2990, 2985 and 2002 genes detected were used for plotting of single cell-1, cell-2 and cell-3, respectively. For 72 h, 441 transcripts detected were used for plotting of all three cells (cell 4, cell 5 and cell 6). Correlation efficient was indicated inside the plots.

satisfactory to calibrate the possible bias during single cell analysis, we further applied a single-cell RT-qPCR analysis to 24 single *Synechocystis* cells randomly selected from each cultivation condition. To verify the heterogeneity revealed by the BaSiC RNA-seq transcriptomic analysis of single *Synechocystis* cells, one gene for a transposase (*slr1684*), which belongs to a functional category of 'Mobile elements', one gene for asparaginyl-tRNA synthetase (*sll0495*), which belongs to a functional category of 'Protein synthesis', as well as the internal control (16S rRNA gene), were selected for single-cell gene expression measurements by RT-qPCR method established previously (Gao *et al.*, 2011). The optimized primer sets (Gao *et al.*, 2011) and their corresponding gene targets were as follows: *slr1684* (forward primer, 5′-GTAAGGTCATGGATAGAATATGGATTTAAG-3′; reverse primer, 5′-TCCCACCATTTTTGAATTTGCT-3′; with an amplicon of 101 bp); *sll0495* (forward primer, 5′-AGGGATGGTTACGCACTAAACG-3′; reverse primer, 5′-ACTTCCATCCAGCACCACTTG-3′; with an amplicon of 101 bp); 16S rRNA (forward primer, 5′-TGTAGCGGTGAAATGCGTAG-3′; reverse primer, 5′-CCACGCCTAGTATCCATCGT-3′; with an amplicon of 152 bp). A total of 24 individual cells were collected from each time point for the analysis. Primers for single-cell RT-qPCR were designed using Primer-BLAST (www.ncbi.nlm.nih.gov/tools/primer-blast/) and evaluated as described previously, briefly, 10 pairs of qPCR primers were designed and evaluated for the efficiency at single-cell level for each of the three genes (Gao *et al.*, 2011; Grun *et al.*, 2014). Briefly, total RNA isolated from a single cell was reverse transcribed into cDNA using the VILO cDNA Synthesis Kit (Invitrogen, Carlsbad, CA, USA). The resulted cDNA in a final volume of 10 µL was diluted 2.5 folds and used as template for SYBR green based RT-qPCR analysis on the StepOne thermal cycler (Life Technology, Carlsbad, CA, USA). At least three

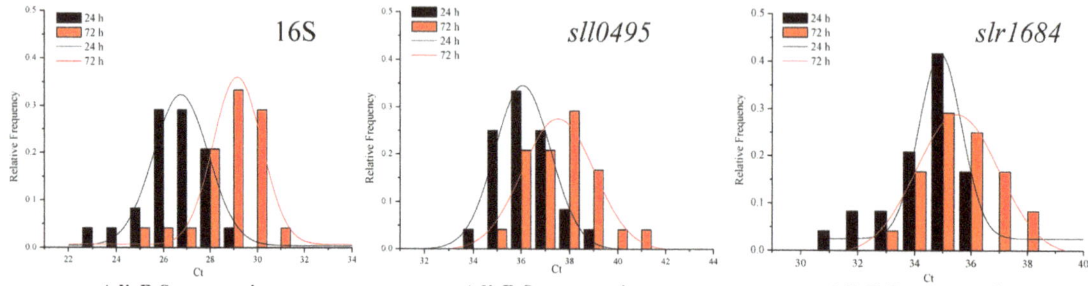

	16S Adj. R-Square	16S sigma	sll0495 Adj. R-Square	sll0495 sigma	slr1684 Adj. R-Square	slr1684 sigma
24 h	0.95977	1.17723	0.98108	1.15355	0.92063	0.76276
72 h	0.95493	1.02529	0.92073	1.48065	0.97386	1.45617

Figure 4.4 RT-qPCR analysis of the selected genes. Population distribution plots (horizontal axis represents Ct value, vertical axis represents percentage of total cell population) and fitting curves reveal that long term nitrogen starvation (72 h) display considerable heterogeneity in transcript expression levels in comparison with that of short term treatment (24 h). A–C represents 16s rRNA, *slr1684* (transposase), and *sll0495* (asparaginyl-tRNA synthetase), respectively.

technical replicates were included for each qPCR analysis. Statistical analyses for single cell RT-qPCR data, such as distribution significance tests, were conducted using the OriginPro software package (V8, OriginLab, Northampton, MA, USA).

In the single-cell transcriptomic data, *slr1684* and *sll0495* were found with increasing heterogeneity after nitrogen starvation treatment. As the RT-qPCR involves no cDNA pre-amplification, we expected it provided more reliable measurements to validate results from the single-cell transcriptomics. Since there is no suitable internal reference gene available for single cell RT-qPCR (Bengtsson *et al.*, 2008; Qi *et al.*, 2014, 2016), the raw Ct values of all single cells were presented in this study. The RT-qPCR results confirmed that the gene expression levels of 16S rRNA, *slr1684* and *sll0495* were slightly down-regulated at 72 h when compared with 24 h after nitrogen starvation stress, consistent with the increased degradation of cyanobacteria total RNA after longer nitrogen starvation observed in this study. More importantly, remarkable differences in terms of single cell gene expression distribution patterns of these selected genes between different time-points after a nitrogen starvation treatment were also observed: higher *sigma* coefficient factors (σ) of curve fitting were detected for both *slr1684* and *sll0495* at 72 h than those at 24 h, 1.46 *vs.* 0.76 and 1.48 *vs.* 1.15, respectively, with $R^2 > 0.92$. In general, lower *sigma* values indicate a narrower distribution. Therefore, the population at 72 h displayed considerable heterogeneity in transcript expression levels compared to that of 24 h for *slr1684* and *sll0495* genes, while no such heterogeneity increase was observed for 16s rRNA gene (Fig. 4.4). The gene expression heterogeneity of selected genes increased at 72 h after the prolonged nitrogen starvation treatment suggesting that increased gene expression heterogeneity may be one of the responses of a genetically identical bacterial population towards challenging environmental conditions.

Technical and biological significance

In this study, using NuGen RNA amplification system as the core, we developed the first RNA-seq based whole transcriptomics method for single bacterial cells, BaSiC RNA-seq, and validated it on *Synechocystis* cells containing only ~5 fg total RNA per cell. The single *Synechocystis* cells under nitrogen starvation condition for 24 and 72 h were randomly isolated and subjected to global transcriptomic profiling using this method. The results demonstrated that the BaSiC RNA-seq method has achieved up to 82–98% coverage of the *Synechocystis* genome which is at the same or higher level of coverage as several other single-cell transcriptomic methods developed for animal cells (61.4–96%) (Tang *et al.*, 2009), fungi (4–7%) (de Bekker *et al.*, 2011) and bacteria (94–96%) (Kang *et al.*, 2011).

As the first RNA-seq transcriptomic method for bacterial single cells, the BaSiC RNA-seq method includes several other advantages:

1. High accessibility: it utilizes a combination of commercially available kits with minor modifications for all procedures from RNA isolation, amplification to sequencing library preparation, no special reagents or instruments was needed so that the method can be accessed by most of the molecular biology laboratories. In addition, its utilization of commercial kits also provided good extra quality controls across the laboratories so that the transcriptomic data can be more comparable.
2. User-friendly and less time-consuming: core parts of the method, RNA isolation and amplification processing took only 10 min and ~5.5 h, respectively, which allows preparation of several dozens of single-cell samples for sequencing within 1–2 days.
3. Sensitivity and reliability: it can be used to sequence transcriptome from a total RNA less than a few *fg*, which makes it suitable for most of prokaryotic cells. For the future application of environmental samples, we expect single-cell isolation using either micromanipulation or other high through-put technologies will allow capture of clear single cells out of environmental samples. In addition, necessary modification and improvement may still needed if we want to extend the application to environmental samples.
4. Generality: the key component used for RNA amplification in the BaSiC RNA-seq method, the One-Direct RNA Amplification System, was also applied in single-cell transcriptome of fungi (de Bekker *et al.*, 2011) and prostate cancer cells (Welty *et al.*, 2013) recently, since both oligo dT and random primers are used for the first stranded cDNA synthesis during RNA amplification, it is thus expected that with little modification, the BaSiC RNA-seq protocol can also be applied to transcriptomic profiling of eukaryotic single cells.

Bacterial cells have developed significant abilities to survive and propagate in highly diverse and changing environments by evolving phenotypic heterogeneity (de Jong *et al.*, 2010). Single-cell analysis of gene expression patterns during nutrient starvation in *Bacillus subtilis* reveals large phenotypic variation (de Jong *et al.*, 2012), which is considered as a risk spreading strategy and can ensure survival of at least a subpopulation of cells in unpredictably fluctuating environments. In the study, preliminary single-cell transcriptomics based on six *Synechocystis* single cells showed a possibility of increased gene expression heterogeneity after long-term nitrogen starvation stress. As phenotypic heterogeneity could be resulted from the gene expression heterogeneity, the increased gene expression heterogeneity could be a benefit to the survival of bacterial population.

In addition, the single-cell transcriptomics and single-cell RT-qPCR analyses showed that not all functional categories contributed equally to the overall increase of gene expression heterogeneity, genes belonged to the functional categories such as 'Mobile elements' seemed to play more roles. Mobile elements are DNA sequences such as plasmids, transposons, insertion sequences, and miniature inverted repeat transposable elements (Kazazian, 2004), among which transposons have been suggested as an important mean of adaptive variation and evolution at the genomic level in many microbes under stress conditions (Feng *et al.*, 2013; Maumus *et al.*, 2009; Zhang and Saier, 2011). Comparative genomics analysis showed an extremely high and hierarchical diversity of mobile elements present in cyanobacteria (Lin *et al.*, 2011; Mikheeva *et al.*, 2013), and transposase genes are involved in genomic rearrangements associated with differentiation process, formation of filaments and the regulation of cell–cell interactions in the colonial way of life in cyanobacteria (Mikheeva *et al.*, 2013). Although more proof is still needed, the preliminary results based on a limited number of single cells implied a possibility that the increased heterogeneity of gene expression of selected functional categories such as 'Mobile elements' could be one of the important means to increase phenotypic variations, which will eventually enhance the survival possibilities of a bacterial population under nutrient starvation and other adverse stress conditions.

Conclusions

Overall, we have developed an integrated workflow for transcriptome analysis from single microbes, and successfully applied it to single *Synechocystis* cells. Moreover, the results reveal significant and even increasing heterogeneity among isogenic populations under stress. Further analysis of gene functional categories indicated that 'Mobile elements' and 'Protein synthesis' category contributed the most for the increased heterogeneity, which might be a strategy for the population to survive under environmental stress. The preliminary results showed that the gene expression levels and the distribution patterns of the selected mobile elements and protein synthesis related genes were highly correlated between the single-cell transcriptomic data and single-cell RT-qPCR data. Together with various noise validation models such as spike-in control RNAs and more various high through-put chip-level RT-qPCR instruments become available in the future, the single-cell bacterial transcriptomics method described here can be further validated.

Considerations and future trends

Ever since the beginning of PCR technology, scientists have never stopped making attempt to analyse cells at single-cell level. In recent years, single-cell sequencing has been quickly developed, accompanying with the keeping dropping of the price of next-generation sequencing. Due to significant amplification and possible bias involved in this single-cell transcriptomic method, for example, DNA amplification also observed in negative controls, caution should be observed when interpreting the single-cell transcriptomic data quantitatively before the method is further evaluated and improved. In our recent study, we only studied RNA-seq in three cells per condition and perhaps the low sampling was not ideal in order to discriminate between a heterogeneous stochastic behaviour and genuine different expression profiles of subset of cells within the population. This is also the major reason for which we did not come to statistical analyses for details about what genes are differentially expressed between 24 and 72 h treatments for a given pair of cells. What's more, spike-in RNA with known copy numbers should also be added at the single cell lysis step, to validate the amplification bias from cell to cell. However, given the experimental limitations, we showed that heterogeneity can be detected with confidence and so as the first RNA-seq based transcriptomic method for single bacterial cells, the method may be valuable in many research areas, such as microbial ecology and environmental sciences. Though, compared with the emerging single-cell sequencing methods for eukaryotic cells in recent years, analysing single prokaryotic cells, especially for transcriptomes, still can be improved.

Some of the important aspects are briefly discussed below:

1. Cell lysis. Microbes usually contain a cell wall. Microbiologists have to be very careful on choosing an appropriate lysis strategy, which should be capable to lysis a cell without damaging the DNA or RNA inside. Although some methodologies appealed to be efficient (Kang *et al.*, 2011; Pamp *et al.*, 2012), the differences between cell wall components of various microbes make it challenging to find a universal lysis method. In addition, if the lysate is directly used without purification, the efficiency of other downstream reactions, such as reverse transcript or PCR, should be carefully validated to make sure not being affected. Moreover, a mild and downstream friendly lysis method is essential for combing single-cell analysis with high-throughput microfluidic devices.
2. mRNA enrichment. mRNA typically constitutes as little as 1–5% of total RNA in a prokaryotic cell. Although not necessary, mRNA enrichment can increase transcript coverage and therefore increase the resolution of the resulting transcriptome maps. mRNAs from eukaryotic cells are usually polyadenylated, which makes them easier to be enriched by using oligo dT primers. For prokaryotic cells, of which mRNAs lack the poly(A) tail, several approaches have been recently developed for RNA-seq (Filiatrault, 2011; Sorek and Cossart, 2010). Capturing rRNA with magnetic beads is the most commonly used approach in commercial bulk cell level prokaryotic RNA-seq workflows, though the depletion efficiency varies between organisms (Passalacqua *et al.*, 2009; Perkins *et al.*,

2009; Yoder-Himes *et al.*, 2009). Another possible approach is to selectively polyadenylate mRNA but not rRNA, following by selectively synthesis cDNA using oligo dT primers (Amara and Vijaya, 1997; Frias-Lopez *et al.*, 2008; Wendisch *et al.*, 2001). The third approach is to selectively degrade rRNA but not mRNA, since most bacterial and archaeal mRNAs carry a 5′ triphosphate, while rRNAs and tRNAs carry a 5′ monophosphate. It has been reported that with a 5′phosphate-dependent exonuclease, the amounts of rRNAs and one tRNA species were significantly reduced after enrichment in a single-cell microarray method (Kang *et al.*, 2015; Kang *et al.*, 2011). The latter two methods might be more suitable for single-cell RNA-seq, especially when the unculturable microbes from nature environments are analysed, where the commercial probes may not efficiently capture the rRNAs.

3 RNA spike-in and unique molecular identifiers (UMI). In RNA-seq experiments, ERCC (External RNA Controls Consortium) RNA spike-in molecules, which contains 96 different synthetic polyadenylated RNAs, are widely used to evaluate technical performance (Baker *et al.*, 2005). It is necessary in the future study to use RNA spike-in, because the fragments being sequenced are typically amplified several orders of magnitude (from ~*fg* level to ~*ug* level in our method) and any little bias at the beginning could finally lead to huge differences in terms of the gene expression levels. Depending on the strategy for cDNA synthesis, a commercial spike-in kit or other similar synthetic RNAs could be used. To minimize amplification bias and technical noise, unique molecular identifiers (UMI) have been applied in some single-cell sequencing protocols (Kivioja *et al.*, 2012). UMIs are random, usually four to ten nucleotides integrated into the primer for reverse transcription and served as a barcode for each mRNA molecule. After sequencing, the amplification bias could be estimated by counting each label, and the number of UMIs can be directly translated to the number of the molecules being sequenced, with a proper mathematical correction (Grun *et al.*, 2014; Kivioja *et al.*, 2012). Technical noise can also arize during library amplification, but can be eliminated by using the UMIs (Fu *et al.*, 2011; Islam *et al.*, 2014). It has been reported that technical noise can be reduced by 2-fold by using UMIs (Grun *et al.*, 2014).

4 Integrated analysis with multiple 'omics'. Methodologies for analysing genomics (Zong *et al.*, 2012), transcriptomics (Goetz and Trimarchi, 2012; Hashimshony *et al.*, 2012; Picelli *et al.*, 2014) or epigenomics (Cheow *et al.*, 2015; Lorthongpanich *et al.*, 2013; Smallwood *et al.*, 2014) for eukaryotic single cells, have been recently developed (Nawy, 2014), and some can even be accessed through commercial services. Since these three omics are strictly connected, it will be ideal to achieve integrated analysis of all these omics in the same single cells. Integrated methods for studying multiple omics have also been developed in the past year for eukaryotic single cells (Angermueller *et al.*, 2016; Dey *et al.*, 2015; Hou *et al.*, 2016; Macaulay *et al.*, 2015). Recently, a new method for analysing the circular RNAs in single mouse embryos (Fan *et al.*, 2015), which also lack the polyadenylation structure, has been developed, and might be useful for studying transcriptomics in single prokaryotic cells.

5 High-throughput microfluidic devices. The emerging microfluidic devices have made it possible to integrate multiple steps, from single-cell isolation to DNA/RNA isolation, amplification, and detection into a fully integrated system. In microfluidic devices, the reaction volume can be minimized to nano litres (nl), which is order of magnitude less than the traditional vial-based methods. As the reaction volume minimized, the reaction efficiency could be improved due to the increasing concentration of the starting molecules, while will also help reduce external contamination (Duncombe *et al.*, 2015; Konry *et al.*, 2016; Sackmann *et al.*, 2014; Saliba *et al.*, 2014). In addition, microfluidic devices make it possible for high-throughput experiments. Comparing with the micromanipulation method, which usually costs couple of minutes to isolate one cell, microfluidic devices can capture over 10,000 cells per hour (Klein *et al.*, 2015). Commercial high-throughput microfluidic devices, such as Fludigm C1™, which is

compatible with the Clontech SMARTer™ library construction kits, have recently been introduced into the single-cell sequencing analysis (Burns *et al.*, 2015; Trapnell *et al.*, 2014), and could play even more important roles in the future.

Acknowledgements

The research was supported by grants from the National Natural Science Foundation of China (project No. 31170043, No. 31270086 and No. 31370115), the National Basic Research Program of China ('973' program, project No. 2014CB745100, 2011CBA00803 and No. 2012CB721101), and the National High-tech R&D Program ('863' program, project No. 2012AA02A707). The funders had no role in study design, data collection and analysis, decision to publish, or preparation of the manuscript. The authors would also like to thank Lianju Gao, Xingrui Li, Yaxing Wang and Guangsheng Pei in our laboratory for their assistance with sampling and statistical analyses. Permission of Figure reprints from *Integr. Biol.* (2015, 7, 1466–1476) was given by the Royal Society of Chemistry.

References

Amara, R.R., and Vijaya, S. (1997). Specific polyadenylation and purification of total messenger RNA from *Escherichia coli*. Nucleic Acids Res. *25*, 3465–3470. http://dx.doi.org/10.1093/nar/25.17.3465.

Angermueller, C., Clark, S.J., Lee, H.J., Macaulay, I.C., Teng, M.J., Hu, T.X., Krueger, F., Smallwood, S.A., Ponting, C.P., Voet, T., *et al.* (2016). Parallel single-cell sequencing links transcriptional and epigenetic heterogeneity. Nat. Methods *13*, 229–232. http://dx.doi.org/10.1038/nmeth.3728.

Baker, S.C., Bauer, S.R., Beyer, R.P., Brenton, J.D., Bromley, B., Burrill, J., Causton, H., Conley, M.P., Elespuru, R., Fero, M., *et al.* (2005). The External RNA Controls Consortium: a progress report. Nat. Methods *2*, 731–734. http://dx.doi.org/10.1038/nmeth1005-731.

Becskei, A., Kaufmann, B.B., and van Oudenaarden, A. (2005). Contributions of low molecule number and chromosomal positioning to stochastic gene expression. Nat. Genet. *37*, 937–944. http://dx.doi.org/10.1038/ng1616.

Bengtsson, M., Hemberg, M., Rorsman, P., and Ståhlberg, A. (2008). Quantification of mRNA in single cells and modelling of RT-qPCR induced noise. BMC Mol. Biol. *9*, 63. http://dx.doi.org/10.1186/1471-2199-9-63.

Brehm-Stecher, B.F., and Johnson, E.A. (2004). Single-cell microbiology: tools, technologies, and applications. Microbiol. Mol. Biol. Rev. *68*, 538–559. http://dx.doi.org/10.1128/MMBR.68.3.538-559.2004.

Burns, J.C., Kelly, M.C., Hoa, M., Morell, R.J., and Kelley, M.W. (2015). Single-cell RNA-Seq resolves cellular complexity in sensory organs from the neonatal inner ear. Nat. Commun. *6*, 8557. http://dx.doi.org/10.1038/ncomms9557.

Cardenas, E., and Tiedje, J.M. (2008). New tools for discovering and characterizing microbial diversity. Curr. Opin. Biotechnol. *19*, 544–549. http://dx.doi.org/10.1016/j.copbio.2008.10.010.

Cheow, L.F., Quake, S.R., Burkholder, W.F., and Messerschmidt, D.M. (2015). Multiplexed locus-specific analysis of DNA methylation in single cells. Nat. Protoc. *10*, 619–631. http://dx.doi.org/10.1038/nprot.2015.041.

Colman-Lerner, A., Gordon, A., Serra, E., Chin, T., Resnekov, O., Endy, D., Pesce, C.G., and Brent, R. (2005). Regulated cell-to-cell variation in a cell-fate decision system. Nature *437*, 699–706. http://dx.doi.org/10.1038/nature03998.

de Bekker, C., Bruning, O., Jonker, M.J., Breit, T.M., and Wösten, H.A. (2011). Single cell transcriptomics of neighboring hyphae of *Aspergillus niger*. Genome Biol. *12*, R71. http://dx.doi.org/10.1186/gb-2011-12-8-r71.

de Jong, I.G., Veening, J.W., and Kuipers, O.P. (2010). Heterochronic phosphorelay gene expression as a source of heterogeneity in *Bacillus subtilis* spore formation. J. Bacteriol. *192*, 2053–2067. http://dx.doi.org/10.1128/JB.01484-09.

de Jong, I.G., Veening, J.W., and Kuipers, O.P. (2012). Single cell analysis of gene expression patterns during carbon starvation in *Bacillus subtilis* reveals large phenotypic variation. Environ. Microbiol. *14*, 3110–3121. http://dx.doi.org/10.1111/j.1462-2920.2012.02892.x.

Dey, S.S., Kester, L., Spanjaard, B., Bienko, M., and van Oudenaarden, A. (2015). Integrated genome and transcriptome sequencing of the same cell. Nat. Biotechnol. *33*, 285–289. http://dx.doi.org/10.1038/nbt.3129.

Duncombe, T.A., Tentori, A.M., and Herr, A.E. (2015). Microfluidics: reframing biological enquiry. Nat. Rev. Mol. Cell. Biol. *16*, 554–567. http://dx.doi.org/10.1038/nrm4041.

Fan, X., Zhang, X., Wu, X., Guo, H., Hu, Y., Tang, F., and Huang, Y. (2015). Single-cell RNA-seq transcriptome analysis of linear and circular RNAs in mouse preimplantation embryos. Genome Biol. *16*, 148. http://dx.doi.org/10.1186/s13059-015-0706-1.

Feng, G., Leem, Y.E., and Levin, H.L. (2013). Transposon integration enhances expression of stress response genes. Nucleic Acids Res. *41*, 775–789. http://dx.doi.org/10.1093/nar/gks1185.

Fenical, W., and Jensen, P.R. (2006). Developing a new resource for drug discovery: marine actinomycete bacteria. Nat. Chem. Biol. *2*, 666–673. http://dx.doi.org/10.1038/nchembio841.

Filiatrault, M.J. (2011). Progress in prokaryotic transcriptomics. Curr. Opin. Microbiol. *14*, 579–586. http://dx.doi.org/10.1016/j.mib.2011.07.023.

Frias-Lopez, J., Shi, Y., Tyson, G.W., Coleman, M.L., Schuster, S.C., Chisholm, S.W., and Delong, E.F. (2008). Microbial community gene expression in ocean surface waters. Proc. Natl. Acad. Sci. U.S.A. *105*, 3805–3810. http://dx.doi.org/10.1073/pnas.0708897105.

Fu, G.K., Hu, J., Wang, P.H., and Fodor, S.P. (2011). Counting individual DNA molecules by the stochastic attachment of diverse labels. Proc. Natl. Acad. Sci.

U.S.A. *108*, 9026–9031. http://dx.doi.org/10.1073/pnas.1017621108.

Gao, W., Zhang, W., and Meldrum, D.R. (2011). RT-qPCR based quantitative analysis of gene expression in single bacterial cells. J. Microbiol. Methods *85*, 221–227. http://dx.doi.org/10.1016/j.mimet.2011.03.008.

Goetz, J.J., and Trimarchi, J.M. (2012). Transcriptome sequencing of single cells with Smart-Seq. Nat. Biotechnol. *30*, 763–765. http://dx.doi.org/10.1038/nbt.2325.

Grün, D., Kester, L., and van Oudenaarden, A. (2014). Validation of noise models for single-cell transcriptomics. Nat. Methods *11*, 637–640. http://dx.doi.org/10.1038/nmeth.2930.

Hashimshony, T., Wagner, F., Sher, N., and Yanai, I. (2012). CEL-Seq: single-cell RNA-Seq by multiplexed linear amplification. Cell. Rep. *2*, 666–673. http://dx.doi.org/10.1016/j.celrep.2012.08.003.

Head, S.R., Komori, H.K., Hart, G.T., Shimashita, J., Schaffer, L., Salomon, D.R., and Ordoukhanian, P.T. (2011). Method for improved Illumina sequencing library preparation using NuGEN Ovation RNA-Seq System. BioTechniques *50*, 177–180. http://dx.doi.org/10.2144/000113613.

Hou, Y., Guo, H., Cao, C., Li, X., Hu, B., Zhu, P., Wu, X., Wen, L., Tang, F., Huang, Y., *et al.* (2016). Single-cell triple omics sequencing reveals genetic, epigenetic, and transcriptomic heterogeneity in hepatocellular carcinomas. Cell Res. *26*, 304–319. http://dx.doi.org/10.1038/cr.2016.23.

Islam, S., Zeisel, A., Joost, S., La Manno, G., Zajac, P., Kasper, M., Lönnerberg, P., and Linnarsson, S. (2014). Quantitative single-cell RNA-seq with unique molecular identifiers. Nat. Methods *11*, 163–166. http://dx.doi.org/10.1038/nmeth.2772.

Jaitin, D.A., Kenigsberg, E., Keren-Shaul, H., Elefant, N., Paul, F., Zaretsky, I., Mildner, A., Cohen, N., Jung, S., Tanay, A., *et al.* (2014). Massively parallel single cell RNA-seq for marker-free decomposition of tissues into cell types. Science *343*, 776–779. http://dx.doi.org/10.1126/science.1247651.

Kang, Y., McMillan, I., Norris, M.H., and Hoang, T.T. (2015). Single prokaryotic cell isolation and total transcript amplification protocol for transcriptomic analysis. Nat. Protoc. *10*, 974–984. http://dx.doi.org/10.1038/nprot.2015.058.

Kang, Y., Norris, M.H., Zarzycki-Siek, J., Nierman, W.C., Donachie, S.P., and Hoang, T.T. (2011). Transcript amplification from single bacterium for transcriptome analysis. Genome Res. *21*, 925–935. http://dx.doi.org/10.1101/gr.116103.110.

Katayama, S., Töhönen, V., Linnarsson, S., and Kere, J. (2013). SAMstrt: statistical test for differential expression in single-cell transcriptome with spike-in normalization. Bioinformatics *29*, 2943–2945. http://dx.doi.org/10.1093/bioinformatics/btt511.

Kazazian, H.H. (2004). Mobile elements: drivers of genome evolution. Science *303*, 1626–1632. http://dx.doi.org/10.1126/science.1089670.

Kivioja, T., Vähärautio, A., Karlsson, K., Bonke, M., Enge, M., Linnarsson, S., and Taipale, J. (2012). Counting absolute numbers of molecules using unique molecular identifiers. Nat. Methods *9*, 72–74. http://dx.doi.org/10.1038/nmeth.1778.

Klein, A.M., Mazutis, L., Akartuna, I., Tallapragada, N., Veres, A., Li, V., Peshkin, L., Weitz, D.A., and Kirschner, M.W. (2015). Droplet barcoding for single-cell transcriptomics applied to embryonic stem cells. Cell *161*, 1187–1201. http://dx.doi.org/10.1016/j.cell.2015.04.044.

Konry, T., Sarkar, S., Sabhachandani, P., and Cohen, N. (2016). Innovative tools and technology for analysis of single cells and cell-cell interaction. Annu. Rev. Biomed. Eng. *18*, 259–284. http://dx.doi.org/10.1146/annurev-bioeng-090215-112735.

Lee, S., and Kemp, P.F. (1994). Single-cell RNA content of natural marine planktonic bacteria measured by hybridization with multiple 16S rRNA-targeted fluorescent probes. Limnol. Oceanog. *39*, 869–879. http://dx.doi.org/10.4319/lo.1994.39.4.0869.

Lee, S., Malone, C., and Kemp, P.F. (1993). Use of multiple 16S rRNA-targeted fluorescent probes to increase signal strength and measure cellular RNA from natural planktonic bacteria. Mar. Ecol. (Progress Series) *101*, 193–193. http://dx.doi.org/10.3354/meps101193.

Lidstrom, M.E., and Meldrum, D.R. (2003). Life-on-a-chip. Nat. Rev. Microbiol. *1*, 158–164. http://dx.doi.org/10.1038/nrmicro755.

Lin, S., Haas, S., Zemojtel, T., Xiao, P., Vingron, M., and Li, R. (2011). Genome-wide comparison of cyanobacterial transposable elements, potential genetic diversity indicators. Gene *473*, 139–149. http://dx.doi.org/10.1016/j.gene.2010.11.011.

Lloyd, K.G., Schreiber, L., Petersen, D.G., Kjeldsen, K.U., Lever, M.A., Steen, A.D., Stepanauskas, R., Richter, M., Kleindienst, S., Lenk, S., *et al.* (2013). Predominant archaea in marine sediments degrade detrital proteins. Nature *496*, 215–218. http://dx.doi.org/10.1038/nature12033.

Lorthongpanich, C., Cheow, L.F., Balu, S., Quake, S.R., Knowles, B.B., Burkholder, W.F., Solter, D., and Messerschmidt, D.M. (2013). Single-cell DNA-methylation analysis reveals epigenetic chimerism in preimplantation embryos. Science *341*, 1110–1112. http://dx.doi.org/10.1126/science.1240617.

Macaulay, I.C., Haerty, W., Kumar, P., Li, Y.I., Hu, T.X., Teng, M.J., Goolam, M., Saurat, N., Coupland, P., Shirley, L.M., *et al.* (2015). G&T-seq: parallel sequencing of single-cell genomes and transcriptomes. Nat. Methods *12*, 519–522. http://dx.doi.org/10.1038/nmeth.3370.

Maumus, F., Allen, A.E., Mhiri, C., Hu, H., Jabbari, K., Vardi, A., Grandbastien, M.A., and Bowler, C. (2009). Potential impact of stress activated retrotransposons on genome evolution in a marine diatom. BMC Genomics *10*, 624. http://dx.doi.org/10.1186/1471-2164-10-624.

Mikheeva, L.E., Karbysheva, E.A., and Shestakov, S.V. (2013). The role of mobile genetic elements in the evolution of cyanobacteria. Russ. J. Genet. 3, 91–101. http://dx.doi.org/10.1134/S2079059713020032.

Nawy, T. (2014). Single-cell sequencing. Nat. Methods *11*, 18. http://dx.doi.org/10.1038/nmeth.2771.

Newman, J.R., Ghaemmaghami, S., Ihmels, J., Breslow, D.K., Noble, M., DeRisi, J.L., and Weissman, J.S. (2006). Single-cell proteomic analysis of *S. cerevisiae* reveals the architecture of biological noise. Nature *441*, 840–846. http://dx.doi.org/10.1038/nature04785.

Pamp, S.J., Harrington, E.D., Quake, S.R., Relman, D.A., and Blainey, P.C. (2012). Single-cell sequencing provides clues about the host interactions of segmented filamentous bacteria (SFB). Genome Res. *22*, 1107–1119. http://dx.doi.org/10.1101/gr.131482.111.

Parkhomchuk, D., Amstislavskiy, V., Soldatov, A., and Ogryzko, V. (2009). Use of high throughput sequencing to observe genome dynamics at a single cell level. Proc. Natl. Acad. Sci. U.S.A. *106*, 20830–20835. http://dx.doi.org/10.1073/pnas.0906681106.

Passalacqua, K.D., Varadarajan, A., Ondov, B.D., Okou, D.T., Zwick, M.E., and Bergman, N.H. (2009). Structure and complexity of a bacterial transcriptome. J. Bacteriol. *191*, 3203–3211. http://dx.doi.org/10.1128/JB.00122-09.

Perkins, T.T., Kingsley, R.A., Fookes, M.C., Gardner, P.P., James, K.D., Yu, L., Assefa, S.A., He, M., Croucher, N.J., Pickard, D.J., *et al.* (2009). A strand-specific RNA-Seq analysis of the transcriptome of the typhoid bacillus *Salmonella typhi*. PLoS Genet. *5*, e1000569. http://dx.doi.org/10.1371/journal.pgen.1000569.

Picelli, S., Faridani, O.R., Björklund, A.K., Winberg, G., Sagasser, S., and Sandberg, R. (2014). Full-length RNA-seq from single cells using Smart-seq2. Nat. Protoc. *9*, 171–181. http://dx.doi.org/10.1038/nprot.2014.006.

Qi, Z., Chen, L., and Zhang, W. (2016). Comparison of transcriptional heterogeneity of eight genes between batch *Desulfovibrio vulgaris* biofilm and planktonic culture at a single-cell level. Front. Microbiol. *7*, 597. http://dx.doi.org/10.3389/fmicb.2016.00597.

Qi, Z., Pei, G., Chen, L., and Zhang, W. (2014). Single-cell analysis reveals gene expression heterogeneity in syntrophic dual-culture of *Desulfovibrio vulgaris* with *Methanosarcina barkeri*. Sci. Rep. *4*, 7478. http://dx.doi.org/10.1038/srep07478.

Richaud, C., Zabulon, G., Joder, A., and Thomas, J.C. (2001). Nitrogen or sulfur starvation differentially affects phycobilisome degradation and expression of the *nblA* gene in *Synechocystis* strain PCC 6803. J. Bacteriol. *183*, 2989–2994. http://dx.doi.org/10.1128/JB.183.10.2989-2994.2001.

Sackmann, E.K., Fulton, A.L., and Beebe, D.J. (2014). The present and future role of microfluidics in biomedical research. Nature *507*, 181–189. http://dx.doi.org/10.1038/nature13118.

Saliba, A.E., Westermann, A.J., Gorski, S.A., and Vogel, J. (2014). Single-cell RNA-seq: advances and future challenges. Nucleic Acids Res. *42*, 8845–8860. http://dx.doi.org/10.1093/nar/gku555.

Schmid, A., Kortmann, H., Dittrich, P.S., and Blank, L.M. (2010). Chemical and biological single cell analysis. Curr. Opin. Biotechnol. *21*, 12–20. http://dx.doi.org/10.1016/j.copbio.2010.01.007.

Shalek, A.K., Satija, R., Adiconis, X., Gertner, R.S., Gaublomme, J.T., Raychowdhury, R., Schwartz, S., Yosef, N., Malboeuf, C., Lu, D., *et al.* (2013). Single-cell transcriptomics reveals bimodality in expression and splicing in immune cells. Nature *498*, 236–240. http://dx.doi.org/10.1038/nature12172.

Shi, X., Gao, W., Chao, S.H., Zhang, W., and Meldrum, D.R. (2013). Monitoring the single-cell stress response of the diatom *Thalassiosira pseudonana* by quantitative real-time reverse transcription-PCR. Appl. Environ. Microbiol. *79*, 1850–1858. http://dx.doi.org/10.1128/AEM.03399-12.

Smallwood, S.A., Lee, H.J., Angermueller, C., Krueger, F., Saadeh, H., Peat, J., Andrews, S.R., Stegle, O., Reik, W., and Kelsey, G. (2014). Single-cell genome-wide bisulfite sequencing for assessing epigenetic heterogeneity. Nat. Methods *11*, 817–820. http://dx.doi.org/10.1038/nmeth.3035.

Sorek, R., and Cossart, P. (2010). Prokaryotic transcriptomics: a new view on regulation, physiology and pathogenicity. Nat. Rev. Genet. *11*, 9–16. http://dx.doi.org/10.1038/nrg2695.

Stewart, P.S., and Franklin, M.J. (2008). Physiological heterogeneity in biofilms. Nat. Rev. Microbiol. *6*, 199–210. http://dx.doi.org/10.1038/nrmicro1838.

Strovas, T.J., Sauter, L.M., Guo, X., and Lidstrom, M.E. (2007). Cell-to-cell heterogeneity in growth rate and gene expression in *Methylobacterium extorquens* AM1. J. Bacteriol. *189*, 7127–7133. http://dx.doi.org/10.1128/JB.00746-07.

Tang, F., Barbacioru, C., Wang, Y., Nordman, E., Lee, C., Xu, N., Wang, X., Bodeau, J., Tuch, B.B., Siddiqui, A., *et al.* (2009). mRNA-Seq whole-transcriptome analysis of a single cell. Nat. Methods *6*, 377–382. http://dx.doi.org/10.1038/nmeth.1315.

Trapnell, C., Cacchiarelli, D., Grimsby, J., Pokharel, P., Li, S., Morse, M., Lennon, N.J., Livak, K.J., Mikkelsen, T.S., and Rinn, J.L. (2014). The dynamics and regulators of cell fate decisions are revealed by pseudotemporal ordering of single cells. Nat. Biotechnol. *32*, 381–386. http://dx.doi.org/10.1038/nbt.2859.

Wang, D., and Bodovitz, S. (2010). Single cell analysis: the new frontier in 'omics'. Trends Biotechnol. *28*, 281–290. http://dx.doi.org/10.1016/j.tibtech.2010.03.002.

Wang, J., Chen, L., Chen, Z., and Zhang, W. (2015). RNA-seq based transcriptomic analysis of single bacterial cells. Integr. Biol. *7*, 1466–1476. http://dx.doi.org/10.1039/c5ib00191a.

Wang, J., Shi, X., Johnson, R.H., Kelbauskas, L., Zhang, W., and Meldrum, D.R. (2013). Single-cell analysis reveals early manifestation of cancerous phenotype in pre-malignant esophageal cells. PLoS One *8*, e75365. http://dx.doi.org/10.1371/journal.pone.0075365.

Welty, C.J., Coleman, I., Coleman, R., Lakely, B., Xia, J., Chen, S., Gulati, R., Larson, S.R., Lange, P.H., Montgomery, B., *et al.* (2013). Single cell transcriptomic analysis of prostate cancer cells. BMC Mol. Biol. *14*, 6. http://dx.doi.org/10.1186/1471-2199-14-6.

Wendisch, V.F., Zimmer, D.P., Khodursky, A., Peter, B., Cozzarelli, N., and Kustu, S. (2001). Isolation of *Escherichia coli* mRNA and comparison of expression using mRNA and total RNA on DNA microarrays. Anal. Biochem. *290*, 205–213. http://dx.doi.org/10.1006/abio.2000.4982.

Westermann, A.J., Gorski, S.A., and Vogel, J. (2012). Dual RNA-seq of pathogen and host. Nat. Rev. Microbiol. *10*, 618–630. http://dx.doi.org/10.1038/nrmicro2852.

Yoder-Himes, D.R., Chain, P.S., Zhu, Y., Wurtzel, O., Rubin, E.M., Tiedje, J.M., and Sorek, R. (2009). Mapping the *Burkholderia cenocepacia* niche response via high-throughput sequencing. Proc. Natl. Acad. Sci.

U.S.A. *106*, 3976–3981. http://dx.doi.org/10.1073/pnas.0813403106.

Zeng, J., Wang, J., Gao, W., Mohammadreza, A., Kelbauskas, L., Zhang, W., Johnson, R.H., and Meldrum, D.R. (2011). Quantitative single-cell gene expression measurements of multiple genes in response to hypoxia treatment. Anal. Bioanal. Chem. *401*, 3–13. http://dx.doi.org/10.1007/s00216-011-5084-2.

Zhang, K., Martiny, A.C., Reppas, N.B., Barry, K.W., Malek, J., Chisholm, S.W., and Church, G.M. (2006). Sequencing genomes from single cells by polymerase cloning. Nat. Biotechnol. *24*, 680–686. http://dx.doi.org/10.1038/nbt1214.

Zhang, Z., and Saier, M.H. (2011). Transposon-mediated adaptive and directed mutations and their potential evolutionary benefits. J. Mol. Microbiol. Biotechnol. *21*, 59–70. http://dx.doi.org/10.1159/000333108.

Zong, C., Lu, S., Chapman, A.R., and Xie, X.S. (2012). Genome-wide detection of single-nucleotide and copy-number variations of a single human cell. Science *338*, 1622–1626. http://dx.doi.org/10.1126/science.1229164.

Transcriptomics of Cyanobacterial Stress Responses: Genes, Sensors and Molecular Triggers

5

Maria A. Sinetova[1], Anna A. Zorina[1], Kirill S. Mironov[1] and Dmitry A. Los[1*]

[1]Institute of Plant Physiology, Russian Academy of Sciences, Moscow, Russian Federation.

*Correspondence: losda@ippras.ru

Abstract

Systemic analysis of transcriptomes of the cyanobacterium *Synechocystis* sp. PCC 6803 revealed that all stress-induced genes can be separated into two groups: one is clustered around heat-shock and another – around cold-shock inducible genes. At the initial stages of stress (in a range of 15–20 min) genes induced by heat and cold stress never overlap. Genes for so-called heat shock proteins (HSPs) are induced by various stressors, e.g. heat, salt, hyperosmotic environment, reactive oxygen species (ROS), changes in light intensity and quality, or in the redox potential of the photosynthetic electron transport chain components. The number of specifically heat-induced genes is rather limited and their functions are mostly unknown. Cold-stress-induced genes also overlap, but with different sets of genes induced by all above mentioned stressors with the exception of heat stress. The analysis suggests that ROS (in particular, H_2O_2) and redox changes of the components of the photosynthetic electron chain may function as universal triggers for stress responses in cyanobacteria.

Introduction

Cyanobacteria belong to a unique group of ancient prokaryotic organisms that perform oxygenic photosynthesis (Zimorski *et al.*, 2014). Studies of fossil microorganisms in Precambrian rocks (3.5–0.5 billion years ago) indicated the temporal morphological changes in fossil cyanobacterial communities caused by the irreversible changes of physicochemical conditions on Earth (Rozanov and Astafieva, 2009). Cyanobacteria adapted to all global changes through several billion years, including reversal changes in temperature associated with cold glacial epochs separated by warmer interglacial periods. Studies on modern extremophilic cyanobacteria demonstrated that they are very conservative and have changed insignificantly both morphologically and physiologically during the past, at least, 2 billion years (Zavarzin, 2008). Therefore, studies on modern and relict cyanobacteria are important for understanding the principles of temperature acclimation and stress resistance.

Cyanobacterial species are widely distributed in nature and they inhabit almost all the environments – from Antarctica, where temperatures exceed −20°C (Chrismas *et al.*, 2015), to hot springs, where temperature reaches +70°C (Klatt *et al.*, 2013). Cyanobacteria of Arctic and Antarctic ice, where temperature is always below 0°C, are metabolically active and perform oxygenic photosynthesis (Varin *et al.*, 2012). Some thermophilic species perform active photosynthesis at + 50–60°C and experience cold stress at +35–45°C (Kiseleva *et al.*, 2000).

The complete nucleotide sequences of the genomes of hundreds cyanobacterial species have been determined. These genomic data provide a molecular background for studies of the stress sensing and transduction mechanisms. Random or targeted mutagenesis with transposons or with antibiotic-resistant cartridges can be applied to many cyanobacterial species. Random or targeted knock-out mutants can be generated, and the function of the corresponding genes can be determined. DNA-microarray (Murata and Suzuki, 2006) and RNA-Seq (Kopf *et al.*, 2014) based transcriptomic analysis of gene expression has been developed. These powerful techniques allow a whole-genome-scale examination of the expression of each individual gene under stress conditions and the effect of mutations on gene expression. The results generated by modern experimental approaches are now handled with the tools of bioinformatics that allow systemic analysis of massive sets of expression data, identification of the putative regulatory circuits, and setup for metabolic re-modelling (Berla *et al.*, 2013; Sinetova and Los, 2016a,b).

Previous studies demonstrated that cyanobacterial cells universally respond to various stressors, e.g. heat shock, oxidative, salt, or hyperosmotic stress, strong light or UV irradiation (Los *et al.*, 2010). These universal integrated responses include a transient decrease in cytoplasmic volume (Bachin *et al.*, 2015; Sinetova *et al.*, 2015), photosynthetic activity (Los *et al.*, 2013), and enhancement of respiration activity (Červený *et al.*, 2015).

Functional genomics, which employs directed mutagenesis together with the analysis of transcriptomes, revealed that cyanobacteria induce similar sets of genes in response to different stressors (Los *et al.*, 2010). In addition, multi-sensor systems have been identified capable of perceiving various abiotic stresses, regardless of their nature. For example, inactivation of sensor histidine kinase Hik33 in the cyanobacterium *Synechocystis* sp. PCC 6803 (hereinafter – *Synechocystis*) interferes with the cell responses to cold, salt, hyperosmotic, and oxidative stress (Murata and Los, 2006). Another multi-sensor histidine kinase Hik31 is involved in the regulation of autotrophic growth (Kahlon *et al.*, 2006), light responses (Nagarajan *et al.*, 2012; 2014), resistance to divalent cations, Cd^{2+}, Cu^{2+} and Zn^{2+} (Giner-Lamia *et al.*, 2012; 2014). Histidine kinase Hik34 acts as a negative regulator of transcription under heat shock (Suzuki *et al.*, 2005), as a positive regulator under salt and hyperosmotic stress, and as an autoregulator – under oxidative stress (Los *et al.*, 2010).

All together these observations suggest that various stressors may generate some universal signal(s), which trigger(s) adaptive cellular responses. Here we will try to analyse systematically stress transcriptomes of *Synechocystis* in order to detect the candidates for the universal signal(s) that trigger(s) stress responses in cyanobacteria.

Heat stress and other abiotic stresses

Systematic analysis of transcriptomes of the cyanobacterium *Synechocystis* revealed that all stress-induced genes can be separated into two groups: one is clustered around heat-shock and another – around cold-shock inducible genes. At the initial stages of stress (in a range of 15–20 min) genes induced by heat and cold stress never overlap (Fig. 5.1). Genes for so-called heat shock proteins (HSPs) are induced by various stressors, e.g. heat, salt, hyperosmotic environment, reactive oxygen species (ROS), changes in light intensity and quality, or in the redox potential of the electron transport chain components. The number of specifically heat-induced genes is rather limited and their functions are mostly unknown. Cold-stress-induced genes also overlap, but with different sets of genes induced by all above mentioned stressors (see below).

Heat shock

The optimum growth temperature for *Synechocystis* cells is in the range 30–35°C, and heat stress occurs under approximately 10°C upwards shift in temperature, i.e. at 40–45°C. *Synechocystis* cells are able to survive after 24 h incubation at 44°C, which indicates their relative heat resistance (Červený *et al.*, 2015).

Transcriptome analysis with DNA microarrays revealed 82 genes that are induced during the first 15–20 min by heat shock with the induction factor (IF) of 2 and higher (Inaba *et al.*, 2003; Suzuki *et al.*, 2005; Prakash *et al.*, 2009) (Table 5.1). However, these genes are induced transiently, and the induction of their transcription is not visible already in 30–60 min after heat shock (Suzuki *et al.*, 2005).

Most strongly induced genes belong to a group of so-called heat-shock protein (HSP) genes, namely genes for chaperones *hspA*, *groES*, *groEL1*, *groEL2*, *dnaJ*, *htpG*, *dnaK*, *clpB1*, protease *HtrA*, sigma factor of RNA polymerase *sigB*, sensory histidine kinase *hik34*, superoxide dismutase *sodB* and some others.

Sigma factor (SigB), whose transcription is

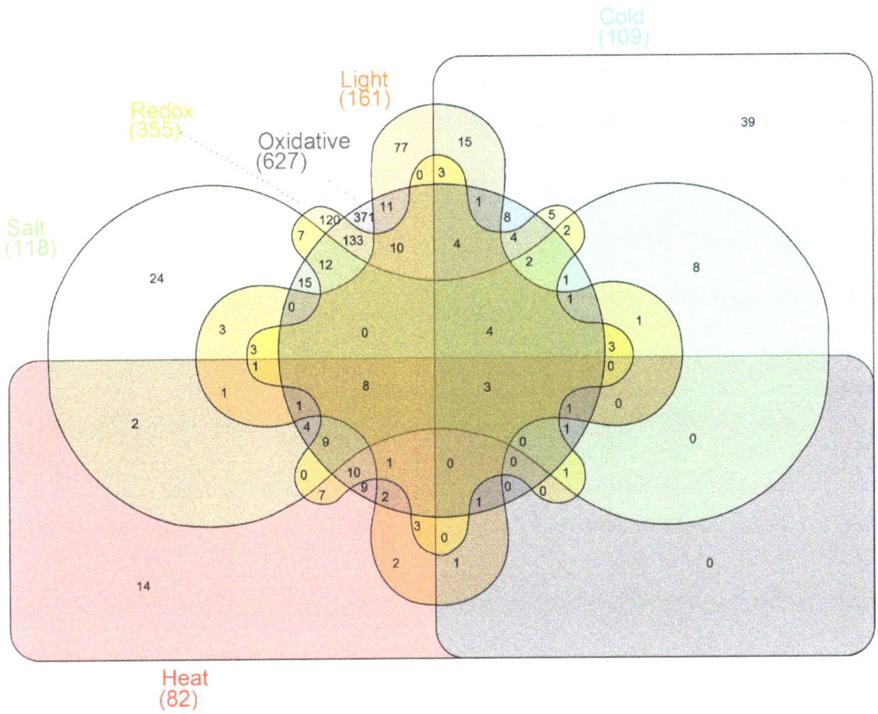

Figure 5.1 Venn diagram of stress-dependent transcription in *Synechocystis*. Stressors of different nature induce transcription of similar or different set of genes. Numbers in brackets indicate total number of genes induced by the individual stressor. Numbers with no brackets indicate number of genes induced by the individual stressor or several stressors. The diagram is the result of the analysis of DNA microarray expression data (http://www.genome.jp/kegg/expression).

controlled by histidine kinase Hik34, is required for normal synthesis of *hspA* mRNA, and for survival under short-term heat (Imamura *et al.*, 2003; Tuominen *et al.*, 2006) and salt (Pollari *et al.*, 2008) stress. Besides the heat shock, the expression of this gene is strongly induced by hyperosmotic and salt stress (Paithoonrangsarid *et al.*, 2004; Shoumskaya *et al.*, 2005), high-light intensity (Hihara *et al.*, 2001), ultraviolet (UV-B) (Huang *et al.*, 2003), oxidative stress (Hihara *et al.*, 2003; Singh *et al.*, 2004), and at low pH (Ohta *et al.*, 2005; Uchiyama *et al.*, 2012). More than a half of genes induced by heat and other stresses belong to a category of genes with unknown function. Among those 82 heat-shock induced genes, only seven genes respond to heat specifically, i.e. they are not induced by other stressors (lower part of Table 5.1).

In fact, HSP genes, are not strictly the genes for heat-shock-induced proteins. All these genes should be rather assigned to a category of general stress-induced genes, since they respond non-specifically to the stressors of different nature: high temperature, salt, hyperosmotic environment, strong and UV light, low pH values. Therefore HSP genes should rather be attributed to a family of core transcriptional response genes, or CTR genes (Singh *et al.*, 2010). Since stress-induced up-regulation of these genes was also confirmed at the protein level (Quiao *et al.*, 2013; Gao *et al.*, 2015), these proteins may be called as core stress proteins (CSPs) or general stress proteins (GSPs). To avoid the confusion with CSPs as accepted definition for Cold Shock Proteins, we prefer to use the abbreviation GSPs.

Strong light

Strong light caused a rapid (within 15 min) induction of more than 100 genes with induction factor of 2 or more (Table 5.1) (Hihara *et al.*, 2001). These are genes for subunits of $NADPH^+$ dehydrogenase subunits (*ndh*), genes encoding the components of the CO_2 concentrating mechanism (*ccm*) and RuBisCo. Probably, the expression of *ndh* and *ccm* genes is induced by certain signals that operate in

Table 5.1 Stress-induced genes of *Synechocystis* sp. PCC 6803: heat stress and other stresses

				Abiotic stressors						Redox		ROS
ORF	Gene	Protein	Function	Heat	Salt	Osmo	Light	UV-B	pH	DBMIB	DCMU	H_2O_2
slr1285	*hik34*	Hik34	Sensor histidine kinase Hik34	+	+	+		+	+	+		+
sll0306	*sigB*	SigB	Sigma factor B of RNA polymerase	+	+	+		+	+	+		+
slr2075	*groES*	GroES	HSP 10 kDa co-chaperone	+	+	+	+	+		+		
slr2076	*groEL1*	GroEL1	HSP 60 kDa chaperone 1	+	+	+	+	+		+		
sll0416	*groEL2*	GroEL2	HSP 60 kDa chaperone 2	+	+	+	+					
sll1514	*hspA*	HspA	HSP 17 kDa	+	+	+	+	+	+	+		+
sll0170	*dnaK2*	DnaK2	HSP 70 kDa	+	+	+	+	+	+	+		+
slr0093	*dnaJ*	DnaJ	HSP 40 kDa	+	+	+		+	+	+		+
sll0430	*htpG*	HtpG	HSP 90 kDa	+	+	+	+	+		+		
slr1641	*clpB1*	ClpB1	HSP 100 kDa chaperone	+	+	+	+	+		+		+
slr1204	*htrA*	HtrA	Serine protease	+	+	+		+	+	+		+
slr0008	*ctpA*	CtpA	C-terminal processing protease	+			+	+	+			+
sll1621	*sll1621*	PrxA	Peroxiredoxin	+	+		+	+		+		+
slr0095	*smtA*	SmtA	S-adenosyl methionine methyltransferase	+	+		+			+		+
slr1674	*slr1674*	Slr1674	Thermoprotector of PS II	+	+		+	+	+	+		+
slr1963	*ocp*	Ocp	Orange carotinoid-binding protein	+		+	+	+		+		
slr1512	*sbtA*	SbtA	Na-dependent bicarbonate transporter	+			+					+
slr1516	*sodB*	SodB	Superoxide dismutase	+	+	+	+	+	+			
sll0528	*sll0528*	Sll0528	Site-2-protease	+	+	+	+	+	+	+		+
slr1675	*hypA1*	HypA1	Hydrogenase	+	+	+	+	+	+	+		+
ssl3044	*ssl3044*	Fdx	Ferredoxin	+	+	+	+	+	+	+		+
slr1687	*nblB1*	NblB1	Phycobilisome degradation protein	+	+	+	+	+	+	+	+	+
sll0939	*sll0939*	Sll0939	Low-pH resistance protein	+	+	+		+	+	+	+	+
slr0967	*slr0967*	Slr0967	Low-pH resistance protein	+	+	+		+	+	+	+	+
slr1686	*slr1686*		Unknown function	+	+		+	+				+
slr1603	*slr1603*		Unknown function	+	+	+		+		+		+
sll1853	*sll1853*		Unknown function	+	+					+		
sll0846	*sll0846*		Unknown function	+	+	+	+	+	+	+		+
sll1884	*sll1884*		Unknown function	+	+	+	+					+
*sll1009**	*frpC**	FrpC	Membrane Fe-regulated protein	+								

Table 5.1 Continued

				Abiotic stressors						Redox		ROS
ORF	Gene	Protein	Function	Heat	Salt	Osmo	Light	UV-B	pH	DBMIB	DCMU	H_2O_2
*sll1501**	*cbiA**	CbiA	Cobalamine biosynthesis protein	+								
*sll0441**	*sll0441**		Unknown function	+								
*sll1892**	*sll1892**		Unknown function	+								
*slr0670**	*slr0670**		Unknown function	+								
*sll0982**	*sll0982**		Unknown function	+								
*slr1127**	*slr1127**		Unknown function	+								

*Genes that are induced specifically by heat stress. ORF – open reading frame. Heat – heat stress (44°C for 15–20 min); salt – salt stress (0.5 M NaCl for 15–20 min); Osmo – hyperosmotic stress (0.5 M sorbitol for 15–20 min); Light – change of light intensity (from 50 to 300 μmol quanta/(m^2 s) for 30 min); UV-B – UV-B light for 30 min; pH – low pH ~4.0 (30 min); DBMIB – 10 μM 2,5-dibromo-6-isopropyl-3-methyl-1,4-benzoquinone (30 min); DCMU – 10 μM 3-(3,4-dichlorphenyl)-1,1-dimethylurea (30 min); H_2O_2 – 250 μM hydrogen peroxide (20 min). Adapted from Sinetova and Los (2016a).

similar manner under an increase of light intensity and a decrease in CO_2 concentration in the culture medium.

Strong light also induces the same set of GSP genes: *hspA*, *clpB1*, *dnaK2*, *htpG*, *groES*, *groEL1* and *groEL2*. Similarly to heat shock, strong light transiently activates transcription of the majority of these genes. The high-light-inducible (*hli*) genes are also activated during the first 15–20 min after exposure to strong light. At the same time, transcription of *groES*, *groEL*, and *groEL2*, remains at a high level even after many hours of exposure to stress (Hihara *et al.*, 2001).

One would assume that the strong light will increase the rate of formation of reactive oxygen species (ROS) and, accordingly, the induction of genes for their neutralization. However, the expected high level of induction was observed only for the *sodB* gene for superoxide dismutase, and, to some degree, for the *gpx1* gene that encodes glutathione peroxidase. The level of transcription of a catalase (*katG*) or thioredoxin peroxidase (*tpx*) genes did not change significantly (Hihara *et al.*, 2001, 2003).

Ultraviolet (UV-B)

The effect of UV-B on transcriptome of *Synechocystis* was estimated with custom-made DNA microarrays (Huang *et al.*, 2003); therefore, the quantitative results cannot be compared to those obtained with commercial arrays. However, qualitative data can be compared. UV-B irradiation of cells for 20 min induced 55 genes with the IF of 2 and more (Table 5.1). UV-B enhanced transcription of two genes encoding the precursors of the main Photosystem II (PS II) protein, D1 (*psbA2* and *psbA3*), genes for proteases (*ctpA* and *ftsH*) that are involved in the maturation and degradation of the D1 protein, and *nblB* gene, whose product is involved in the disassembly phycobilisomes, for example, under nitrogen starvation (Hsiao *et al.*, 2004). The list of induced genes includes heat-shock induced regulatory gens *hik34*, *sigB*, and *sigD*, as well as GSP genes: *groES*, *groEL1*, *cpn60*, *hspA*, *clpB*, *htpG*, *dnaK2*, *dnaJ*, *htpG*) and some others.

Salt (NaCl) stress

Salt stress (0.5 M NaCl for 15 min) up-regulates 118 genes in *Synechocystis* cells with the IF of 2 and higher (Shoumskaya *et al.*, 2005). Evidently, the list includes all the representatives of gens for GSP. NaCl specifically induced only a small set of genes that includes several genes for ABC-type transporters (*cbiQ*, *norM*, *ycf85*); genes responsible for the synthesis of quinones (*menB*, *menG*, *ubiH*), biotin metabolism (*birA*), NAD (*nadC*) and riboflavin (*ribF*), acetylation of proteins (*rimI*), protease *HhoB*, one of translation initiation factors (*infC*) and several genes of unknown function.

Hyperosmotic stress

Similarly to NaCl, hyperosmotic stress (0.5 M sorbitol for 15 min) causes rapid and strong induction GSP genes (*hspA*, *htpG*, *dnaK2*, *dnaJ*, *clpB1*, and *cpn60*), as well as genes for a protease (*htrA*), the regulatory genes (*sigB* and *sigD*), and a gene for superoxide dismutase (*sodB*) (Paithoonrangsarid *et al.*, 2004; Shapiguzov *et al.*, 2005). Short-term (15–20 min) salt and hyperosmotic stresses induce essentially the same set of genes, probably, because of a rapid decrease in the cytoplasmic volume due to water loss at the initial stages of both stresses (Paithoonrangsarid *et al.*, 2004; Shumskaya *et al.*, 2005). However, there is a possibility of regulation that has emerged due to stress-induced ROS formation, or due to redox changes in the main pools of photosynthetic electron transport chain (ETC).

Genes specifically induced by sorbitol include those for phosphate (*pstS*), nitrate/nitrite transporters (*ntrABC*), ribosomal proteins (*rplI* and *rplY*), periplasmic protease HhoA, genes involved in lipid metabolism (*sll0208* and *sll1377*), and several genes of unknown function.

Redox regulation

Maintaining an optimal redox status within the cells is important for regulation of gene expression. Changes in redox status in the thylakoid membranes as a result of stress factors directly affect the functioning of the photosynthetic ETC. The crucial roles in the maintenance of the redox homeostasis play thiol molecules: glutathione, thioredoxin and glutaredoxin peroxiredoxins.

Table 5.1 shows that, for the most part, the GSP genes are also regulated via changes in the redox state of plastoquinone (PQ) pool of PS II. Such changes can be simulated by the addition of inhibitors that interrupt with a flow of electrons before (DCMU) or after (DBMIB) the PQ pool. The addition of DBMIB strongly induces the transcription of HSP (GSP) genes (Hihara *et al.*, 2003; Singh *et al.*, 2004). Changes in the redox state PQ pool may be also caused by an increase in light intensity (see above) or by changing the pH of the environment.

Oxidative stress

The induction of GSP genes may be controlled by the reactive oxygen species (ROS) as well. The formation of ROS provoked by different stressors may influence the redox state of a cell. ROS are present in several forms: singlet oxygen ($^1O_2$), superoxide anion ($O_2{\cdot}^-$), hydrogen peroxide (H_2O_2), or hydroxyl radical (OH·) (Schmitt *et al.*, 2014; 2016). Singlet oxygen is formed by the energization of O_2. It is characterized by an extremely short life-time and high degree of reactivity with surrounding proteins, lipids, and pigments. Other three forms of reduced oxygen intermediates are characterized by different properties and, therefore, by different mechanisms of action on cellular processes. Superoxide anion and hydroxyl radical have both an unpaired electron, which determines a high potential of their interaction with surrounding molecules. Superoxide radical is negatively charged and, therefore, can not diffuse through the membrane. It oxidizes clusters $[4Fe–4S]^{2+}$ to $[3Fe–4S]^{1+}$ with the release of ferrous iron (Fe^{2+}). The hydroxyl radical is also characterized by ultra-high reactivity and low life-time. H_2O_2 is less reactive, but may recover to hydroxyl radical via the Fenton reaction and, thus, become destructive for the cells (Latifi *et al.*, 2009). In the experimental system with photosynthetic organisms oxidative stress can be induced by the addition of H_2O_2, methyl viologen (paraquat) (Prosecka *et al.*, 2009) or by the exposure to strong light that oversaturates the photosynthetic apparatus and leads to the formation of ROS (Hakkila *et al.*, 2014).

The addition of 0.25 mM H_2O_2 to a culture medium induces the transcription of 118 genes with the IF of 3 and above (Kanesaki *et al.*, 2007). Among these genes there are GSP genes (*hik34*, *sigB*, *hspA*, *dnaJ*, *clpB1*, *dnaK2*, *ctpA*, *hypA1*, *htpG*) that, as already indicated, are induced by heat, strong light, ultraviolet light, salt, and sorbitol (Table 5.1).

Similarly to heat shock and other stressors, ROS induce genes that are necessary for stress neutralization. These include genes for peroxiredoxin *aphC* (*sll1621*), catalase *katG* (*sll1987*), (Vlasits *et al.*, 2010) and thioredoxin *trx* (sll0755) (Pérez-Pérez *et al.*, 2009). ROS induce transcriptional regulator *perR* (*slr1738*) (Li *et al.*, 2004), which is directly regulated by H_2O_2 via oxidation of His-37 and His-91 residues, which leads to dissociation of PerR•DNA complex (Lee and Helmann, 2006).

The stress-transcriptome analysis suggests that ROS (in particular, H_2O_2) and redox changes of the components of the photosynthetic electron chain may function as universal triggers for stress responses in cyanobacteria.

Cold stress and other abiotic stresses

It is known that a limited number of genes are induced by both high and low temperatures. This concerns, for example, the GSP operon, *groESL1*, which encodes the multiple-stress-induced 60 kDa chaperonin, GroEL1, and 10 kDa co-chaperonin, GroES (Prakash *et al.*, 2010). Such induction, however, was never detected in short-term (15–30 min) cold-shock experiments that allow a dissection of early stages of stress signal perception and transduction. Therefore, based on the stress transcriptomic analysis, we will consider that the lists of heat- and cold-stress-induced genes do not overlap (Fig. 5.1).

Cold-stress

The genes that are induced by cold stress were identified by whole-genome scale transcriptomic analysis (Suzuki *et al.*, 2001; Inaba *et al.*, 2003; Prakash *et al.*, 2009; Liu *et al.*, 2014). Cold stress was applied by exposing the cells of motile (Liu *et al.*, 2014) or non-motile (Suzuki *et al.*, 2001) *Synechocystis* strains grown at 28–34°C to 22°C or 10°C for 20–30 min. Cold stress transcriptomics shows more than 100 cold-inducible genes with the IF 2 or higher. These genes may be grouped into major functional categories, such as signal perception and transduction; transcription and translation; cell wall and membrane maintenance; photosynthesis and respiration; various cellular functions (cofactor biosynthesis, nucleotide metabolism); unknown functions (Table 5.2).

Cold-induced regulatory genes are limited to two genes for the histidine kinases (*hik3* and *hik31*) and two genes for the DNA-binding transcriptional regulators (*rre5* and *sfsA*). Hik3 is equipped with PAS/PAC and GAF sensory domains (Takala *et al.*, 2014). Hik31 (or CopS) is a transmembrane multisensory histidine kinase that regulates autotrophic growth (Kahlon *et al.*, 2006), responses to light (Nagarajan *et al.*, 2012, 2014), and resistance to bivalent cations, Cd^{2+}, Cu^{2+} and Zn^{2+} (Giner-Lamia *et al.*, 2012). Hik31 controls genes that are involved in photosynthesis and genes for ribosomal proteins. Fluctuations in photosynthetic electron transport (addition of DBMIB, nitrogen limitation, etc.) oxidize plastocyanin and promote the release of Cu^{2+}, which binds to Hik31 and trigger the regulatory cascade (Giner-Lamia *et al.*, 2012; 2014). Rre5 belongs to a PatA subfamily of two component response regulators, which is involved in CO_2 uptake and associated with pH homeostasis. SfsA-like protein is involved in regulation of sugar catabolism, but its exact function in cyanobacteria is unknown.

Cold-induced *sigD* (*rpoD*) encodes a Group 2 RNA polymerase σ factor, which is the only σ factor abundant in the dark (Tuominen *et al.*, 2003). Transcription of *sigD* is induced by UV-B and strong light (Koskinen *et al.*, 2016), by the inhibitor of photosynthetic electron transfer chain, DCMU (Tuominen *et al.*, 2003) and by H_2O_2 (Li *et al.*, 2004). These data suggest that SigD may be involved in light-dependent redox regulation of transcription.

The cofactor of Rho transcriptional terminator, NusG operates in combination with histone-like nucleoid-structuring protein H-NS and Rho-dependent transcriptional terminators to diminish genome-wide antisense transcription (Peters *et al.*, 2012). Under normal growth conditions, the contribution of Rho-NusG modulation of sense and antisense transcription should be limited. Under cold stress, however, antisense transcription may cause a serious additional problem together with difficulties in the maintenance of a proper RNA secondary structure(s), and loss of speed, efficiency, and fidelity of transcription and translation. Thus, the activation of NusG expression may help cells to silence the global antisense transcription under cold stress.

Low temperatures also activate genes for ribosomal machinery: 50S ribosomal proteins L20 and L11, 30S proteins S12, S21, and ribosome chaperone trigger factor (Tig). The latter supports early folding events and prevents misfolding and aggregation of proteins. The *smpB* gene encodes the protein that is required for rescue of ribosomes stalled on defective messages. SmpB binds to SsrA RNA to mimic tRNA that mediates the addition of a short peptide tag to the C-terminus of the partially synthesized nascent polypeptide chain (Karzai *et al.*, 1999). The SsrA-tagged proteins are then degraded by C-terminal proteases. Among cold-induced genes there are genes for tRNA- and ribosomal protein methyltransferases, translation elongation factor, and peptide chain release factor 2 that participates in termination of protein synthesis at UAA and UGA stop-codons.

Table 5.2 Stress-induced genes of *Synechocystis* sp. PCC 6803: cold stress and other stresses

				Abiotic stressors						Redox		ROS
OPC	Gene	Protein	Function	Cold	Salt	Osmo	Light	UV	pH	DBMIB	DCMU	H_2O_2
*sll1124**	*hik3**	Hik3	Phytochrome-like sensor histidine kinase	+								
sll0790	*hik31*	Hik31	Sensor histidine kinase	+						+		
slr1594	*rre5*	Rre5	Response regulator	+			+			+		
slr0083	*crhR*	CrhR	RNA helicase	+	+	+	+			+		
sll0517	*rbpA1*	RbpA1	RNA binding protein A1	+								+
*slr1639**	*smpB**	SmtB	SsrA-binding protein	+								
*sll0767**	*rplT**	Rpl20	50S ribosomal protein L20	+								
*sll1865**	*pfbB**	PfbB	Peptide chain release factor 2	+								
sll0533	*tig*	Tig	Ribosome trigger factor	+		+	+					
sll1096	*rpsL*	Rps12	30S ribosomal protein S12	+		+	+					
slr0082	*rimO*	RimO	Methyltransferase of ribosomal protein S12	+	+	+	+			+		
slr1105	*typA*	TypA	GTP-binding protein TypA/BipA	+		+	+			+		+
*slr0321**	*era**		GTP-binding protein, ERA homologue	+								
*slr1803**	*mbpA**		Adenine-specific DNA methylase	+								
*sll0398**	*dgt**		dGTP triphosphohydrolase	+								
*sll1258**	*dcd**		dCTP deaminase	+								
slr1512	*sbtA*	SbtA	Na-dependent bicarbonate transporter	+			+					+
sll1541	*syc2*	Syc2	Carotenoid oxygenase	+	+		+	+		+		+
slr1291	*ndhD2*	NdhD2	NADH dehydrogenase subunit 4	+			+	+		+		+
ssl2542	*hliA*	HliA	High light inducible protein	+	+	+	+	+		+		+
ssr2595	*hliB*	HliB	High light inducible protein	+	+	+	+	+	+	+	+	
slr1544	*lilA*	LilA	Light-harvesting protein LilA	+	+	+	+	+	+	+	+	+
slr1687	*nblB*	NblB	Phycocyanobilin lyase NblB	+	+	+	+	+	+	+	+	+
sll1742	*nusG*	NusG	Transcription antitermination protein	+	+	+	+					+
sll1818	*rpoA*	RpoA	RNA polymerase alpha subunit σ^{70}	+		+	+					
sll2012	*sigD*	SigD	RNA polymerase sigma factor SigD	+	+		+	+	+	+	+	+
*ssl3044**	*ssl3044**	Fdx	Ferredoxin	+	+	+	+	+	+	+		+
slr1350[1]	*desA*[1]	DesA	Δ12 fatty acid desaturase	+			+					
sll1441[1]	*desB*[1]	DesB	ω3 fatty acid desaturase	+			+					
slr1992	*gpx2*	Gpx2	Hydroperoxy fatty acid reductase	+			+	+				

Table 5.2 Continued

				Abiotic stressors						Redox		ROS
OPC	Gene	Protein	Function	Cold	Salt	Osmo	Light	UV	pH	DBMIB	DCMU	H_2O_2
*slr1238**	*gshB**	GshB	Glutathione synthetase	+								
sll1483	*sll1483*		Salt-induced periplasmic protein	+	+	+	+	+	+	+	+	+
*slr0401**	*potD**	PotD	Polyamine-binding ABC transporter	+								
slr1392	*feoB*	FeoB	Ferrous iron transport protein	+	+					+		+
sll0157	*sll0157*		Zn-dependent hydrolase	+	+	+		+		+		
sll0383	*cbiM*	CbiM	Cobalamin biosynthesis protein M	+	+						+	+
*slr0239**	*cbiF**	CbiF	Cobalamin biosynthesis protein F	+								
*slr0323**	*ams1**		α-Mannosidase	+								
*slr0017**	*murA**	MurA	UDP-N-acetylglucosamine 1-carboxyvinyltransferase	+		+						
sll1863	*sll1863*		Unknown protein	+	+	+						+
sll1862	*sll1862*		Unknown protein	+	+	+						+
sll1853	*sll1853*		Unknown protein	+	+					+		
slr0551	*slr0551*		Unknown protein	+		+	+	+		+	+	
slr0959	*slr0959*		Unknown protein	+	+		+	+				
slr1686	*slr1686*		Unknown protein	+	+		+	+				+

[1]Genes that have the induction factor 2 or less in the microarray experiments, and that have been characterized as strongly cold-inducible by Northern or RT-PCR analysis.
*Genes that are induced specifically by cold stress (Sinetova and Los, 2016b).

Liu *et al.* (2014) listed 36 genes for tRNAs and 15 genes for ribosomal proteins that were up-regulated by cold stress. However, in those experiments, cold treatment was provided by placing *Synechocystis* cells grown at 28°C to 10°C or 4°C. It is therefore unlikely, that an increase in the amounts of the specific mRNAs reflects physiological cellular responses at such low temperatures (Inoue *et al.*, 2001). Such increase might rather result from the arrest of all physiological processes, including mRNA degradation, caused by a severe cold shock.

At low temperatures, structure of RNA (e.g. formation of RNA helices, tertiary structures, and complex ribonucleoproteins (Jarmoskaite and Russell, 2014), becomes critical for proper translation. Therefore, it is not surprizing that genes for RNA helicases are induced by cold stress (Khemici and Linder, 2016). The most studied is the *crhR* gene that encodes a DEAD (Asp-Glu-Ala-Asp) box family RNA helicase. The *crhR* gene is rapidly and transiently induced by cold stress. Inactivation of this gene leads to severe changes in the physiology of *Synechocystis* at both normal (30–34°C) and low (20–24°C) growth temperatures (Prakash *et al.*, 2010; Rosana *et al.*, 2012). Unlike wild-type cells, cold stressed CrhR-deficient mutant cells could not perform light-induced state transitions. While wild-type cells maintain reduced PQ pool under cold stress, mutant cells have PQ oxidized (Sireesha *et al.*, 2012).

The ability of cells to acclimate to low temperatures is explained, at least in part, by their capacity to desaturate the fatty acids (FAs) in the membrane lipids and to adjust the fluidity of the membranes to new cold environment. In cyanobacterial cells, the desaturation reactions occur in FAs that are bound to glycerolipids. Under cold stress, the *de novo* synthesis of FAs ceases for several hours. During this period, the existing FAs of lipids are desaturated to a certain level, after which cells begin to grow again. An increase in cold-induced FA desaturation is caused by induction of the expression of three known genes for the fatty acid desaturases, FADs (*desA* for Δ12-desaturase, *desB* for ω3-desaturase, and *desD* for Δ6-desaturase), accompanied by an increase in their mRNA stability (Los *et al.*, 1997).

The stability of three desaturase mRNAs (*desA*, *desB*, and *desD*) at low temperatures is ensured by RbpA3, a member of a family of RNA binding proteins (Tang *et al.*, 2010). The *rbpA3*-deficient mutant demonstrated a decrease in the amounts of those specific desaturase mRNAs and in total absence of polyunsaturated FAs in membrane lipids of *Synechocystis* under cold stress. At the same time, the transcripts of *rbpA1* (encodes another cold-induced RNA-binding protein of Rpb-family) and *crhR* (cold-induced RNA helicase, see below) were not affected by the knock-out of *rbpA3*.

It should be also noted that the *rbpA3*-deficient mutant appeared to be salt-sensitive, possibly, because of a decrease in polyunsaturated FAs in the membranes due to mutation in *rbpA3* (Wang and Xu, 2013). The effect of FA unsaturation on salt tolerance was previously demonstrated in the double mutant of *Synechocystis* deficient in *desA* and *desD* genes for Δ12- and Δ6 desaturases (Allakhverdiev *et al.*, 1999). Now it becomes clear that any stimulus that causes a decrease in FA unsaturation (knock-out of desaturase genes or a decrease in *des* mRNA stability) leads to sensitivity of cells to salt at elevated concentrations. This conclusion was also confirmed by the overexpression of *rbpA3* or *desA* in the *rbpA3*-deficient strain: both restored the amount of polyunsaturated FAs and salt-tolerant phenotype (Wang and Xu, 2013).

Although the specific targets for RbpA1 in cyanobacteria are still unknown, the inactivation of *rbpA1* led to a great reduction in the acquired cold and high-light tolerance of *Synechocystis*, while overexpression facilitated stress- resistance (Tan *et al.*, 2011).

Some cold-induced genes are light-dependent. Two of four *hli* (high light induced) genes, *hliA* and *hliB*, are strongly induced by cold, while *hliC* and *hliD* are not (He *et al.*, 2001). At the same time, the *ycf39* gene that encodes a putative PS II assembly factor, which interacts with HliD during the assembly of light harvesting complexes (Chidgey *et al.*, 2014), was found to be cold-inducible. Another light and cold-inducible gene (*slr1544*) also encodes light-harvesting protein, LilA (Kufryk *et al.*, 2008). The *ndhD2* gene for a subunit of NADH dehydrogenase is a part of PedR regulon consisting of eight genes: *ndhD2, chlB, chlL, chlN* – for protochlorophyllide reductase subunits, *rpe* – for ribulose-phosphate 3-epimerase, *slr1957* and *sll0296* – for two unknown proteins, and *pedR* itself – a redox-sensitive DNA-binding transcription regulator (Nakamura and Hihara, 2006; Horiuchi *et al.*, 2010). It should be noted that the *gshB* gene for glutathione synthase, which is involved in redox regulation and acclimation under changing light (Cameron and Pakrasi, 2010), is also induced by cold.

Low temperatures induce genes for the cell wall biosynthesis (*murA, ams1*), for cofactor biosynthesis (*cbiM, cbiF, moaA*) and for several transporters (*potD, feoB*). The latter, *feoB*, codes for a high affinity ferrous iron (Fe^{2+}) transport protein (Hantke, 2003). It should be noted that fatty acid desaturases, including those induced by low temperatures, require Fe ions for their activity (Los and Murata, 1998; Guy *et al.*, 2011).

Another group of cold-induced genes includes a family that encodes proteins for nucleotide binding and modification. Their functions in thermal acclimation are barely known. About a half of cold-induced genes is not functionally characterized and their role in cold stress responses and cold acclimation is absolutely unknown.

Specific and non-specific cold-induced genes

Environmental stressors induce sets of genes required for cell protection against adverse conditions and acclimation. Among 109 cold-induced genes, only 38 genes are specifically induced by cold (marked by asterisks in Table 5.2). This list is limited to genes of *rbpA*-family that encode RNA-binding proteins (RbpA2 and RbpA3); *smpB, rplT*, and *pfbB* involved in translation; *era* and *mbpA* for GTP-binding protein and DNA methylase; *cbiF* for the first step of cobalamin biosynthesis; *ams1* for α-mannosidase; and several genes of unknown function.

Other cold-induced genes respond to two or more abiotic stresses, including salt, hyperosmotic, oxidative, acidic stresses, strong light and UV, chemical agents that alter the redox status of the PQ pool (Table 5.2). The induction of the same genes by different stressors suggests the existence of some universal trigger(s) of stress responses. In the photosynthetic cells or organisms, the primary candidates for such universal triggers might be electric signalling via redox-responsive systems or chemical signalling with ROS or NO-derived reactive species.

Redox regulation system consists of components of photosynthetic electron transport chain and the regulatory proteins that may perceive the difference of electric potentials, for example, transcription factor PedR (Ssl0564) (Nakamura and Hihara, 2006; Horiuchi *et al.*, 2010) or sensor kinase Hik31 (Kahlon *et al.*, 2006; Giner-Lamia *et al.*, 2012). Stress-induced ROS should have sufficient life-time to perform their regulatory functions (Schmitt *et al.*, 2014). Electron transport inhibitors of photosynthetic electron transport (DCMU, DBMIB or methylviologen) and the generator of ROS response (H_2O_2) evidence in favour of the hypothesis of their involvement in the initiation of stress responses as universal triggers. However, the systematic analysis of the stress-induced transcription in *Synechocystis* indicates that the overall regulatory scheme is somewhat more complicated.

Systematic analysis of transcriptome shows that redox- and ROS-regulatory pathways play important role in stress-induced gene expression (Fig. 5.1). A redox potential of cells may change due to the emergence of ROS (Liu *et al.*, 2005). In such cases, about a quarter of genes induced by H_2O_2 and by a derivative of benzoquinone, DBMIB, overlaps. Some part of genes falls under the control of the ROS, and another part – under redox control (Fig. 5.1). In addition, there is a relatively minor set of stress-induced genes that does not depend on the ROS or redox regulatory systems.

Cold sensing

The ability of organisms to detect changes in temperature and to escape from harmful temperatures is crucial for the survival. In animals and human, several transient receptor potential (TRP) ion channels serve as temperature sensors (Peier *et al.*, 2003; McKemy *et al.*, 2002). Some taste sensors, e.g. menthol (McKemy *et al.*, 2002), wasabi (Story *et al.*, 2003) and the chili pepper (Caterina *et al.*, 1997), have been characterized as temperature-sensitive ion channels. The initial cold-induced constriction of blood vessels is mediated via TRPA1-dependent ROS (superoxide) production (Moparthi *et al.*, 2014) that stimulates α-adrenoceptors and Rho-kinase-mediated myosin light chain phosphorylation cascade. The restoration of blood flow also depends on TRPA1 activation mediated by dilator neuropeptides and nitric oxide derived from the activity of neuronal NO-synthase (Aubdool *et al.*, 2014).

In plants, cold sensors have not been characterized, whereas heat sensors are represented by calcium-permeable cyclic nucleotide-gated ion channels (Saidi *et al.*, 2009; Gao *et al.*, 2012) that mediate heat-induced Ca^{2+} influx, facilitate expression of heat shock protein (HSP) genes and the acquisition of thermotolerance (Finka *et al.*, 2012). Similarly to cold stress events, membrane lipid composition affects plant heat sensing and modulated Ca^{2+}-dependent heat shock response (Saidi *et al.*, 2010).

In bacteria, including cyanobacteria, thermosensor proteins represented by histidine kinases (Los *et al.*, 2010; de Mendoza, 2014) and operate as two-component regulatory systems composed of a sensory kinase and a cognate response regulator. Thus, a vegetable channel-based thermosensory system is much closer to human one than to an ancient kinase-based sensing system of cyanobacteria.

Cytoplasmic membrane as a primary temperature-sensing compartment

The intriguing question is how cells detect the temperature or a change in temperature, which leads to the cold-induced gene expression and, ultimately, to acclimation. First attempts to answer this question have been made in the studies of expression of FAD genes in relation to dynamics of the membrane structure in *Synechocystis*. It is well known that cyanobacteria desaturate FAs of their membrane lipids in response to a decrease in the ambient temperature. Insertion of the additional double bonds into the acyl chains of membrane lipids by FADs compensates for a decrease in membrane fluidity caused by a drop in temperature (Los and Murata, 2004; Los *et al.*, 2013). In cells grown at 36°C, the induction of *desA* transcription started at 28°C, whereas in cells grown at 32°C the induction of *desA* started at 26°C (Fig. 5.2A). This observation indicated

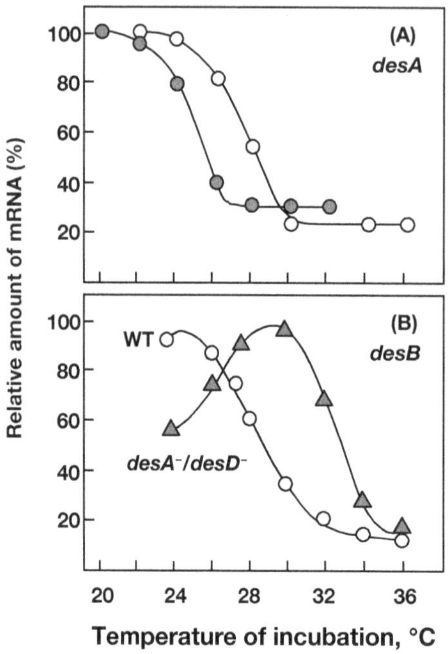

Figure 5.2 Cold-induced transcription of *desA* (A) and *desB* (B) genes in *Synechocystis*. Cells of wild type (open circles) or *desA*⁻/*desD*⁻ double mutant (grey triangles) were grown at 36°C and then incubated at designate temperatures for 30 min (B) (Mironov *et al*., 2012a). Alternatively, wild type cells (grey circles) were grown at 32°C, and then treated at designated temperatures for 30 min (A) (Los *et al*., 1993). The efficiency of transcription was estimated by Northern blotting with the gene-specific probes.

that 1) cells somehow sense the change in temperature, but not the absolute temperature; and 2) cells sense the change in temperature only when it exceeds a certain threshold (5–6°C). Such small change in ambient temperature represents only a 2% reduction in molecular motion of membrane lipids (Los *et al.*, 1993).

Evidence supporting the primary role of the membranes in perception of the temperature signals came from a series of experiments that simulated a cold-induced increase in membrane viscosity by catalytic hydrogenation of FAs in the cytoplasmic membrane of *Synechocystis* cells grown at optimal growth temperature (Fig. 5.3). Partial (5%) catalytic saturation of FAs in the cytoplasmic membrane lipids caused its rigidification *in vivo* and activated transcription of *desA* under isothermal conditions (Fig. 5.3B). Thus, a decrease in membrane fluidity (in other words – an increase in viscosity or rigidification) due to cold stress or catalytic hydrogenation resulted in a rapid activation of the expression of the *desA* gene for the FAD that functions as an enzymatic membrane fluidizer. These findings strengthened the hypothesis about a membrane as a primary site of temperature perception and suggested that the primary signal in temperature sensing might be a change in the fluidity/viscosity of a cytoplasmic membrane (Vigh *et*

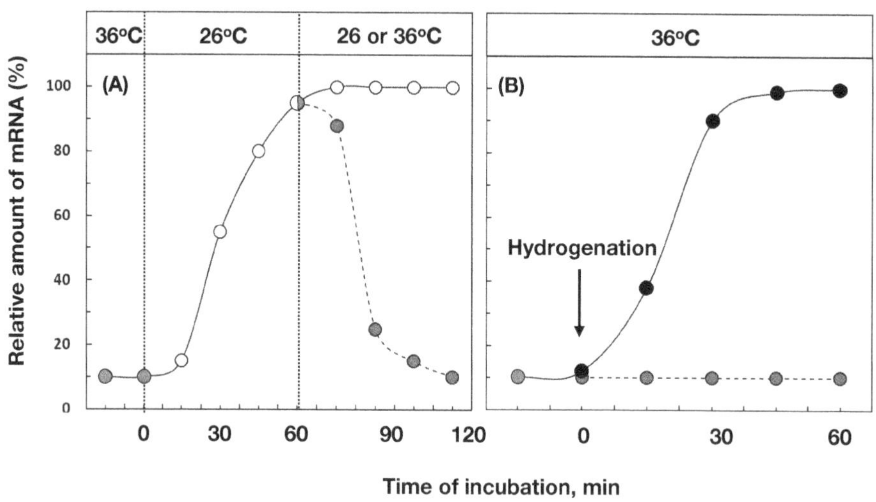

Figure 5.3 Cold-induced and hydrogenation-induced transcription of *desA* gene in *Synechocystis*. (A) Cells were grown at 36°C, subjected to 26°C for 1 h, and then transferred back to 36°C or further incubated at 26°C. (B) Cells were grown at 36°C and hydrogenated for 10 min at 36°C (Vigh *et al*., 1993). The efficiency of transcription was estimated by Northern blotting with the gene-specific probe.

al., 1993). These experiments demonstrated the existence of a feedback link between the physical state of the cytoplasmic membrane and the expression of a gene that encodes a protein, which is necessary for the maintenance of membrane physical properties.

Third line of evidence was provided by the experiments with the double knock-out mutant of *Synechocystis* deficient in *desA* and *desD* genes coding for Δ12- and Δ6-FADs. In such cells, characterized by the rigidified membranes, the induction of the terminal FAD gene, *desB*, started at significantly higher temperatures than in wild-type cells (Mironov *et al.*, 2012a) (Fig. 5.2B).

Thus, it became evident that a putative cold sensor may be rather located in a cytoplasmic membrane of a cell. It might sense temperature-induced changes in the fluidity/viscosity of the membrane, which is regulated via a feedback loop between the fluidity and transcription of FADs regulating that fluidity.

Two-component cold sensor and transducer

To search for a putative cold sensor, a reporter strain of *Synechocystis* was constructed that carried bacterial luciferase cassette under control of the cold-inducible promoter of *desB* gene (Los *et al.*, 1997). Random mutagenesis of this strain with an antibiotic-resistance cartridge revealed that cold inducibility of this promoter is controlled by the transmembrane protein, histidine kinase Hik33 (Suzuki *et al.*, 2000). Among 46 potential two-component regulatory systems of *Synechocystis*, histidine kinase Hik33 (Suzuki *et al.*, 2000) and response regulator Rre26 (RpaB) (Kappell and van Waasbergen, 2007) have been identified as the components of cold perception and transduction system. As pointed above, the genome-wide analysis of gene expression in *Synechocystis* with DNA microarrays revealed more than 100 genes that are highly induced by cold. It appeared that about a half of these genes is controlled by Hik33 and may be induced due to the changes in membrane fluidity (Murata and Los, 2006; Mironov *et al.*, 2012b). Another part of cold-inducible genes is regulated by some other cold-sensing system that is fluidity-independent and differs from Hik33 (Los *et al.*, 2010).

Molecular mechanisms of cold sensing by thermosensory histidine kinases

Physical mechanisms by which Hik33 recognize stress factors are still not known. It was assumed that Hik33 may detect changes in the membrane physical properties during cold stress and at the initial stages of salt or hyperosmotic stress (Murata and Los, 2006), when a rapid decrease of cytoplasmic volume and shrinkage of cells occurs. However, it is still unclear whether autophosphorylation and activation of Hik33 results from a change in physical motion of membrane lipids or from a change in the surface charge cause by rigidification of a cytoplasmic membrane. It is also possible, that Hik33 may sense redox changes or ROS via its PAS domain (Vogt and Schippers, 2015).

A clue for understanding the possible molecular mechanisms of temperature sensing by transmembrane sensors via changes in the membrane properties recently came from studies of the bacillary cold sensor, histidine kinase DesK (Saita and de Mendoza, 2015). DesK of *Bacillus subtilis* carries five TM-spanning segments that sense an increase in the order of membrane lipids and promote a kinase dominant state of DesK, which undergoes autophosphorylation, and transfers the phosphate group to the response regulator DesR. Phosphorylated DesR activates expression of the gene for Δ5-desaturase – the only FAD in *B. subtilis* – that is involved in the maintenance of the membrane fluidity (de Mendoza, 2014). A truncated DesK, in which four transmembrane domains were deleted, and the N-terminus of the transmembrane domain TM1 is fused to the C-terminus of TM5, still performs the membrane-mediated DesK signalling (Saita *et al.*, 2016). The cold signalling by the engineered sensor protein with a single transmembrane domain implies that lateral movements of the transmembrane domains are not important for sensing the membrane physical state (Fig. 5.4). Thus, the temperature sensing requires a match between hydrophobic thickness of the membrane and the hydrophobic thickness of a transmembrane segment of integral membrane protein. Any discrepancy alters the structure of the sensor kinase and triggers the cold stress response (Saita and de Mendoza, 2015). Resent mutagenesis and modelling experiments revealed that structural

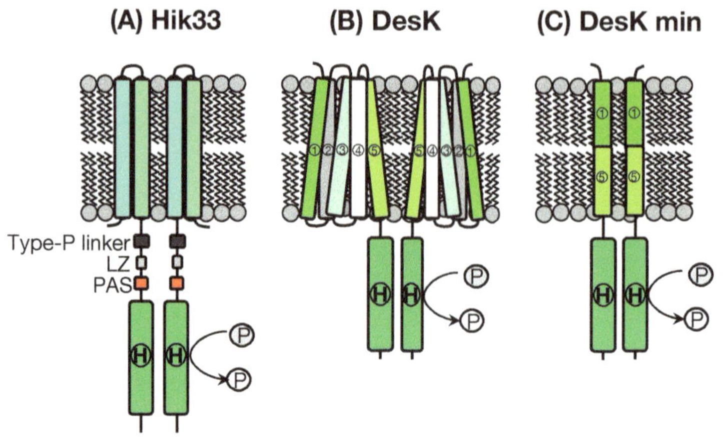

Figure 5.4 Cold sensors. (A) The histidine kinase Hik33 of *Synechocystis* consists of two transmembrane domains, type-P linker, leucine zipper (LZ), and PAS domain. (B) The histidine kinase DesK of *Bacillus subtilis* includes five transmembrane segments. The rigidification of the membrane promotes autophosporylation and transfer the phosphate group to the response regulator DesR, which activates expression of the *des* gene coding for a Δ5-desaturase. (C) A truncated chimera of DesK represents the '*minimum sensor*' (DesK min), in which the N-terminal domain of TM1 is fused to the C-terminal domain of TM5. This engineered sensor allows near-normal control of membrane-mediated DesK signalling (Los *et al.*, 2013).

changes in a two-helix coiled coil, which connects the transmembrane sensory domain to the cytosolic catalytic region, control the signalling state of DesK (Saita *et al.*, 2015). It is now accepted that membrane thickening is the main driving force for cold signal sensing: it forces helix stretching and rotation prompting an asymmetric sensor kinase-competent state.

It should be noted, however, that, unlike Hik33 of *Synechocystis*, DesK of *Bacillus* does not carry a PAS domain. Therefore, the regulation through Hik33 may be more complicated and include light-dependent events. DesK of *Bacillus* was unable to substitute the inactivated Hik33 in *Synechocystis*, while Hik33 homologue from another cyanobacterium, *Synechococcus elongatus* PCC 7942, could successfully complement the knock-out of Hik33 in *Synechocystis* in terms of cold sensing and turning on the responsive gene expression (Kiseleva, Suzuki, Los, Murata – unpublished). This may reflect the differences in membrane properties of cyanobacteria and *Bacillus*, as well as light-dependent manner of regulation via Hik33 sensor.

Multifunctional sensory histidine kinases

Different signalling pathways of a cell may share particular elements involved in transduction of the corresponding signal. One amazing phenomenon found in cyanobacteria, is that the same molecular sensor can perceive different external environment stimuli and trigger several distinct cascades of responses interacting with different response regulators. This is true for Hik33 of *Synechocystis*, which was first described as a protein that ensures resistance to chemical inhibitors of photosynthesis, difunon and diuron (Bartsevich and Shestakov, 1995). Later it was found that Hik33 senses low temperature and interacts with Rre26 response regulator thus transmitting a cold signal to activate a set of the responsive genes (Suzuki *et al.*, 2000, 2001). In addition, Hik33 perceives the signals from sorbitol (hyperosmotic) and NaCl (salt) stresses and transmits this signal(s) to induce a different set of responsive genes via another response regulator, Rre31 (Shoumskaya *et al.*, 2005). Hik33 also perceives oxidative stress and operates via yet unknown response regulator and PerR transcription factor (Kanesaki *et al.*, 2007).

Hik33 plays a role in signal perception in cold, salt, osmotic, and oxidative stress (Murata and Los, 2006). The cooperation of a sensor with different response regulators under different stress conditions suggests the presence of additional factors interacting with the known components. A yeast two-hybrid screen revealed a specific interaction between Hik33 and a small SipA-like protein, Ssl3451 (Sato *et al.*, 2009). Functional analysis demonstrated that Ssl3451 interacts with Hik33 and enhances its phosphorylation at low temperature (Sakayori *et al.*, 2009). Thus, such mode of interaction, may indeed forward the reciprocal preference of a histidine kinase to a certain response regulator.

Another example of the multifunctional sensor is transmembrane Hik31, which senses Cu^{2+} released from plastocyanin oxidized by various abiotic stressors, or chemicals, including those interacting with photosynthetic electron transport chain (Giner-Lamia *et al.*, 2012).

Light regulation of cold sensing

Fluidity-dependent Hik33-regulated genes (*desB*, *ndhD2*, *hliB*) are induced by cold only when cells are illuminated, but never in the dark (Fig. 5.5; Mironov *et al.*, 2012b). A link between membrane fluidity and light regulation of the cold-induced *desB* gene was demonstrated in the mutant defective in Hik33 and in the double *desA*$^-$/*desD*$^-$ mutant with rigidified membranes (Mironov *et al.*, 2012a). Light-dependence of *desB* transcription explains well the light-dependent appearance of ω3-unsaturated FAs in cold-treated cells of *Synechocystis*. Transcription of Hik33-independent genes (*crhR*, *rbpA1*, etc.) does not depend on changes in the fluidity of the membranes, and these genes are equally induced by cold stress both in the light and in darkness (Mironov *et al.*, 2012b).

Light-regulated responses usually depend on the state of different photoreceptors, especially, phytochromes (Kreslavski *et al.*, 2012). In higher plants, phytochromes are usually transformed into their active form by absorption of red light with the maximum of 630–670 nm. Far-red light with the maximum of 730 nm transforms phytochromes back to their inactive form. Phytochrome-like receptors with two photoreversible states have been also found in cyanobacteria (Russel, 2004). Cyanobacterial phytochrome-like proteins are highly homologous to phytochromes of higher plants, and they are represented by the sensor histidine kinases (Hughes *et al.*, 1997).

Investigation of the action spectra that is important for cold inducibility of genes revealed that transcription of Hik33-dependent genes (e.g. *desB*) is enhanced by cold under red light (~700 nm). Moreover, a brief flash of red light provided along with cold stress, could trigger the synthesis of ω3 unsaturated FAs (Mironov *et al.*, 2014). This observation suggests that red light is indeed a triggering regulator obligatory for the cold induction of Hik33-dependent genes. The investigation of cold inducibility of the *desB* and *hliB* genes in the Hik33-deficient mutant of *Synechocystis* revealed that their transcription did not respond either to light or temperature (Fig. 5.6), whereas Hik33-independent *rbpA* gene was induced by cold in a light-independent manner.

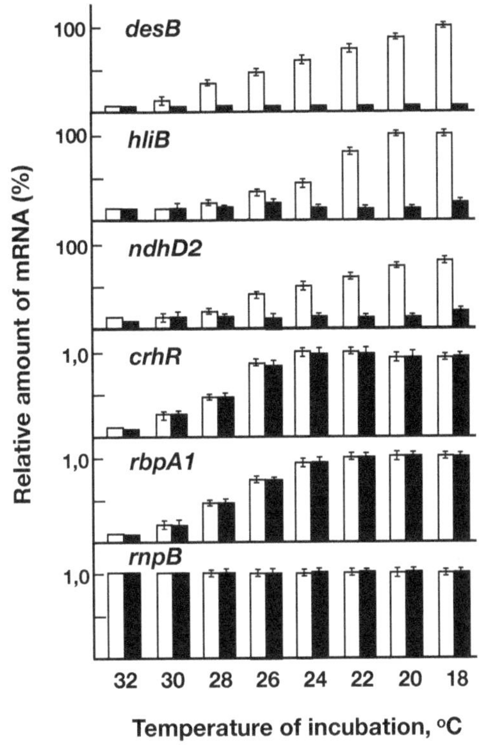

Figure 5.5 Light-dependent and light-independent cold-induced transcription of genes in *Synechocystis* sp. PCC 6803. Cells of wild-type were at 32°C under continuous illumination and incubated at 30°C, 28°C, 26°C, 24°C, 22°C, 20°C or 18°C for 30 min in the light or in darkness. The changes in the levels of the specific mRNAs were determined by the RT-PCR method. The *desB*, *hliB*, and *ndhD2* represent Hik33-dependent cold-inducible genes. The *crhR* and *rbpA1* genes represent Hik33-independent cold-inducible genes. The *rnpB* gene represents the constitutive gene for RNase P. Quantification of the results of RT-PCR is presented: open bars represent samples exposed to white light; dark bars represent samples incubated in the dark.

A number of sensor histidine kinases are involved in light perception in *Synechocystis*. The receptors of blue (Zirak *et al.*, 2007; Moon *et al.*, 2010), green (Hirose *et al.*, 2008), and red (Yeh *et*

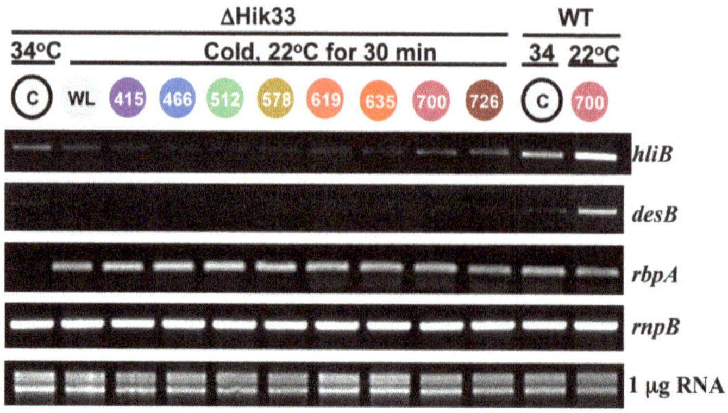

Figure 5.6 Light-quality-dependent cold-induced transcription of genes in the mutant of *Synechocystis*, deficient in histidine kinase Hik33. Wild-type cells are shown for comparison to the Hik33-deficient cells (ΔHik33). Wild-type cells were incubated at 34°C under white light and treated at 22°C for 30 min under 700 nm red light. The *hliB* and *desB* represent Hik33-dependent cold-inducible genes. The *rbpA* gene represents Hik33-independent cold-inducible gene. The *rnpB* gene represents the constitutive gene for RNase P. Ribosomal (23S and 16S) RNAs were stained with ethidium bromide (Mironov *et al.*, 2014).

al., 1997; Russel, 2004) light have been characterized. Green and red light receptors operate in a phytochrome-like mode, implying green – red or red – far red switching cycles, respectively.

In regard to cold stress, pulses of red (635 nm) and far red light (726 nm) of various duration and combinations did not reveal any notable changes in gene expression in wild-type cells and had no effect in the Hi33-deficient mutant. Disappearance of cold induction of *desB* and *hliB* genes in the Hik33-deficient mutant, which are strongly induced by cold under red light of 700 nm in wild-type cells, suggests that Hik33 itself may sense red light, as suggested previously (Kreslavski *et al.*, 2012), although its mode of operation differs from that of a phytochrome system.

Other candidates for cold sensors: changes in genomic DNA supercoiling

Changes to the supercoiling of genomic DNA play an important role in the regulation of gene expression in response to environmental stress in bacteria (Dorman, 2006). Temperature-dependent alterations in DNA supercoiling were proposed to be one of the sensory mechanisms that enhance the expression of genes involved in cold acclimation. Cold stress caused an increase in the negative supercoiling of the promoter region of the *desB* gene and controls its induction at low temperatures (Los, 2004). Genome-wide transcriptome analysis demonstrated that transcription of many genes depends on cold-induced changes in DNA supercoiling, including Hik33-independent genes (*crhR, rlpA, rpbA1*, and some others) (Prakash *et al.*, 2009). The molecular basis for such activation of genes under cold stress is still unknown. It is possible that an increase in negative supercoiling of DNA due to cold stress changes the promoter spacing or availability of regulatory sequences that are recognized by RNA polymerase or cold-induced DNA-binding transcription factors (Fig. 5.7).

Other candidates for cold sensors: Ser/Thr protein kinases

Serine/threonine protein kinases (Ser/Thr kinases) are known as the participants of signal-transduction pathways in eukaryotes. Whole-genome sequencing indicates that Ser/Thr kinases are also widely present in prokaryotes, including cyanobacteria. Among twelve putative genes for Ser/Thr kinases in *Synechocystis*, seven of these genes encode proteins that belong to the PKN2 subfamily of Ser/Thr kinases and five encode proteins that belong to the ABC1 subfamily of Ser/Thr kinases (Zorina, 2013). The genes for kinases of the PKN2 type are designated *spkA, spkB, spkC, spkD, spkE, spkF* and *spkG* and those for kinases of the ABC1-type are designated *spkH, spkI, spkJ, spkK* and *spkL* (CyanoBase, http://genome.microbedb.jp/cyanobase/Synechocystis). The *spkA* and *spkB* genes appear to

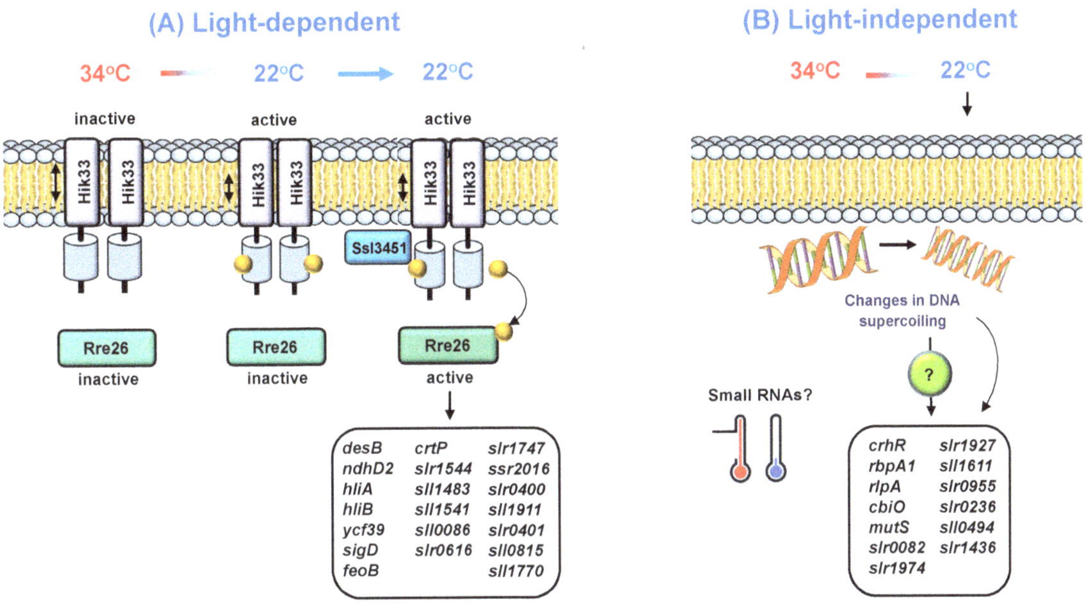

Figure 5.7 A scheme of light-dependent and light-independent regulation of cold-induced gene expression in *Synechocystis*. Light-dependent cold-induced gene expression operates via red-light-dependent transmembrane histidine kinase Hik33, which senses the cold-induced changed in membrane fluidity. Hik33 and Rre26 form a two-component regulatory system based on classical phosphorelay. Cold-induced autophosphorylation of Hik33 is facilitated by interaction with a small protein, Ssl3451, which may target Hik33 to Rre26. As a result, a group of genes is induced by cold in the light (A). Light-independent group of genes is activated via cold-induced changes in genomic DNA supercoiling (B). The impact of small RNAs into regulation of cold-induced gene expression requires further investigation.

encode proteins that participate in the control of cell motility, whereas the *spkC* gene encodes a kinase that might be involved in the regulation of nitrogen metabolism. Inactivation of *spkD* gene results in growth retardation at low inorganic carbon (C*i*) concentrations, although it does not affect the C*i* uptake system (Zorina, 2013). SpkC, SpkF, and SpkK have been shown to participate in phosphorylation of GroES co-chaperone (Zorina *et al.*, 2011). Filamentous cyanobacteria with complex cellular organization usually have much more genes for eukaryotic-type Ser/Thr protein kinases: e.g. 43 in *Arthrospira platensis* and 52 in *Anabaena variabilis*.

Systematic analysis of possible involvement of Ser/Thr kinases in cold responses in *Synechocystis* revealed that the membrane located Ser/Thr kinases (SpkC, SpkD, SpkF, SpkH, SpkI, or SpkL) are not involved in the regulation of transcription of cold-inducible genes (*hliB*, *ndhD2*, *rbpA*, and *crhR*). Screening of the collection of *Synechocystis* mutants, defective in Ser/Thr kinases, identified four enzymes – SpkB, SpkD, SpkE, and SpkG – as possible transcriptional regulators of cold stress responses (Zorina *et al.*, 2014). The mutant lacking SpkE displayed the transcription profile similar to the Hik33-deficient mutant in terms of a spectrum of cold-inducible genes.

The SpkE mutant also substantially differed from wild-type cells in a pattern of proteins. This mutation is not lethal for cyanobacterial cells, although a complete absence of several protein groups indicates that SpkE makes a substantial contribution to the protein pattern in *Synechocystis*. A phosphorylation signal from high-molecular-weight proteins was completely undetectable in the *spkE*-deficient mutant. The intracellular substrates of SpkE are still to be identified. In contrast to earlier assumptions of its nonfunctional character, SpkE appeared as a functionally active kinase and utilizes basic cell proteins as preferential substrates.

Thus, SpkE participates in regulation of cold stress responses by modulating the level of transcription of some genes (e.g. *ndhD2*, *crhR*, *desB*, *hliB*), as well as the extent of protein phosphorylation (Zorina *et al.*, 2014). The mutant deficient in SpkE cannot induce transcription of both Hik33-dependent (*ndhD2, desB, hliB*) and Hik33-independent (*crhR*) genes. This suggests a regulatory cross-talk between His and Ser/Thr kinases in the regulation of the responses to cold stress in cyanobacteria.

Conclusions: Sensors and universal triggers of stress responses

Environmental stressors induce sets of genes required for acclimation or protection of cells against adverse conditions (Los *et al.*, 2010). Systematic analysis of the stress transcriptomes of *Synechocystis* shows that among such genes there is a distinct group of genes for core or general stress proteins (GSPs), formerly known as genes for heat shock proteins (HSPs). These genes are induced by various stressors of different physical nature – high temperatures, strong and UV light, salt and osmotic stress, high acidity of the medium (Table 5.1).

The induction of the same set of genes by different stressors suggests the existence of universal signal(s) that trigger(s) stress responses at the level of transcription. In photosynthetic cells, the candidates for the role of such triggers are as follows.

First, it might be the redox regulation system, consisting of components of PQ pool and ETC. This system should be able to perceive the difference in electric potential similarly to the redox-sensitive transcription factor, PedR (Ssl0564) (Nakamura and Hihara, 2006; Horiuchi *et al.*, 2010) or redox sensitive transmembrane histidine kinase, Hik31, which triggers the regulatory cascade by binding Cu^{2+} ions released from oxidized plastocyanin (Giner-Lamia *et al.*, 2012, 2014).

Second candidate for the role of universal triggers might be stress-induced ROS that have sufficient lifetime to carry out their regulatory functions (Schmitt *et al.*, 2014; 2016). H_2O_2-generated signals are perceived by four histidine kinases – Hik33, Hik34, Hik16 and Hik41 (Kanesaki *et al.*, 2007).

ROS sensing under heat stress may be performed by Hik34, which is also able to perceive heat, salt, and hyperosmotic stimuli, or by Hik31. Hik34, paired with transcriptional Rre1 factor, controls transcription of a sigma factor of RNA polymerase, SigB, which, in turn, is involved in the regulation of cell responses to heat, salt and hyperosmotic stress.

Cold stress may be perceived by transmembrane domains of Hik33. Apart from sensing a physical change in the membrane thickness, Hik33 may also sense ROS by the PAS-domain. Hik33 is light dependent, at least under cold stress. It may perceive cold, salt, hyperosmotic and oxidative stresses.

Transcriptomic analysis shows that some universal chemical or electric stimuli may exist that trigger the expression of stress-dependent genes regardless the physical nature of a stressor. The candidates for such triggers are reactive oxygen species and changes in redox status of plastoquinone pool. Application of the inhibitor(s) of photosynthetic ETC (DBMIB, DCMU, or methylviologen), or ROS (H_2O_2) evidence that redox status and ROS may be involved in the initiation of stress responses as universal triggers. Changing the redox potential of cells may be due to the emergence of ROS. In this case, about one-third of genes induced by H_2O_2 and DBMIB intersect. ROS control more than one-third of stress-induced genes, and the remaining one-third are under redox control (Fig. 5.1). In addition, about one-quarter of stress-induced genes do not depend either on the ROS or on the redox potential. We have previously found that some stress-dependent genes can be induced by changing the degree of supercoiling of the genomic DNA. This type of regulation does not require any regulators with SH-groups that can undergo oxidation or reduction cycles. Further experimental studies are necessary to directly confirm or refute the hypothesis about the universal triggers of stress responses in photosynthetic cells.

Acknowledgement

This work was supported by a grant from the Russian Science Foundation to D.A.L. (no. 14-24-00020).

References

Allakhverdiev, S.I., Nishiyama, Y., Suzuki, I., Tasaka, Y., and Murata, N. (1999). Genetic engineering of the unsaturation of fatty acids in membrane lipids alters the tolerance of *Synechocystis* to salt stress. Proc. Natl. Acad. Sci. U.S.A. *96*, 5862–5867. http://dx.doi.org/10.1073/pnas.96.10.5862.

Aubdool, A.A., Graepel, R., Kodji, X., Alawi, K.M., Bodkin, J.V., Srivastava, S., Gentry, C., Heads, R., Grant, A.D., Fernandes, E.S., *et al.* (2014). TRPA1 is essential for the vascular response to environmental cold exposure. Nat. Commun. *5*, 5732. http://dx.doi.org/10.1038/ncomms6732.

Bachin, D., Nazarenko, L.V., Mironov, K.S., Pisareva, T., Allakhverdiev, S.I., and Los, D.A. (2015). Mechanosensitive ion channel MscL controls ionic fluxes during cold and heat stress in *Synechocystis*. FEMS Microbiol. Lett. *362*, fnv090. http://dx.doi.org/10.1093/femsle/fnv090.

Bartsevich, V.V., and Shestakov, S.V. (1995). The *dspA* gene product of the cyanobacterium *Synechocystis* sp. strain PCC 6803 influences sensitivity to chemically different growth inhibitors and has amino acid similarity to histidine protein kinases. Microbiology UK *141*, 2915–2920. http://dx.doi.org/10.1099/13500872-141-11-2915.

Berla, B.M., Saha, R., Immethun, C.M., Maranas, C.D., Moon, T.S., and Pakrasi, H.B. (2013). Synthetic biology of cyanobacteria: unique challenges and opportunities. Front. Microbiol. *4*, 246. http://dx.doi.org/10.3389/fmicb.2013.00246.

Cameron, J.C., and Pakrasi, H.B. (2010). Essential role of glutathione in acclimation to environmental and redox perturbations in the cyanobacterium *Synechocystis* sp. PCC 6803. Plant Physiol. *154*, 1672–1685. http://dx.doi.org/10.1104/pp.110.162990.

Caterina, M.J., Schumacher, M.A., Tominaga, M., Rosen, T.A., Levine, J.D., and Julius, D. (1997). The capsaicin receptor: a heat-activated ion channel in the pain pathway. Nature *389*, 816–824. http://dx.doi.org/10.1038/39807.

Červený, J., Sinetova, M.A., Zavřel, T., and Los, D.A. (2015). Mechanisms of high temperature resistance of *Synechocystis* sp. PCC 6803: an impact of histidine kinase 34. Life *5*, 676–699. http://dx.doi.org/10.3390/life5010676.

Chidgey, J.W., Linhartová, M., Komenda, J., Jackson, P.J., Dickman, M.J., Canniffe, D.P., Koník, P., Pilný, J., Hunter, C.N., and Sobotka, R. (2014). A cyanobacterial chlorophyll synthase-HliD complex associates with the Ycf39 protein and the YidC/Alb3 insertase. Plant Cell *26*, 1267–1279. http://dx.doi.org/10.1105/tpc.114.124495.

Chrismas, N.A., Anesio, A.M., and Sánchez-Baracaldo, P. (2015). Multiple adaptations to polar and alpine environments within cyanobacteria: a phylogenomic and Bayesian approach. Front. Microbiol. *6*, 1070. http://dx.doi.org/10.3389/fmicb.2015.01070.

de Mendoza, D. (2014). Temperature sensing by membranes. Annu. Rev. Microbiol. *68*, 101–116. http://dx.doi.org/10.1146/annurev-micro-091313-103612.

Dorman, C.J. (2006). DNA supercoiling and bacterial gene expression. Sci. Prog. *89*, 151–166. http://dx.doi.org/10.3184/003685006783238317.

Finka, A., Cuendet, A.F., Maathuis, F.J., Saidi, Y., and Goloubinoff, P. (2012). Plasma membrane cyclic nucleotide gated calcium channels control land plant thermal sensing and acquired thermotolerance. Plant Cell *24*, 3333–3348. http://dx.doi.org/10.1105/tpc.112.095844.

Gao, F., Han, X., Wu, J., Zheng, S., Shang, Z., Sun, D., Zhou, R., and Li, B. (2012). A heat-activated calcium-permeable channel – Arabidopsis cyclic nucleotide-gated ion channel 6 – is involved in heat shock responses. Plant. J. *70*, 1056–1069. http://dx.doi.org/10.1111/j.1365-313X.2012.04969.x.

Gao, L., Wang, J., Ge, H., Fang, L., Zhang, Y., Huang, X., and Wang, Y. (2015). Toward the complete proteome of *Synechocystis* sp. PCC 6803. Photosynth. Res. *126*, 203–219. http://dx.doi.org/10.1007/s11120-015-0140-y.

Giner-Lamia, J., López-Maury, L., and Florencio, F.J. (2014). Global transcriptional profiles of the copper responses in the cyanobacterium *Synechocystis* sp. PCC 6803. PLoS One *9*, e108912. http://dx.doi.org/10.1371/journal.pone.0108912.

Giner-Lamia, J., López-Maury, L., Reyes, J.C., and Florencio, F.J. (2012). The CopRS two-component system is responsible for resistance to copper in the cyanobacterium *Synechocystis* sp. PCC 6803. Plant Physiol. *159*, 1806–1818. http://dx.doi.org/10.1104/pp.112.200659.

Guy, J.E., Whittle, E., Moche, M., Lengqvist, J., Lindqvist, Y., and Shanklin, J. (2011). Remote control of regioselectivity in acyl-acyl carrier protein-desaturases. Proc. Natl. Acad. Sci. U.S.A. *108*, 16594–16599. http://dx.doi.org/10.1073/pnas.1110221108.

Hakkila, K., Antal, T., Rehman, A.U., Kurkela, J., Wada, H., Vass, I., Tyystjärvi, E., and Tyystjärvi, T. (2014). Oxidative stress and photoinhibition can be separated in the cyanobacterium *Synechocystis* sp. PCC 6803. Biochim. Biophys. Acta *1837*, 217–225. http://dx.doi.org/10.1016/j.bbabio.2013.11.011.

Hantke, K. (2003). Is the bacterial ferrous iron transporter FeoB a living fossil? Trends Microbiol. *11*, 192–195. http://dx.doi.org/10.1016/S0966-842X(03)00100-8.

He, Q., Dolganov, N., Bjorkman, O., and Grossman, A.R. (2001). The high light-inducible polypeptides in *Synechocystis* PCC6803. Expression and function in high light. J. Biol. Chem. *276*, 306–314. http://dx.doi.org/10.1074/jbc.M008686200.

Hihara, Y., Kamei, A., Kanehisa, M., Kaplan, A., and Ikeuchi, M. (2001). DNA microarray analysis of cyanobacterial gene expression during acclimation to high light. Plant Cell *13*, 793–806. http://dx.doi.org/10.1105/tpc.13.4.793.

Hihara, Y., Sonoike, K., Kanehisa, M., and Ikeuchi, M. (2003). DNA microarray analysis of redox-responsive genes in the genome of the cyanobacterium *Synechocystis* sp. strain PCC 6803. J. Bacteriol. *185*, 1719–1725. http://dx.doi.org/10.1128/JB.185.5.1719-1725.2003.

Hirose, Y., Shimada, T., Narikawa, R., Katayama, M., and Ikeuchi, M. (2008). Cyanobacteriochrome CcaS is the green light receptor that induces the expression of phycobilisome linker protein. Proc. Natl. Acad. Sci. U.S.A. *105*, 9528–9533. http://dx.doi.org/10.1073/pnas.0801826105.

Horiuchi, M., Nakamura, K., Kojima, K., Nishiyama, Y., Hatakeyama, W., Hisabori, T., and Hihara, Y. (2010). The PedR transcriptional regulator interacts with thioredoxin to connect photosynthesis with gene expression in cyanobacteria. Biochem. J. *431*, 135–140. http://dx.doi.org/10.1042/BJ20100789.

Hsiao, H.Y., He, Q., Van Waasbergen, L.G., and Grossman, A.R. (2004). Control of photosynthetic and high-light-responsive genes by the histidine kinase DspA: negative and positive regulation and interactions between signal transduction pathways. J. Bacteriol. *186*, 3882–3888. http://dx.doi.org/10.1128/JB.186.12.3882-3888.2004.

Huang, L., McCluskey, M.P., Ni, H., and LaRossa, R.A. (2002). Global gene expression profiles of the cyanobacterium *Synechocystis* sp. strain PCC 6803 in response to irradiation with UV-B and white light. J. Bacteriol. *184*, 6845–6858. http://dx.doi.org/10.1128/JB.184.24.6845-6858.2002

Hughes, J., Lamparter, T., Mittmann, F., Hartmann, E., Gärtner, W., Wilde, A., and Börner, T. (1997). A prokaryotic phytochrome. Nature *386*, 663. http://dx.doi.org/10.1038/386663a0.

Imamura, S., Asayama, M., Takahashi, H., Tanaka, K., Takahashi, H., and Shirai, M. (2003). Antagonistic dark/light-induced SigB/SigD, group 2 sigma factors, expression through redox potential and their roles in cyanobacteria. FEBS Lett. *554*, 357–362. http://dx.doi.org/10.1016/S0014-5793(03)01188-8.

Inaba, M., Suzuki, I., Szalontai, B., Kanesaki, Y., Los, D.A., Hayashi, H., and Murata, N. (2003). Gene-engineered rigidification of membrane lipids enhances the cold inducibility of gene expression in *Synechocystis*. J. Biol. Chem. *278*, 12191–12198. http://dx.doi.org/10.1074/jbc.M212204200.

Inoue, N., Taira, Y., Emi, T., Yamane, Y., Kashino, Y., Koike, H., and Satoh, K. (2001). Acclimation to the growth temperature and the high-temperature effects on photosystem II and plasma membranes in a mesophilic cyanobacterium, *Synechocystis* sp. PCC6803. Plant Cell Physiol. *42*, 1140–1148. http://dx.doi.org/10.1093/pcp/pce147.

Jarmoskaite, I., and Russell, R. (2014). RNA helicase proteins as chaperones and remodelers. Annu. Rev. Biochem. *83*, 697–725. http://dx.doi.org/10.1146/annurev-biochem-060713-035546.

Kahlon, S., Beeri, K., Ohkawa, H., Hihara, Y., Murik, O., Suzuki, I., Ogawa, T., and Kaplan, A. (2006). A putative sensor kinase, Hik31, is involved in the response of *Synechocystis* sp. strain PCC 6803 to the presence of glucose. Microbiology *152*, 647–655.

Kanesaki, Y., Yamamoto, H., Paithoonrangsarid, K., Shoumskaya, M., Suzuki, I., Hayashi, H., and Murata, N. (2007). Histidine kinases play important roles in the perception and signal transduction of hydrogen peroxide in the cyanobacterium, *Synechocystis* sp. PCC 6803. Plant J. *49*, 313–324. http://dx.doi.org/10.1111/j.1365-313X.2006.02959.x

Kappell, A.D., and van Waasbergen, L.G. (2007). The response regulator RpaB binds the high light regulatory 1 sequence upstream of the high-light-inducible *hliB* gene from the cyanobacterium *Synechocystis* PCC 6803. Arch. Microbiol. *187*, 337–342. http://dx.doi.org/10.1007/s00203-007-0213-1.

Karzai, A.W., Susskind, M.M., and Sauer, R.T. (1999). SmpB, a unique RNA-binding protein essential for the peptide-tagging activity of SsrA (tmRNA). EMBO J. *18*, 3793–3799. http://dx.doi.org/10.1093/emboj/18.13.3793.

Khemici, V., and Linder, P. (2016). RNA helicases in bacteria. Curr. Opin. Microbiol. *30*, 58–66. http://dx.doi.org/10.1016/j.mib.2016.01.002.

Kiseleva, L.L., Serebriiskaya, T.S., Horvàth, I., Vigh, L., Lyukevich, A.A., and Los, D.A. (2000). Expression of the gene for the Δ9 acyl-lipid desaturase in the thermophilic cyanobacterium. J. Mol. Microbiol. Biotechnol. *2*, 331–338.

Klatt, C.G., Inskeep, W.P., Herrgard, M.J., Jay, Z.J., Rusch, D.B., Tringe, S.G., Parenteau, M.N., Ward, D.M., Boomer, S.M., Bryant, D.A., and Miller, S.R. (2013). Community structure and function of high-temperature chlorophototrophic microbial mats inhabiting diverse geothermal environments. Front. Microbiol. *4*, 106. http://dx.doi.org/10.3389/fmicb.2013.00106

Kopf, M., Klähn, S., Scholz, I., Matthiessen, J.K., Hess, W.R., and Voß, B. (2014). Comparative analysis of the primary transcriptome of *Synechocystis* sp. PCC 6803. DNA Res. *21*, 527–539. http://dx.doi.org/10.1093/dnares/dsu018.

Koskinen, S., Hakkila, K., Gunnelius, L., Kurkela, J., Wada, H., and Tyystjärvi, T. (2016). *In vivo* recruitment analysis and a mutant strain without any group 2 σ factor reveal roles of different σ factors in cyanobacteria. Mol. Microbiol. *99*, 43–54. http://dx.doi.org/10.1111/mmi.13214.

Kreslavski, V.D., Fomina, I.R., Los, D.A., Carpentier, R., Kuznetsov, V.V., and Allakhverdiev, S.I. (2012). Red and near infra-red signaling: Hypothesis and perspectives. J. Photochem. Photobiol. C: Photochem. Rev. *13*, 190–203. http://dx.doi.org/10.1016/j.jphotochemrev.2012.01.002.

Kufryk, G., Hernandez-Prieto, M.A., Kieselbach, T., Miranda, H., Vermaas, W., and Funk, C. (2008). Association of small CAB-like proteins (SCPs) of *Synechocystis* sp. PCC 6803 with Photosystem II. Photosynth. Res. *95*, 135–145. http://dx.doi.org/10.1007/s11120-007-9244-3.

Latifi, A., Ruiz, M., and Zhang, C.C. (2009). Oxidative stress in cyanobacteria. FEMS Microbiol. Rev. *33*, 258–278. http://dx.doi.org/10.1111/j.1574-6976.2008.00134.x.

Lee, J.W., and Helmann, J.D. (2006). The PerR transcription factor senses H2O2 by metal-catalysed histidine oxidation. Nature *440*, 363–367. http://dx.doi.org/10.1038/nature04537.

Li, H., Singh, A.K., McIntyre, L.M., and Sherman, L.A. (2004). Differential gene expression in response to hydrogen peroxide and the putative PerR regulon of *Synechocystis* sp. strain PCC 6803. J. Bacteriol. *186*, 3331–3345. http://dx.doi.org/10.1128/JB.186.11.3331-3345.2004.

Liu, H., Colavitti, R., Rovira, I.I., and Finkel, T. (2005). Redox-dependent transcriptional regulation. Circ. Res. *97*, 967–974. http://dx.doi.org/10.1161/01.RES.0000188210.72062.10.

Liu, Z., Cui, H., Liu, Z., Wang, Y., Cui, Y., Liu, Z., and Qin, S. (2014). Transcriptomic analysis of *Synechocystis* sp. PCC6803 under low-temperature stress. Chin. J. Oceanol. Limnol. 32, 403–418. http://dx.doi.org/10.1007/s00343-014-3174-7.

Los, D.A. (2004). The effect of low-temperature-induced DNA supercoiling on the expression of the desaturase genes in *Synechocystis*. Cell. Mol. Biol. *50*, 605–612. http://dx.doi.org/10.1170/T550.

Los, D., Horvath, I., Vigh, L., and Murata, N. (1993). The temperature-dependent expression of the desaturase gene *desA* in *Synechocystis* PCC6803. FEBS Lett. *318*, 57–60. http://dx.doi.org/10.1016/0014-5793(93)81327-V.

Los, D.A., Ray, M.K., and Murata, N. (1997). Differences in the control of the temperature-dependent expression of four genes for desaturases in *Synechocystis* sp. PCC 6803. Mol. Microbiol. *25*, 1167–1175. http://dx.doi.org/10.1046/j.1365-2958.1997.5641912.x

Los, D.A., and Murata, N. (1998). Structure and expression of fatty acid desaturases. Biochim. Biophys. Acta *1394*, 3–15. http://dx.doi.org/10.1016/S0005-2760(98)00091-5.

Los, D.A., and Murata, N. (2004). Membrane fluidity and its roles in the perception of environmental signals. Biochim. Biophys. Acta *1666*, 142–157. http://dx.doi.org/10.1016/j.bbamem.2004.08.002.

Los, D.A., Zorina, A., Sinetova, M., Kryazhov, S., Mironov, K., and Zinchenko, V.V. (2010). Stress sensors and signal transducers in cyanobacteria. Sensors *10*, 2386–2415. http://dx.doi.org/10.3390/s100302386.

Los, D.A., Mironov, K.S., and Allakhverdiev, S.I. (2013). Regulatory role of membrane fluidity in gene expression and physiological functions. Photosynth. Res. *116*, 489–509. http://dx.doi.org/10.1007/s11120-013-9823-4.

McKemy, D.D., Neuhausser, W.M., and Julius, D. (2002). Identification of a cold receptor reveals a general role for TRP channels in thermosensation. Nature *416*, 52–58. http://dx.doi.org/10.1038/nature719.

Mironov, K.S., Maksimov, E.G., Maksimov, G.V., and Los, D.A. (2012a). Feedback between membrane fluidity and transcription of the *desB* gene for the ω3-desaturase in the cyanobacterium *Synechocystis*. Mol. Biol. *46*, 134–141. http://dx.doi.org/10.1134/S002689331201013X.

Mironov, K.S., Sidorov, R.A., Trofimova, M.S., Bedbenov, V.S., Tsydendambaev, V.D., Allakhverdiev, S.I., and Los, D.A. (2012b). Light-dependent cold-induced fatty acid unsaturation, changes in membrane fluidity, and alterations in gene expression in *Synechocystis*. Biochim. Biophys. Acta *1817*, 1352–1359. http://dx.doi.org/10.1016/j.bbabio.2011.12.011.

Mironov, K.S., Sidorov, R.A., Kreslavski, V.D., Bedbenov, V.S., Tsydendambaev, V.D., and Los, D.A. (2014). Cold-induced gene expression and ω^3 fatty acid unsaturation is controlled by red light in *Synechocystis*. J. Photochem. Photobiol. B. Biol. *137*, 84–88. http://dx.doi.org/10.1016/j.jphotobiol.2014.03.001.

Moon, Y.J., Lee, E.M., Park, Y.M., Park, Y.S., Chung, W.I., and Chung, Y.H. (2010). The role of cyanopterin in UV/blue light signal transduction of cyanobacterium *Synechocystis* sp. PCC 6803 phototaxis. Plant Cell Physiol. *51*, 969–980. http://dx.doi.org/10.1093/pcp/pcq059.

Moparthi, L., Survery, S., Kreir, M., Simonsen, C., Kjellbom, P., Högestätt, E.D., Johanson, U., and Zygmunt, P.M. (2014). Human TRPA1 is intrinsically cold- and chemosensitive with and without its N-terminal ankyrin repeat domain. Proc. Natl. Acad. Sci. U.S.A. *111*, 16901–16906. http://dx.doi.org/10.1073/pnas.1412689111.

Murata, N., and Los, D.A. (2006). Histidine kinase Hik33 is an important participant in cold signal transduction in cyanobacteria. Physiol. Plant. *126*, 17–27. http://dx.doi.org/10.1111/j.1399-3054.2006.00608.x.

Murata, N., and Suzuki, I. (2006). Exploitation of genomic sequences in a systematic analysis to access how cyanobacteria sense environmental stress. J. Exp. Bot. *57*, 235–247. http://dx.doi.org/10.1093/jxb/erj005.

Nagarajan, S., Sherman, D.M., Shaw, I., and Sherman, L.A. (2012). Functions of the duplicated hik31 operons in central metabolism and responses to light, dark, and carbon sources in Synechocystis sp. strain PCC 6803. J. Bacteriol. *194*, 448–459. http://dx.doi.org/10.1128/JB.06207-11.

Nagarajan, S., Srivastava, S., and Sherman, L.A. (2014). Essential role of the plasmid *hik31* operon in regulating central metabolism in the dark in *Synechocystis* sp. PCC 6803. Mol. Microbiol. *91*, 79–97. http://dx.doi.org/10.1111/mmi.12442.

Nakamura, K., and Hihara, Y. (2006). Photon flux density-dependent gene expression in *Synechocystis* sp. PCC 6803 is regulated by a small, redox-responsive, LuxR-type regulator. J. Biol. Chem. *281*, 36758–36766. http://dx.doi.org/10.1074/jbc.M606797200.

Ohta, H., Shibata, Y., Haseyama, Y., Yoshino, Y., Suzuki, T., Kagasawa, T., Kamei, A., Ikeuchi, M., and Enami, I. (2005). Identification of genes expressed in response to acid stress in *Synechocystis* sp. PCC 6803 using DNA microarrays. Photosynth. Res. *84*, 225–230. http://dx.doi.org/10.1007/s11120-004-7761-x.

Paithoonrangsarid, K., Shoumskaya, M.A., Kanesaki, Y., Satoh, S., Tabata, S., Los, D.A., Zinchenko, V.V., Hayashi, H., Tanticharoen, M., Suzuki, I., *et al.* (2004). Five histidine kinases perceive osmotic stress and regulate distinct sets of genes in *Synechocystis*. J. Biol. Chem. *279*, 53078–53086. http://dx.doi.org/10.1074/jbc.M410162200.

Peier, A.M., Moqrich, A., Hergarden, A.C., Reeve, A.J., Andersson, D.A., Story, G.M., Earley, T.J., Dragoni, I., McIntyre, P., Bevan, S., *et al.* (2002). A TRP channel that senses cold stimuli and menthol. Cell *108*, 705–715. http://dx.doi.org/10.1016/S0092-8674(02)00652-9.

Peters, J.M., Mooney, R.A., Grass, J.A., Jessen, E.D., Tran, F., and Landick, R. (2012). Rho and NusG suppress pervasive antisense transcription in *Escherichia coli*. Genes Dev. *26*, 2621–2633. http://dx.doi.org/10.1101/gad.196741.112.

Pérez-Pérez, M.E., Martín-Figueroa, E., and Florencio, F.J. (2009). Photosynthetic regulation of the cyanobacterium *Synechocystis* sp. PCC 6803 thioredoxin system and functional analysis of TrxB (Trx x) and TrxQ (Trx y) thioredoxins. Mol. Plant. *2*, 270–283. http://dx.doi.org/10.1093/mp/ssn070.

Pollari, M., Gunnelius, L., Tuominen, I., Ruotsalainen, V., Tyystjärvi, E., Salminen, T., and Tyystjärvi, T. (2008). Characterization of single and double inactivation strains reveals new physiological roles for group 2 sigma factors in the cyanobacterium *Synechocystis* sp. PCC 6803. Plant Physiol. *147*, 1994–2005. http://dx.doi.org/10.1104/pp.108.122713.

Prakash, J.S., Sinetova, M., Zorina, A., Kupriyanova, E., Suzuki, I., Murata, N., and Los, D.A. (2009). DNA supercoiling regulates the stress-inducible expression of genes in the cyanobacterium *Synechocystis*. Mol. Biosyst. *5*, 1904–1912. http://dx.doi.org/10.1039/B903022k.

Prakash, J.S., Krishna, P.S., Sirisha, K., Kanesaki, Y., Suzuki, I., Shivaji, S., and Murata, N. (2010). An RNA helicase, CrhR, regulates the low-temperature-inducible expression of heat-shock genes *groES*, *groEL1* and *groEL2* in *Synechocystis* sp. PCC 6803. Microbiology *156*, 442–451. http://dx.doi.org/10.1099/mic.0.031823-0.

Prosecka, J., Orlov, A.V., Fantin, Y.S., Zinchenko, V.V., Babykin, M.M., and Tichy, M. (2009). A novel

ATP-binding cassette transporter is responsible for resistance to viologen herbicides in the cyanobacterium *Synechocystis* sp. PCC 6803. FEBS J. *276*, 4001–4011. http://dx.doi.org/10.1111/j.1742-4658.2009.07109.x.

Qiao, J., Shao, M., Chen, L., Wang, J., Wu, G., Tian, X., Liu, J., Huang, S., and Zhang, W. (2013). Systematic characterization of hypothetical proteins in *Synechocystis* sp. PCC 6803 reveals proteins functionally relevant to stress responses. Gene *512*, 6–15. http://dx.doi.org/10.1016/j.gene.2012.10.004.

Rosana, A.R., Ventakesh, M., Chamot, D., Patterson-Fortin, L.M., Tarassova, O., Espie, G.S., and Owttrim, G.W. (2012). Inactivation of a low temperature-induced RNA helicase in *Synechocystis* sp. PCC 6803: physiological and morphological consequences. Plant Cell Physiol. *53*, 646–658. http://dx.doi.org/10.1093/pcp/pcs020.

Rozanov, A. Yu., and Astafieva, M.M. (2009). The evolution of the early precambrian geobiological systems. Paleontol. J. *43*, 911–927. http://dx.doi.org/10.1134/S0031030109080103.

Lamparter, T. (2004). Evolution of cyanobacterial and plant phytochromes. FEBS Lett. *573*, 1–5. http://dx.doi.org/10.1016/j.febslet.2004.07.050.

Saidi, Y., Finka, A., Muriset, M., Bromberg, Z., Weiss, Y.G., Maathuis, F.J., and Goloubinoff, P. (2009). The heat shock response in moss plants is regulated by specific calcium-permeable channels in the plasma membrane. Plant Cell *21*, 2829–2843. http://dx.doi.org/10.1105/tpc.108.065318.

Saidi, Y., Peter, M., Finka, A., Cicekli, C., Vigh, L., and Goloubinoff, P. (2010). Membrane lipid composition affects plant heat sensing and modulates Ca^{2+}-dependent heat shock response. Plant Signal. Behav. *5*, 1530–1533.

Saita, E.A., and de Mendoza, D. (2015). Thermosensing via transmembrane protein-lipid interactions. Biochim. Biophys. Acta *1848*, 1757–1764. http://dx.doi.org/10.1016/j.bbamem.2015.04.005.

Saita, E., Abriata, L.A., Tsai, Y.T., Trajtenberg, F., Lemmin, T., Buschiazzo, A., Dal Peraro, M., de Mendoza, D., and Albanesi, D. (2015). A coiled coil switch mediates cold sensing by the thermosensory protein DesK. Mol. Microbiol. *98*, 258–271. http://dx.doi.org/10.1111/mmi.13118.

Saita, E., Albanesi, D., and de Mendoza, D. (2016). Sensing membrane thickness: Lessons learned from cold stress. Biochim. Biophys. Acta *1861*, 837–846. http://dx.doi.org/10.1016/j.bbalip.2016.01.003.

Sakayori, T., Shiraiwa, Y., and Suzuki, I. (2009). A *Synechocystis* homolog of SipA protein, Ssl3451, enhances the activity of the histidine kinase Hik33. Plant Cell Physiol. *50*, 1439–1448. http://dx.doi.org/10.1093/pcp/pcp089.

Sato, S., Shimoda, Y., Muraki, A., Kohara, M., Nakamura, Y., and Tabata, S. (2007). A large-scale protein protein interaction analysis in *Synechocystis* sp. PCC6803. DNA Res. *14*, 207–216.

Schmitt, F.J., Renger, G., Friedrich, T., Kreslavski, V.D., Zharmukhamedov, S.K., Los, D.A., Kuznetsov, V.V., and Allakhverdiev, S.I. (2014). Reactive oxygen species: re-evaluation of generation, monitoring and role in stress-signaling in phototrophic organisms. Biochim. Biophys. Acta *1837*, 835–848. http://dx.doi.org/10.1016/j.bbabio.2014.02.005.

Schmitt, F.-J., Kreslavki, V.D., Zharmukhamedov, S.K., Friedrich, T., Renger, G., Los, D.A., Kuznetsov, V.V., and Allakhverdiev, S.I. (2016). The multiple roles of various reactive oxygen species (ROS) in photosynthetic organisms. In Photosynthesis. A New Approach to the Molecular, Cellular, and Organismal Level. S.I. Allakhverdiev, ed. (Beverly, MA, USA, Scrivener Publishing LLC), pp. 1–84. http://dx.doi.org/10.1002/9781119084150.ch1.

Shapiguzov, A., Lyukevich, A.A., Allakhverdiev, S.I., Sergeyenko, T.V., Suzuki, I., Murata, N., and Los, D.A. (2005). Osmotic shrinkage of cells of *Synechocystis* sp. PCC 6803 by water efflux via aquaporins regulates osmostress-inducible gene expression. Microbiology *151*, 447–455. http://dx.doi.org/10.1099/mic.0.27530-0.

Shoumskaya, M.A., Paithoonrangsarid, K., Kanesaki, Y., Los, D.A., Zinchenko, V.V., Tanticharoen, M., Suzuki, I., and Murata, N. (2005). Identical Hik-Rre systems are involved in perception and transduction of salt signals and hyperosmotic signals but regulate the expression of individual genes to different extents in *Synechocystis*. J. Biol. Chem. *280*, 21531–21538. http://dx.doi.org/10.1074/jbc.M412174200.

Sinetova, M.A., Mironov, K.S., Mustardy, L., Shapiguzov, A., Bachin, D., Allakhverdiev, S.I., and Los, D.A. (2015). Aquaporin-deficient mutant of *Synechocystis* is sensitive to salt and high-light stress. J. Photochem. Photobiol. B. Biol. *152*, 377–382. http://dx.doi.org/10.1016/j.jphotobiol.2015.07.012.

Sinetova, M.A., and Los, D.A. (2016a). Lessons from the cyanobacterial transcriptomics: universal genes and triggers of stress responses. Mol. Biol. (Moscow) *50*, 606–614. http://dx.doi.org/10.1134/S0026893316040117.

Sinetova, M.A., and Los, D.A. (2016b). New insights in cyanobacterial cold stress responses: genes, sensors, and molecular triggers. Biochim. Biophys. Acta – General Subjects. *1860*, 2391–2403. http://dx.doi.org/10.1016/j.bbagen.2016.07.006.

Singh, A.K., Li, H., and Sherman, L.A. (2004). Microarray analysis and redox control of gene expression in the cyanobacterium *Synechocystis* sp. PCC 6803. Physiol. Plant. *120*, 27–35. http://dx.doi.org/10.1111/j.0031-9317.2004.0232.x.

Singh, A.K., Elvitigala, T., Cameron, J.C., Ghosh, B.K., Bhattacharyya-Pakrasi, M., and Pakrasi, H.B. (2010). Integrative analysis of large scale expression profiles reveals core transcriptional response and coordination between multiple cellular processes in a cyanobacterium. BMC Syst. Biol. *4*, 105. http://www.biomedcentral.com/1752–0509/4/105.

Sireesha, K., Radharani, B., Krishna, P.S., Sreedhar, N., Subramanyam, R., Mohanty, P., and Prakash, J.S. (2012). RNA helicase, CrhR is indispensable for the energy redistribution and the regulation of photosystem stoichiometry at low temperature in *Synechocystis* sp. PCC6803. Biochim. Biophys. Acta *1817*, 1525–1536. http://dx.doi.org/10.1016/j.bbabio.2012.04.016.

Story, G.M., Peier, A.M., Reeve, A.J., Eid, S.R., Mosbacher, J., Hricik, T.R., Earley, T.J., Hergarden, A.C., Andersson, D.A., Hwang, S.W., McIntyre, P., Jegla, T., Bevan, S., and Patapoutian, A. (2003). ANKTM1, a TRP-like channel expressed in nociceptive neurons, is activated by cold temperatures. Cell *21*, 819–829. http://dx.doi.org/10.1016/S0092-8674(03)00158-2.

Suzuki, I., Los, D.A., Kanesaki, Y., Mikami, K., and Murata, N. (2000). The pathway for perception and transduction of low-temperature signals in *Synechocystis*. EMBO J. *19*, 1327–1334. http://dx.doi.org/10.1093/emboj/19.6.1327.

Suzuki, I., Kanesaki, Y., Mikami, K., Kanehisa, M., and Murata, N. (2001). Cold-regulated genes under control of the cold sensor Hik33 in *Synechocystis*. Mol. Microbiol. *40*, 235–244. http://dx.doi.org/10.1046/j.1365-2958.2001.02379.x

Suzuki, I., Kanesaki, Y., Hayashi, H., Hall, J.J., Simon, W.J., Slabas, A.R., and Murata, N. (2005). The histidine kinase Hik34 is involved in thermotolerance by regulating the expression of heat shock genes in *Synechocystis*. Plant Physiol. *138*, 1409–1421. http://dx.doi.org/10.1104/pp.104.059097.

Takala, H., Björling, A., Berntsson, O., Lehtivuori, H., Niebling, S., Hoernke, M., Kosheleva, I., Henning, R., Menzel, A., Ihalainen, J.A., *et al.* (2014). Signal amplification and transduction in phytochrome photosensors. Nature *509*, 245–248. http://dx.doi.org/10.1038/nature13310.

Tan, X., Zhu, T., Shen, S., Yin, C., Gao, H., and Xu, X. (2011). Role of Rbp1 in the acquired chill-light tolerance of cyanobacteria. J. Bacteriol. *193*, 2675–2683. http://dx.doi.org/10.1128/JB.01454-10.

Tang, Q., Tan, X., and Xu, X. (2010). Effects of a type-II RNA-binding protein on fatty acid composition in *Synechocystis* sp. PCC 6803. Chin. Sci. Bull. *55*, 2416–2421. http://dx.doi.org/10.1007/s11434-010-3254-9.

Tuominen, I., Tyystjärvi, E., and Tyystjärvi, T. (2003). Expression of primary sigma factor (PSF) and PSF-like sigma factors in the cyanobacterium *Synechocystis* sp. strain PCC 6803. J. Bacteriol. *185*, 1116–1119. http://dx.doi.org/10.1128/JB.185.3.1116-1119.2003.

Tuominen, I., Pollari, M., Tyystjärvi, E., and Tyystjärvi, T. (2006). The SigB σ factor mediates high temperature responses in the cyanobacterium *Synechocystis* sp. PCC6803. FEBS Lett. *580*, 319–323. http://dx.doi.org/10.1016/j.febslet.2005.11.082

Uchiyama, J., Asakura, R., Kimura, M., Moriyama, A., Tahara, H., Kobayashi, Y., Kubo, Y., Yoshihara, T., and Ohta, H. (2012). Slr0967 and Sll0939 induced by the SphR response regulator in *Synechocystis* sp. PCC 6803 are essential for growth under acid stress conditions. Biochim. Biophys. Acta *1817*, 1270–1276. http://dx.doi.org/10.1016/j.bbabio.2012.03.028.

Varin, T., Lovejoy, C., Jungblut, A.D., Vincent, W.F., and Corbeil, J. (2012). Metagenomic analysis of stress genes in microbial mat communities from Antarctica and the High Arctic. Appl. Environ. Microbiol. *78*, 549–559. http://dx.doi.org/10.1128/AEM.06354-11.

Vigh, L., Los, D.A., Horváth, I., and Murata, N. (1993). The primary signal in the biological perception of temperature: Pd-catalyzed hydrogenation of membrane lipids stimulated the expression of the *desA* gene in *Synechocystis* PCC6803. Proc. Natl. Acad. Sci. U.S.A. *90*, 9090–9094.

Vlasits, J., Furtmüller, P.G., Jakopitsch, C., Zamocky, M., and Obinger, C. (2010). Probing hydrogen peroxide oxidation kinetics of wild-type *Synechocystis* catalase-peroxidase (KatG) and selected variants. Biochim. Biophys. Acta *1804*, 799–805. http://dx.doi.org/10.1016/j.bbapap.2009.12.007.

Vogt, J.H., and Schippers, J.H. (2015). Setting the PAS, the role of circadian PAS domain proteins during environmental adaptation in plants. Front. Plant Sci. *6*, 513. http://dx.doi.org/10.3389/fpls.2015.00513.

Wang, Y., and Xu, X. (2013). Effects of Rbp3 on lipid peroxidation and salt tolerance in *Synechocystis* sp. PCC 6803. FEBS Lett. *587*, 1446–1451. http://dx.doi.org/10.1016/j.febslet.2013.03.028.

Yeh, K.C., Wu, S.H., Murphy, J.T., and Lagarias, J.C. (1997). A cyanobacterial phytochrome two-component light sensory system. Science *277*, 1505–1508. http://dx.doi.org/10.1126/science.277.5331.1505.

Zavarzin, G.A. (2008). Microbial biosphere. In Biosphere Origin and Evolution, N. Dobretsov, N. Kolchanov, A. Rozanov, G. Zavarzin, eds. (Dortdrecht, Netherlands, Springer), pp. 25–42. http://dx.doi.org/10.1007/978-0-387-68656-1_2.

Zimorski, V., Ku, C., Martin, W.F., and Gould, S.B. (2014). Endosymbiotic theory for organelle origins. Curr. Opin. Microbiol. *22*, 38–48. http://dx.doi.org/10.1016/j.mib.2014.09.008.

Zirak, P., Penzkofer, A., Lehmpfuhl, C., Mathes, T., and Hegemann, P. (2007). Absorption and emission spectroscopic characterization of blue-light receptor Slr1694 from Synechocystis sp. PCC6803. J. Photochem. Photobiol. B. Biol. *86*, 22–34. http://dx.doi.org/10.1016/j.jphotobiol.2006.08.007.

Zorina, A., Stepanchenko, N., Novikova, G.V., Sinetova, M., Panichkin, V.B., Moshkov, I.E., Zinchenko, V.V., Shestakov, S.V., Suzuki, I., Murata, N., *et al.* (2011). Eukaryotic-like Ser/Thr protein kinases SpkC/F/K are involved in phosphorylation of GroES in the cyanobacterium *Synechocystis*. DNA Res. *18*, 137–151. http://dx.doi.org/10.1093/dnares/dsr006.

Zorina, A.A. (2013). Eukaryotic-type protein kinases of cyanobacteria. Russ. J. Plant Physiol. *60*, 589–596. http://dx.doi.org/10.1134/S1021443713040195.

Zorina, A.A., Bedbenov, V.S., Novikova, G.V., Panichkin, V.B., and Los, D.A. (2014). Involvement of serine/threonine protein kinases in cold stress response in the cyanobacterium *Synechocystis* sp. PCC 6803: functional characterization of a protein kinase SpkE. Mol. Biol. (Moscow) *48*, 390–398. http://dx.doi.org/10.1134/S0026893314030212.

Transcriptomic and Proteomic Analysis to Understand Systems-level Properties of Diurnal Cycles in Nitrogen-fixing Cyanobacteria

6

Uma K. Aryal[1]* and Louis A. Sherman[2]*

[1]Purdue Proteomics Facility, Bindley Bioscience Center, West Lafayette, IN, USA.
[2]Department of Biological Sciences, Purdue University, West Lafayette, IN, USA.

*Correspondence: uaryal@purdue.edu and lsherman@purdue.edu

Abstract

Photosynthetic cyanobacteria are important model organisms for many studies, and like plants, they have developed complex strategies for regulating metabolism during alternating light–dark cycles. Transcriptomics and proteomics technologies are now routinely applied for systems-level analysis of cyanobacteria. Although transcriptional regulation is a mature field, analysis of the cyanobacterial proteome is in its infancy. Thus, this review will concentrate on proteomics and focus on new technological improvements that have been applied to cyanobacteria. Proteins are the key functional units of the cell and their functions are controlled and driven by post-translational modifications (PTMs), and protein–protein interactions. We will emphasize research on the diazotrophic, unicellular cyanobacteria of the genus *Cyanothece*, since these strains perform two disparate types of metabolism: photosynthesis, which functions during the light, and nitrogen fixation, which functions during the dark. Transcriptional features varied dramatically throughout the 24 h diel cycle. However, proteome analysis revealed that proteins are much more stable throughout the light and dark and provides clues as to how cells conserve energy by keeping proteins within a modest range during the light–dark cycle. We emphasize that the development of new methods and mass spectrometers that enhances the performance of the proteomic workflow will be critical for future research, which should focus on PTMs, endogenous protein complexes and protein–protein interactions.

Introduction

Cyanobacteria and their ancestors have been around some 3 billion years and developed the two-photosystem electron transport apparatus that led to oxygenic photosynthesis. During this long period, cells have had to contend with the light–dark diel period, in which light for photosynthesis was high for part of the day and absent at other parts. One adaptation to this large change in light intensity was the circadian clock to control circadian rhythms. Circadian rhythms are self-sustaining and adjustable cycles that are entrained by light (L)/dark (D) (or temperature) cycles and that occur within about a 24 h period. The structure and functioning of the circadian clock in cyanobacteria has been studied extensively in the unicellular strain *Synechococcus elongatus* PCC 7942 (Dong *et al.*, 2010; Johnson *et al.*, 2011; Mackey and Golden, 2007; Shultzaberger *et al.*, 2015). This strain was chosen because of its superior genetic system and the choice was fortuitous. This organism contains one set of genes, termed *kaiABC*, that encode the clock proteins and they were quickly identified by mutagenesis. Subsequent work has resulted in the cyanobacterial clock becoming the best-characterized clock among all organisms. The cyanobacteria are currently the only bacteria known to have such a clock and a recent study has provided an interesting twist to the story. Pattanayak and co-workers (2016) have shown that the fundamental metabolite-sensing aspects of the clock Kai proteins can track daily cycles without a dedicated light sensor. Such metabolic sensing may be important to the system to be discussed in this review.

Nitrogen fixation is another function that began over 3 billion years ago and one that is still found exclusively in bacteria. It is possible that oxygenic photosynthesis formed in a cell already capable of N_2-fixation and that strategies to protect nitrogenase from oxygen toxicity have been in place for 3 billion years (Boyd *et al.*, 2015; Larsson *et al.*, 2011; Latysheva *et al.*, 2012; Stueken *et al.*, 2015). We have been interested in the strategy employed by unicellular cyanobacteria of the genus *Cyanothece* (Bandyopadhyay *et al.*, 2011; Sherman *et al.*, 2010; Welsh *et al.*, 2008). We have determined that these cells perform photosynthesis during the day and restrict N_2-fixation to the night time. They also use an ingenious strategy to produce energy at night and to remove any traces of oxygen in the cytoplasm. During photosynthesis and CO_2 fixation in the light, carbon is stored as glycogen in granules, which is then used as a substrate for respiration in the dark. This oxygen scavenging protects the nitrogenase for many hours in the dark (usually 6–8 h), at which time nitrogenase is degraded. A great many studies have been performed on *Cyanothece* sp. ATCC 51142 (henceforth *Cyanothece* 51142), a benthic strain, and *Cyanothece* sp. PCC 7822 (henceforth *Cyanothece* 7822), a fresh-water strain. Of the various *Cyanothece* spp. studied, only *Cyanothece* 7822 has been found to have a useable genetic system (Min and Sherman, 2010), so working with the two strains has been beneficial.

Synechococcus has many advantages for the study of the clock structure and function, but it is less metabolically robust than the *Cyanothece* spp. It does not fix nitrogen and it does not grow naturally on an external carbon source; thus, the types of physiological experiments performed have been limited. For this reason, the *Cyanothece* strains provide an excellent opportunity to integrate the physiology into a circadian rhythm mechanism that was designed for two key global processes – photosynthesis and N_2 fixation. Nonetheless, it is also noteworthy that *Synechococcus* has a number of other attributes that are more suitable for large-scale genomics, transcriptomics and proteomics experiments (Rubin *et al.*, 2015). *Synechococcus* has a smaller genome than *Cyanothece* 51142 and 7822 (2.72 Mb vs. 5.4 Mb and 6.37 Mb, respectively) and a less complex genome (one circular chromosome and two small plasmids, vs. six scaffolds and seven scaffolds, respectively). In addition, *Cyanothece* 51142 has 7 *kai* genes and *Cyanothece* 7822 has 11 *kai* genes and this makes the study of the clock mechanism and regulation much more difficult. Nonetheless, we will summarize the many studies with microarrays, RNA-seq and proteomics that have provided an excellent understanding of how gene expression is integrated in *Cyanothece* 51142 and *Cyanothece* 7822 into a daily light–dark strategy to cope with too much light or too much oxygen. We will concentrate on proteomics and provide a detailed background on the current approaches and technology at the forefront of this field.

Systems biology and proteomics

Much research in the biological sciences focuses on the complex relationship between genotypes and phenotypes. Traditionally, such studies were carried out with the 'one gene – one protein – one function' concept (Bensimon *et al.*, 2012), which was daunting and time consuming, and yielded incomplete data sets that provided little quantitative information. This approach has been superseded by the development of powerful genome sequencing technologies for mapping protein-coding genes in the genome, and functional genomic technologies, such as transcriptomics, proteomics and metabolomics, that are used to discover the dynamic aspect of the genomes including gene transcription, translation, protein modifications, and protein–protein interactions.

Life science research now focuses on understanding the biological system as a whole, and integrates experimental data sets with computational methods to predict and characterize the dynamic properties of cellular components and networks. High-throughput transcriptomic microarray data are readily available for systems-wide analysis. However, recent advances in mass spectrometry (MS)-based proteomic techniques, combined with robust protein or peptide separation techniques and database search algorithms, now allow researchers to simultaneously analyse thousands of proteins in a single experiment (Gstaiger and Aebersold, 2009). This synergy has rapidly accelerated the pace of biological research, and in recent years, we have seen an explosion of data from bottom-up proteomics studies. The global analysis of proteins arguably forms the principle level of information required to understand how cells function and adapt to external perturbations. Below, we outline current proteomic technologies, highlight successful applications of proteomics in cyanobacterial research, and discuss the challenges for future applications.

Proteomics connects genomics and transcriptomics to cellular structures, functions, and pathways that underpin the phenotypic traits of an organism. Cellular functions and interactions of many proteins are driven by their post-translational modifications (PTMs), and interactions with other cellular components including proteins, lipids, and metabolites. So, the proteome is highly dynamic, and more complex than the genome, and any alteration of modifications and interactions can impact phenotypic traits. Proteomics studies focus on comprehensive analysis of all the expressed proteins (proteome) in the cell, including their quantitation, interactions, subcellular localization and PTMs. Proteomics technologies are now commonly applied to complement other high-throughput analyses to generate complex, multi-faceted datasets for systems-level analysis (Moore and Weeks, 2011). Because of the diverse physicochemical properties of the amino acids, numerous diverse modifications and degradations, proteome level analysis is highly challenging compared to genomic and transcriptomic analyses. The complexity is further increased by the physical interactions of multiple protein subunits to form complexes and signalling networks (Altelaar *et al.*, 2013). During the last two decades, proteomics technologies have advanced immensely through technological advances in instrumentation, sample preparation, and computational and bioinformatics analyses (Cox and Mann, 2011). Most proteomic studies reply on liquid chromatography (LC) coupled with tandem mass spectrometry (LC-MS/MS) as the core technology, which is referred to as bottom-up proteomics.

Gel-based proteomics

The general proteomics workflow is depicted in Fig. 6.1. First introduced by O'Farrell in 1975 (O'Farrell, 1975), two-dimensional polyacrylamide gel electrophoresis (2D PAGE) was a key technology in proteome research (Gorg *et al.*, 2004; Hu *et al.*, 2003). 2D-PAGE resolves proteins in the first dimension based on isoelectric point, and in the second dimension based on molecular weight. 2D gels are particularly well-suited for detecting changes in protein phosphorylation, since each phosphorylation event produces an incremental reduction in the pI, resulting in characteristic horizontal rows of protein spots on 2D gels (Tang *et al.*, 2008). Good sample preparation is the key to achieving reproducible, high quality gels, and improvements in sample preparation, as well as protein visualization and MS techniques led to a significant increase in the number of proteomic studies employing 2D-PAGE (Rabilloud, 2002).

Gel-based proteomics still remains popular, but the technique has several limitations, such as a very limited dynamic range, so that it is less effective in detecting low abundance proteins (Kosako *et al.*, 2009). Proteins in biological systems can vary in dynamic range from six to nine orders of magnitude (Anderson and Anderson, 2002), but staining of 2D gels has a dynamic range of only a hundred-fold, so that only more abundant proteins are detected (Patton, 2002). The difference gel electrophoresis (DIGE) technique is more sensitive than conventional 2D-PAGE, but is still not sufficient to detect low abundance proteins (Deng *et al.*, 2007). The technique is very low throughput, and downstream sample processing is complicated and cumbersome. These limitations have motivated investigators to use 'gel-free' shotgun proteomics techniques, which will be discussed below.

Shotgun proteomics

Shotgun proteomics refers to the use of bottom-up proteomics techniques in characterizing proteins in a complex mixture. It typically consists of LC-MS/MS analysis of peptides derived from digested proteins, followed by database searching for matching MS spectra to the theoretical peptide spectra generated from protein sequences. Trypsin is the most commonly used (serine) protease for protein digestion. Other proteases, such as Glu-C, Lys-C, chemotrypsin or Asp-N protease are particularly useful for proteins with regions of high lysine (R) or arginine (K) content because these high RK regions are extensively cleaved by trypsin to yield very short

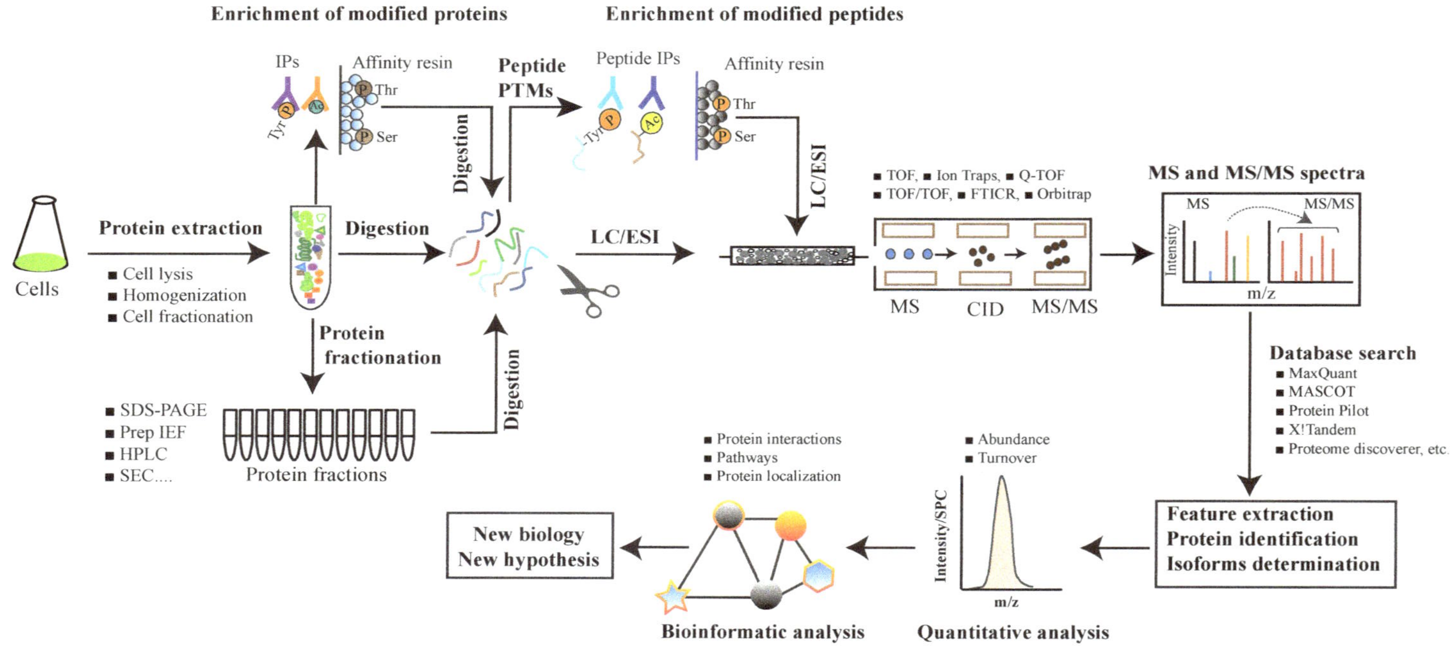

Figure 6.1 Proteomics workflow. Cell or tissue lysates are divided into different cell fractions by differential centrifugation. Proteins are digested, usually by trypsin, to generate peptides before mass spectrometry (MS) analysis. To reduce sample complexity, proteins are fractionated based on physicochemical properties, such as molecular weight (SDS-PAGE, SEC), charge (prep IEF), or hydrophobicity (RP-HPLC). Alternatively, subsets of the proteins can be targeted for MS analysis through their enrichment such as phosphorylation (P) or acetylation (Ac) using affinity-based resins or antibody-based immunoprecipitation (IP). Enzymatic digestion of the whole proteome generates thousands of peptides, but only a subset of those peptides can be detected by the MS. Therefore, peptide sample complexity is also reduced by fractionation based on the charge, isoelectric point, hydrophobicity. Modified peptides can be targeted through their enrichment using affinity-based resins or antibody-based IPs. These prefractionated or enriched peptide samples are introduced to the LC, and peptides eluting through the LC are detected by the MS and MS/MS analysis. Many peptides co-elute through the LC and ionization of highly abundant peptide species can suppress the signals of co-eluting less abundant peptide species, and prevent their detection in the MS. After ionization, peptide precursor ions (MS1) are introduced into the mass spectrometer, which records their mass-to-charge (*m/z*) ratio with high accuracy. For identification, single precursors are selected on the basis of observed intensity and subjected to CID for fragmentation and tandem MS (MS/MS) event. The MS and MS/MS spectra are then matched to known peptide sequence database using search algorithms such as Mascot, SEQUEST, Protein Pilot or X!Tandem. The identified peptides and protein data are quantified (either relatively or absolutely) using quantitative tools such as MaxQuant, Progenesis, etc. The protein abundance information is used to determine changes in protein expression. The results are further analysed using bioinformatic tools to interpret results, determine localization, related pathways, or protein–protein interactions. ESI, electrospray ionization; Prep IEF, preparatory isoelectric focusing; SEC, size exclusion chromatography; CID, collision-induced dissociation; RP-HPLC, reverse phase-high-performance liquid chromatography; PTM, post-translational modification.

peptides which provide little or no useful MS/MS sequence context. Protein sample preparation is the most critical for any proteomic experiment. Extraction methods should be efficient to recover as much of the protein as possible with little contamination. Various reagents are used to optimize methods for extracting proteins, and it is important to know that some of these reagents interfere with LC-MS/MS analysis. For example, residual amount of phenylmethylsulfonyl fluoride (PMSF), a serine protease inhibitor commonly used to prevent protein degradation, inhibits trypsin digestion (Matthiesen and Mutenda, 2007). Detergents such as SDS are frequently used to extract membrane proteins, but residual SDS interferes both with proteolytic digestion and LC-MS/MS analysis.

For complex protein sample, fractionation methods are implemented to separate proteins (e.g. 1D gels, gel filtration or ion exchange columns) or peptides (e.g. strong cation exchange or reverse phase columns) before LC-MS/MS analysis. Despite these technological advances, there are still many challenges in proteomic analysis that must be addressed. Reproducibility, sensitivity, and comprehensive analyses are important and should be considered in order to design successful MS-based proteomic experiments. Modern hybrid mass spectrometers can detect proteins in the low femtomole or attomole range, and this is sufficient to detect most proteins in their pure state. However, modern mass spectrometers cannot deal with the wide dynamic range of proteins. When complex protein or peptide mixtures are ionized, ion signal suppression from high abundant proteins/peptide ions prevents the detection of low abundant species. Sample pre-fraction at the protein or peptide levels is therefore, routinely used to reduce sample complexity and improve dynamic range.

Proteolytic digestion of a complex sample produces thousands of peptides. Only a small fractions of those peptides can be analysed by a modern tandem MS (Matthiesen and Mutenda, 2007; Rubin *et al.*, 2015). This 'under- sampling' issue is a huge challenge for quantifying low abundance proteins in a reproducible manner. This is further complicated by PTMs such as phosphorylation, acetylation, glycosylation, and lipidation. Sample pre-fractionation, repeated analysis and targeted proteomics methods are used to partially improve the under-sampling problem. Additionally, only a small fraction of the MS/MS spectra collected during the proteomics experiment can be reliably matched to peptides. PTMs are among several factors contributing to the low spectral matching. Current computational algorithms are not able to identify many unsuspected modifications, such as lipid-modified peptides. PTMs may reduce ionization efficiency and, may also alter or impair MS/MS fragmentation patterns. Developing an algorithm to reliably detect the fragmentation pattern of modified peptides is necessary.

A typical mass spectrometer consists of an ion source, mass analysers and a detector (Fig. 6.2). The type of mass analysers differentiates the MS instrument platforms. Ion trap, Orbitrap, and ion cyclotron resonance mass analysers separate ions based on m/z resonance frequency. Quadrupole mass analysers use m/z stability as a method of ion separation, and a time-of-flight (TOF) analyser separates ions based on flight time. Other properties of the MS are mass accuracy, resolution, sensitivity, mass scan range, scan speed, dynamic range, and the type of MS/MS fragmentation method. Many of the mass spectrometers available in the market today are hybrid instruments; they combine more than one analyser to achieve higher sensitivity, accuracy, speed and flexibility.

A number of methods have been developed for quantitative analysis of proteins using MS which can be broadly divided into stable isotope labelling and label-free methods. Label-free methods employ either spectral counts (SPC) or extracted ion current (XIC). Spectral counts are the numbers of non-redundant MS/MS spectra matching to a protein. This approach is based on the observation that more abundant proteins will produce more peptides than the low abundance proteins, and are sampled more frequently in fragment ion scans, and generate more MS/MS spectra than the low abundance peptides. Relative quantitation by spectral count thus involves comparing the number of spectra from the same protein between different samples (Liu *et al.*, 2004). Spectral count is a protein-centric approach, and works reasonably well for proteins with high spectral counts, but the reliability decreases dramatically for proteins with low spectral counts (Zhou *et al.*, 2010). The number of spectra increases almost linearly with protein abundance after normalizing. Many peptides often match to multiple proteins or protein

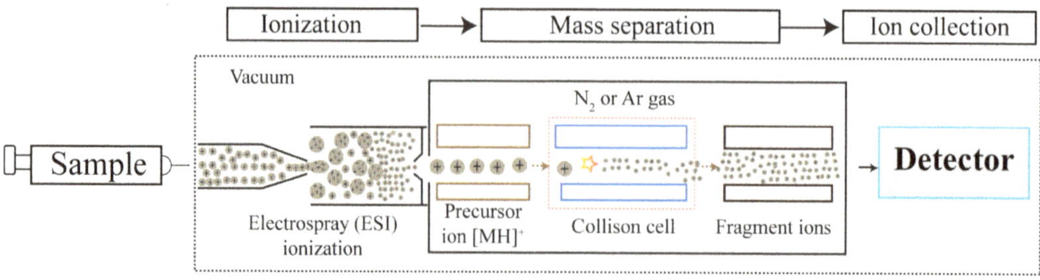

Figure 6.2 Elements of mass spectrometer. Peptides are separated by HPLC and eluted into an electrospray ion source where they are nebulized in small, highly charged droplets. A high voltage applied at the front end of the MS generates charged droplets which pass through the heated capillary and/or a curtain N_2 gas across the spray to help the desolvation process. The peptide ions pass from the source into the mass analyser, where the bulk solvent from the droplets is pumped away by the vacuum, and ions are drawn into the mass analyser. In tandem MS, the first mass analyser isolates a single peptide ion (m/z) species, and directs this ion to collision-induced dissociation (CID) in the collision cell, which induces fragmentation of the peptide into fragment ions and neutral fragments. The fragment ions are then analysed based on m/z to produce a product ion spectrum in the second mass analyser, and then recorded by the detector. The information contained in the tandem or MS/MS spectrum is used to deduce peptide sequence by database searches.

isoforms, which are referred as 'shared peptides', and assigning the shared peptides counts to multiple proteins is an issue. This is generally achieved by calculating the adjusted spectral counts, which is the sum of the unique spectral counts and a fraction of the spectral counts from peptides shared among multiple proteins and distributed in proportion to their unique spectral counts (Olinares *et al.*, 2010; Zybailov *et al.*, 2006). Low dynamic range of the spectral counts measurement, compared to the XIC MS1 intensity measurement, is also a problem of the method.

Relative quantitation using MS1 intensity measurements involves comparing the XIC of peptide ion belonging to a given protein (Wang *et al.*, 2003). The XICs are independent of the MS/MS or product ion scans. They are generated from multiple scans of the same peptide in an LC-MS run, and their peak areas are integrated over the chromatographic time scale. Higher numbers of scans (data points) of the same peptide generate better XIC MS1 peaks. A prerequisite for accurate label-free MS1 quantitation is the reproducible peptide quantitation by MS and the reproducible peptide retention with m/z ion features. Label-free techniques have been performed in many studies and are promising alternatives to stable isotope labelling. They are fast, easy to perform, and inexpensive, since no special techniques or reagents are required. Label-free methods allow for higher dynamic range measurements, and analysis of an unlimited number of samples. However, sample analysis under identical settings is critical for accurate measurements, and peptide alignment is crucial for defining MS1 signals.

Stable isotope labelling techniques are based on the introduction of a differential mass tag, which affects only the mass of a protein or peptide without changing the chemical properties during chromatography or MS (Aebersold and Mann, 2003). Stable isotopes are introduced metabolically (SILAC, ^{15}N) or chemically (iTRAQ, ICAT, TMT, ^{18}O) at either the protein or peptide level. This approach helps to resolve the problem associated with LC-MS/MS variations in label-free quantitation. Relative or absolute quantitation can be achieved by comparing XIC MS1 intensity of the labelled 'heavy' peptides against the 'light' peptides (Ong *et al.*, 2002).

In metabolic labelling, stable isotopes are incorporated into the cells through the growth medium during normal cell growth. Differently labelled samples can be pooled together for subsequent preparation and analysis, which avoids variability of sample preparation. However, this method is applicable to organisms or cells that can be grown in the growth medium, and is not applicable to samples that are not metabolically active (e.g. plasma) (Elliott *et al.*, 2009). While the original ^{15}N labelling can only compare two samples in one experiment, high-throughput quantitation was developed in the form of stable isotope labelling by amino acids in cell cultures (Ong *et al.*, 2002). SILAC incorporates heavy and light forms of arginine or lysine *in vivo* and allows detection of very small changes in protein levels or protein modifications. In chemical labelling, label is introduced to proteins or peptides by a chemical reaction, such as with isotope-coded affinity tags (ICAT) (Gygi *et al.*, 1999). ICAT labels are introduced at the protein level, and specifically bind to thiol groups of cysteine. This chemical labelling strategy effectively reduces sample complexity, but also limits its application to only quantify cysteine-containing peptides. Another limitation of ICAT is that only two samples can be compared in a single analysis, and the D0/D8 ICAT reagents are known to cause large shift in chromatographic retention of peptides. After digestion, the labelled peptides were purified on Avidin affinity columns.

The development of isobaric mass tags such as tandem mass tag (TMT) (Thompson *et al.*, 2003) and isobaric tags for relative and absolute quantification (iTRAQ) (Gygi *et al.*, 1999; Ross *et al.*, 2004) now allow for the comparison of up to eight samples in parallel. iTRAQ utilizes isobaric reagents to label the primary amines of peptides and proteins. It contains a reporter group of an *N*-hydroxy succinimide ester group (masses = 113, 114, 115, 116, 117, 118, 119, 121), a balance group and an anime specific peptide reactive *N*-hydroxy succinimide ester group (Ross *et al.*, 2004). The balance group functions to make the labelled peptides from each sample isobaric (same mass), and quantitation is performed based on the analysis of reporter groups that are generated upon fragmentation during LC-MS/MS analysis. TMT and iTRAQ labelling are attractive methods for quantitative proteomic analysis such as biomarker discovery due to their ability to multiplex up to eight samples.

For highly accurate protein or peptide quantitation, multiple reaction monitoring (MRM) is routinely applied. MRM targets specific peptides in complex samples by typically using a triple quadrupole (QQQ) mass spectrometer or hybrid QQQ/linear ion trap mass spectrometer. These instruments have two mass filters that can select a predefined peptide ion and a combination of its specific fragment ions to analyse and monitor over time for accurate quantitation. The acquisition of data only for a selected peptides allows measurements with higher precision, sensitivity, and throughput (Liebler and Zimmerman, 2013). For absolute quantitation, MRM assays are performed by incorporating isotopically labelled synthetic peptides as internal standards (Gerber *et al.*, 2003). The concentration the target peptide in the sample can be calculated based on the known concentration of the isotopically labelled synthetic peptide spiked into the sample (Lange *et al.*, 2008). The MRM method allows us to achieve greater sensitivity towards low abundance peptides, and is capable of detecting attomole level of peptides across a dynamic range of up to 10^5 (Qian *et al.*, 2006; Wolf-Yadlin *et al.*, 2007).

Triple quadrupole and quadrupole-ion trap (Q-Trap) MS instruments are used for MRM assays. A limitation of QQQ instruments is the relatively low resolution of precursor ion measurement, and limited duty cycle (the rate at which transitions can be sampled with an acceptable signal-to-noise ratio). The hydrid quadrupole-time-of-flight (QqTOF) instrument can also be used for similar quantitative applications. This MS contains a mass-resolving quadrupole (Q) and a collision cell (q) similar to a triple quadrupole; however, the third quadrupole (Q3) is replaced with a time-of-flight (TOF) mass analyser offering high sensitivity, mass resolution, and mass accuracy for both precursor and product ion spectra (Chernushevich *et al.*, 2001). The ability of both the Q Trap and QqTOF instruments to acquire full scan product ion data with high mass accuracy allows these platforms to be alternative instrumentation for performing quantitation, in addition to the more traditionally used QQQ MS.

Cyanobacterial proteomics

Proteomics has gained increasing popularity in cyanobacterial research. Both gel and gel-free proteomic approaches have been applied to a diverse group of unicellular and filamentous cyanobacteria including *Synechocystis* 6803, *Anabaena*, *Cyanothece* sp. (ATCC 51142, PCC 7822, PCC 7424, PCC 7425, PCC8801, PCC8802) (Aryal *et al.*, 2011, 2012b, 2013; Stöckel *et al.*, 2011; Wegener *et al.*, 2010), *Arthrospira* 8005, and others (Battchikova *et al.*, 2015). Proteomics studies in cyanobacteria have primarily focused on identifying differentially expressed proteins during normal or stressed conditions. With the development of new technology and analytical methods, cyanobacterial proteomics has also shifted from identification of rather small sets of proteins to large-scale analyses of the whole proteomes, including post-translational modifications (PTMs), and protein–protein interactions. One major application of proteomic technology for cyanobacterial research has been to study environmental stress responses, and photosynthesis and nitrogen fixation during the diurnal life cycle. More generally, *Synechocystis* sp. 6803 has become the model organism for investigating the stress responses of photosynthetic organisms, and is one of the most extensively studied cyanobacteria (Huang *et al.*, 2006).

Proteomic analysis of stress responses

Proteomic analysis have been applied to examine the changes in protein expression in cyanobacteria under salt, heat, cold, metal, acid or nutrients (carbon or nitrogen) stresses (Aryal *et al.*, 2014; Castielli *et al.*, 2009; Eriksson and Clarke, 1996; Fulda *et al.*, 2006; Guo *et al.*, 2014; Kurian *et al.*, 2006; Sato *et al.*, 2008; Tanaka *et al.*, 1997; Tanaka and Nakamoto, 1999; Tuominen *et al.*, 2008; Wegener *et al.*, 2010). Analysis of periplasmic and soluble proteins of *Synechocystis* sp. 6803 after cold osmotic shock by 2-DE and MALDI-TOF MS identified enzymes involved in the synthesis of glucosylglycerol, heat shock proteins (DnaK, GroEL, GroES, Hsp) and reactive oxygen species (ROS) scavenging proteins (superoxide dismutase, peroxiredoxin, etc.) (Fulda *et al.*, 2006). Expression of a number of proteins involved in cell wall biosynthesis was also enhanced by osmotic stress. Plasma membrane proteomics of *Synechocystis* sp. 6803 after salt stress resulted in the identification of many ABC transporters, proteins involved in PSII protection, thylakoid membrane formation, and multiple regulatory proteins (Huang *et al.*, 2006). Cyanobacteria exposed to heat stress showed increased expression of two chaperones (GroEL1 and GroEL2) (Sato *et al.*, 2008; Tanaka *et al.*, 1997). Inactivation of *clpB* protease genes, and genes encoding chaperones and heat shock proteins (*hspA, htpG, dnaK2*), sigma factors (*sigB* and *sigC*) induced thermal sensitivity (Eriksson and Clarke, 1996; Tanaka and Nakamoto, 1999; Tuominen *et al.*, 2008; Varvasovszki *et al.*, 2003). The Hik34 histidine kinase mutation in *Synechocystis* also down regulated the expression of several heat shock genes as the *Synechocystis* *Δhik34* deletion mutants showed increased thermal tolerance under short-term heat stress (Slabas *et al.*, 2006; Zorina *et al.*, 2011; Červený *et al.* 2015).

Thermal (heat) shock increased the expression of chaperones, several proteases, and proteins related to metabolic processes (ribosomal subunits, tRNA synthetases, elongation factors) in *Synechocystis* sp. 6803 (Suzuki *et al.*, 2006). Proteins involved in amino acid biosynthesis (D-3-phosphoglycerate dehydrogenase, arginosuccinate synthetase, aspartate kinase), proteins from the water-splitting complex (PsbO), the reaction centre (PsbW) in PSII, and energy metabolism (ATP synthase subunits, components of the glycolytic/pentose phosphate pathway) also showed increased expression, whereas phycobilisomes and Rubisco proteins showed decreased expression in response to heat stress (Suzuki *et al.*, 2006). Subcellular proteomic analysis of high temperature stress in *Spirulina platensis,* a cyanobacterium commonly grown in outdoor ponds, showed up-regulation of proteins involved in two-component response systems, and indicated a connection between thermal stress and with nitrogen and ammonia assimilation, and rapid inactivation of photosynthesis (Hongsthong *et al.*, 2009). A comparative proteomic analysis of the responses of *Synechocystis* sp. PCC 6803 to acid stress, using 2-DE coupled with MALDI-MS and LC-MS/MS, showed increased expression of cytoplasmic malate dehydrogenase and decreased expression of elongation factor EF-G, polyribonucleotide nucleotidyl transferase, and Rubisco small subunit (RbcS) (Kurian *et al.*, 2006). The authors reported striking changes in the expression of periplasmic proteins due to acid treatment, including oxalate decarboxylase, carbonic anhydrase, and periplasmic iron-binding proteins.

In one of the most comprehensive analyses, Pakrasi and co-workers performed proteomic analysis of *Synechocystis* sp. *PCC 6803* under 33 different environmental conditions (without nitrogen, iron, sulfur, phosphorus, ammonia, NaCl, glucose, cold stress, etc.) (Wegener *et al.*, 2010). They applied gel-free and label free MS-based quantitative proteomic profiling to identify 1198 differentially regulated proteins. Many housekeeping proteins such as key amino acid biosynthesis were up-regulated in most of the nutrient depleted conditions, but photosynthetic proteins were mostly down-regulated, particularly under nitrogen depletion. They postulated that urea and arginine are the preferred carbon and nitrogen sources when cells experienced C and N depletion. Under limited carbon and nitrogen sources, *Synechocystis* cells actively utilize internal and external carbon sources as evident from the up-regulation of proteins involved in glycolysis, glucose transport, and pentose phosphate pathways (Wegener *et al.*, 2010). Mota and co-workers (Mota *et al.*, 2015) studied the effects of several heavy metals on protein profiles of *Cyanothece* sp. CCY 0110, the highly efficient extracellular polymeric substances (EPS) -producing cyanobacterium. This study reported distinct specific effects related to the time of exposure and/or the concentration of an essential (Cu^{2+}) and a non-essential (Cd^{2+}) metal, and the majority of the differentially regulated proteins were associated with photosynthesis (PsaA, PsbU, ferredoxin 2Fe–2S, ferredoxin NADP, plastocyanin, etc.), CO_2 fixation (RbcL, Pgk, Fructose bisphosphate aldolase, glyceraldehyde 3-phosphate dehydrogenase), and carbohydrate metabolism, translation, and nitrogen and amino acid metabolism.

Proteomic analysis of post-translational modifications

Posttranslational modifications (PTMs) such as phosphorylation, acetylation, glycosylation, and thiol oxidation are key components of the signalling networks, and play important regulatory roles (Aryal *et al.*, 2012a; Gstaiger and Aebersold, 2009; Xiong *et al.*, 2016). PTMs control the activity, localization and stability of proteins, and their interactions with other macromolecules (Gstaiger and Aebersold, 2009). Since functional genomics cannot provide experimental evidence for protein modifications, mass spectrometry-based proteomics is a useful technique for analysing PTMs. There are more than 300 different types of PTMs, and only a handful of those have been extensively studied (Jensen, 2006; Zhao and Jensen, 2009; see also Chapter 7 of this book).

Phosphorylation

Phosphorylation is one of the most important and widespread PTMs. It may affect approximately one-third of the proteome and plays an important role in many cellular processes (Kosako *et al.*, 2009). The identification of phosphorylated proteins and phosphorylation sites is challenging, because cellular levels of such proteins are often too low for detection using conventional methods (Jensen, 2004; Tang *et al.*, 2008). The size, charge and hydrophobicity of tryptic phosphopeptides also make it difficult to identify and determine the stoichiometry of phosphorylation for individual proteins using MS (Aryal *et al.*, 2012a; Jensen, 2004).

Over the years, several methods have been developed for the analysis of the protein PTMs. Antibody-based affinity enrichment is a popular and useful technique for enriching the corresponding modified proteins. For example, phosphotyrosine (pTyr)-specific antibodies are used for the enrichment of pTyr-modified proteins (Pandey *et al.*, 2000). Phosphoserine (pSer)- and phosphothreonine (pThr)-specific antibodies are also available, and can now be routinely applied (Grønborg *et al.*, 2002). Other affinity-based phosphopeptide enrichment techniques include immobilized metal-affinity chromatography (IMAC) (Posewitz and Tempst, 1999; Stensballe *et al.*, 2001), and metal oxide affinity chromatography using TiO_2 or ZrO_2 columns (Larsen *et al.*, 2005; Pinkse *et al.*, 2004). IMAC takes advantage of the affinity of the phosphate moiety for Fe (III) or Ga (III), and has been successfully applied to identify thousands of phosphorylated residues in proteins from various species (Aryal *et al.*, 2012a; Aryal and Ross, 2010). Under optimized conditions, IMAC is also useful for the enrichment of intact phosphoproteins (Aryal *et al.*, 2012a; Link *et al.*, 1999). The TiO_2-based enrichment is more selective and more specific towards phosphopeptides, particularly singly phosphorylated peptides than the IMAC-based method (Larsen *et al.*, 2005).

Phosphorylation-based signal transduction mechanism in cyanobacteria sometime involve two-component signal transduction systems, consisting of histidine kinases and the corresponding response regulators (Los *et al.*, 2010). But, recent phosphoproteomic analyses using MS have identified many Ser/Thr/Tyr kinases and phosphatases that play important roles in regulating numerous biological processes in cyanobacteria, such as stress adaptation, cell differentiation, cell motility, carbon and nitrogen metabolism and photosynthesis (Zhang *et al.*, 2005; Zorina *et al.*, 2011). Genes encoding putative Ser/Thr/Tyr kinases range from none to 52 in cyanobacterial strains (Zhang *et al.*, 2005), and indicate the importance of phosphorylation-based signalling systems in cyanobacterial physiology and adaptations (Baniulis *et al.*, 2009; Sobczyk *et al.*, 1993). Protein phosphorylation in cyanobacteria has been confirmed using [^{32}P] phosphate labelling since the 1980s (Spat *et al.*, 2015), although only a few phosphorylated proteins have been characterized. Phosphoproteins were shown to be involved in environmental adaptation in cyanobacteria (Spat *et al.*, 2015). The nitrogen regulatory protein PII (GlnB), which plays a critical role in sensing changes in the internal N/C ratio, is the most intensively studied and best elucidated phosphoprotein in cyanobacteria (Forchhammer, 2008). Phycobiliproteins and phycobilisome linker proteins also are known to be phosphorylated in cyanobacteria (Harrison *et al.*, 1991; Piven *et al.*, 2005). The circadian clock protein, KaiC undergoes circadian oscillations through autocatalytic phosphorylation/de-phosphorylation cycles (Axmann *et al.*, 2014; Williams, 2007). Yang and co-workers (Yang *et al.*, 2013) used biochemical enrichment of phosphopeptides combined with MS to identify 410 unique phosphoryl sites, and 280 unique phosphopeptides from 245 proteins in *Synechococcus* PCC 7002 cells. Several groups have also investigated phosphoproteome dynamics in *Synechocystis* sp. PCC 6803 under autotrophic to heterotrophic growth using both gel-based and gel-free MS methods (Lee *et al.*, 2015; Mikkat *et al.*, 2014; Spat *et al.*, 2015). These observations reveal that reversible protein phosphorylation in cyanobacteria plays an essential role in many cellular processes such as photosynthesis and carbon/nitrogen metabolism.

Glycosylation

Protein glycosylation provides a means for cell–cell recognition and communication, and has provided the basis for several proteomic studies (Dube and Bertozzi, 2005; Wang *et al.*, 2012). Glycoprotein analysis by MS is in its infancy compared to phosphoproteins, but several promising methods, including lectin-affinity chromatography followed by MS, have been developed (Harvey, 2005; Wang *et al.*, 2012; Wilson *et al.*, 2002). Lectins have distinct specificities towards various classes of glycan and can be used to enrich both glycoproteins and glycopeptides (Gabius *et al.*, 2002). Combining different types of lectins can be used to investigate *N*-glycosylated and *O*-glycosylated proteins and peptides (Yang and Hancock, 2004). Lectin-affinity chromatography can be coupled with the hydrophilic-interaction liquid chromatography (HILIC) for the isolation of the enriched glycopeptides prior to MS analysis (Hagglund *et al.*, 2004). Tagging proteins modified by *O*-linked *N*-acetylglucosamine and farnesylation by metabolic incorporation of PTM precursor analogues, and activation of such analogue to react with biotin enables for selective recovery of the modified glycoproteins by streptavidin-affinity enrichment and identification by MS.

Acetylation

PTMs by lysine acetylation is also emerging as a major regulatory mechanism in both eukaryotes and prokaryotes (Verdin and Ott, 2015). Protein acetylation is the addition of an acetyl group to either the ε-amines of Lys side chains (Lysine acetylation) or the α-amines of protein N-termini (N-terminal acetylation). However, compared to other PTMs, the physiological role of *N*-terminal acetylation is relatively unclear. Owing to the technical challenges in identifying acetylation sites, the study of protein acetylation has been challenging (Choudhary *et al.*, 2014). It took more than 20 years from the discovery of histone acetylation until the first non-histone acetylation of tubulin was discovered (L'Hernault and Rosenbaum, 1985). Immunoprecipitation with anti-acetyl lysine antibody, followed by MS, is a commonly used technique to identify lysine-acetylated proteins. Enrichment of N(α)-acetylated peptides via reductive dimethyl labelling, followed by identification with MS, has been developed (Chen *et al.*, 2013).

For many years, acetylation was synonymous with histone acetylation, since it was considered to be an epigenetic modification that occurred mainly or exclusively on chromatin-associated proteins (Choudhary *et al.*, 2014). However, several recent proteomic studies have reported that the majority of acetylation events occur on non-nuclear proteins (Choudhary *et al.*, 2009; Kim *et al.*, 2006; Zhao *et al.*, 2010). Acetylation has turned out to be widespread in the mitochondrial proteome and is more prevalent than phosphorylation (Weinert *et al.*, 2011). An analysis of a global acetylome in *Synechocystis* sp. PCC 6803 exposed to three different stresses (nitrogen-deficiency, heterotrophic growth, and high light) was performed by peptide prefractionation followed by antibody enrichment, and MS. This study identified 776 acetylation sites on 513 acetylated proteins (Mo *et al.*, 2015). Many proteins involved in cellular metabolism and in photosynthesis, including the subunits of phycocyanin (CpcA, CpcB, CpcC, and CpcG) and allophycocyanin (ApcA, ApcB, ApcD, ApcE, and ApcF), were found to be lysine acetylated (Mo *et al.*, 2015). Relatively smaller number of cyanobacterial proteins have been reported as *N*-terminal acetylation (Xiong *et al.*, 2016). These proteins included the Rieske iron–sulfur protein (PetC), Cytochrome b_{559} subunit beta (PsbF) and PS II reaction centre protein J (PsbJ) (Baniulis *et al.*, 2009; Guskov *et al.*, 2009). Such results are promising and need to be repeated and extended to more strains and additional environmental perturbations.

Thiol oxidation

Protein modification by reversible thiol oxidation plays an essential role in many biological processes in cyanobacteria (Foyer and Noctor, 2009). Protein thiol oxidation forms diverse modifications such as disulfide bond formation (Sato and Inaba, 2012), *S*-nitrosylation (Derakhshan *et al.*, 2007; Hess *et al.*, 2005), *S*-glutathionylation (Hess *et al.*, 2005), and *S*-sulfenylation (Paulsen *et al.*, 2011). Redox regulation of photosynthesis is associated with the photosynthetic electron transport (PET) chain and regulation of gene expression. In light/dark cycles, thiol oxidation (e.g. disulfide formation), through the thioredoxin and glutaredoxin systems modulates the activation status of many enzymes linked to PS I and PS II, as well as many other redox-related processes in plants (Buchanan and Balmer, 2005) and cyanobacteria (Lindahl and Florencio, 2003). Little is known about how the redox process is involved beyond photosynthesis; more importantly, what specific cysteine (Cys) residues serve as redox switches in photosynthetic organisms. Measurement of *in vivo* thiol redox status has been challenging, because free thiols are unstable and prone to oxidation during sample preparation with labile, reversibly oxidized Cys-residues (Guo *et al.*, 2014). Most initial methods used differential chemical or isotopic labelling followed by biotin-based affinity enrichment. In these methods, free protein thiols were initially blocked by alkylation, and then specific oxidized Cys-residues were selectively reduced and labelled with biotin tags (Jaffrey *et al.*, 2001), fluorophores (Nelson *et al.*, 2010), radioactive compounds (Leichert *et al.*, 2008), or isotope-coded affinity tags (Sethuraman *et al.*, 2004) and MS analysis. Thiol-reactive chemical probes coupled with click chemistry have also been applied to identify redox-sensitive proteins or Cys-residues (Deng *et al.*, 2013; Weerapana *et al.*, 2010), and *in vivo* thiol redox changes (Sadler *et al.*, 2014). An alternative iodoTMT or cysTMT switch approach for multiplexed quantification of reversible thiol modifications has also been reported (Murray *et al.*, 2012). However, the specificity of the anti-TMT-based enrichment is relatively poor; in one recent report, only 21% of enriched peptides were observed to be TMT-labelled cysteine containing peptides (Qu *et al.*, 2014). Recently, resin-assisted enrichment of Cys-containing peptides and Cys-based modifications also have been developed (Forrester *et al.*, 2009; Liu *et al.*, 2004). Applying resin-assisted enrichment with TMT labelling and MS analysis to *Synechocystis* PCC 6803, Qian and co-workers (Guo *et al.*, 2014) reported ~2100 Cys-sites from 1060 proteins under light, dark, and 3-(3,4-dichlorophenyl)-1,1-dimethylurea (a PS II inhibitor) conditions. They also confirmed the functional significance of redox sensitive cysteine modifications in NADP-dependent glyceraldehyde-3-phosphate dehydrogenase, peroxiredoxin, and glucose 6-phosphate dehydrogenase, using site-specific mutagenesis and biochemical studies. This highlighted the importance of redox-mediated changes in the thiol proteome in cyanobacteria. These proteomic studies clearly indicated that PTMs are ubiquitous in cyanobacteria, and play essential roles in maintaining cell physiology and function (Gallon *et al.*, 2000).

Protein–protein interactions

Many proteins form complexes with other proteins or macromolecules to execute their functions. As such, protein complexes are the cornerstones of many cellular control mechanisms (Alberts, 1998; Srere, 2000), and can be viewed as fundamental building blocks of cell biology that enable precise control within and between cellular pathways (Hartwell *et al.*, 1999). One way to deducing a protein function is to identify the interacting partners of that protein. Subunits of protein complexes are generally involved in the same cellular pathways (Phizicky *et al.*, 2003). Different techniques for pairwise detection of protein–protein interactions have been developed, such as phage display (Lindmark *et al.*, 1996), protein chips (Zhu *et al.*, 2001), Bimolecular fluorescence complementation (BiFC) (Kerppola, 2006), and the yeast two-hybrid (Y2H) system (Fields and Song, 1989; Fields and Sternglanz, 1994). Y2H screen is a genetic approach to the identification of protein–protein interactions. It relies on the modular nature of many eukaryotic transcription factors, which normally consist of two functionally and structurally independent domains: DNA-binding (DB) and activation (AD) domains that recruits the transcriptional machinery (Phizicky *et al.*, 2003). Expression of a reporter gene requires the binding of a TF, and the DB domain binds to the particular DNA sequence upstream of the reporter gene, while the AD domain activates reporter gene expression. The protein fused to the DBD is referred to as the 'bait', and the protein fused to the AD is referred as the 'prey'. In practice, a protein of interest or 'bait' is screened against a library of 'preys' (Phizicky *et al.*, 2003). Y2H assays are sensitive, very flexible for automation and high-throughput analysis, and have been successfully applied to different model organisms (Krogan *et al.*, 2006; Mukhtar *et al.*, 2011; Uetz *et al.*, 2000; Yu *et al.*, 2008). However, there are technical limitations to the yeast two-hybrid system (Jansen and Gerstein, 2004; Wodak *et al.*, 2009). For example, it often gives high false positive results, and interaction partners have to be identified by colony PCR analysis and sequencing, making such screens more expensive and time-consuming. It is only capable of detecting binary interactions, and detection of interacting proteins only possible in living cells. Fluorescence-based interaction assays have also been developed, but have not been used on a high-throughput basis.

Tandem affinity purification (TAP) (Rigaut *et al.*, 1999) followed by MS technique has been used to systematically purify multiprotein complexes on a large-scale (Gavin *et al.*, 2002). TAP method involves the fusion of the TAP tag to the target protein and the introduction of the construct to the host cell or organism (Puig *et al.*, 2001). It has been successfully applied to many different model organisms (Braun *et al.*, 2013; Gavin *et al.*, 2002; Uetz *et al.*, 2000). A disadvantage of TAP is that each tagged protein must be analysed individually, and broad proteome coverage would require the generation of thousands of transgenic lines and greater numbers of parallel purifications and MS analyses. The expression level of the tagged proteins and the effect of the affinity tag on protein function are important technical considerations for this technology (Wodak *et al.*, 2009). TAP is not possible in many organisms because it is not feasible to generate the large number of high-quality transgenic lines that would be needed. An alternative approach to protein complex analysis is to capitalize on the parallel protein separation, identification and quantification using MS (Aryal *et al.*, 2014; Kristensen *et al.*, 2012).

Recently, native gel electrophoresis and size exclusion chromatography which separate proteins based on their mass, and ion exchange chromatography (IEX) which separates proteins based on the charge are combined with quantitative LC-MS/MS profiling and cluster analysis to predict protein complexes (Dong *et al.*, 2008; Liu *et al.*, 2008). These methods have been successfully applied to analyse the size and subunit composition of endogenous protein complexes in the *Arabidopsis* chloroplast stroma (Olinares *et al.*, 2010; Peltier *et al.*, 2006), Arabidopsis leaf cytosol (Aryal *et al.*, 2014). This approach requires only a well-annotated genome sequence and retains useful information on subcellular localization. This powerful approach was also applied to predict putative protein complexes in *Arabidopsis*. In order to use profiling-based methods to predict protein complex composition from a cell fraction, hundreds of LC-MS/MS samples from different column fractions would be needed to sufficiently reduce the confounding effects of chance co-elution. Such strategies are being developed in human cell culture systems, in which MS-based abundance profiling of native complexes has been coupled with different types of column chromatography and bioinformatic analyses to make global predictions about protein complex composition (Havugimana *et al.*, 2012; Kristensen *et al.*, 2012) or to analyse how protein complexes can change as a function of growth factor treatment (Kristensen *et al.*, 2012) or posttranscriptional control (Kirkwood *et al.*, 2013). Alternatively, computational methods are used to predict genome-wide protein–protein interactions (PPIs) (Rhodes *et al.*, 2005; von Mering *et al.*, 2002).

Many protein complexes have been identified in different cyanobacterial strains including a giant chlorophyll–protein complex induced by iron deficiency (Boekema *et al.*, 2001), photosynthetic complexes (Dühring *et al.*, 2006; Tucker and Sherman, 2000) Kai protein complex (Kageyama *et al.*, 2006), a light-harvesting phycobilisome protein complex (Grossman *et al.*, 1993). These studies relied on the isolation and analysis of individual protein complexes using techniques such as co-immunoprecipitation, affinity tagging and pull down assays, fluorescence microscopy, and others. However, at present, proteome-wide analysis of protein complexes using multidimensional chromatographic separations combined with MS approaches have not been applied to discover novel protein complexes in cyanobacteria. In the future, development and applications of these powerful tools to broadly analyse the composition, localization, and dynamics of protein complex systems in cyanobacteria under various environmental growth conditions is expected.

Growth of cyanobacteria in the light and dark

The day-night cycle is obviously of critical importance to cyanobacteria and it is necessary to understand how different genes are expressed throughout this diel cycle. Thus, many experiments have been performed to analyse expression differences in the light and dark. For example, Wattiez and co-workers (Matallana-Surget *et al.*, 2014) performed diel proteomic-profiling of the non-N_2 fixing cyanobacterium *Arthrospira platensis* PCC 8005 using gel-based and gel-free proteomic approach. They identified more than 1300 proteins, 500 of which were identified as metal binding proteins using immobilized metal (Cu^{2+}) affinity chromatography, and a total of 482 proteins during a diel cycle of which 30 proteins were differentially regulated under the light–dark transitions. *A. platensis* possesses several psbA genes encoding the protein D1 from PSII (psbA1, psbA2, psbA3, psbA4), and only one D1 protein, encoded by the *psbA1* gene was down-regulated upon light–dark cycle (Matallana-Surget *et al.*, 2014). Several phycocyanin and allophycocyanin proteins, cytochrome f, the FtsZ cell division protein, and a protein homologous to glycine rich RNA binding were differentially regulated.

A number of labs have performed proteomics analysis in *Cyanothece* 51142 using both metabolic labelling and label free approaches. Proteomic analysis of *Cyanothece* 51142 using partial ('unsaturated') metabolic labelling identified 721 actively synthesized proteins with significant temporal changes in expression throughout the 48-h light–dark diurnal cycles (Aryal *et al.*, 2011). Of those, 425 proteins matched with previously characterized cycling transcripts (Stöckel *et al.*, 2008). Nitrogenase, an enzyme involved in nitrogen fixation was expressed only in the dark period, and proteins involved in respiration and glycogen metabolism also showed increased expression in the dark cycle, which suggested that N_2-fixation is mediated by higher respiration and glycogen metabolism in the dark. We further applied the same partial metabolic labelling strategy for proteomic analysis of *Cyanothece* 51142 under nitrogen sufficient and constant light conditions (Aryal *et al.*, 2012b), revealing dynamic changes in abundances of 414 proteins including those involved in carbon fixation, cellular protection, and redox regulation. In another study, we conducted comparative proteome analysis of *Cyanothece* 51142 and *Cyanothece* 7822 under 8 different growth conditions (with and without nitrogen, with or without glycerol, and continuous light or light–dark cycle) using label free spectral counting. This study also revealed that nitrogenase expression was linked to the dark period under N_2-fixing condition in the both strains but exogenous addition of glycerol induced expression of nitrogenase during part of the light period, together with cytochrome *c* oxidase (Cox), glycogen phosphorylase (Glp), and glycolytic and pentose phosphate pathway (PPP) enzymes (Aryal *et al.*, 2013; Bandyopadhyay *et al.*, 2011). This proved that higher levels of respiration and glycogen breakdown aided nitrogenase expression in the light. Key enzymes of the Calvin cycle were inhibited in *Cyanothece* 51142 in the presence of glycerol and under N_2-producing conditions, but in *Cyanothece* 7822, the Calvin cycle still played a role in cofactor recycling during N_2-fixation.

Cyanothece spp. exhibited considerable variation in physiological and biochemical characteristics. The comparative assessment of the proteomes of five *Cyanothece* strains using a label free shotgun proteomics approach identified 642 orthologous proteins common to all strains (core proteome) (Aryal *et al.*, 2014). This study provided direct evidence of significant evolutionary differences of *Cyanothece* PCC 7425 from the other *Cyanothece* strains, as revealed previously by comparative genomics (Bandyopadhyay *et al.*, 2011). Recently, Guerreiro and co-workers (Guerreiro *et al.*, 2016) applied size exclusion chromatography (SEC) combined with LC-MS analysis of *Synechococcus elongatus* PCC 7942 to study diurnal and circadian rhythms at the protein complex level, and reported that RNA polymerase, the ribosome and photosynthetic protein complexes and several other known complexes showed changes in abundances as well as in their assembly.

A very recent proteomics analysis of *Synechocystis* 6803 by Pakrasi and co-workers (Liberton *et al.*, 2016) found a very distinct separation of the thylakoid and plasma membranes. A larger number of proteins uniquely localized to the plasma membrane (PM) (459 proteins) compared to thylakoid membrane (TM) (176 proteins). PM proteins were characterized by those associated with the transport, secretion, and trafficking, and the TM proteins were those involved in the energetics of electron transport. Not only is the plasma membrane protein composition more diverse, but this work clearly identified two very separate membrane systems. Thus, large-scale proteomics answered long-standing questions about how separate these membrane systems were and whether or not there was any interaction at the biogenesis level. Indeed, the membrane systems now always seem to be quite distinct.

Transcriptional regulation in *Cyanothece* spp.

Due to their robust metabolic periodicities, the *Cyanothece* strains are excellent for analysis of transcriptional and translational regulation. Previous work in this area was performed on the model organisms, *Synechococcus elongatus* (Ruffing, 2013; Vijayan *et al.*, 2011; see also Chaper 8 of this book) and *Synechocystis* sp. PCC 6803 (Li *et al.*, 2004; Singh *et al.*, 2010; Singh *et al.*, 2003; Summerfield *et al.*, 2008) and has provided the tools for the current research. We have used both microarray and RNA-seq experiments to analyse transcriptional regulation in *Cyanothece* 51142 and *Cyanothece* 7822. The main questions that we will pursue in the remaining sections of this review revolve around how the mRNA and the protein levels vary throughout the diurnal cycle and what percentage of all proteins actually cycle during the 24 h period. We will concentrate on the integration of the transcriptomics and proteomics experiments, all of which were performed under nitrogen fixing, LD conditions, unless mentioned otherwise.

We first grew *Cyanothece* 51142 under a variety of different LD cycles (Toepel *et al.*, 2008; Toepel *et al.*, 2009). This enabled us to determine the response of the cell to these conditions and to differentiate between diurnal and circadian-regulated genes. *Cyanothece* 51142 became acclimated in several ways to growth under short LD conditions (Toepel *et al.*, 2009). Nitrogen was fixed in every second dark period and only once in each 24 h period. Not surprisingly, N_2 fixation was strongly correlated to the energy status of the cells and to glycogen breakdown, and high respiration rates were necessary to provide appropriate energy and anoxic conditions for this process. We demonstrated that the main metabolic genes involved in photosynthesis, respiration, nitrogen fixation, and central carbohydrate metabolism have strong or total circadian-regulated components. The short LD cycles enabled us to identify transcriptional differences among the family of *psbA* genes, as well as the differing patterns of the uptake hydrogenase (*hup*) genes, which followed the same pattern as nitrogenase genes. On the other hand, the bidirectional hydrogenase (*hox*) genes displayed a diurnal, dark-dependent gene expression. This was the first experimental evidence of the transcriptional relationship of the *nif* and *hup* genes in *Cyanothece* 51142. We later demonstrated the physiological importance of *hup* to *nif* in *Cyanothece* 7822 (Zhang *et al.*, 2014).

Transcriptomic profiles were then analysed using context-likelihood of relatedness analysis (CLR) over four separate experiments using different LD periodicities (Toepel *et al.*, 2009). Fig. 6.3 depicts this analysis as a network of relationships between genes that approximates a 24 h clock. The tightness of a cluster in the large number of contacts indicated groups of genes that are transcribed in a similar fashion. We particularly analysed the network using a hierarchical clustering approach that identified clusters of genes that are tightly related by CLR, and identified four large clusters and a number of smaller clusters. There are three main clusters that displayed circadian behaviour and that are circled: a photosynthesis cluster, a nitrogenase cluster and a respiratory/carbohydrate

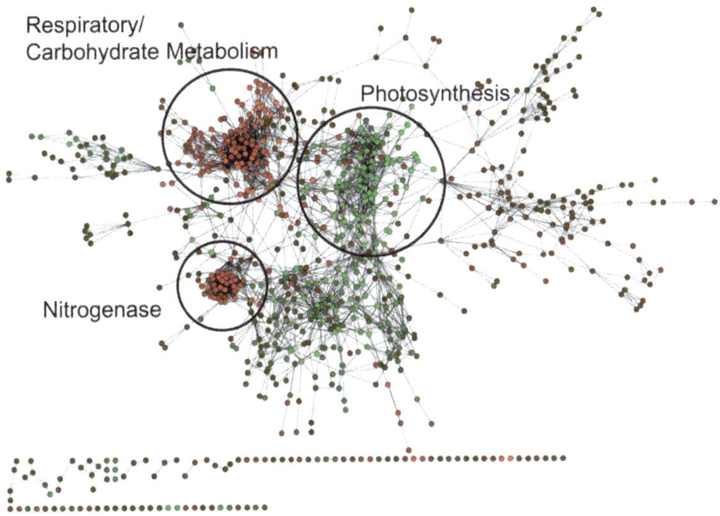

Figure 6.3 Inferred network of diurnal and circadian associations. The context likelihood of relatedness (CLR) algorithm was used to infer statistically significant gene–to-gene associations, based on the combined expression profiles from growth under different light–dark conditions: growth in 12h LD, growth in 12h LD followed by continuous light, growth in 6h LD, and growth following transition into anaerobic conditions. In the network, circles represent genes and edges between them an association with a CLR z-score of 5.0 or higher. The genes are coloured according to the transcriptional levels at 5h Dark (red, high; green, low). Three clusters of interest are labelled and correspond to important metabolic groupings. The majority of genes with inferred associations are included in the large network, but there are a number of smaller unconnected networks (including isolated pairs of genes) displayed at the bottom of the figure (Toepel *et al*. 2009).

metabolism cluster (Fig. 6.3). The photosynthesis cluster contained genes encoding proteins for all the major photosynthesis structures. This is a complicated set of genes that are scattered throughout the chromosome and it was satisfying to find them form a reasonable cluster. The nitrogenase cluster was the tightest and most cohesive cluster, as was expected from previous work. The CLR analysis determined that all of the genes in the nitrogenase grouping fell within this cluster. The cluster at 11 o'clock included genes involved in respiration and in central metabolism, such as glycolysis, the pentose phosphate pathway, and glycogen degradation. This is a most interesting cluster because it includes some gene sets that encode metabolic processes that peak at the end of the light period and some with peaks towards the end of the dark. Interestingly, this cluster includes circadian-regulated genes that are needed to prepare for the next light or dark phase. These results not only support the hypotheses of diurnal control of the major processes inside *Cyanothece* 51142, but also suggested a strong feedback regulation regarding the energy status of the cells. In general, these results showed a very tight regulation at the transcriptional level of key metabolic processes, but one that had both circadian and diurnal regulation.

This work was followed by a more detailed systems biology approach to the disparate data on the N_2-fixing growth of *Cyanothece* 51142. We represented the transcript on the data from *Cyanothece* 51142 under diurnal LD cycles as a high level functional abstraction and described development of a predictive *in silico* model of diurnal and circadian behaviour in terms of regulatory and metabolic processes. Incorporating network topology into the model improved our predictive ability of the system under new conditions. A model that robustly described the transcriptional behaviour of *Cyanothece* 51142 under different cyclic and non-cyclic growth conditions was an important advance towards understanding gene regulation under these conditions. The computational and modelling techniques used were described in our recent publication (McDermott *et al.*, 2011) and the model utilized all of the transcriptomic data that was generated on this organism over a number of years.

A model of cyclic transcript behaviour in *Cyanothece* 51142 is shown in Fig. 6.4. During the early part of the light period, the Photosystem II genes

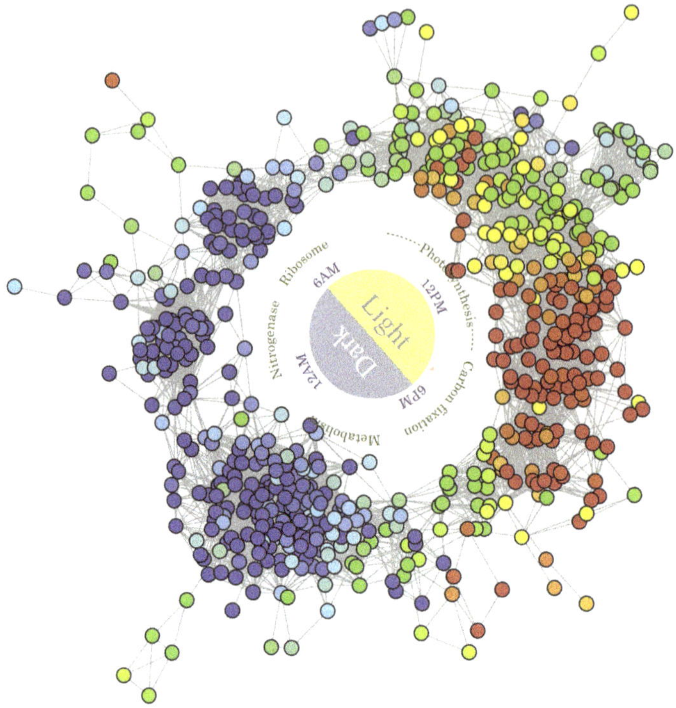

Figure 6.4 A wreath diagram that represents the topology of inferred networks used to identify system control points and to construct a predictive model of cyclic functional behaviour in *Cyanothece* sp. ATCC 51142. The data is represented on a 24h clock and is based on transcriptional experiments performed on 12h LD cycles. The temporal ordering of gene expression is represented in the network by the location of a gene at its peak expression. The genes expressed in the early light (green and yellow dots) primarily encode proteins for photosynthesis (Photosystem II), light-harvesting proteins, ATP synthase, CO_2 fixation and central metabolism. The genes transcribed in the late light (red dots) encode Photosystem I proteins and many key regulatory proteins (e.g. SigD and RpaA) that are involved in control of early dark processes. The genes expressed in the dark (light blue and purple dots) are primarily circadian and encode the nitrogenase proteins and proteins involved in respiration. The figure is based on data from McDermott *et al*. (2011).

were transcribed, followed by those involved with carbon fixation. In the late light period, genes encoding PSI proteins were transcribed, as were some key regulatory genes. During the early dark, transcription of the nitrogenase and respiratory cluster predominated, and was followed by other housekeeping activities, such as ribosome gene transcription in the late dark. During this late dark phase, the cell is preoccupied with transcribing genes involved with important functions during the early light phase. The complete cyclic wreath network for *Cyanothece* 51142 permitted the identification of the peak transcription time of specific regulatory proteins (McDermott *et al*., 2011), including transcription factors involved in the regulation of specific light or dark activities. From the studies, we concluded that transcription in *Cyanothece* 51142 was highly cyclic, and based on the specific metabolic role of the genes. The model identified a number of important regulatory proteins and these predictions could have been tested via mutational analysis.

However, although *Cyanothece* 51142 is a wonderful organism for physiological and regulatory studies, we were unable to generate a stable genetic system in this organism. Thus, we investigated other *Cyanothece* species and determined that *Cyanothece* 7822 was a genetically tractable system (Min and Sherman, 2010b). This system had a relatively large genome, but also demonstrated excellent oscillations of nitrogen fixation and photosynthesis. These oscillations were not as highly synchronized as those in *Cyanothece* 51142 (Aryal *et al*., 2013), and we determined an important difference between these two organisms. *Cyanothece* 51142 has large starch-like granules that formed during

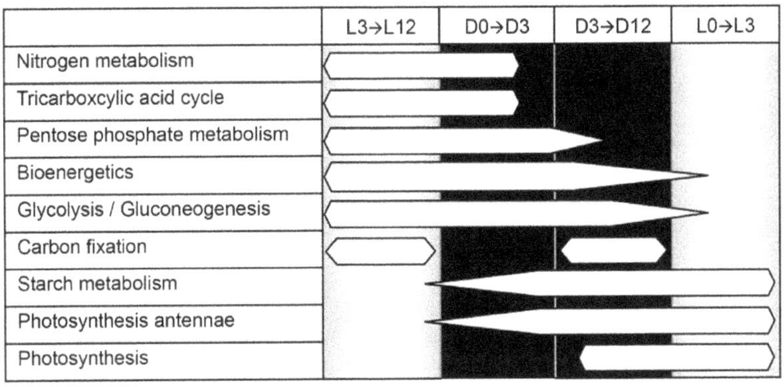

Figure 6.5 Time scale for differential gene expression in *Cyanothece* sp. PCC 7822 grown in 12 h LD cycles. The bar thickness is a qualitative measure of the mRNA levels at a particular time point. The two dark segments (early and late dark) are highlighted in black. The diagram begins with the late light period (L3 to L12) on the left, to best represent the pattern of gene expression throughout the day, to the early light period (L0 to L3) on the right. Thus, nitrogenase genes (in nitrogen metabolism) begin transcription in the light and continue into the early dark. (From Welkie *et al.*, 2014 with permission.)

photosynthesis, whereas *Cyanothece* 7822 formed much smaller glycogen granules (Welkie *et al.*, 2015; Welkie *et al.*, 2013). This difference in carbohydrate storage may impact on the strategy used to protect nitrogenase when oxygen is present and may require that more granules need to be degraded over a longer period of time to quench cytoplasmic oxygen. This led us to perform detailed transcriptional and proteomic studies on *Cyanothece* 7822 (Welkie *et al.*, 2014). In parallel to the physiological results, the transcription patterns were not as highly synchronized in *Cyanothece* 7822 and it was more likely that a particular metabolic system demonstrated transcription over the light–dark or dark–light boundaries (Fig. 6.5). Thus, photosynthesis gene transcription typically began in the late dark and continued into the early light, whereas the genes for nitrogenase were transcribed in the late light into the early dark. In all other regards, the two systems were quite similar in terms of transcriptional patterns. The transcriptional patterns are still cyclic, but not as constrained to the light or the dark. At the same time, the degradation of the smaller glycogen granules and corresponding rates of respiration begin in the late light and continue well into the dark (Welkie *et al.*, 2014).

Integration of proteomics with transcriptomics in *Cyanothece*

The significant oscillations in transcript levels provided the impetus to understand protein changes throughout the diurnal cycle. Thus, we utilized the techniques described earlier to analyse proteome profiles of multiple strains of *Cyanothece* (Aryal *et al.*, 2014). We also analysed the proteome under a variety of different physiological conditions, including changes in light, nitrogen, and glycerol (Aryal *et al.*, 2013). The proteome of both *Cyanothece* 51142 and *Cyanothece* 7822 were then analysed under light–dark, nitrogen-fixing conditions. These results indicated that far fewer proteins oscillated throughout the diurnal cycle. In *Cyanothece* 51142, proteome analysis identified 68% of the protein-including ORF's predicted from the *Cyanothece* 51142 genome (Stöckel *et al.*, 2011). However, of the 1232 proteins identified under LD N_2-fixing conditions, only 250 proteins exhibited cyclic changes in abundance over the 2-day diurnal periods of 12 h LD. Of these 250 cyclic proteins, 175 overlapped with genes that were previously identified as cyclic transcripts. It is clear from these observations that post-translational regulation is of great significance in the nitrogen fixing *Cyanothece* 51142. The functional categories for the 250 oscillating proteins were also quite illuminating. About 40% of the proteins belonged to central-metabolism based pathways (central, intermediary and energy), such as nitrogen fixation and glycogen degradation. In addition, close to 20% of the proteins were involved in photosynthesis and respiration and another 20% were associated with the biosynthesis of cofactors and the cell envelope (Stöckel *et al.*, 2011). Thus, analysis of the *Cyanothece* 51142 proteome indicated that the abundance of proteins related to central metabolic processes generally followed the expression pattern of their appropriate transcripts, although changes at the protein level were not as tightly regulated. For example, eight proteins involved in N_2 fixation were highest during the dark, consistent with the usual observed peak of nitrogen fixation during this phase. Indeed, an overview of diurnal changes in protein abundance involved in central metabolism showed a clear light–dark pattern, consistent with physiological measurements. This analysis also revealed some important regulatory proteins associated with either photosynthesis or nitrogen fixation.

Similar results were obtained for *Cyanothece* 7822: in this case, some 6766 mRNAs and 1322 proteins were identified at four time points across a 24-hour LD cycle (Welkie *et al.*, 2014). Once again, some 200 cycling proteins were identified, once again with the majority in the photosynthesis and nitrogen fixing categories. It was quite evident that in *Cyanothece* 7822 most proteins do not cycle. The analysis of proteins in central metabolism is shown in Fig. 6.6. Although most transcripts vary substantially during these time points, most of the proteins stay within a 1.5–2.0 fold range. Nonetheless, many proteins involved in key metabolic functions demonstrated expression to a greater extent in either the light or the dark, depending upon the need for that function. The dynamic range of the protein levels under these conditions was typically far less than that shown at the mRNA level, thus implying that statements relating to absolute levels of induction/repression at the transcription level may not be important at the protein level.

Those proteins involved with bioenergetics were mainly present in the dark. This was particularly true of proteins from the main cytochrome oxidase operon, as well as other proteins involved in respiratory energy production. Most importantly, proteins encoded by the nitrogenase cluster genes were differentially expressed with a peak at D3 for all proteins except for NifD, which peaked at the end of the dark period. As shown in Fig. 6.7, the protein and mRNA patterns were far more similar than for those genes in central metabolism. The fairly high levels of protein in the late light period were reflected by an enhancement in mRNA transcription in the late light and demonstrated the circadian cycling pattern of these genes.

Conclusion

The development of new and better proteomic tools has been one of the important developments in current biological research. These techniques provide important information on protein levels, protein modifications and provide a much better understanding of the cell's response to environmental alterations. We have used these tools to determine the relationship of transcriptional to protein levels utilizing cyanobacteria with an oscillating metabolism that has shown a very high degree of transcriptional cycling during LD growth under nitrogen-fixing conditions. However, many fewer proteins demonstrated cycling in both *Cyanothece* strains and it is clear that cells do not want to expend energy on making and degrading all cellular proteins during each 24 h period. Similar results were obtained from the simpler, but ecologically important, unicellular cyanobacterium, *Prochlorococcus* (Waldbauer *et al.*, 2012) and indicated that this is likely to be a common trait among cyanobacteria and other bacteria. Therefore, we believe that proteomics analysis is an important aspect of any future analysis of cellular regulation. The advent of procedures for identifying protein modifications will further enhance the use of proteomics and will be of vital importance for systems like the metabolic oscillations in *Cyanothece* spp. It will be of great interest to determine how many proteins are modified in *Cyanothece* spp. and at what time of the day or night. Such results might generate valuable new information about the uses and evolution of various types of protein modifications and determine the strategies used by cells to permit oscillatory metabolism while conserving the energy needed to degrade and re-synthesize thousands of cellular proteins.

Future research directions

Proteins are the key functional units of the cell. Many protein functions are modulated by PTMs, interactions, cellular localization, and spatial and temporal dynamics. The proteomics field is rapidly evolving, so that researchers now can identify and quantify proteins with high sensitivity and at high

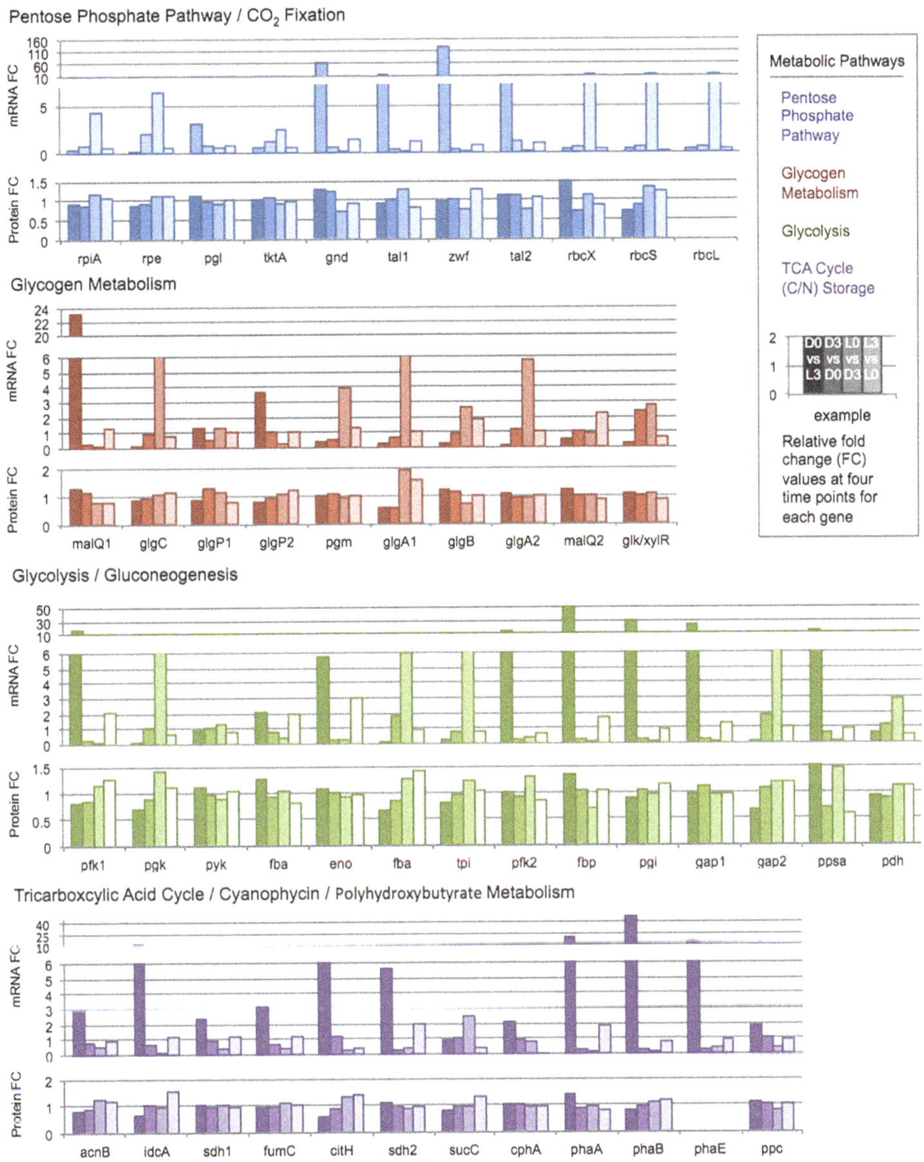

Figure 6.6 The relative fold change expression levels of both mRNA and protein for 47 genes involved with central metabolism in *Cyanothece* 7822 are shown. The y-axis represents the relative fold change in mRNA level or in protein level. The x-axis represents the four comparisons performed in this experiment: D0 vs. L3 (mid-day to sunset); D3 vs. D0 (mid-night to sunset); D3 vs. L0 (mid-night to sunrise); and, L3 vs. L0 (sunrise to early morning). The figure is coloured coded as follows for the protein-mRNA expression dynamics. PPP, blue; glycogen metabolism, red; glycolysis, green; and TCA, purple. Genes within each groups displayed are as follows: PPP and Carbon Fixation: Cyan7822_0018, *rpiA*; Cyan7822_0124, *rpe*; Cyan7822_0172, *pgl*; Cyan7822_1076, *tktA*; Cyan7822_3017, *gnd*; Cyan7822_4172, *tal1*; Cyan7822_4173, *zwf*; Cyan7822_5424, *tal2*; Cyan7822_2900, *rbcX*; Cyan7822_2901, *rbcS*; Cyan7822_2899, *rbcL*. Glycogen Metabolism: Cyan7822_1182, *malQ1*; Cyan7822_2049, *glgC*; Cyan7822_2294, *glgP2*; Cyan7822_2322, *glgP1*; Cyan7822_2491, *pgm*; Cyan7822_2570, *glgA1*; Cyan7822_2889, *glgB*; Cyan7822_4734, *glgA2*; Cyan7822_5145, *malQ2*; Cyan7822_5605, *glk/xylR*. Glycolysis: Cyan7822_0159, *pfk1*; Cyan7822_1455, *pgk*; Cyan7822_1523, *pyk*; Cyan7822_1612,*fba1*; Cyan7822_1633, *eno*; Cyan7822_1659, *fba2* Cyan7822_1688, *tpi*; Cyan7822_3033, *pfk2*; Cyan7822_4171, *fbp*; Cyan7822_4412, *pgi*; Cyan7822_4423, *gap1*; Cyan7822_4638, *gap2*; Cyan7822_1627, *ppsA*; Cyan7822_2730, *pdh*. TCA and C/N Storage: Cyan7822_0752, *acnB*; Cyan7822_1400, *idcA*; Cyan7822_1551, *sdh1*; Cyan7822_2824, *fumC*; Cyan7822_3004, *citH*; Cyan7822_3071, *sdh2*; Cyan7822_3782, *sucC*; Cyan7822_2816, *cphB*; Cyan7822_1326, *phaA*; Cyan7822_1327, *phaB*; Cyan7822_1330, *phaE*; Cyan7822_2645, *pcc*. (From Welkie *et al.*, 2014 with permission.)

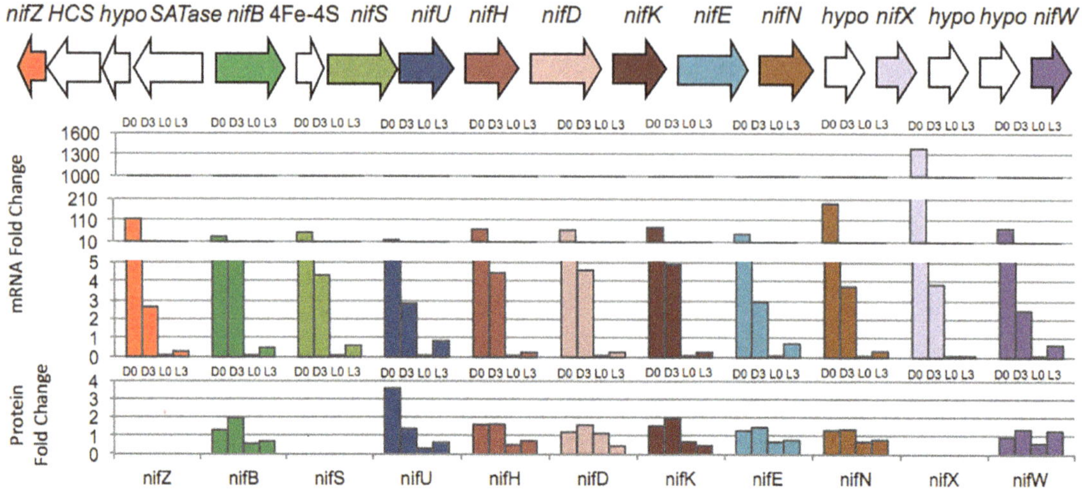

Figure 6.7 Genes in the *nif* cassette in *Cyanothece* 7822 display consistent expression in both mRNA and protein during the dark phase. It is notable that *nifD* show anti-phase abundance maxima with the respectable transcripts. Peptides for the genes *nifZ* and *nifX* were not detected in the analysis. The x-axis represents the four comparisons: D0 vs. L3 (mid-day to sunset); D3 vs. D0 (mid-night to sunset); D3 vs. L0 (mid-night to sunrise); and, L3 vs. L0 (sunrise to early morning). (From Welkie *et al.*, 2014 with permission.)

throughput. Since functional genomics cannot provide experimental evidence for PTMs, new proteomic technologies are essential for analysis of PTMs. There are more than 300 different types of PTMs, and only a handful of those modifications, such as phosphorylation and glycosylation, have been extensively studied (Jensen, 2006; Zhao and Jensen, 2009). Therefore, future research should focus on developing new proteomic technologies for the characterization of other modifications. New and improved computational and mass spectrometry-based methods are also needed for unbiased and high-throughput analysis of the structure, assembly and dynamics of large multi-protein complexes and protein–protein interactions.

Modern proteomics relies on mass spectrometry for high sensitivity proteome detection, and builds upon the continued improvement of increasingly sophisticated hybrid instruments for the comprehensive characterization of the cellular proteome. Proteins in biological systems can vary in dynamic range from 6 to 9 orders of magnitude (Anderson and Anderson, 2002), but the dynamic range of the modern highly sophisticated mass spectrometers can vary from 4 to 5 orders of magnitude. Thus, identification of many low abundant proteins in a cell is still a huge challenge. This has been addressed by protein or peptide level fractionations and depletion of high abundant proteins, yet many very low abundant proteins still remain undetected. Development of more sensitive hybrid mass spectrometers that enhance the performance of the modern proteomic workflow will be critical in the future research. We also argue that the large-scale study of endogenous protein complexes, their modifications, interactions and dynamic behaviour will be the core of the future proteomics research, including in cyanobacteria. We expect that these challenges will inform those working to improve the technology, so that new methods will provide researchers with a complete understanding of protein dynamics throughout the complex life cycle of organisms like cyanobacteria.

References

Aebersold, R., and Mann, M. (2003). Mass spectrometry-based proteomics. Nature *422*, 198–207. http://dx.doi.org/10.1038/nature01511

Alberts, B. (1998). The cell as a collection of protein machines: Preparing the next generation of molecular biologists. Cell 92, 291–294. http://dx.doi.org/10.1016/S0092-8674(00)80922-8.

Altelaar, A.F., Munoz, J., and Heck, A.J. (2013). Next-generation proteomics: towards an integrative view of proteome dynamics. Nat. Rev. Genet. *14*, 35–48. http://dx.doi.org/10.1038/nrg3356.

Anderson, N.L., and Anderson, N.G. (2002). The human plasma proteome: history, character, and diagnostic prospects. Mol. Cell. Proteomics. *1*, 845–867. http://dx.doi.org/10.1074/mcp.R200007-MCP200.

Aryal, U.K., Callister, S.J., McMahon, B.H., McCue, L.A., Brown, J., Stöckel, J., Liberton, M., Mishra, S., Zhang,

X., Nicora, C.D., *et al.* (2014). Proteomic profiles of five strains of oxygenic photosynthetic cyanobacteria of the genus *Cyanothece*. J. Proteome. Res. *13*, 3262–3276. http://dx.doi.org/10.1021/pr5000889.

Aryal, U.K., Callister, S.J., Mishra, S., Zhang, X., Shutthanandan, J.I., Angel, T.E., Shukla, A.K., Monroe, M.E., Moore, R.J., Koppenaal, D.W., *et al.* (2013). Proteome analyses of strains ATCC 51142 and PCC 7822 of the diazotrophic cyanobacterium *Cyanothece* sp. under culture conditions resulting in enhanced H_2 production. Appl. Environ. Microbiol. 79, 1070–1077. http://dx.doi.org/10.1128/AEM.02864-12.

Aryal, U.K., Krochko, J.E., and Ross, A.R. (2012a). Identification of phosphoproteins in Arabidopsis thaliana leaves using polyethylene glycol fractionation, immobilized metal–ion affinity chromatography, two-dimensional gel electrophoresis and mass spectrometry. J. Proteome Res. *11*, 425–437. http://dx.doi.org/10.1021/pr200917t.

Aryal, U.K., and Ross, A.R.S. (2010). Enrichment and analysis of phosphopeptides under different experimental conditions using titanium dioxide affinity chromatography and mass spectrometry. Rapid Commun. Mass Spectrom. *24*, 219–231. http://dx.doi.org/10.1002/rcm.4377.

Aryal, U.K., Stöckel, J., Krovvidi, R.K., Gritsenko, M.A., Monroe, M.E., Moore, R.J., Koppenaal, D.W., Smith, R.D., Pakrasi, H.B., and Jacobs, J.M. (2011). Dynamic proteomic profiling of a unicellular cyanobacterium *Cyanothece* ATCC51142 across light–dark diurnal cycles. BMC Syst. Biol. *5*, 194. http://dx.doi.org/10.1186/1752-0509-5-194.

Aryal, U.K., Stöckel, J., Welsh, E.A., Gritsenko, M.A., Nicora, C.D., Koppenaal, D.W., Smith, R.D., Pakrasi, H.B., and Jacobs, J.M. (2012b). Dynamic proteome analysis of *Cyanothece* sp. ATCC 51142 under constant light. J. Proteome. Res. *11*, 609–619. http://dx.doi.org/10.1021/pr200959x.

Axmann, I.M., Hertel, S., Wiegard, A., Dörrich, A.K., and Wilde, A. (2014). Diversity of KaiC-based timing systems in marine Cyanobacteria. Mar. Genomics *14*, 3–16. http://dx.doi.org/10.1016/j.margen.2013.12.006.

Bandyopadhyay, A., Elvitigala, T., Welsh, E., Stöckel, J., Liberton, M., Min, H., Sherman, L.A., and Pakrasi, H.B. (2011). Novel metabolic attributes of the genus cyanothece, comprising a group of unicellular nitrogen-fixing *Cyanothece*. MBio 2, e00214–11. http://dx.doi.org/10.1128/mBio.00214-11.

Baniulis, D., Yamashita, E., Whitelegge, J.P., Zatsman, A.I., Hendrich, M.P., Hasan, S.S., Ryan, C.M., and Cramer, W.A. (2009). Structure-function, stability, and chemical modification of the cyanobacterial cytochrome b_6f complex from *Nostoc* sp. PCC 7120. J. Biol. Chem. *284*, 9861–9869. http://dx.doi.org/10.1074/jbc.M809196200.

Battchikova, N., Angeleri, M., and Aro, E.M. (2015). Proteomic approaches in research of cyanobacterial photosynthesis. Photosynth. Res. *126*, 47–70. http://dx.doi.org/10.1007/s11120-014-0050-4.

Bensimon, A., Heck, A.J., and Aebersold, R. (2012). Mass spectrometry-based proteomics and network biology. Annu. Rev. Biochem. *81*, 379–405. http://dx.doi.org/10.1146/annurev-biochem-072909-100424.

Boekema, E.J., Hifney, A., Yakushevska, A.E., Piotrowski, M., Keegstra, W., Berry, S., Michel, K.P., Pistorius, E.K., and Kruip, J. (2001). A giant chlorophyll–protein complex induced by iron deficiency in cyanobacteria. Nature *412*, 745–748. http://dx.doi.org/10.1038/35089104.

Boyd, E.S., Costas, A.M., Hamilton, T.L., Mus, F., and Peters, J.W. (2015). Evolution of molybdenum nitrogenase during the transition from anaerobic to aerobic metabolism. J. Bacteriol. *197*, 1690–1699. http://dx.doi.org/10.1128/JB.02611-14.

Braun, J.M., Sathyanarayana, S., and Hauser, R. (2013). Phthalate exposure and children's health. Curr. Opin. Pediatr. *25*, 247–254. http://dx.doi.org/10.1097/MOP.0b013e32835e1eb6.

Buchanan, B.B., and Balmer, Y. (2005). Redox regulation: a broadening horizon. Annu. Rev. Plant Biol. *56*, 187–220. http://dx.doi.org/10.1146/annurev.arplant.56.032604.144246.

Castielli, O., De la Cerda, B., Navarro, J.A., Hervás, M., and De la Rosa, M.A. (2009). Proteomic analyses of the response of cyanobacteria to different stress conditions. FEBS Lett. *583*, 1753–1758. http://dx.doi.org/10.1016/j.febslet.2009.03.069.

Červený, J., Sinetova, M.A., Zavřel, T., and Los, D.A. (2015). Mechanisms of high temperature resistance of *Synechocystis* sp. PCC 6803: an impact of histidine kinase 34. Life 5, 676–699. http://dx.doi.org/10.3390/life5010676.

Chen, S.H., Chen, C.R., Chen, S.H., Li, D.T., and Hsu, J.L. (2013). Improved N(α)-acetylated peptide enrichment following dimethyl labeling and SCX. J. Proteome. Res. *12*, 3277–3287. http://dx.doi.org/10.1021/pr400127j.

Chernushevich, I.V., Loboda, A.V., and Thomson, B.A. (2001). An introduction to quadrupole-time-of-flight mass spectrometry. J. Mass. Spectrom. *36*, 849–865. http://dx.doi.org/10.1002/jms.207.

Choudhary, C., Kumar, C., Gnad, F., Nielsen, M.L., Rehman, M., Walther, T.C., Olsen, J.V., and Mann, M. (2009). Lysine acetylation targets protein complexes and co-regulates major cellular functions. Science *325*, 834–840. http://dx.doi.org/10.1126/science.1175371.

Choudhary, C., Weinert, B.T., Nishida, Y., Verdin, E., and Mann, M. (2014). The growing landscape of lysine acetylation links metabolism and cell signalling. Nat. Rev. Mol. Cell. Biol. *15*, 536–550. http://dx.doi.org/10.1038/nrm3841.

Cox, J., and Mann, M. (2011). Quantitative, high-resolution proteomics for data-driven systems biology. Annu. Rev. Biochem. *80*, 273–299. http://dx.doi.org/10.1146/annurev-biochem-061308-093216.

Deng, X., Weerapana, E., Ulanovskaya, O., Sun, F., Liang, H., Ji, Q., Ye, Y., Fu, Y., Zhou, L., Li, J., *et al.* (2013). Proteome-wide quantification and characterization of oxidation-sensitive cysteines in pathogenic bacteria. Cell. Host. Microbe. *13*, 358–370. http://dx.doi.org/10.1016/j.chom.2013.02.004.

Deng, Z., Zhang, X., Tang, W., Oses-Prieto, J.A., Suzuki, N., Gendron, J.M., Chen, H., Guan, S., Chalkley, R.J., Peterman, T.K., *et al.* (2007). A proteomics study of brassinosteroid response in Arabidopsis. Mol. Cell. Proteomics. *6*, 2058–2071. http://dx.doi.org/10.1074/mcp.M700123-MCP200.

Derakhshan, B., Wille, P.C., and Gross, S.S. (2007). Unbiased identification of cysteine *S*-nitrosylation sites on proteins. Nat. Protoc. *2*, 1685–1691. http://dx.doi.org/10.1038/nprot.2007.210.

Dong, G., Yang, Q., Wang, Q., Kim, Y.I., Wood, T.L., Osteryoung, K.W., van Oudenaarden, A., and Golden, S.S. (2010). Elevated ATPase activity of KaiC applies a circadian checkpoint on cell division in *Synechococcus elongatus*. Cell *140*, 529–539. http://dx.doi.org/10.1016/j.cell.2009.12.042.

Dong, M., Yang, L.L., Williams, K., Fisher, S.J., Hall, S.C., Biggin, M.D., Jin, J., and Witkowska, H.E. (2008). A "tagless" strategy for identification of stable protein complexes genome-wide by multidimensional orthogonal chromatographic separation and iTRAQ reagent tracking. J. Proteome Res. 7, 1836–1849. http://dx.doi.org/10.1021/pr700624e.

Dube, D.H., and Bertozzi, C.R. (2005). Glycans in cancer and inflammation – potential for therapeutics and diagnostics. Nat. Rev. Drug Discov. *4*, 477–488. http://dx.doi.org/10.1038/nrd1751.

Dühring, U., Axmann, I.M., Hess, W.R., and Wilde, A. (2006). An internal antisense RNA regulates expression of the photosynthesis gene *isiA*. Proc. Natl. Acad. Sci. U.S.A. *103*, 7054–7058. http://dx.doi.org/10.1073/pnas.0600927103.

Elliott, M.H., Smith, D.S., Parker, C.E., and Borchers, C. (2009). Current trends in quantitative proteomics. J. Mass. Spectrom. *44*, 1637–1660. http://dx.doi.org/10.1002/jms.1692.

Eriksson, M.J., and Clarke, A.K. (1996). The heat shock protein ClpB mediates the development of thermotolerance in the cyanobacterium *Synechococcus* sp. strain PCC 7942. J. Bacteriol. *178*, 4839–4846.

Fields, S., and Song, O. (1989). A novel genetic system to detect protein-protein interactions. Nature *340*, 245–246. http://dx.doi.org/10.1038/340245a0.

Fields, S., and Sternglanz, R. (1994). The two-hybrid system: an assay for protein-protein interactions. Trends Genet. *10*, 286–292. http://dx.doi.org/10.1016/0168-9525(90)90012-U.

Forchhammer, K. (2008). P(II) signal transducers: novel functional and structural insights. Trends Microbiol. *16*, 65–72. http://dx.doi.org/10.1016/j.tim.2007.11.004.

Forrester, M.T., Thompson, J.W., Foster, M.W., Nogueira, L., Moseley, M.A., and Stamler, J.S. (2009). Proteomic analysis of *S*-nitrosylation and denitrosylation by resin-assisted capture. Nat. Biotechnol. *27*, 557–559. http://dx.doi.org/10.1038/nbt.1545.

Foyer, C.H., and Noctor, G. (2009). Redox regulation in photosynthetic organisms: signaling, acclimation, and practical implications. Antioxid. Redox. Signal. *11*, 861–905. http://dx.doi.org/10.1089/ars.2008.2177.

Fulda, S., Mikkat, S., Huang, F., Huckauf, J., Marin, K., Norling, B., and Hagemann, M. (2006). Proteome analysis of salt stress response in the cyanobacterium *Synechocystis* sp. strain PCC 6803. Proteomics *6*, 2733–2745. http://dx.doi.org/10.1002/pmic.200500538.

Gabius, H.J., André, S., Kaltner, H., and Siebert, H.C. (2002). The sugar code: functional lectinomics. Biochim. Biophys. Acta *1572*, 165–177.

Gallon, J.R., Cheng, J., Dougherty, L.J., Gallon, V.A., Hilz, H., Pederson, D.M., Richards, H.M., Rüggeberg, S., and Smith, C.J. (2000). A novel covalent modification of nitrogenase in a cyanobacterium. FEBS Lett. *468*, 231–233. http://dx.doi.org/10.1016/S0014-5793(00)01229-1.

Gavin, A.C., Bösche, M., Krause, R., Grandi, P., Marzioch, M., Bauer, A., Schultz, J., Rick, J.M., Michon, A.M., Cruciat, C.M., *et al.* (2002). Functional organization of the yeast proteome by systematic analysis of protein complexes. Nature *415*, 141–147. http://dx.doi.org/10.1038/415141a.

Gerber, S.A., Rush, J., Stemman, O., Kirschner, M.W., and Gygi, S.P. (2003). Absolute quantification of proteins and phosphoproteins from cell lysates by tandem MS. Proc. Natl. Acad. Sci. U.S.A. *100*, 6940–6945. http://dx.doi.org/10.1073/pnas.0832254100.

Görg, A., Weiss, W., and Dunn, M.J. (2004). Current two-dimensional electrophoresis technology for proteomics. Proteomics *4*, 3665–3685. http://dx.doi.org/10.1002/pmic.200401031.

Grønborg, M., Kristiansen, T.Z., Stensballe, A., Andersen, J.S., Ohara, O., Mann, M., Jensen, O.N., and Pandey, A. (2002). A mass spectrometry-based proteomic approach for identification of Serine/Threonine-phosphorylated proteins by enrichment with phospho-specific antibodies: Identification of a novel protein, Frigg, as a protein kinase A substrate. Mol. Cell. Proteomics *1*, 517–527. http://dx.doi.org/10.1074/mcp.M200010-MCP200.

Grossman, A.R., Schaefer, M.R., Chiang, G.G., and Collier, J.L. (1993). The phycobilisome, a light-harvesting complex responsive to environmental conditions. Microbiol. Rev. *57*, 725–749.

Gstaiger, M., and Aebersold, R. (2009). Applying mass spectrometry-based proteomics to genetics, genomics and network biology. Nat. Rev. Genet. *10*, 617–627. http://dx.doi.org/10.1038/nrg2633.

Guerreiro, A.C., Penning, R., Raaijmakers, L.M., Axman, I.M., Heck, A.J., and Altelaar, A.F. (2016). Monitoring light/dark association dynamics of multi–protein complexes in cyanobacteria using size exclusion chromatography-based proteomics. J. Proteomics *142*, 33–44. http://dx.doi.org/10.1016/j.jprot.2016.04.030.

Guo, J., Nguyen, A.Y., Dai, Z., Su, D., Gaffrey, M.J., Moore, R.J., Jacobs, J.M., Monroe, M.E., Smith, R.D., Koppenaal, D.W., *et al.* (2014). Proteome-wide light/dark modulation of thiol oxidation in cyanobacteria revealed by quantitative site-specific redox proteomics. Mol. Cell. Proteomics. *13*, 3270–3285. http://dx.doi.org/10.1074/mcp.M114.041160.

Guskov, A., Kern, J., Gabdulkhakov, A., Broser, M., Zouni, A., and Saenger, W. (2009). Cyanobacterial photosystem II at 2.9-Å resolution and the role of quinones, lipids, channels and chloride. Nat. Struct. Mol. Biol. *16*, 334–342. http://dx.doi.org/10.1038/nsmb.1559.

Gygi, S.P., Rist, B., Gerber, S.A., Turecek, F., Gelb, M.H., and Aebersold, R. (1999). Quantitative analysis of complex protein mixtures using isotope-coded affinity tags. Nat. Biotechnol. *17*, 994–999. http://dx.doi.org/10.1038/13690.

Hagglund, P., Bunkenborg, J., Elortza, F., Jensen, O.N., and Roepstorff, P. (2004). A new strategy for identification of N–glycosylated proteins and unambiguous assignment of their glycosylation sites using HILIC enrichment and partial deglycosylation. J. Proteome Res. *3*, 556–566. http://dx.doi.org/10.1021/pr034112b.

Harrison, M.A., Tsinoremas, N.F., and Allen, J.F. (1991). Cyanobacterial thylakoid membrane proteins are reversiblyphosphorylatedunderplastoquinone-reducing conditions *in vitro*. FEBS Lett. *282*, 295–299. http://dx.doi.org/10.1016/0014-5793(91)80499-S.

Hartwell, L.H., Hopfield, J.J., Leibler, S., and Murray, A.W. (1999). From molecular to modular cell biology. Nature *402* (Suppl. 6761), C47–52. http://dx.doi.org/10.1038/35011540.

Harvey, D.J. (2005). Proteomic analysis of glycosylation: structural determination of *N*- and *O*-linked glycans by mass spectrometry. Expert. Rev. Proteomics. *2*, 87–101. http://dx.doi.org/10.1586/14789450.2.1.87.

Havugimana, P.C., Hart, G.T., Nepusz, T., Yang, H., Turinsky, A.L., Li, Z., Wang, P.I., Boutz, D.R., Fong, V., Phanse, S., *et al.* (2012). A census of human soluble protein complexes. Cell *150*, 1068–1081. http://dx.doi.org/10.1016/j.cell.2012.08.011.

Hess, D.T., Matsumoto, A., Kim, S.O., Marshall, H.E., and Stamler, J.S. (2005). Protein *S*-nitrosylation: purview and parameters. Nat. Rev. Mol. Cell. Biol. *6*, 150–166. http://dx.doi.org/10.1038/nrm1569.

Hongsthong, A., Sirijuntarut, M., Yutthanasirikul, R., Senachak, J., Kurdrid, P., Cheevadhanarak, S., and Tanticharoen, M. (2009). Subcellular proteomic characterization of the high-temperature stress response of the cyanobacterium *Spirulina platensis*. Proteome Sci. *7*, 33. http://dx.doi.org/10.1186/1477-5956-7-33.

Hu, Y., Wang, G., Chen, G.Y.J., Fu, X., and Yao, S.Q. (2003). Proteome analysis of *Saccharomyces cerevisiae* under metal stress by two-dimensional differential gel electrophoresis. Electrophoresis *24*, 1458–1470. http://dx.doi.org/10.1002/elps.200390188

Huang, F., Fulda, S., Hagemann, M., and Norling, B. (2006). Proteomic screening of salt-stress-induced changes in plasma membranes of *Synechocystis* sp. strain PCC 6803. Proteomics *6*, 910–920. http://dx.doi.org/10.1002/pmic.200500114.

Jaffrey, S.R., Erdjument Bromage, H., Ferris, C.D., Tempst, P., and Snyder, S.H. (2001). Protein *S*-nitrosylation: a physiological signal for neuronal nitric oxide. Nat. Cell Biol. *3*, 193–197. http://dx.doi.org/10.1038/35055104.

Jansen, R., and Gerstein, M. (2004). Analyzing protein function on a genomic scale: the importance of gold-standard positives and negatives for network prediction. Curr. Opin. Microbiol. *7*, 535–545. http://dx.doi.org/10.1016/j.mib.2004.08.012

Jensen, O.N. (2004). Modification-specific proteomics: characterization of post-translational modifications by mass spectrometry. Curr. Opin. Chem. Biol. *8*, 33–41. http://dx.doi.org/10.1016/j.cbpa.2003.12.009.

Jensen, O.N. (2006). Interpreting the protein language using proteomics. Nat. Rev. Mol. Cell. Biol. *7*, 391–403.

Johnson, C.H., Stewart, P.L., and Egli, M. (2011). The cyanobacterial circadian system: from biophysics to bioevolution. Annu. Rev. Biophys. *40*, 143–167. http://dx.doi.org/10.1146/annurev-biophys-042910-155317.

Kageyama, H., Nishiwaki, T., Nakajima, M., Iwasaki, H., Oyama, T., and Kondo, T. (2006). Cyanobacterial circadian pacemaker: Kai protein complex dynamics in the KaiC phosphorylation cycle *in vitro*. Mol. Cell. *23*, 161–171. http://dx.doi.org/10.1016/j.molcel.2006.05.039.

Kerppola, T.K. (2006). Design and implementation of bimolecular fluorescence complementation (BiFC) assays for the visualization of protein interactions in living cells. Nat. Protoc. *1*, 1278–1286. http://dx.doi.org/10.1038/nprot.2006.201

Kim, S.C., Sprung, R., Chen, Y., Xu, Y., Ball, H., Pei, J., Cheng, T., Kho, Y., Xiao, H., Xiao, L., *et al.* (2006). Substrate and functional diversity of lysine acetylation revealed by a proteomics survey. Mol. Cell *23*, 607–618. http://dx.doi.org/10.1016/j.molcel.2006.06.026.

Kirkwood, K.J., Ahmad, Y., Larance, M., and Lamond, A.I. (2013). Characterization of native protein complexes and protein isoform variation using size-fractionation-based quantitative proteomics. Mol. Cell. Proteomics. *12*, 3851–3873. http://dx.doi.org/10.1074/mcp.M113.032367.

Kosako, H., Yamaguchi, N., Aranami, C., Ushiyama, M., Kose, S., Imamoto, N., Taniguchi, H., Nishida, E., and Hattori, S. (2009). Phosphoproteomics reveals new ERK MAP kinase targets and links ERK to nucleoporin-mediated nuclear transport. Nat. Struct. Mol. Biol. *16*, 1026–1035. http://dx.doi.org/10.1038/nsmb.1656.

Kristensen, A.R., Gsponer, J., and Foster, L.J. (2012). A high-throughput approach for measuring temporal changes in the interactome. Nat. Methods *9*, 907–909. http://dx.doi.org/10.1038/nmeth.2131.

Krogan, N.J., Cagney, G., Yu, H., Zhong, G., Guo, X., Ignatchenko, A., Li, J., Pu, S., Datta, N., Tikuisis, A.P., *et al.* (2006). Global landscape of protein complexes in the yeast *Saccharomyces cerevisiae*. Nature *440*, 637–643. http://dx.doi.org/10.1038/nature04670

Kurian, D., Jansèn, T., and Mäenpää, P. (2006). Proteomic analysis of heterotrophy in *Synechocystis* sp. PCC 6803. Proteomics *6*, 1483–1494. http://dx.doi.org/10.1002/pmic.200500413

L'Hernault, S.W., and Rosenbaum, J.L. (1985). *Chlamydomonas* alpha-tubulin is posttranslationally modified by acetylation on the epsilon-amino group of a lysine. Biochemistry *24*, 473–478. http://dx.doi.org/10.1021/bi00323a034

Lange, V., Picotti, P., Domon, B., and Aebersold, R. (2008). Selected reaction monitoring for quantitative proteomics: a tutorial. Mol. Syst. Biol. *4*, 222. http://dx.doi.org/10.1038/msb.2008.61.

Larsen, M.R., Thingholm, T.E., Jensen, O.N., Roepstorff, P., and Jørgensen, T.J. (2005). Highly selective enrichment of phosphorylated peptides from peptide mixtures using titanium dioxide microcolumns. Mol. Cell. Proteomics. *4*, 873–886. http://dx.doi.org/10.1074/mcp.T500007-MCP200

Larsson, J., Nylander, J.A., and Bergman, B. (2011). Genome fluctuations in cyanobacteria reflect evolutionary, developmental and adaptive traits. BMC Evol. Biol. *11*, 187. http://dx.doi.org/10.1186/1471-2148-11-187.

Latysheva, N., Junker, V.L., Palmer, W.J., Codd, G.A., and Barker, D. (2012). The evolution of nitrogen fixation in cyanobacteria. Bioinformatics *28*, 603–606. http://dx.doi.org/10.1093/bioinformatics/bts008.

Lee, D.-G., Kwon, J., Eom, C.-Y., Kang, Y.-M., Roh, S.W., Lee, K.-B., and Choi, J.-S. (2015). Directed analysis of cyanobacterial membrane phosphoproteome using stained phosphoproteins and titanium-enriched phosphopeptides. J. Microbiol. *53*, 279–287. http://dx.doi.org/10.1007/s12275-015-5021-8.

Leichert, L.I., Gehrke, F., Gudiseva, H.V., Blackwell, T., Ilbert, M., Walker, A.K., Strahler, J.R., Andrews, P.C., and Jakob, U. (2008). Quantifying changes *in* the thiol redox proteome upon oxidative stress *in vivo*. Proc. Natl. Acad. Sci. U.S.A. *105*, 8197–8202. http://dx.doi.org/10.1073/pnas.0707723105.

Li, H., Singh, A.K., McIntyre, L.M., and Sherman, L.A. (2004). Differential gene expression in response to hydrogen peroxide and the putative PerR regulon of *Synechocystis* sp. strain PCC 6803. J. Bacteriol. *186*, 3331–3345. http://dx.doi.org/10.1128/JB.186.11.3331-3345.2004

Liebler, D.C., and Zimmerman, L.J. (2013). Targeted quantitation of proteins by mass spectrometry. Biochemistry *52*, 3797–3806. http://dx.doi.org/10.1021/bi400110b.

Liberton, M., Saha, R., Jacobs, J.M., Nguyen, A.Y., Gritsenko, M.A., Smith, R.D., Koppenaal, D.W., and Pakrasi, H.B. (2016). Global proteomic analysis reveals an exclusive role of thylakoid membranes in bioenergetics of a model cyanobacterium. Mol. Cell. Proteomics *15*, 2021–2032. http://dx.doi.org/10.1074/mcp.M115.057240.

Lindahl, M., and Florencio, F.J. (2003). Thioredoxin-linked processes in cyanobacteria are as numerous as in chloroplasts, but targets are different. Proc. Natl. Acad. Sci. U.S.A. *100*, 16107–16112. http://dx.doi.org/10.1073/pnas.2534397100

Lindmark, H., Jacobsson, K., Frykberg, L., and Guss, B. (1996). Fibronectin-binding protein of *Streptococcus equi* subsp. *zooepidemicus*. Infect. Immun. *64*, 3993–3999.

Link, A.J., Eng, J., Schieltz, D.M., Carmack, E., Mize, G.J., Morris, D.R., Garvik, B.M., and Yates, J.R., 3rd (1999). Direct analysis of protein complexes using mass spectrometry. Nat. Biotechnol. *17*, 676–682. http://dx.doi.org/10.1038/10890.

Liu, H., Sadygov, R.G., and Yates, J.R. (2004). A model for random sampling and estimation of relative protein abundance in shotgun proteomics. Anal. Chem. *76*, 4193–4201. http://dx.doi.org/10.1021/ac0498563

Liu, X., Yang, W.C., Gao, Q., and Regnier, F. (2008). Toward chromatographic analysis of interacting protein networks. J. Chromatogr. A. *1178*, 24–32. http://dx.doi.org/10.1016/j.chroma.2007.10.067

Los, D.A., Zorina, A., Sinetova, M., Kryazhov, S., Mironov, K., and Zinchenko, V.V. (2010). Stress sensors and signal transducers in cyanobacteria. Sensors *10*, 2386–2415. http://dx.doi.org/10.3390/s100302386.

Mackey, S.R., and Golden, S.S. (2007). Winding up the cyanobacterial circadian clock. Trends Microbiol. *15*, 381–388. http://dx.doi.org/10.1016/j.tim.2007.08.005.

Matallana-Surget, S., Derock, J., Leroy, B., Badri, H., Deschoenmaeker, F., and Wattiez, R. (2014). Proteome-wide analysis and diel proteomic profiling of the cyanobacterium *Arthrospira platensis* PCC 8005. PLoS One *9*, e99076. http://dx.doi.org/10.1371/journal.pone.0099076.

Matthiesen, R., and Mutenda, K.E. (2007). Introduction to proteomics. Methods Mol. Biol. *367*, 1–35. http://dx.doi.org/10.1385/1-59745-275-0:1.

McDermott, J.E., Oehmen, C.S., McCue, L.A., Hill, E., Choi, D.M., Stöckel, J., Liberton, M., Pakrasi, H.B., and Sherman, L.A. (2011). A model of cyclic transcriptomic behavior in the cyanobacterium *Cyanothece* sp. ATCC 51142. Mol. Biosyst. *7*, 2407–2418. http://dx.doi.org/10.1039/c1mb05006k.

Mikkat, S., Fulda, S., and Hagemann, M. (2014). A 2D gel electrophoresis-based snapshot of the phosphoproteome in the cyanobacterium *Synechocystis* sp. strain PCC 6803. Microbiology *160*, 296–306. http://dx.doi.org/10.1099/mic.0.074443-0.

Min, H., and Sherman, L.A. (2010). Genetic transformation and mutagenesis via single-stranded DNA in the unicellular, diazotrophic cyanobacteria of the genus *Cyanothece*. Appl. Environ. Microbiol. *76*, 7641–7645. http://dx.doi.org/10.1128/AEM.01456-10.

Mo, R., Yang, M., Chen, Z., Cheng, Z., Yi, X., Li, C., He, C., Xiong, Q., Chen, H., Wang, Q., *et al.* (2015). Acetylome analysis reveals the involvement of lysine acetylation in photosynthesis and carbon metabolism in the model cyanobacterium *Synechocystis* sp. PCC 6803. J. Proteome Res. *14*, 1275–1286. http://dx.doi.org/10.1021/pr501275a.

Moore, J.B., and Weeks, M.E. (2011). Proteomics and systems biology: current and future applications in the nutritional sciences. Adv. Nutr. *2*, 355–364. http://dx.doi.org/10.3945/an.111.000554.

Mota, R., Pereira, S.B., Meazzini, M., Fernandes, R., Santos, A., Evans, C.A., De Philippis, R., Wright, P.C., and Tamagnini, P. (2015). Effects of heavy metals on *Cyanothece* sp. CCY 0110 growth, extracellular polymeric substances (EPS) production, ultrastructure and protein profiles. J. Proteom. *120*, 75–94. http://dx.doi.org/10.1016/j.jprot.2015.03.004.

Mukhtar, M.S., Carvunis, A.R., Dreze, M., Epple, P., Steinbrenner, J., Moore, J., Tasan, M., Galli, M., Hao, T., Nishimura, M.T., *et al.* (2011). Independently evolved virulence effectors converge onto hubs in a plant immune system network. Science *333*, 596–601. http://dx.doi.org/10.1126/science.1203659.

Murray, C.I., Uhrigshardt, H., O'Meally, R.N., Cole, R.N., and Van Eyk, J.E. (2012). Identification and quantification of *S*-nitrosylation by cysteine reactive tandem mass tag switch assay. Mol. Cell. Proteomics. *11*, M111.013441. http://dx.doi.org/10.1074/mcp.M111.013441.

Nelson, K.J., Klomsiri, C., Codreanu, S.G., Soito, L., Liebler, D.C., Rogers, L.C., Daniel, L.W., and Poole, L.B. (2010). Use of dimedone-based chemical probes for sulfenic acid detection methods to visualize and identify labeled proteins. Methods Enzymol. *473*, 95–115. http://dx.doi.org/10.1016/S0076-6879(10)73004-4.

O'Farrell, P.H. (1975). High resolution two-dimensional electrophoresis of proteins. J. Biol. Chem. *250*, 4007–4021.

Olinares, P.D., Ponnala, L., and van Wijk, K.J. (2010). Megadalton complexes in the chloroplast stroma of *Arabidopsis thaliana* characterized by size exclusion chromatography, mass spectrometry, and hierarchical clustering. Mol. Cell. Proteomics *9*, 1594–1615. http://dx.doi.org/10.1074/mcp.M000038-MCP201.

Ong, S.E., Blagoev, B., Kratchmarova, I., Kristensen, D.B., Steen, H., Pandey, A., and Mann, M. (2002). Stable isotope labeling by amino acids in cell culture, SILAC, as a simple and accurate approach to expression proteomics. Mol. Cell. Proteomics *1*, 376–386.

Pandey, A., Podtelejnikov, A.V., Blagoev, B., Bustelo, X.R., Mann, M., and Lodish, H.F. (2000). Analysis of receptor signaling pathways by mass spectrometry: identification of vav-2 as a substrate of the epidermal and platelet-derived growth factor receptors. Proc. Natl. Acad. Sci. U.S.A. *97*, 179–184. http://dx.doi.org/10.1073/pnas.97.1.179.

Pattanayak, G.K., Lambert, G., Bernat, K., and Rust, M.J. (2015). Controlling the cyanobacterial clock by synthetically rewiring metabolism. Cell. Rep. *13*, 2362–2367. http://dx.doi.org/10.1016/j.celrep.2015.11.031.

Patton, W.F. (2002). Detection technologies in proteome analysis. J. Chromatogr. B. Analyt. Technol. Biomed. Life. Sci. *771*, 3–31. http://dx.doi.org/10.1016/S1570-0232(02)00043-0.

Paulsen, C.E., Truong, T.H., Garcia, F.J., Homann, A., Gupta, V., Leonard, S.E., and Carroll, K.S. (2012). Peroxide-dependent sulfenylation of the EGFR catalytic site enhances kinase activity. Nat. Chem. Biol. *8*, 57–64. http://dx.doi.org/10.1038/nchembio.736.

Peltier, J.-B., Cai, Y., Sun, Q., Zabrouskov, V., Giacomelli, L., Rudella, A., Ytterberg, A.J., Rutschow, H., and van Wijk, K.J. (2006). The oligomeric stromal proteome of *Arabidopsis thaliana* chloroplasts. Mol. Cell. Proteomics *5*, 114–133. http://dx.doi.org/10.1074/mcp.M500180-MCP200.

Phizicky, E., Bastiaens, P.I., Zhu, H., Snyder, M., and Fields, S. (2003). Protein analysis on a proteomic scale. Nature *422*, 208–215. http://dx.doi.org/10.1038/nature01512.

Pinkse, M.W., Uitto, P.M., Hilhorst, M.J., Ooms, B., and Heck, A.J. (2004). Selective isolation at the femtomole level of phosphopeptides from proteolytic digests using 2D-NanoLC-ESI-MS/MS and titanium oxide precolumns. Anal. Chem. *76*, 3935–3943. http://dx.doi.org/10.1021/ac0498617.

Piven, I., Ajlani, G., and Sokolenko, A. (2005). Phycobilisome linker proteins are phosphorylated in *Synechocystis* sp. PCC 6803. J. Biol. Chem. *280*, 21667–21672. http://dx.doi.org/10.1074/jbc.M412967200.

Posewitz, M.C., and Tempst, P. (1999). Immobilized gallium(III) affinity chromatography of phosphopeptides. Anal. Chem. *71*, 2883–2892. http://dx.doi.org/10.1021/ac981409y.

Puig, O., Caspary, F., Rigaut, G., Rutz, B., Bouveret, E., Bragado-Nilsson, E., Wilm, M., and Séraphin, B. (2001). The tandem affinity purification (TAP) method: a general procedure of protein complex purification. Methods *24*, 218–229. http://dx.doi.org/10.1006/meth.2001.1183.

Qian, W.J., Jacobs, J.M., Liu, T., Camp, D.G., and Smith, R.D. (2006). Advances and challenges in liquid chromatography-mass spectrometry-based proteomics profiling for clinical applications. Mol. Cell. Proteomics *5*, 1727–1744. http://dx.doi.org/10.1074/mcp.M600162-MCP200.

Qu, Z., Meng, F., Bomgarden, R.D., Viner, R.I., Li, J., Rogers, J.C., Cheng, J., Greenlief, C.M., Cui, J., Lubahn, D.B., *et al.* (2014). Proteomic quantification and site-mapping of S-nitrosylated proteins using isobaric iodoTMT reagents. J. Proteome Res. *13*, 3200–3211. http://dx.doi.org/10.1021/pr401179v.

Rabilloud, T. (2002). Two-dimensional gel electrophoresis in proteomics: old, old fashioned, but it still climbs up the mountains. Proteomics *2*, 3–10. http://dx.doi.org/10.1002/1615-9861(200201)2:1<3::AID-PROT3>3.0.CO;2-R.

Rhodes, D.R., Tomlins, S.A., Varambally, S., Mahavisno, V., Barrette, T., Kalyana-Sundaram, S., Ghosh, D., Pandey, A., and Chinnaiyan, A.M. (2005). Probabilistic model of the human protein-protein interaction network. Nat. Biotechnol. *23*, 951–959. http://dx.doi.org/10.1038/nbt1103.

Rigaut, G., Shevchenko, A., Rutz, B., Wilm, M., Mann, M., and Séraphin, B. (1999). A generic protein purification method for protein complex characterization and proteome exploration. Nat. Biotechnol. *17*, 1030–1032. http://dx.doi.org/10.1038/13732

Ross, P.L., Huang, Y.N., Marchese, J.N., Williamson, B., Parker, K., Hattan, S., Khainovski, N., Pillai, S., Dey, S., Daniels, S., *et al.* (2004). Multiplexed protein quantitation in *Saccharomyces cerevisiae* using amine-reactive isobaric tagging reagents. Mol. Cell. Proteomics *3*, 1154–1169. http://dx.doi.org/10.1074/mcp.M400129-MCP200.

Rubin, B.E., Wetmore, K.M., Price, M.N., Diamond, S., Shultzaberger, R.K., Lowe, L.C., Curtin, G., Arkin, A.P., Deutschbauer, A., and Golden, S.S. (2015). The essential gene set of a photosynthetic organism. Proc. Natl. Acad. Sci. U.S.A. *112*, E6634–43. http://dx.doi.org/10.1073/pnas.1519220112.

Ruffing, A.M. (2013). RNA-Seq analysis and targeted mutagenesis for improved free fatty acid production in an engineered cyanobacterium. Biotechnol. Biofuels *6*, 113. http://dx.doi.org/10.1186/1754-6834-6-113.

Sadler, N.C., Melnicki, M.R., Serres, M.H., Merkley, E.D., Chrisler, W.B., Hill, E.A., Romine, M.F., Kim, S., Zink, E.M., Datta, S., *et al.* (2014). Live cell chemical profiling of temporal redox dynamics in a photoautotrophic cyanobacterium. ACS Chem. Biol. *9*, 291–300. http://dx.doi.org/10.1021/cb400769v.

Sato, S., Ikeuchi, M., and Nakamoto, H. (2008). Expression and function of a *groEL* paralog in the thermophilic cyanobacterium *Thermosynechococcus elongatus* under heat and cold stress. FEBS Lett. *582*, 3389–3395. http://dx.doi.org/10.1016/j.febslet.2008.08.034.

Sato, Y., and Inaba, K. (2012). Disulfide bond formation network in the three biological kingdoms, bacteria, fungi and mammals. FEBS J. *279*, 2262–2271. http://dx.doi.org/10.1111/j.1742-4658.2012.08593.x.

Sethuraman, M., McComb, M.E., Heibeck, T., Costello, C.E., and Cohen, R.A. (2004). Isotope-coded affinity tag approach to identify and quantify oxidant-sensitive protein thiols. Mol. Cell. Proteomics *3*, 273–278. http://dx.doi.org/10.1074/mcp.T300011-MCP200.

Sherman, L.A., Min, H., Toepel, J., and Pakrasi, H.B. (2010). Better living through *Cyanothece* – unicellular diazotrophic cyanobacteria with highly versatile metabolic systems. Adv. Exp. Med. Biol. *675*, 275–290. http://dx.doi.org/10.1007/978-1-4419-1528-3_16.

Shultzaberger, R.K., Boyd, J.S., Diamond, S., Greenspan, R.J., and Golden, S.S. (2015). Giving time purpose: the *Synechococcus elongatus* clock in a broader network context. Annu. Rev. Genet. *49*, 485–505. http://dx.doi.org/10.1146/annurev-genet-111212-133227.

Singh, A.K., Elvitigala, T., Cameron, J.C., Ghosh, B.K., Bhattacharyya-Pakrasi, M., and Pakrasi, H.B. (2010). Integrative analysis of large scale expression profiles reveals core transcriptional response and coordination between multiple cellular processes in a cyanobacterium. BMC Syst. Biol. *4*, 1–14. http://dx.doi.org/10.1186/1752-0509-4-105.

Singh, A.K., McIntyre, L.M., and Sherman, L.A. (2003). Microarray analysis of the genome-wide response to iron deficiency and iron reconstitution in the cyanobacterium *Synechocystis* sp. PCC 6803. Plant Physiol. *132*, 1825–1839. http://dx.doi.org/10.1104/pp.103.024018.

Slabas, A.R., Suzuki, I., Murata, N., Simon, W.J., and Hall, J.J. (2006). Proteomic analysis of the heat shock response in *Synechocystis* PCC6803 and a thermally tolerant knockout strain lacking the histidine kinase 34 gene. Proteomics *6*, 845–864. http://dx.doi.org/10.1002/pmic.200500196.

Sobczyk, A., Schyns, G., Tandeau de Marsac, N., and Houmard, J. (1993). Transduction of the light signal during complementary chromatic adaptation in the cyanobacterium *Calothrix* sp. PCC 7601: DNA-binding proteins and modulation by phosphorylation. EMBO. J. *12*, 997–1004.

Spät, P., Maček, B., and Forchhammer, K. (2015). Phosphoproteome of the cyanobacterium *Synechocystis* sp. PCC 6803 and its dynamics during nitrogen starvation. Front. Microbiol. *6*, 248. http://dx.doi.org/10.3389/fmicb.2015.00248.

Srere, P.A. (2000). Macromolecular interactions: tracing the roots. Trends Biochem. Sci *25*, 150–153. http://dx.doi.org/10.1016/S0968-0004(00)01550-4.

Stensballe, A., Andersen, S., and Jensen, O.N. (2001). Characterization of phosphoproteins from electrophoretic gels by nanoscale Fe(III) affinity chromatography with off-line mass spectrometry analysis. Proteomics *1*, 207–222. http://dx.doi.org/10.1002/1615-9861(200102)1:2<207::AID-PROT207>3.0.CO;2-3.

Stöckel, J., Jacobs, J.M., Elvitigala, T.R., Liberton, M., Welsh, E.A., Polpitiya, A.D., Gritsenko, M.A., Nicora, C.D., Koppenaal, D.W., Smith, R.D., *et al.* (2011). Diurnal rhythms result in significant changes in the cellular protein complement in the cyanobacterium *Cyanothece* 51142. PLoS One *6*, e16680. http://dx.doi.org/10.1371/journal.pone.0016680.

Stöckel, J., Welsh, E.A., Liberton, M., Kunnvakkam, R., Aurora, R., and Pakrasi, H.B. (2008). Global transcriptomic analysis of *Cyanothece* 51142 reveals robust diurnal oscillation of central metabolic processes. Proc. Natl. Acad. Sci. U.S.A. *105*, 6156–6161. http://dx.doi.org/10.1073/pnas.0711068105.

Stüeken, E.E., Buick, R., Guy, B.M., and Koehler, M.C. (2015). Isotopic evidence for biological nitrogen fixation by molybdenum-nitrogenase from 3.2 Gyr. Nature *520*, 666–669. http://dx.doi.org/10.1038/nature14180.

Summerfield, T.C., Toepel, J., and Sherman, L.A. (2008). Low-oxygen induction of normally cryptic *psbA* genes in cyanobacteria. Biochemistry *47*, 12939–12941. http://dx.doi.org/10.1021/bi8018916.

Suzuki, I., Simon, W.J., and Slabas, A.R. (2006). The heat shock response of *Synechocystis* sp. PCC 6803 analysed by transcriptomics and proteomics. J. Exp. Bot. *57*, 1573–1578. http://dx.doi.org/10.1093/jxb/erj148.

Tanaka, N., Hiyama, T., and Nakamoto, H. (1997). Cloning, characterization and functional analysis of *groESL* operon from thermophilic cyanobacterium *Synechococcus vulcanus*. Biochim. Biophys. Acta *1343*, 335–348. http://dx.doi.org/10.1016/S0167-4838(97)00159-3

Tanaka, N., and Nakamoto, H. (1999). HtpG is essential for the thermal stress management in cyanobacteria. FEBS Lett. *458*, 117–123. http://dx.doi.org/10.1016/S0014-5793(99)01134-5.

Tang, W., Deng, Z., Oses-Prieto, J.A., Suzuki, N., Zhu, S., Zhang, X., Burlingame, A.L., and Wang, Z.Y. (2008). Proteomics studies of brassinosteroid signal transduction using prefractionation and two-dimensional DIGE. Mol. Cell. Proteomics *7*, 728–738. http://dx.doi.org/10.1074/mcp.M700358-MCP200.

Thompson, A., Schäfer, J., Kuhn, K., Kienle, S., Schwarz, J., Schmidt, G., Neumann, T., Johnstone, R., Mohammed, A.K., and Hamon, C. (2003). Tandem mass tags: a novel quantification strategy for comparative analysis of complex protein mixtures by MS/MS. Anal. Chem. *75*, 1895–1904. http://dx.doi.org/10.1021/ac0262560.

Toepel, J., Welsh, E., Summerfield, T.C., Pakrasi, H.B., and Sherman, L.A. (2008). Differential transcriptional analysis of the cyanobacterium *Cyanothece* sp. strain ATCC 51142 during light–dark and continuous-light growth. J. Bacteriol. *190*, 3904–3913. http://dx.doi.org/10.1128/JB.00206-08.

Toepel, J.R., McDermott, J.E., Summerfield, T.C., and Sherman, L.A. (2009). Transcriptional analysis of the unicellular, diazotrophic cyanobacterium *Cyanothece* sp. ATCC 51142 grown under short day/night cycles. J. Phycol. *45*, 610–620. http://dx.doi.org/10.1111/j.1529-8817.2009.00674.x.

Tucker, D.L., and Sherman, L.A. (2000). Analysis of chlorophyll-protein complexes from the cyanobacterium *Cyanothece* sp. ATCC 51142 by non-denaturing gel electrophoresis. Biochim. Biophys. Acta *1468*, 150–160. http://dx.doi.org/10.1016/S0005-2736(00)00263-7.

Tuominen, I., Pollari, M., Aguirre von Wobeser, E., Tyystjärvi, E., Ibelings, B.W., Matthijs, H.C., and Tyystjärvi, T. (2008). Sigma factor SigC is required for heat acclimation of the cyanobacterium *Synechocystis* sp. strain PCC 6803. FEBS Lett. *582*, 346–350. http://dx.doi.org/10.1016/j.febslet.2007.12.030.

Uetz, P., Giot, L., Cagney, G., Mansfield, T.A., Judson, R.S., Knight, J.R., Lockshon, D., Narayan, V., Srinivasan, M., Pochart, P., *et al.* (2000). A comprehensive analysis of protein–protein interactions in *Saccharomyces cerevisiae*. Nature *403*, 623–627. http://dx.doi.org/10.1038/35001009.

Varvasovszki, V., Glatz, A., Shigapova, N., Jósvay, K., Vígh, L., and Horváth, I. (2003). Only one *dnaK* homolog, *dnaK2*, is active transcriptionally and is essential in *Synechocystis*. Biochem. Biophys. Res. Commun. *305*, 641–648. http://dx.doi.org/10.1016/S0006-291X(03)00822-2.

Verdin, E., and Ott, M. (2015). 50 years of protein acetylation: from gene regulation to epigenetics, metabolism and beyond. Nat. Rev. Mol. Cell. Biol. *16*, 258–264. http://dx.doi.org/10.1038/nrm3931.

Vijayan, V., Jain, I.H., and O'Shea, E.K. (2011). A high resolution map of a cyanobacterial transcriptome. Genome Biol. *12*, R47. http://dx.doi.org/10.1186/gb-2011-12-5-r47.

von Mering, C., Krause, R., Snel, B., Cornell, M., Oliver, S.G., Fields, S., and Bork, P. (2002). Comparative assessment of large-scale data sets of protein-protein interactions. Nature *417*, 399–403. http://dx.doi.org/10.1038/nature750.

Wang, L., Aryal, U.K., Dai, Z., Mason, A.C., Monroe, M.E., Tian, Z.X., Zhou, J.Y., Su, D., Weitz, K.K., Liu, T., *et al.* (2012). Mapping N-linked glycosylation sites in the secretome and whole cells of *Aspergillus niger* using hydrazide chemistry and mass spectrometry. J. Proteome Res. *11*, 143–156. http://dx.doi.org/10.1021/pr200916k.

Wang, W., Zhou, H., Lin, H., Roy, S., Shaler, T.A., Hill, L.R., Norton, S., Kumar, P., Anderle, M., and Becker, C.H. (2003). Quantification of proteins and metabolites by mass spectrometry without isotopic labeling or spiked standards. Anal. Chem. *75*, 4818–4826. http://dx.doi.org/10.1021/ac026468x.

Weerapana, E., Wang, C., Simon, G.M., Richter, F., Khare, S., Dillon, M.B., Bachovchin, D.A., Mowen, K., Baker, D., and Cravatt, B.F. (2010). Quantitative reactivity profiling predicts functional cysteines in proteomes. Nature *468*, 790–795. http://dx.doi.org/10.1038/nature09472.

Wegener, K.M., Singh, A.K., Jacobs, J.M., Elvitigala, T., Welsh, E.A., Keren, N., Gritsenko, M.A., Ghosh, B.K., Camp, D.G., Smith, R.D., *et al.* (2010). Global proteomics reveal an atypical strategy for carbon/nitrogen assimilation by a cyanobacterium under diverse environmental perturbations. Mol. Cell. Proteomics *9*, 2678–2689. http://dx.doi.org/10.1074/mcp.M110.000109.

Weinert, B.T., Wagner, S.A., Horn, H., Henriksen, P., Liu, W.R., Olsen, J.V., Jensen, L.J., and Choudhary, C. (2011). Proteome-wide mapping of the *Drosophila* acetylome demonstrates a high degree of conservation of lysine acetylation. Sci. Signaling *4*, ra48. http://dx.doi.org/10.1126/scisignal.2001902.

Welkie, D., Zhang, X., Markillie, M.L., Taylor, R., Orr, G., Jacobs, J., Bhide, K., Thimmapuram, J., Gritsenko, M., Mitchell, H., *et al.* (2014). Transcriptomic and proteomic dynamics in the metabolism of a diazotrophic cyanobacterium, *Cyanothece* sp. PCC 7822 during a diurnal light–dark cycle. BMC Genomics *15*, 1185. http://dx.doi.org/10.1186/1471-2164-15-1185.

Welkie, D.G., Lee, B.H., and Sherman, L.A. (2016). Altering the Structure of Carbohydrate Storage Granules in the Cyanobacterium *Synechocystis* sp. Strain PCC 6803 through Branching-Enzyme Truncations. J. Bacteriol. *198*, 701–710. http://dx.doi.org/10.1128/JB.00830-15.

Welkie, D.G., Sherman, D.M., Chrisler, W.B., Orr, G., and Sherman, L.A. (2013). Analysis of carbohydrate storage granules in the diazotrophic cyanobacterium *Cyanothece* sp. PCC 7822. Photosynth. Res. *118*, 25–36. http://dx.doi.org/10.1007/s11120-013-9941-z.

Welsh, E.A., Liberton, M., Stöckel, J., Loh, T., Elvitigala, T., Wang, C., Wollam, A., Fulton, R.S., Clifton, S.W., Jacobs, J.M., *et al.* (2008). The genome of *Cyanothece* 51142, a unicellular diazotrophic cyanobacterium important in the marine nitrogen cycle. Proc. Natl. Acad. Sci. U.S.A. *105*, 15094–15099. http://dx.doi.org/10.1073/pnas.0805418105.

Williams, S.B. (2007). A circadian timing mechanism in the cyanobacteria. Adv. Microb. Physiol. *52*, 229–296. http://dx.doi.org/10.1016/S0065-2911(06)52004-1.

Wilson, N.L., Schulz, B.L., Karlsson, N.G., and Packer, N.H. (2002). Sequential analysis of *N*- and *O*-linked glycosylation of 2D-PAGE separated glycoproteins. J. Proteome Res. *1*, 521–529. http://dx.doi.org/10.1021/pr025538d.

Wodak, S.J., Pu, S., Vlasblom, J., and Séraphin, B. (2009). Challenges and rewards of interaction proteomics. Mol. Cell. Proteomics *8*, 3–18. http://dx.doi.org/10.1074/mcp.R800014-MCP200.

Wolf-Yadlin, A., Hautaniemi, S., Lauffenburger, D.A., and White, F.M. (2007). Multiple reaction monitoring for robust quantitative proteomic analysis of cellular signaling networks. Proc. Natl. Acad. Sci. U.S.A. *104*, 5860–5865. http://dx.doi.org/10.1073/pnas.0608638104.

Xiong, Q., Chen, Z., and Ge, F. (2016). Proteomic analysis of post translational modifications in cyanobacteria. J. Proteomics *134*, 57–64. http://dx.doi.org/10.1016/j.jprot.2015.07.037.

Yang, M.K., Qiao, Z.X., Zhang, W.Y., Xiong, Q., Zhang, J., Li, T., Ge, F., and Zhao, J.D. (2013). Global phosphoproteomic analysis reveals diverse functions of serine/threonine/tyrosine phosphorylation in the model cyanobacterium *Synechococcus* sp. strain PCC 7002. J. Proteome Res. *12*, 1909–1923. http://dx.doi.org/10.1021/pr4000043.

Yang, Z., and Hancock, W.S. (2004). Approach to the comprehensive analysis of glycoproteins isolated from human serum using a multi-lectin affinity column. J. Chromatogr. A. *1053*, 79–88. http://dx.doi.org/10.1016/j.chroma.2004.08.150.

Yu, H., Braun, P., Yildirim, M.A., Lemmens, I., Venkatesan, K., Sahalie, J., Hirozane Kishikawa, T., Gebreab, F., Li, N., Simonis, N., *et al.* (2008). High-quality binary protein interaction map of the yeast interactome network. Science *322*, 104–110. http://dx.doi.org/10.1126/science.1158684.

Zhang, C.C., Jang, J., Sakr, S., and Wang, L. (2005). Protein phosphorylation on Ser, Thr and Tyr residues in cyanobacteria. J. Mol. Microbiol. Biotechnol. *9*, 154–166. http://dx.doi.org/10.1159/000089644

Zhang, X., Sherman, D.M., and Sherman, L.A. (2014). The uptake hydrogenase in the unicellular diazotrophic cyanobacterium *Cyanothece* sp. strain PCC 7822 protects nitrogenase from oxygen toxicity. J. Bacteriol. *196*, 840–849. http://dx.doi.org/10.1128/JB.01248-13.

Zhao, S., Xu, W., Jiang, W., Yu, W., Lin, Y., Zhang, T., Yao, J., Zhou, L., Zeng, Y., Li, H., *et al.* (2010). Regulation of cellular metabolism by protein lysine acetylation. Science *327*, 1000–1004. http://dx.doi.org/10.1126/science.1179689.

Zhao, Y., and Jensen, O.N. (2009). Modification–specific proteomics: Strategies for characterization of post–translational modifications using enrichment techniques. Proteomics *9*, 4632–4641. http://dx.doi.org/10.1002/pmic.200900398.

Zhou, J.Y., Afjehi-Sadat, L., Asress, S., Duong, D.M., Cudkowicz, M., Glass, J.D., and Peng, J. (2010). Galectin-3 is a candidate biomarker for amyotrophic lateral sclerosis: discovery by a proteomics approach. J. Proteome Res. *9*, 5133–5141. http://dx.doi.org/10.1021/pr100409r.

Zhu, H., Bilgin, M., Bangham, R., Hall, D., Casamayor, A., Bertone, P., Lan, N., Jansen, R., Bidlingmaier, S., Houfek, T., *et al.* (2001). Global analysis of protein activities using proteome chips. Science *293*, 2101–2105.

Zorina, A., Stepanchenko, N., Novikova, G.V., Sinetova, M., Panichkin, V.B., Moshkov, I.E., Zinchenko, V.V., Shestakov, S.V., Suzuki, I., Murata, N., *et al.* (2011). Eukaryotic-like Ser/Thr protein kinases SpkC/F/K are involved in phosphorylation of GroES in the cyanobacterium *Synechocystis*. DNA Res. *18*, 137–151. http://dx.doi.org/10.1093/dnares/dsr006.

Zybailov, B., Mosley, A.L., Sardiu, M.E., Coleman, M.K., Florens, L., and Washburn, M.P. (2006). Statistical analysis of membrane proteome expression changes in *Saccharomyces cerevisiae*. J. Proteome. Res. *5*, 2339–2347. http://dx.doi.org/10.1021/pr060161n.

Proteomic Analysis of Post Translational Modifications in Cyanobacteria

7

Qian Xiong[1], Zhuo Chen[1] and Feng Ge[1*]

[1]Key Laboratory of Algal Biology, Institute of Hydrobiology, Chinese Academy of Sciences, Wuhan, China.

*Correspondence: gefeng@ihb.ac.cn

Abstract

Cyanobacteria are a diverse group of Gram-negative bacteria and the only prokaryotes capable of oxygenic photosynthesis. Recently, cyanobacteria have attracted great interest due to their crucial roles in global carbon and nitrogen cycles and their ability to produce clean and renewable biofuels. To survive at various environmental conditions, cyanobacteria have developed a complex signal transduction network to sense environmental signals and implement adaptive changes. The post-translational modifications (PTMs) systems play important regulatory roles in the signalling networks of cyanobacteria. Mass spectrometry (MS)-based proteomic technologies have been applied to identify PTMs in a systematic manner. Although the proteomic studies of PTMs carried out in cyanobacteria were limited, these data have provided clues to elucidate their sophisticated sensing mechanisms that contribute to their evolutionary and ecological success. This chapter aims to summarize the current status of PTM studies and recent publications regarding PTM proteomics in cyanobacteria, and discuss the novel developments and applications for the analysis of PTMs in cyanobacteria. Challenges, opportunities and future perspectives in the proteomics studies of PTMs in cyanobacteria are also discussed.

Introduction

Cyanobacteria are a morphologically diverse and widespread group of Gram-negative bacteria (Beck *et al.*, 2012). These ancient microorganisms (≈2.6 billion years) are the only prokaryotes capable of oxygenic photosynthesis, they are believed to be the progenitors of chloroplasts in higher plants and make substantial contribution to global CO_2 assimilation, O_2 production, and N_2 fixation (Chellamuthu *et al.*, 2013; Fay, 1992). Cyanobacteria are responsible for a substantial part of the annual primary production in the oceans (Partensky *et al.*, 1999). Recently, cyanobacteria have attracted great attention due to their essential roles in global nitrogen and carbon cycles and their ability to produce clean and renewable biofuels (Lubner *et al.*, 2011; Parmar *et al.*, 2011; Rosgaard *et al.*, 2012).

Cyanobacteria have successfully adapted to all habitats with extraordinary capacity to survive in varieties of extreme environments, such as high temperatures, high salinity, and extreme pH (Crispim and Gaylarde, 2005). To adapt to changing environmental conditions and survive in extreme environments, cyanobacteria have developed a delicate signal transduction network to sense external signals and implement adaptive changes (Mann, 1994; Piven *et al.*, 2005).

Post-translational modifications (PTMs) are defined as the covalent processing events which change the properties of a protein by proteolytic cleavage or by addition of a modifying group(e.g. phosphate, acetyl, or methyl) to specific amino acid residues (Mann and Jensen, 2003). PTMs are dynamic cellular processes occurring in various cell compartments that modulate diverse protein properties, including folding, conformation, activity, and consequently, regulation of protein function (Hart and Ball, 2013; Olsen and Mann, 2013).

The PTM systems are key components of the signal transduction network and play important regulatory roles (Ubersax and Ferrell, 2007; Yang, 2004). A large number of PTMs events have already been reported in cyanobacteria, which are extensively involved in cellular processes and metabolic homeostasis (which will be further discussed in section 'PTMs and proteomic analysis of PTMs in cyanobacteria'). However, most existing studies are sporadic. Systematic investigation of PTMs will be greatly helpful to obtain a comprehensive understanding of the signal transduction network and regulatory mechanisms in cyanobacteria. Proteomic investigations of PTMs have started only at the beginning of the last decade (Beausoleil *et al.*, 2004; Ficarro *et al.*, 2002; Gruhler *et al.*, 2005a). Now the study of PTMs via MS-based proteomic strategies constitutes a vibrant and large field of research. Although the proteomics studies of PTMs carried out in cyanobacteria were limited (Chardonnet *et al.*, 2015; Chen *et al.*, 2015; Lee *et al.*, 2015; Mikkat *et al.*, 2014; Mo *et al.*, 2015; Spät *et al.*, 2015; Yang *et al.*, 2013; 2014), these studies provided clues to elucidate the complex signal transduction network and the biochemical processes that contribute to their ecological success. Here we provide a snapshot of the PTM proteomic studies, including the current status, the progresses have been made, and the main challenges.

PTMs and proteomic analysis of PTMs in cyanobacteria

PTM studies in cyanobacteria

PTMs participate in numerous biological processes in cyanobacteria, such as circadian rhythms (Axmann *et al.*, 2014; Williams, 2007), global carbon/nitrogen control (Forchhammer, 2004), nitrogen fixation (Gallon *et al.*, 2000), complementary chromatic adaptation (Sobczyk *et al.*, 1993; 1994). PTM events are also reported to exist in photosynthesis apparatus, including photosystem I (Xu *et al.*, 2001), photosystem II (Michel *et al.*, 1988), phycobiliproteins (Miller *et al.*, 2008; Schluchter *et al.*, 2010) and cytochrome b_6f complex (Baniulis *et al.*, 2009). Types or classes of PTMs that have already been reported in cyanobacteria include protein splicing (Caspi *et al.*, 2003; Nichols and Evans, 2004; Shah *et al.*, 2012; Wei *et al.*, 2006; Yang *et al.*, 2004), phosphorylation (Lee *et al.*, 2015; Mikkat *et al.*, 2014; Spät *et al.*, 2015; Yang *et al.*, 2013), acetylation (Mo *et al.*, 2015), methylation (Kim *et al.*, 2011; Swanson and Glazer, 1990; Yang *et al.*, 2014), biotinylation (Gornicki *et al.*, 1993; von Leoprechting *et al.*, 1998; Xu *et al.*, 1994), persulfide formation (Behshad and Bollinger, 2009; Behshad *et al.*, 2004; Campanini *et al.*, 2006), ADP-ribosylation (Silman *et al.*, 1995), glutathionylation (Chardonnet *et al.*, 2015), prenylation (McIntosh *et al.*, 2011), palmitoylation (Gallon *et al.*, 2000), Histidine hydroxyl modification (Sugiura *et al.*, 2013) and carboxyl-terminal processing (Inagaki *et al.*, 2001; Kanervo *et al.*, 1997).

Protein splicing is a post-translational process catalysed by a large family of proteins called split inteins (Mills and Perler, 2005). The split DnaE inteins family in cyanobacteria have been well studied, including the structure, distribution, and splicing efficiencies (Caspi *et al.*, 2003; Nichols and Evans, 2004; Shah *et al.*, 2012; Wei *et al.*, 2006; Yang *et al.*, 2004). These comprehensive surveys of split inteins provide valuable information for the development of intein-based tools for chemical biology.

Cyanobacteria are prolific sources of secondary metabolites that exhibit antimicrobial, anticancer, and immunosuppressive activities that could be used in drug development (Burja *et al.*, 2001). A major part of cyanobacterial secondary metabolites are cyclic or linear peptides or process peptidic substructures produced through both nonribosomal (such as microcystin and nodularin) and ribosomal biosynthetic pathways (such as cyanobactins and microviridin) (Welker and von Dohren, 2006). The majority of these cyanobacterial peptides can be heavily modified, including epimerization, glycosylation, methylation, acylation, formylation, halogenation, sulfation, heterocyclization, oxidation and prenylation (reviewed by Welker, 2005; Welker and von Dohren, 2006).

The above studies indicate that PTMs are ubiquitous in cyanobacteria and play key roles in maintaining the normal physiological function. Although informative, these studies are not able to provide a global view of the PTM (s) status in cyanobacteria. In the following sections, we will summarize and discuss the recent progresses in several PTM proteomic studies in cyanobacteria.

Phosphoproteome

Reversible protein phosphorylation, one of the most widespread PTMs, is a pivotal mechanism by which environmental signals are transmitted to modify protein expression or activity in both prokaryotic and eukaryotic cells (Jers *et al.*, 2008; Rogers and Foster, 2009; Soufi *et al.*, 2008). The two-component signal transduction systems, consisting of histidine kinases and the corresponding response regulators, was previously considered as the predominant phosphorylation based signal transduction mechanism in cyanobacteria (Los *et al.*, 2010). However, recently, comprehensive functional analyses have revealed that Ser/Thr/Tyr kinases and phosphatases play vital roles in regulating many critical processes in cyanobacteria, such as stress adaptation, cell differentiation, cell motility, carbon and nitrogen metabolism and photosynthesis (Zhang *et al.*, 2005). The number of genes encode putative Ser/Thr/Tyr kinases range from none to 52 in the fully sequenced cyanobacteria strains (Zhang *et al.*, 2005), suggesting that both the two phospho-based signalling systems are important in cyanobacteria.

The existence of protein phosphorylation events in cyanobacteria has been confirmed using [^{32}P] phosphate labelling strategies since 1980s (reviewed by Mann, 1994; Spät *et al.*, 2015). In cyanobacteria, protein phosphorylation was shown to be involved in environmental acclimation (Hagemann *et al.*, 1993; Sanders *et al.*, 1989; Spät *et al.*, 2015). However, only a very few phosphoproteins have been identified and characterized. The nitrogen regulatory protein PII (GlnB), which plays a critical role in sensing changes in the internal N/C ratio, is the most intensively studied and best elucidated phosphoprotein in cyanobacteria (reviewed by Forchhammer, 2004, 2008). Another example is the phosphorylation of phycobiliproteins and phycobilisome linker proteins (Harrison *et al.*, 1991; Piven *et al.*, 2005). KaiC, a central clock protein in cyanobacteria, was also reported to undergo circadian oscillations through autocatalytic phosphorylation/de-phosphorylation cycle (reviewed by Williams, 2007; Axmann *et al.*, 2014).

Our group performed the first genome-wide and site-specific phosphoproteome study in the unicellular, marine cyanobacterium *Synechococcus* sp. strain PCC 7002 (hereafter *Synechococcus* 7002) (Yang *et al.*, 2013). *Synechococcus* 7002 is a model organism for studying photosynthetic carbon fixation and the development of biofuels (Ludwig and Bryant, 2011, 2012), it can tolerate extremely high-light intensities and high salinity (Nomura *et al.*, 2006). By using biochemical enrichment of phosphopeptides in combination with MS, we identified 280 unique phosphopeptides from 245 proteins with the determination of 410 Ser/Thr/Tyr phosphorylation sites in *Synechococcus* 7002 cells grown to early/mid-exponential phase under standard conditions (38°C, 1% (v/v) CO_2 in air, 250 µmol photons/m^2/s). This set of cyanobacterial phosphoproteins constitute a significant advance in understanding the physiological functions underlying Ser/Thr/Tyr phosphorylation and facilitate the elucidation of the global signalling networks in *Synechococcus* 7002 and cyanobacteria in general.

After that, phosphoproteomic studies in *Synechocystis* sp. strain PCC 6803 (hereafter *Synechocystis* 6803) have been successively reported (Chen *et al.*, 2015; Lee *et al.*, 2015; Mikkat *et al.*, 2014 Spät *et al.*, 2015). *Synechocystis* 6803 can grow under conditions ranging from photoautotrophic to total heterotrophic, making it an ideal model cyanobacterium for studying photosynthesis, adaptation to environmental changes, carbon metabolism, and other metabolic processes (Varman *et al.*, 2013; Wegener *et al.*, 2010; Zhang *et al.*, 2013).

Mikkat *et al.* (2014) studied the protein phosphorylation pattern in *Synechocystis* 6803 using a 2D gel electrophoresis (2DE)-based approach and a phosphoprotein-specific dye, Pro-Q Diamond. Thirty-two phosphoproteins were identified from a mixture of *Synechocystis* cultivated under high- and low-salt concentrations using MALDI-TOF MS, and for eight of them the phosphorylated amino acid residues were mapped by subsequent MS investigations of enriched phosphopeptides. Although gel-free workflows allow to identify a larger number of phosphoproteins and phosphorylation sites (Spät *et al.*, 2015; Yang *et al.*, 2013), gel-based approach offers the advantage of visualizing phosphoproteins. Furthermore, in the 2-DE gels, intact phosphoproteins can be separated from non-phosphorylated or differently modified forms of the same protein, and thus the detected phosphoprotein can be structurally analysed.

Spät *et al.* (2015) also reported the global phosphoproteome analysis of *Synechocystis* 6803: they employed a gel-free phosphoproteome strategy based on TiO_2 enrichment of phosphopeptides, followed by MS analysis. A total of 301 phosphorylation events were identified, of which

262 could be localized to specific Ser, Thr, or Tyr residue. Phosphoproteome dynamics in response to nitrogen starvation were also investigated in a quantitative manner using stable isotope dimethylation labelling. The dynamics of 2111 proteins and 148 phosphorylation events from two independent experiments were quantified. This is the first quantitative PTM proteomic analysis in cyanobacteria and opens the way for more quantitative proteomics studies referring to more PTM classes.

Lee *et al.* (2015) present the first directed shotgun membrane phosphoproteome in *Synechocystis* based on gel-based phosphoprotein staining and in-gel digestion. Total membrane proteins of *Synechocystis* were separated by SDS-PAGE and phosphoprotein-stained gel bands were selectively subjected to in-gel trypsin digestion. The phosphopeptides were identified by using TiO_2-based phosphopeptides enrichment coupled with high resolution liquid chromatography-Fourier transform tandem mass spectrometry (LC-FT-MS) analysis. The phosphorylation sites of the resulting peptides were determined by assigning the neutral loss of [$M\text{-}H_3PO_4$] to Ser, Thr, and Tyr residues. Altogether, they identified 33 phosphoproteins, including 11 integral membrane proteins.

Our group reported 194 *in vivo* phosphorylation sites from 149 proteins in *Synechocystis*, which were identified using a combination of peptides pre-fractionation, TiO_2 enrichment and liquid chromatography-tandem mass spectrometry (LC-MS/MS) analysis (Chen *et al.*, 2015). These phosphorylated proteins are implicated in diverse biological processes, such as photosynthesis. Further functional studies revealed that phosphorylation at the serine or/and threonine residues of CpcB is important post-translational modifications, which is directly or indirectly involved in efficient energy transfer and state transitions. This study provides novel insights into the molecular mechanisms of protein phosphorylation in the regulation of photosynthesis in cyanobacteria and may facilitate the elucidation of the entire regulatory network by linking kinases to their physiological substrates.

All these phosphoproteomic data revealed that reversible protein phosphorylation in cyanobacteria plays an essential role in fundamental biological processes such as photosynthesis and carbon/nitrogen metabolism. These datasets will be useful for elucidating the comprehensive phosphorylation signalling networks in cyanobacteria.

Acetylome

Protein acetylation is the addition of an acetyl group to either the ε-amines of Lys side chains (lysine acetylation) or the α-amines of protein N-termini (N-terminal acetylation). In cyanobacteria, only a few protein N-terminal acetylation events have been reported, for example, N-terminal acetylation of several proteins such as the Rieske iron–sulfur protein (PetC), Cytochrome b_{559} subunit beta (psbF), and Photosystem II reaction centre protein J (PsbJ) (Baniulis *et al.*, 2009; Guskov *et al.*, 2009). N-terminal acetylation of some cyanobacterial peptides (Philmus *et al.*, 2008) or toxins were also reported (Fewer *et al.*, 2013; Shalev-Malul *et al.*, 2008).

Lysine acetylation is a conserved and prevalent PTM in both eukaryotes and prokaryotes, which regulate diverse cellular processes (Jones and O'Connor, 2011; Kim and Yang, 2011). The first system-wide study of bacterial acetylome was conducted in *Escherichia coli*, by employing lysine-acetylated peptide immunoprecipitation and LC-MS/MS (Yu *et al.*, 2008; Zhang *et al.*, 2009). Then this approach has become a standard procedure in bacterial acetylome analyses (Jones and O'Connor, 2011; Smith and Workman, 2009; Yu *et al.*, 2008; Zhang *et al.*, 2009). Our group provide the first extensive data on lysine acetylation in the cyanobacterium *Synechocystis* 6803 based on this method, which is also the first study of protein lysine acetylation reported in cyanobacteria (Mo *et al.*, 2015). *Synechocystis* 6803 cells exposed to three different stresses (nitrogen-deficient, heterotrophic condition, and high light condition) were harvested and pooled before protein extraction. A total of 776 unique lysine acetylation sites on 513 proteins were identified. These acetylated proteins are involved in diverse biological processes and render particular enrichment to photosynthesis and metabolic processes, which are closely linked to cellular energy status. Lysine acetylation is remarkably observed at the active sites of many components of the photosynthetic apparatus, including subunits of photosystem I (PsaA, PsaB, PsaD, and PsaF), photosystem II (PsbA, PsbB, PsabC, PsbO, PsbP, PsbV, and PsbU), phycocyanin (CpcA, CpcB, CpcC, and CpcG) and allophycocyanin (ApcA, ApcB, ApcD,

ApcF, and ApcE), suggesting that lysine acetylation is another critical mechanism of photosynthetic functional regulation (Mann, 1994; Piven *et al.*, 2005). This study would be an important advance in understanding the physiological functions of lysine acetylation in *Synechocystis* and in cyanobacteria in general.

Glutathionylome

Cyanobacteria are continuously challenged with reactive oxygen species (ROS) generated by photosynthesis and respiration. The ROS can be scavenged by cellular antioxidant defence systems, which maintain the redox homeostasis of cellular thiols (Foyer and Noctor, 2011). Glutathione, a highly abundant tripeptide (γ-Glu-Cys-Gly), occurring mostly in its reduced form (GSH), is one of the major cellular antioxidant redox buffers in most organisms (Zhang and Forman, 2012). Moreover, glutathione can form a mixed disulfide bridge between its cysteine thiol and an accessible free thiol on a protein, which termed as protein S-glutathionylation (Dalle-Donne *et al.*, 2009; Zaffagnini *et al.*, 2012). Glutathionylation is a reversible PTM that seems to occur mostly under oxidative or nitrosative stress conditions. It is a crucial mechanism of redox signalling by protecting specific cysteine residues or by modulating protein activity (Dalle-Donne *et al.*, 2009; Grek *et al.*, 2013).

Glutathionylation has been poorly studied in cyanobacteria. Recently glutathionylation was reported to regulate the activity of two proteins from *Synechocystis* 6803 (Marteyn *et al.*, 2013; Sakr *et al.*, 2013). *Synechocystis* 6803 possesses only one monothiol (Picciocchi *et al.*, 2007) and two dithiol glutaredoxins (Grxs), which have been shown to specifically interact with various protein partners that confer the resistance to toxic metals and various other stresses (Lopez-Maury *et al.*, 2009; Marteyn *et al.*, 2009; 2013; Sanchez-Riego *et al.*, 2013). In addition, several targets of one dithiol Grx (Ssr2061) from *Synechocystis* have been identified by proteomic methods (Li *et al.*, 2007).

Large-scale proteomic approaches have been developed to identify proteins undergoing glutathionylation either *in vivo* or *in vitro* as reviewed by Gao *et al.* (2009). These methods, varying from analysis of purified proteins *in vitro* to large-scale proteomic analysis *in vivo*, allowed identification of nearly 200 targets of glutathionylation in mammals (Gao *et al.*, 2009). Chardonnet *et al.* (2015) presented the first systematic study of *in vitro* glutathionylation in *Synechocystis* 6803 by using biotinylated-oxidized glutathione (BioGSSG) and streptavidin-affinity purification as previously described (Brennan *et al.*, 2006; Sullivan *et al.*, 2000), followed by nano LC-MS/MS analysis. Potential sites of glutathionylation were also determined for 125 proteins following tryptic cleavage, streptavidin-affinity purification, and MS analysis. Taken together the two approaches allowed the identification of 383 glutathionylatable proteins that participate in a wide range of cellular processes and metabolic pathways such as carbon and nitrogen metabolisms, cell division, stress responses, and H_2 production (Chardonnet *et al.*, 2015). In addition, the glutathionylation of two putative targets, namely, peroxiredoxin (Sll1621) involved in oxidative stress tolerance and 3-phosphoglycerate dehydrogenase (Sll1908) acting on amino acid metabolism, was confirmed by biochemical studies on the purified recombinant proteins. These results suggest that glutathionylation is a major mechanism used by cyanobacteria to protect themselves against the frequently encountered (photo) oxidative stresses triggered by light and/or metal availabilities. Their studies will stimulate the development of the *in vivo* study of glutathionylation and also pave the way for further works aimed at determining the molecular effects of glutathionylation on activity and/or structure of a high number of potential targets.

Global discovery of PTMs

Many cellular proteins undergo appreciable amounts of PTMs in response to certain stimuli, and this dynamic process occurs in various cell compartments to dictate the fate and activity of the modified proteins (Cain *et al.*, 2014). However, system-wide identification of multiple PTMs remains a highly challenging task, especially in situations where some reversible PTMs are induced by a particular stimulus and are present for only a short period (Olsen and Mann, 2013).

Proteogenomics, referring to the correlation of MS-derived proteomic data to refine genome annotation (Pandey and Pevzner, 2014), can greatly improve genome annotation quality. Moreover, the same experimental proteomic datasets are also useful for identifying PTMs on a proteome-wide level (Cao *et al.*, 2010; Gupta *et al.*, 2007). Our group developed a proteogenomic approach to carry out genome annotation and whole-proteome analysis of PTMs in prokaryotes by employing high resolution and high accuracy MS data, and we used this approach to study the global PTMs in the model cyanobacterium *Synechococcus* 7002 (Yang *et al.*, 2014). We pooled *Synechococcus* cells that had been grown to different phases (exponential phase, stationary phase) under different stresses (iron deficiency, phosphate deficiency, nitrogen deficiency, and vitamin A5 deficiency) before protein extraction. The protein mixture was then subjected to Glu-C and Trypsin digestion, and peptide mixture was analysed by nano-LC-MS/MS with an LTQ-Orbitrap Elite mass spectrometer. For the first time, we analysed the mass spectra using the MODa (MODification via alignment) algorithm (Na *et al.*, 2012), which not only allows for unrestrictive PTM searches with no limitation on the number of modifications per peptide, but also incorporates several unique features to improve the sensitivity and accuracy of peptide identification and eliminate the increases in false positives and false negatives. Using this approach, we identified 23 different types of PTMs on 6704 unique peptides from 2230 *Synechococcus* 7002 proteins with high confidence (FDR < 1%), and confirmed the localization accuracies of these modification sites using probability-based PTM scores. It is remarkably that our results provide a level of coverage similar to that of individual PTM analyses.

The PTMs dataset revealed that numerous proteins with diverse PTM events were involved in the photosynthetic system (Fig. 7.1). Mapping PTM events to the major components of the photosynthetic apparatus will facilitate the integration of proteogenomic data with biological function and may thereby provide insight into the potential functional relevance of the identified PTMs. Particularly, certain sets of modified proteins were statistically overrepresented ($P < 0.05$) in specific functional categories. For example, the proteins identified with methylation, persulfide modification, farnesylation, and hydroxymethylation were overrepresented in photosynthesis and respiration processes, indicating that these PTM events are extensively involved in photosynthesis in *Synechococcus* 7002 (Fig. 7.1 and Table 7.1). The major impact of our methods is highlighted in the unique capability to provide a holistic view of PTM events in cyanobacteria, as well as any other sequenced prokaryotes.

Challenge and perspectives in PTM proteomics of cyanobacteria

In recent years, MS-based proteomic approach has already been applied to the large-scale discovery of PTMs in cyanobacteria and shows its powerful potential in revealing their physiological and metabolic characteristics, as well as the signal transduction networks. However, PTM proteomics studies in cyanobacteria have just taken off and challenges still exist.

Although there are now robust MS-based workflows for PTM proteomic studies, efficient enrichment strategies do not exist for most PTMs, which leads to a major obstacle in the PTM proteomic studies of cyanobacteria. So, the development of efficient and specific enriching approaches for more types of PTMs remains an urgent work, especially for those PTMs have already been identified by global PTMs screening (Yang *et al.*, 2014). In the future, it is imaginable that rapidly improved MS technology could allow the identification of PTM-bearing peptides without prior enrichment. The novel strategy for whole-proteome analysis of PTMs in prokaryote developed by our group already opens the way for the global profiling of PTMs in any other sequenced cyanobacterium or prokaryote without specific enrichment (Yang *et al.*, 2014). This would make it possible to comprehensive analysis of different PTMs simultaneously and will be a hot spot in future PTM study. Cellular processes are dynamic events, and it is highly desirable to investigate PTM proteomic dynamics in response to different environmental stresses in a quantitative manner. Until now, there is only one quantitative PTM proteomic study in cyanobacteria (Spät *et al.*, 2015). At present, the development of methods for accurate protein quantitation is one of the most challenging areas of proteomics. Quantify the PTM-bearing peptides is more challenging because it is based on single peptides, often of very low abundance.

Quantitative proteomic approaches are generally subdivided into three categories: *in vitro* chemical labelling approaches such as isobaric tags for relative and absolute quantification (iTRAQ)

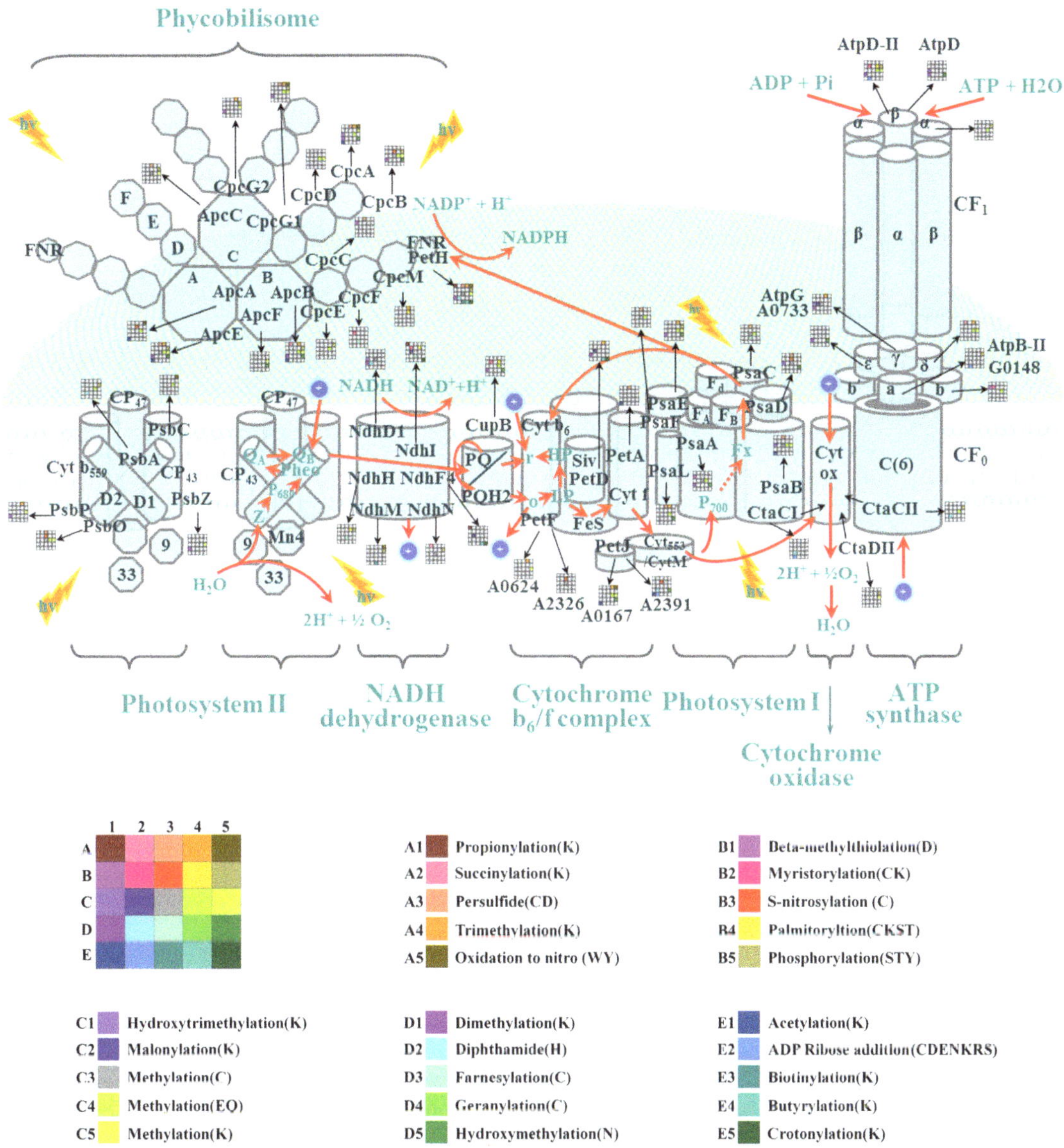

Figure 7.1 Working scheme of the holistic PTM events in photosynthesis pathways in cyanobacteria. Modified proteins are shown using a black arrow. Different PTMs are shown as squares with different colours (Yang *et al.*, 2014). For detailed description, see Table 7.1.

and isobaric tandem mass tag (TMT); metabolic labelling strategies, including stable isotope labelling with amino acids in cell culture (SILAC), ^{15}N labelling; and label-free approaches. The iTRAQ/TMT labelling techniques have been used in cyanobacteria and proved to be efficient and reliable methods for quantitative proteomics (D'Agostino *et al.*, 2014; Ren *et al.*, 2014; Wase *et al.*, 2014; Xiong *et al.*, 2015). However, this reporter-ion-based quantification is better suited for quantitative analysis of the protein expression profiles rather than for the quantification of PTM levels (Elliott *et al.*, 2009; Lemeer *et al.*, 2012; Mueller *et al.*, 2008; Treumann and Thiede, 2010).

The SILAC technique, which based on the incorporation of isotopically labelled amino acids into proteins formed by the growing organism, has become the gold standard because it eliminates many workflow-induced sources for error (Blagoev *et al.*, 2003; Ong *et al.*, 2002). However,

Table 7.1 Post-translational modifications in the cyanobacterial proteome

PTM	No.[a]	Proteins/peptides exhibiting this PTM[b]	Species	Reference
Phosphorylation:* Ser, Thr, Tyr	859	Ubiquitous	*Synechocystis* 6803 *Synechococcus* 7002	Lee *et al.* (2015), Mikkat *et al.* (2014), Spät *et al.* (2015), Yang *et al.* (2013), Yang *et al.* (2014), Zorina *et al.* (2011)
Acetylation:* Lys	259	Ubiquitous	*Synechocystis* 6803	Mo *et al.* (2015), Yang *et al.* (2014)
ADP ribosylation	362	Glutamine synthetase	*Synechocystis* 6803	Silman *et al.* (1995), Yang *et al.* (2014)
Biotinylation	119	PsaD, PsaE, biotin carboxyl carrier protein (BCCP)	*Synechococcus* 7942 *Synechocystis* 6803 *Anabaena* 7120	Gornicki *et al.* (1993), von Leoprechting *et al.* (1998),Xu *et al.* (1994), Yang *et al.* (2014)
Methylation	896	A conserved Asn-72 in the β-subunits of all phycobiliproteins	Ubiquitous	Swanson and Glazer (1990), Yang *et al.* (2014)
Trimethylation	379	C-terminal Lys of pilin	*Synechocystis* 6803	Kim *et al.* (2011), Yang *et al.* (2014)
Persulfide formation	219		Ubiquitous	Behshad and Bollinger (2009), Behshad *et al.* (2004), Campanini *et al.* (2006), Yang *et al.* (2014)
Prenylation	57	Cyanobactins		McIntosh *et al.* (2011), Yang *et al.* (2014)
Palmitoylation	201	Nitrogenase	*Gloeothece* ATCC 27152	Gallon *et al.* (2000),; Yang *et al.* (2014)
Dimethylation	576		*Synechococcus* 7002	Yang *et al.* (2014)
S-nitrosylation	115		*Synechococcus* 7002	Yang *et al.* (2014)
Beta-methylthiolation	146		*Synechococcus* 7002	Yang *et al.* (2014)
Butyrylation	168		*Synechococcus* 7002	Yang *et al.* (2014)
Crotonylation	172		*Synechococcus* 7002	Yang *et al.* (2014)
Diphthamide	136		*Synechococcus* 7002	Yang *et al.* (2014)
Geranylation	41		*Synechococcus* 7002	Yang *et al.* (2014)
Hydroxymethylation	206		*Synechococcus* 7002	Yang *et al.* (2014)
Hydroxytrimethylation	207		*Synechococcus* 7002	Yang *et al.* (2014)
Malonylation	173		*Synechococcus* 7002	Yang *et al.* (2014)
Myristoylation	131		*Synechococcus* 7002	Yang *et al.* (2014)
Oxidation to nitro	209		*Synechococcus* 7002	Yang *et al.* (2014)
Propionylation	157		*Synechococcus* 7002	Yang *et al.* (2014)
Succinylation	292		*Synechococcus* 7002	Yang *et al.* (2014)
Glutathionylation*		Ubiquitous	*Synechocystis* 6803	Chardonnet *et al.* (2015)
His hydroxyl modification		D2-His336 in Photosystem II	*Thermosynechoccocus vulcanus Thermosynechoccocus elongatus*	Sugiura *et al.* (2013)
C-terminal processing		D1 protein	*Synechocystis* 6803	Inagaki *et al.* (2001), Kanervo *et al.* (1997)
O-linked glycan		Pilins	*Synechocystis* 6803	Kim *et al.* (2009)

*Proteomic studies have been performed referring to these PTMs.
[a]The numbers of proteins with PTM reported in Yang *et al.* (2014).
[b]Protein/peptides exhibiting certain kinds of PTMs, which are experimentally verified.

it works well for mammalian cell lines (Ge *et al.*, 2011; Xiong *et al.*, 2012; Zheng *et al.*, 2015), but it is not so effective in autotrophic organisms such as cyanobacteria due to their ability to synthesize all amino acids (Gruhler *et al.*, 2005b; Schutz *et al.*, 2011). The second primary method in stable isotope metabolic labelling relies on the introduction of a heavy isotope such as ^{15}N, often in the form of $Na^{15}NO_3$ or $K^{15}NO_3$ as a nitrogen source, is a powerful tool for the quantitative analysis in plant and bacteria (Engelsberger *et al.*, 2006; Huttlin *et al.*, 2007). Quantitative proteomics based on 'metabolic labelling' has already been successfully used in some cyanobacterial strains (Aryal *et al.*, 2011; 2012; Pandhal *et al.*, 2009). So it has great potential to be applied in the future quantitative PTM studies in cyanobacteria.

Label-free quantitative proteomic approach is the most straightforward method operationally. It is applicable to any source material and allows direct comparison of MS signals between any number of samples (Cox *et al.*, 2014). Label-free approaches are generally considered less accurate than the isotope labelling methods (Pham *et al.*, 2012). However, remarkably improved MS performance and algorithms make label-free quantification extremely attractive in the PTM field (Amoresano *et al.*, 2009).

Multiple PTMs within one protein can coordinately lead to a functional outcome, which is called PTM cross-talk (Beltrao *et al.*, 2013). The interplay between two PTMs was studied in both prokaryote and eukaryote (Swaney *et al.*, 2013; van Noort *et al.*, 2012), which suggested that cross-talk between different PTMs was extensive and remained to be discovered. The large-scale PTM proteomics data revealed that PTM cross-talk may also exist in cyanobacteria since many proteins were discovered to have more than one PTM (Mo *et al.*, 2015; Yang *et al.*, 2013; 2014). Histone PTM cross-talk networks have been systematically discovered in *Saccharomyces cerevisiae* (Guan *et al.*, 2013). However, the method was hard to generalize to the whole proteome. The identification of PTM cross-talk proteome wide remains a great challenge.

Large-scale PTM (s) identification or quantification often simply serves as an initial screen, followed by detailed functional study of key PTM sites, which is also one of the major challenges in the PTM study of cyanobacteria. Mutational study is the traditional method to explore the function of a PTM site. However, there may be more than one PTM on a specific amino acid, so mutational study alone may not be able to fully elucidate the function of a PTM site. Traditional functional studies in combination with molecular simulation methods will facilitate the understanding of the function of key PTM sites in the future.

Cyanobacteria are considered to be a rich source of novel metabolites which is of great importance from a biotechnological and industrial point of view. The majority of cyanobacterial peptides can be heavily modified (Welker and von Dohren, 2006). The isolation/purification and large-scale discovery of the bioactive peptides and their PTMs using proteomics strategy will be great help to the better understanding and future application of these valuable bioactive peptides.

Conclusions

This chapter summarizes the current status of PTM studies as well as latest progress in the PTM proteomics in cyanobacteria. MS-based systems-wide proteomics are proved to have great potential in characterization of the global PTM(s) network. Many fundamental biological processes in cyanobacteria such as photosynthesis are extensively regulated by different PTMs. However, at present, there are still many challenges such as lack of efficient enrichment strategies for most PTMs, restrictions in high accurate quantitative methods and the following functional studies after initial PTM screening. Quantitative proteomic approaches are highly desired in dynamic PTM(s) profiling of cyanobacteria under different environmental stresses, which will lead to a better understanding of the cellular processes and metabolic pathways that resulted in the amazingly adaptations to changing environment. We can foresee that areas such as PTM cross-talk, quantitative PTM profiling, and global discovery of PTMs without specific enrichment will become hotspots in cyanobacteria PTM studies.

Acknowledgements

This work was supported by the Strategic Priority Research Program of the Chinese Academy of

Sciences (Grant No. XDB14030202), the National Natural Science Foundation of China (Grant No. 21207153 To Q. X.) and the Knowledge Innovation Program of the Chinese Academy of Sciences (Grant No. Y35E04-1-501 to Q. X.).

References

Amoresano, A., Carpentieri, A., Giangrande, C., Palmese, A., Chiappetta, G., Marino, G., and Pucci, P. (2009). Technical advances in proteomics mass spectrometry: identification of post-translational modifications. Clin. Chem. Lab. Med. *47*, 647–665. http://dx.doi.org/10.1515/CCLM.2009.154.

Aryal, U.K., Stöckel, J., Krovvidi, R.K., Gritsenko, M.A., Monroe, M.E., Moore, R.J., Koppenaal, D.W., Smith, R.D., Pakrasi, H.B., and Jacobs, J.M. (2011). Dynamic proteomic profiling of a unicellular cyanobacterium *Cyanothece* ATCC51142 across light–dark diurnal cycles. BMC Syst. Biol. *5*, 194. http://dx.doi.org/10.1186/1752-0509-5-194.

Aryal, U.K., Stöckel, J., Welsh, E.A., Gritsenko, M.A., Nicora, C.D., Koppenaal, D.W., Smith, R.D., Pakrasi, H.B., and Jacobs, J.M. (2012). Dynamic proteome analysis of *Cyanothece* sp. ATCC 51142 under constant light. J. Proteome. Res. *11*, 609–619. http://dx.doi.org/10.1021/pr200959x.

Axmann, I.M., Hertel, S., Wiegard, A., Dörrich, A.K., and Wilde, A. (2014). Diversity of KaiC-based timing systems in marine Cyanobacteria. Mar. Geonomics *14*, 3–16. http://dx.doi.org/10.1016/j.margen.2013.12.006.

Baniulis, D., Yamashita, E., Whitelegge, J.P., Zatsman, A.I., Hendrich, M.P., Hasan, S.S., Ryan, C.M., and Cramer, W.A. (2009). Structure-function, stability, and chemical modification of the cyanobacterial cytochrome b_6f complex from *Nostoc* sp. PCC 7120. J. Biol. Chem. *284*, 9861–9869. http://dx.doi.org/10.1074/jbc.M809196200.

Beausoleil, S.A., Jedrychowski, M., Schwartz, D., Elias, J.E., Villén, J., Li, J., Cohn, M.A., Cantley, L.C., and Gygi, S.P. (2004). Large-scale characterization of HeLa cell nuclear phosphoproteins. Proc. Natl. Acad. Sci. U.S.A. *101*, 12130–12135. http://dx.doi.org/10.1073/pnas.0404720101.

Beck, C., Knoop, H., Axmann, I.M., and Steuer, R. (2012). The diversity of cyanobacterial metabolism: genome analysis of multiple phototrophic microorganisms. BMC Genomics *13*, 56. http://dx.doi.org/10.1186/1471-2164-13-56.

Behshad, E., and Bollinger, J.M. (2009). Kinetic analysis of cysteine desulfurase CD0387 from *Synechocystis* sp. PCC 6803: formation of the persulfide intermediate. Biochemistry *48*, 12014–12023. http://dx.doi.org/10.1021/bi802161u.

Behshad, E., Parkin, S.E., and Bollinger, J.M. (2004). Mechanism of cysteine desulfurase Slr0387 from *Synechocystis* sp. PCC 6803: kinetic analysis of cleavage of the persulfide intermediate by chemical reductants. Biochemistry *43*, 12220–12226. http://dx.doi.org/10.1021/bi049143e.

Beltrao, P., Bork, P., Krogan, N.J., and van Noort, V. (2013). Evolution and functional cross-talk of protein post-translational modifications. Mol. Syst. Biol. *9*, 714. http://dx.doi.org/10.1002/msb.201304521.

Blagoev, B., Kratchmarova, I., Ong, S.E., Nielsen, M., Foster, L.J., and Mann, M. (2003). A proteomics strategy to elucidate functional protein–protein interactions applied to EGF signaling. Nat. Biotechnol. *21*, 315–318. http://dx.doi.org/10.1038/nbt790.

Brennan, J.P., Miller, J.I., Fuller, W., Wait, R., Begum, S., Dunn, M.J., and Eaton, P. (2006). The utility of *N,N*-biotinyl glutathione disulfide in the study of protein *S*-glutathiolation. Mol. Cell. Proteomics *5*, 215–225. http://dx.doi.org/10.1074/mcp.M500212-MCP200.

Burja, A.M., Banaigs, B., Abou-Mansour, E., Grant Burgess, J., and Wright, P.C. (2001). Marine cyanobacteria – a prolific source of natural products. Tetrahedron *57*, 9347–9377. http://dx.doi.org/10.1016/S0040-4020(01)00931-0.

Cain, J.A., Solis, N., and Cordwell, S.J. (2014). Beyond gene expression: the impact of protein post-translational modifications in bacteria. J. Proteomics *97*, 265–286. http://dx.doi.org/10.1016/j.jprot.2013.08.012.

Campanini, B., Schiaretti, F., Abbruzzetti, S., Kessler, D., and Mozzarelli, A. (2006). Sulfur mobilization in cyanobacteria: the catalytic mechanism of L-cystine C-S lyase (C-DES) from *Synechocystis*. J. Biol. Chem. *281*, 38769–38780. http://dx.doi.org/10.1074/jbc.M607098200.

Cao, X.J., Dai, J., Xu, H., Nie, S., Chang, X., Hu, B.Y., Sheng, Q.H., Wang, L.S., Ning, Z.B., Li, Y.X., *et al.* (2010). High-coverage proteome analysis reveals the first insight of protein modification systems in the pathogenic spirochete *Leptospira interrogans*. Cell Res. *20*, 197–210. http://dx.doi.org/10.1038/cr.2009.127.

Caspi, J., Amitai, G., Belenkiy, O., and Pietrokovski, S. (2003). Distribution of split DnaE inteins in cyanobacteria. Mol. Microbiol. *50*, 1569–1577. http://dx.doi.org/10.1046/j.1365-2958.2003.03825.x.

Chardonnet, S., Sakr, S., Cassier-Chauvat, C., Le Maréchal, P., Chauvat, F., Lemaire, S.D., and Decottignies, P. (2015). First proteomic study of *S*-glutathionylation in cyanobacteria. J. Proteome Res. *14*, 59–71. http://dx.doi.org/10.1021/pr500625a.

Chellamuthu, V.R., Alva, V., and Forchhammer, K. (2013). From cyanobacteria to plants: conservation of PII functions during plastid evolution. Planta *237*, 451–462. http://dx.doi.org/10.1007/s00425-012-1801-0.

Chen, Z., Zhan, J., Chen, Y., Yang, M., He, C., Ge, F., and Wang, Q. (2015). Effects of phosphorylation of β subunits of phycocyanins on state transition in the model cyanobacterium *Synechocystis* sp. PCC 6803. Plant Cell. Physiol. *56*, 1997–2013. http://dx.doi.org/10.1093/pcp/pcv118.

Cox, J., Hein, M.Y., Luber, C.A., Paron, I., Nagaraj, N., and Mann, M. (2014). Accurate proteome-wide label-free quantification by delayed normalization and maximal peptide ratio extraction, termed MaxLFQ. Mol. Cell. Proteomics *13*, 2513–2526. http://dx.doi.org/10.1074/mcp.M113.031591.

Crispim, C.A., and Gaylarde, C.C. (2005). Cyanobacteria and biodeterioration of cultural heritage: a review. Microb. Ecol. *49*, 1–9. http://dx.doi.org/10.1007/s00248-003-1052-5.

D'Agostino, P.M., Song, X., Neilan, B.A., and Moffitt, M.C. (2014). Comparative proteomics reveals that a saxitoxin-producing and a nontoxic strain of *Anabaena circinalis* are two different ecotypes. J. Proteome. Res. *13*, 1474–1484. http://dx.doi.org/10.1021/pr401007k.

Dalle-Donne, I., Rossi, R., Colombo, G., Giustarini, D., and Milzani, A. (2009). Protein S-glutathionylation: a regulatory device from bacteria to humans. Trends Biochem. Sci*34*, 85–96. http://dx.doi.org/10.1016/j.tibs.2008.11.002.

Elliott, M.H., Smith, D.S., Parker, C.E., and Borchers, C. (2009). Current trends in quantitative proteomics. J. Mass. Spectrom. *44*, 1637–1660. http://dx.doi.org/10.1002/jms.1692.

Engelsberger, W.R., Erban, A., Kopka, J., and Schulze, W.X. (2006). Metabolic labeling of plant cell cultures with $K^{15}NO_3$ as a tool for quantitative analysis of proteins and metabolites. Plant Methods *2*, 14. http://dx.doi.org/10.1186/1746-4811-2-14.

Fay, P. (1992). Oxygen relations of nitrogen fixation in cyanobacteria. Microbiol. Rev. *56*, 340–373.

Fewer, D.P., Wahlsten, M., Österholm, J., Jokela, J., Rouhiainen, L., Kaasalainen, U., Rikkinen, J., and Sivonen, K. (2013). The genetic basis for *O*-acetylation of the microcystin toxin in cyanobacteria. Chem. Biol. *20*, 861–869. http://dx.doi.org/10.1016/j.chembiol.2013.04.020.

Ficarro, S.B., McCleland, M.L., Stukenberg, P.T., Burke, D.J., Ross, M.M., Shabanowitz, J., Hunt, D.F., and White, F.M. (2002). Phosphoproteome analysis by mass spectrometry and its application to *Saccharomyces cerevisiae*. Nat. Biotechnol. *20*, 301–305. http://dx.doi.org/10.1038/nbt0302-301.

Forchhammer, K. (2004). Global carbon/nitrogen control by PII signal transduction in cyanobacteria: from signals to targets. FEMS Microbiol. Rev. *28*, 319–333. http://dx.doi.org/10.1016/j.femsre.2003.11.001.

Forchhammer, K. (2008). P(II) signal transducers: novel functional and structural insights. Trends Microbiol. *16*, 65–72. http://dx.doi.org/10.1016/j.tim.2007.11.004.

Foyer, C.H., and Noctor, G. (2011). Ascorbate and glutathione: the heart of the redox hub. Plant Physiol. *155*, 2–18. http://dx.doi.org/10.1104/pp.110.167569.

Gallon, J.R., Cheng, J., Dougherty, L.J., Gallon, V.A., Hilz, H., Pederson, D.M., Richards, H.M., Rüggeberg, S., and Smith, C.J. (2000). A novel covalent modification of nitrogenase in a cyanobacterium. FEBS Lett. *468*, 231–233. http://dx.doi.org/10.1016/S0014-5793(00)01229-1.

Gao, X.H., Bedhomme, M., Veyel, D., Zaffagnini, M., and Lemaire, S.D. (2009). Methods for analysis of protein glutathionylation and their application to photosynthetic organisms. Mol. Plant *2*, 218–235. http://dx.doi.org/10.1093/mp/ssn072.

Ge, F., Zhang, L., Tao, S.C., Kitazato, K., Zhang, Z.P., Zhang, X.E., and Bi, L.J. (2011). Quantitative proteomic analysis of tumor reversion in multiple myeloma cells. J. Proteome Res. *10*, 845–855. http://dx.doi.org/10.1021/pr100992e.

Gornicki, P., Scappino, L.A., and Haselkorn, R. (1993). Genes for two subunits of acetyl coenzyme A carboxylase of *Anabaena* sp. strain PCC 7120: biotin carboxylase and biotin carboxyl carrier protein. J. Bacteriol. *175*, 5268–5272.

Grek, C.L., Zhang, J., Manevich, Y., Townsend, D.M., and Tew, K.D. (2013). Causes and consequences of cysteine *S*-glutathionylation. J. Biol. Chem. *288*, 26497–26504. http://dx.doi.org/10.1074/jbc.R113.461368.

Gruhler, A., Olsen, J.V., Mohammed, S., Mortensen, P., Faergeman, N.J., Mann, M., and Jensen, O.N. (2005a). Quantitative phosphoproteomics applied to the yeast pheromone signaling pathway. Mol. Cell. Proteomics *4*, 310–327. http://dx.doi.org/10.1074/mcp.M400219-MCP200.

Gruhler, A., Schulze, W.X., Matthiesen, R., Mann, M., and Jensen, O.N. (2005b). Stable isotope labeling of *Arabidopsis thaliana* cells and quantitative proteomics by mass spectrometry. Mol. Cell. Proteomics *4*, 1697–1709. http://dx.doi.org/10.1074/mcp.M500190-MCP200.

Guan, X., Rastogi, N., Parthun, M.R., and Freitas, M.A. (2013). Discovery of histone modification crosstalk networks by stable isotope labeling of amino acids in cell culture mass spectrometry (SILAC MS). Mol. Cell. Proteomics *12*, 2048–2059. http://dx.doi.org/10.1074/mcp.M112.026716.

Gupta, N., Tanner, S., Jaitly, N., Adkins, J.N., Lipton, M., Edwards, R., Romine, M., Osterman, A., Bafna, V., Smith, R.D., *et al.* (2007). Whole proteome analysis of post-translational modifications: applications of mass-spectrometry for proteogenomic annotation. Genome Res. *17*, 1362–1377. http://dx.doi.org/10.1101/gr.6427907.

Guskov, A., Kern, J., Gabdulkhakov, A., Broser, M., Zouni, A., and Saenger, W. (2009). Cyanobacterial photosystem II at 2.9-A resolution and the role of quinones, lipids, channels and chloride. Nat. Struct. Mol. Biol. *16*, 334–342. http://dx.doi.org/10.1038/nsmb.1559.

Hagemann, M., Golldack, D., Biggins, J., and Erdmann, N. (1993). Salt-dependent protein phosphorylation in the cyanobacterium *Synechocystis* PCC 6803. FEMS Microbiol. Lett. *113*, 205–209. http://dx.doi.org/10.1111/j.1574-6968.1993.tb06515.x.

Harrison, M.A., Tsinoremas, N.F., and Allen, J.F. (1991). Cyanobacterial thylakoid membrane proteins are reversibly phosphorylated under plastoquinone-reducing conditions *in vitro*. FEBS Lett. *282*, 295–299. http://dx.doi.org/10.1016/0014-5793(91)80499-S.

Hart, G.W., and Ball, L.E. (2013). Post-translational modifications: a major focus for the future of proteomics. Mol. Cell. Proteomics *12*, 3443. http://dx.doi.org/10.1074/mcp.E113.036491.

Huttlin, E.L., Hegeman, A.D., Harms, A.C., and Sussman, M.R. (2007). Comparison of full versus partial metabolic labeling for quantitative proteomics analysis in *Arabidopsis thaliana*. Mol. Cell. Proteomics *6*, 860–881. http://dx.doi.org/10.1074/mcp.M600347-MCP200.

Inagaki, N., Yamamoto, Y., and Satoh, K. (2001). A sequential two-step proteolytic process in the carboxyl-terminal truncation of precursor D1 protein in *Synechocystis* sp. PCC6803. FEBS Lett. *509*, 197–201. http://dx.doi.org/10.1016/S0014-5793(01)03180-5.

Jers, C., Soufi, B., Grangeasse, C., Deutscher, J., and Mijakovic, I. (2008). Phosphoproteomics in bacteria: towards a systemic understanding of bacterial phosphorylation networks. Expert Rev. Proteomics 5, 619–627. http://dx.doi.org/10.1586/14789450.5.4.619.

Jones, J.D., and O'Connor, C.D. (2011). Protein acetylation in prokaryotes. Proteomics *11*, 3012–3022. http://dx.doi.org/10.1002/pmic.201000812.

Kanervo, E., Tasaka, Y., Murata, N., and Aro, E.M. (1997). Membrane lipid unsaturation modulates processing of the photosystem II reaction-center protein D1 at low temperatures. Plant Physiol. *114*, 841–849. http://dx.doi.org/10.1104/pp.114.3.841.

Kim, G.W., and Yang, X.J. (2011). Comprehensive lysine acetylomes emerging from bacteria to humans. Trends Biochem. Sci36, 211–220. http://dx.doi.org/10.1016/j.tibs.2010.10.001.

Kim, Y.H., Kim, J.Y., Kim, S.Y., Lee, J.H., Lee, J.S., Chung, Y.H., Yoo, J.S., and Park, Y.M. (2009). Alteration in the glycan pattern of pilin in a nonmotile mutant of *Synechocystis* sp. PCC 6803. Proteomics *9*, 1075–1086. http://dx.doi.org/10.1002/pmic.200800372.

Kim, Y.H., Park, K.H., Kim, S.Y., Ji, E.S., Kim, J.Y., Lee, S.K., Yoo, J.S., Kim, H.S., and Park, Y.M. (2011). Identification of trimethylation at C-terminal lysine of pilin in the cyanobacterium *Synechocystis* PCC 6803. Biochem. Biophys. Res. Commun. *404*, 587–592. http://dx.doi.org/10.1016/j.bbrc.2010.11.133.

Lee, D.G., Kwon, J., Eom, C.Y., Kang, Y.M., Roh, S.W., Lee, K.B., and Choi, J.S. (2015). Directed analysis of cyanobacterial membrane phosphoproteome using stained phosphoproteins and titanium-enriched phosphopeptides. J. Microbiol. *53*, 279–287. http://dx.doi.org/10.1007/s12275-015-5021-8.

Lemeer, S., Hahne, H., Pachl, F., and Kuster, B. (2012). Software tools for MS-based quantitative proteomics: a brief overview. Methods Mol. Biol. *893*, 489–499. http://dx.doi.org/10.1007/978-1-61779-885-6_29.

Li, M., Yang, Q., Zhang, L., Li, H., Cui, Y., and Wu, Q. (2007). Identification of novel targets of cyanobacterial glutaredoxin. Arch. Biochem. Biophys. *458*, 220–228. http://dx.doi.org/10.1016/j.abb.2006.12.010.

López-Maury, L., Sánchez-Riego, A.M., Reyes, J.C., and Florencio, F.J. (2009). The glutathione/glutaredoxin system is essential for arsenate reduction in *Synechocystis* sp. strain PCC 6803. J. Bacteriol. *191*, 3534–3543. http://dx.doi.org/10.1128/JB.01798-08.

Los, D.A., Zorina, A., Sinetova, M., Kryazhov, S., Mironov, K., and Zinchenko, V.V. (2010). Stress sensors and signal transducers in cyanobacteria. Sensors *10*, 2386–2415. http://dx.doi.org/10.3390/s100302386.

Lubner, C.E., Applegate, A.M., Knörzer, P., Ganago, A., Bryant, D.A., Happe, T., and Golbeck, J.H. (2011). Solar hydrogen-producing bionanodevice outperforms natural photosynthesis. Proc. Natl. Acad. Sci. U.S.A. *108*, 20988–20991. http://dx.doi.org/10.1073/pnas.1114660108.

Ludwig, M., and Bryant, D.A. (2011). Transcription profiling of the model cyanobacterium *Synechococcus* sp. strain PCC 7002 by Next-Gen (SOLiD™) sequencing of cDNA. Front. Microbiol. *2*, 41. http://dx.doi.org/10.3389/fmicb.2011.00041.

Ludwig, M., and Bryant, D.A. (2012). Acclimation of the global transcriptome of the cyanobacterium *Synechococcus* sp. strain PCC 7002 to nutrient limitations and different nitrogen sources. Front. Microbiol. *3*, 145. http://dx.doi.org/10.3389/fmicb.2012.00145.

Mann, M., and Jensen, O.N. (2003). Proteomic analysis of post-translational modifications. Nat. Biotechnol. *21*, 255–261. http://dx.doi.org/10.1038/nbt0303-255.

Mann, N.H. (1994). Protein phosphorylation in cyanobacteria. Microbiology *140*, 3207–3215.

Marteyn, B., Domain, F., Legrain, P., Chauvat, F., and Cassier-Chauvat, C. (2009). The thioredoxin reductas e-glutaredoxins-ferredoxin crossroad pathway for selenate tolerance in *Synechocystis* PCC6803. Mol. Microbiol. *71*, 520–532. http://dx.doi.org/10.1111/j.1365-2958.2008.06550.x.

Marteyn, B., Sakr, S., Farci, S., Bedhomme, M., Chardonnet, S., Decottignies, P., Lemaire, S.D., Cassier-Chauvat, C., and Chauvat, F. (2013). The *Synechocystis* PCC6803 MerA-like enzyme operates in the reduction of both mercury and uranium under the control of the glutaredoxin 1 enzyme. J. Bacteriol. *195*, 4138–4145. http://dx.doi.org/10.1128/JB.00272-13.

McIntosh, J.A., Donia, M.S., Nair, S.K., and Schmidt, E.W. (2011). Enzymatic basis of ribosomal peptide prenylation in cyanobacteria. J. Am. Chem. Soc. *133*, 13698–13705. http://dx.doi.org/10.1021/ja205458h.

Michel, H., Hunt, D.F., Shabanowitz, J., and Bennett, J. (1988). Tandem mass spectrometry reveals that three photosystem II proteins of spinach chloroplasts contain *N*-acetyl-*O*-phosphothreonine at their NH_2 termini. J. Biol. Chem. *263*, 1123–1130.

Mikkat, S., Fulda, S., and Hagemann, M. (2014). A 2D gel electrophoresis-based snapshot of the phosphoproteome in the cyanobacterium *Synechocystis* sp. strain PCC 6803. Microbiology *160*, 296–306. http://dx.doi.org/10.1099/mic.0.074443-0.

Miller, C.A., Leonard, H.S., Pinsky, I.G., Turner, B.M., Williams, S.R., Harrison, L., Fletcher, A.F., Shen, G., Bryant, D.A., and Schluchter, W.M. (2008). Biogenesis of phycobiliproteins. III. CpcM is the asparagine methyltransferase for phycobiliprotein beta-subunits in cyanobacteria. J. Biol. Chem. *283*, 19293–19300. http://dx.doi.org/10.1074/jbc.M802734200.

Mills, K.V., and Perler, F.B. (2005). The mechanism of intein-mediated protein splicing: variations on a theme. Protein. Pept. Lett. *12*, 751–755.

Mo, R., Yang, M., Chen, Z., Cheng, Z., Yi, X., Li, C., He, C., Xiong, Q., Chen, H., Wang, Q., *et al.* (2015). Acetylome analysis reveals the involvement of lysine acetylation in photosynthesis and carbon metabolism in the model cyanobacterium *Synechocystis* sp. PCC 6803. J. Proteome Res. *14*, 1275–1286. http://dx.doi.org/10.1021/pr501275a.

Mueller, L.N., Brusniak, M.Y., Mani, D.R., and Aebersold, R. (2008). An assessment of software solutions for the analysis of mass spectrometry based quantitative proteomics data. J. Proteome. Res. *7*, 51–61. http://dx.doi.org/10.1021/pr700758r.

Na, S., Bandeira, N., and Paek, E. (2012). Fast multi-blind modification search through tandem mass spectrometry. Mol. Cell. Proteomics *11*, M111.010199. http://dx.doi.org/10.1074/mcp.M111.010199.

Nichols, N.M., and Evans, T.C. (2004). Mutational analysis of protein splicing, cleavage, and self-association reactions mediated by the naturally split Ssp DnaE intein. Biochemistry *43*, 10265–10276. http://dx.doi.org/10.1021/bi0494065.

Nomura, C.T., Sakamoto, T., and Bryant, D.A. (2006). Roles for heme-copper oxidases in extreme high-light and oxidative stress response in the cyanobacterium *Synechococcus* sp. PCC 7002. Arch. Microbiol. *185*, 471–479. http://dx.doi.org/10.1007/s00203-006-0107-7.

Olsen, J.V., and Mann, M. (2013). Status of large-scale analysis of post-translational modifications by mass spectrometry. Mol. Cell. Proteomics *12*, 3444–3452. http://dx.doi.org/10.1074/mcp.O113.034181.

Ong, S.E., Blagoev, B., Kratchmarova, I., Kristensen, D.B., Steen, H., Pandey, A., and Mann, M. (2002). Stable isotope labeling by amino acids in cell culture, SILAC, as a simple and accurate approach to expression proteomics. Mol. Cell. Proteomics *1*, 376–386. http://dx.doi.org/10.1074/mcp.M200025-MCP200.

Pandey, A., and Pevzner, P.A. (2014). Proteogenomics. Proteomics *14*, 2631–2632. http://dx.doi.org/10.1002/pmic.201470173.

Pandhal, J., Ow, S.Y., Wright, P.C., and Biggs, C.A. (2009). Comparative proteomics study of salt tolerance between a nonsequenced extremely halotolerant cyanobacterium and its mildly halotolerant relative using in vivo metabolic labeling and in vitro isobaric labeling. J. Proteome Res. *8*, 818–828. http://dx.doi.org/10.1021/pr800283q.

Parmar, A., Singh, N.K., Pandey, A., Gnansounou, E., and Madamwar, D. (2011). Cyanobacteria and microalgae: a positive prospect for biofuels. Bioresour. Technol. *102*, 10163–10172. http://dx.doi.org/10.1016/j.biortech.2011.08.030.

Partensky, F., Hess, W.R., and Vaulot, D. (1999). *Prochlorococcus*, a marine photosynthetic prokaryote of global significance. Microbiol. Mol. Biol. Rev. *63*, 106–127.

Pham, T.V., Piersma, S.R., Oudgenoeg, G., and Jimenez, C.R. (2012). Label-free mass spectrometry-based proteomics for biomarker discovery and validation. Expert Rev. Mol. Diagn. *12*, 343–359. http://dx.doi.org/10.1586/erm.12.31.

Philmus, B., Christiansen, G., Yoshida, W.Y., and Hemscheidt, T.K. (2008). Post-translational modification in microviridin biosynthesis. Chembiochem *9*, 3066–3073. http://dx.doi.org/10.1002/cbic.200800560.

Picciocchi, A., Saguez, C., Boussac, A., Cassier-Chauvat, C., and Chauvat, F. (2007). CGFS-type monothiol glutaredoxins from the cyanobacterium *Synechocystis* PCC6803 and other evolutionary distant model organisms possess a glutathione-ligated [2Fe–2S] cluster. Biochemistry *46*, 15018–15026. http://dx.doi.org/10.1021/bi7013272.

Piven, I., Ajlani, G., and Sokolenko, A. (2005). Phycobilisome linker proteins are phosphorylated in *Synechocystis* sp. PCC 6803. J. Biol. Chem. *280*, 21667–21672. http://dx.doi.org/10.1074/jbc.M412967200.

Ren, Q., Shi, M., Chen, L., Wang, J., and Zhang, W. (2014). Integrated proteomic and metabolomic characterization of a novel two-component response regulator Slr1909 involved in acid tolerance in *Synechocystis* sp. PCC 6803. J. Proteomics *109*, 76–89. http://dx.doi.org/10.1016/j.jprot.2014.06.021.

Rogers, L.D., and Foster, L.J. (2009). Phosphoproteomics – finally fulfilling the promise? Mol. Biosyst. *5*, 1122–1129. http://dx.doi.org/10.1039/b905580k.

Rosgaard, L., de Porcellinis, A.J., Jacobsen, J.H., Frigaard, N.U., and Sakuragi, Y. (2012). Bioengineering of carbon fixation, biofuels, and biochemicals in cyanobacteria and plants. J. Biotechnol. *162*, 134–147. http://dx.doi.org/10.1016/j.jbiotec.2012.05.006.

Sakr, S., Dutheil, J., Saenkham, P., Bottin, H., Leplat, C., Ortega-Ramos, M., Aude, J.-C., Chapuis, V., Guedeney, G., and Decottignies, P. (2013). The activity of the *Synechocystis* PCC6803 AbrB2 regulator of hydrogen production can be post-translationally controlled through glutathionylation. Int. J. Energ. *38*, 13547–13555. http://dx.doi.org/10.1016/j.ijhydene.2013.07.124.

Sánchez-Riego, A.M., López-Maury, L., and Florencio, F.J. (2013). Glutaredoxins are essential for stress adaptation in the cyanobacterium *Synechocystis* sp. PCC 6803. Front. Plant Sci. *4*, 428. http://dx.doi.org/10.3389/fpls.2013.00428.

Sanders, C.E., Melis, A., and Allen, J.F. (1989). In vivo phosphorylation of proteins in the cyanobacterium *Synechococcus* 6301 after chromatic acclimation to Photosystem I or Photosystem II light. Biochim. Biophys. Acta *976*, 168–172. http://dx.doi.org/10.1016/S0005-2728(89)80226-9.

Schluchter, W.M., Shen, G., Alvey, R.M., Biswas, A., Saunée, N.A., Williams, S.R., Mille, C.A., and Bryant, D.A. (2010). Phycobiliprotein biosynthesis in cyanobacteria: structure and function of enzymes involved in post-translational modification. Adv. Exp. Med. Biol. *675*, 211–228. http://dx.doi.org/10.1007/978-1-4419-1528-3_12.

Schütz, W., Hausmann, N., Krug, K., Hampp, R., and Macek, B. (2011). Extending SILAC to proteomics of plant cell lines. Plant Cell 23, 1701–1705. http://dx.doi.org/10.1105/tpc.110.082016.

Shah, N.H., Dann, G.P., Vila-Perelló, M., Liu, Z., and Muir, T.W. (2012). Ultrafast protein splicing is common among cyanobacterial split inteins: implications for protein engineering. J. Am. Chem. Soc. *134*, 11338–11341. http://dx.doi.org/10.1021/ja303226x.

Shalev-Malul, G., Lieman-Hurwitz, J., Viner-Mozzini, Y., Sukenik, A., Gaathon, A., Lebendiker, M., and Kaplan, A. (2008). An AbrB-like protein might be involved in the regulation of cylindrospermopsin production by *Aphanizomenon ovalisporum*. Environ. Microbiol. *10*, 988–999. http://dx.doi.org/10.1111/j.1462-2920.2007.01519.x.

Silman, N.J., Carr, N.G., and Mann, N.H. (1995). ADP-ribosylation of glutamine synthetase in the cyanobacterium *Synechocystis* sp. strain PCC 6803. J. Bacteriol. *177*, 3527–3533.

Smith, K.T., and Workman, J.L. (2009). Introducing the acetylome. Nat. Biotechnol. 27, 917–919. http://dx.doi.org/10.1038/nbt1009-917.

Sobczyk, A., Bely, A., Tandeau de Marsac, N., and Houmard, J. (1994). A phosphorylated DNA-binding protein is specific for the red-light signal during complementary chromatic adaptation in cyanobacteria. Mol. Microbiol. *13*, 875–885. http://dx.doi.org/10.1111/j.1365-2958.1994.tb00479.x.

Sobczyk, A., Schyns, G., Tandeau de Marsac, N., and Houmard, J. (1993). Transduction of the light signal during complementary chromatic adaptation in the cyanobacterium *Calothrix* sp. PCC 7601: DNA-binding proteins and modulation by phosphorylation. EMBO. J. *12*, 997–1004.

Soufi, B., Jers, C., Hansen, M.E., Petranovic, D., and Mijakovic, I. (2008). Insights from site-specific phosphoproteomics in bacteria. Biochim. Biophys. Acta *1784*, 186–192. http://dx.doi.org/10.1016/j.bbapap.2007.07.018.

Spät, P., Maček, B., and Forchhammer, K. (2015). Phosphoproteome of the cyanobacterium *Synechocystis* sp. PCC 6803 and its dynamics during nitrogen starvation. Front. Microbiol. *6*, 248. http://dx.doi.org/10.3389/fmicb.2015.00248.

Sugiura, M., Koyama, K., Umena, Y., Kawakami, K., Shen, J.R., Kamiya, N., and Boussac, A. (2013). Evidence for an unprecedented histidine hydroxyl modification on D2-His336 in Photosystem II of *Thermosynechoccocus vulcanus* and *Thermosynechoccocus elongatus*. Biochemistry *52*, 9426–9431. http://dx.doi.org/10.1021/bi401213m.

Sullivan, D.M., Wehr, N.B., Fergusson, M.M., Levine, R.L., and Finkel, T. (2000). Identification of oxidant-sensitive proteins: TNF-alpha induces protein glutathiolation. Biochemistry *39*, 11121–11128. http://dx.doi.org/10.1021/bi0007674.

Swaney, D.L., Beltrao, P., Starita, L., Guo, A., Rush, J., Fields, S., Krogan, N.J., and Villén, J. (2013). Global analysis of phosphorylation and ubiquitylation cross-talk in protein degradation. Nat. Methods *10*, 676–682. http://dx.doi.org/10.1038/nmeth.2519.

Swanson, R.V., and Glazer, A.N. (1990). Phycobiliprotein methylation. Effect of the gamma-N-methylasparagine residue on energy transfer in phycocyanin and the phycobilisome. J. Mol. Biol. *214*, 787–796. http://dx.doi.org/10.1016/0022-2836(90)90293-U.

Treumann, A., and Thiede, B. (2010). Isobaric protein and peptide quantification: perspectives and issues. Expert Rev. Proteomics *7*, 647–653. http://dx.doi.org/10.1586/epr.10.29.

Ubersax, J.A., and Ferrell, J.E. (2007). Mechanisms of specificity in protein phosphorylation. Nat. Rev. Mol. Cell. Biol. *8*, 530–541. http://dx.doi.org/10.1038/nrm2203.

van Noort, V., Seebacher, J., Bader, S., Mohammed, S., Vonkova, I., Betts, M.J., Kühner, S., Kumar, R., Maier, T., O'Flaherty, M., *et al.* (2012). Cross-talk between phosphorylation and lysine acetylation in a genome-reduced bacterium. Mol. Syst. Biol. *8*, 571. http://dx.doi.org/10.1038/msb.2012.4.

Varman, A.M., Xiao, Y., Pakrasi, H.B., and Tang, Y.J. (2013). Metabolic engineering of *Synechocystis* sp. strain PCC 6803 for isobutanol production. Appl. Environ. Microbiol. *79*, 908–914. http://dx.doi.org/10.1128/AEM.02827-12.

von Leoprechting, A., Horth, P., Haehnel, W., Schilz, E., and Muhlenhoff, U. (1998). Identification of biotinylation sites on proteins by selective retrieval of 2-iminobiotinylated peptides from proteolytic peptide mixtures: localization of the accessible lysine residues on the photosystem I subunits PsaD and PsaE. Anal. Biochem. *262*, 110–121. http://dx.doi.org/10.1006/abio.1998.2751.

Wase, N., Pham, T.K., Ow, S.Y., and Wright, P.C. (2014). Quantitative analysis of UV-A shock and short term stress using iTRAQ, pseudo selective reaction monitoring (pSRM) and GC-MS based metabolite analysis of the cyanobacterium *Nostoc punctiforme* ATCC 29133. J. Proteomics *109*, 332–355. http://dx.doi.org/10.1016/j.jprot.2014.06.024.

Wegener, K.M., Singh, A.K., Jacobs, J.M., Elvitigala, T., Welsh, E.A., Keren, N., Gritsenko, M.A., Ghosh, B.K., Camp, D.G., 2nd, Smith, R.D., *et al.* (2010). Global proteomics reveal an atypical strategy for carbon/nitrogen assimilation by a cyanobacterium under diverse environmental perturbations. Mol. Cell. Proteomics *9*, 2678–2689. http://dx.doi.org/10.1074/mcp.M110.000109.

Wei, X.Y., Sakr, S., Li, J.H., Wang, L., Chen, W.L., and Zhang, C.C. (2006). Expression of split *dnaE* genes and trans-splicing of DnaE intein in the developmental cyanobacterium *Anabaena* sp. PCC 7120. Res. Microbiol. *157*, 227–234. http://dx.doi.org/10.1016/j.resmic.2005.08.004.

Welker, M., and von Döhren, H. (2006). Cyanobacterial peptides– nature's own combinatorial biosynthesis. FEMS Microbiol. Rev. *30*, 530–563. http://dx.doi.org/10.1111/j.1574-6976.2006.00022.x.

Williams, S.B. (2007). A circadian timing mechanism in the cyanobacteria. Adv. Microb. Physiol. *52*, 229–296. http://dx.doi.org/10.1016/S0065-2911(06)52004-1.

Xiong, Q., Feng, J., Li, S.T., Zhang, G.Y., Qiao, Z.X., Chen, Z., Wu, Y., Lin, Y., Li, T., Ge, F., *et al.* (2015). Integrated transcriptomic and proteomic analysis of the global response of *Synechococcus* to high light stress. Mol. Cell. Proteomics *14*, 1038–1053. http://dx.doi.org/10.1074/mcp.M114.046003.

Xiong, Q., Zhong, Q., Zhang, J., Yang, M., Li, C., Zheng, P., Bi, L.J., and Ge, F. (2012). Identification of novel miR-21 target proteins in multiple myeloma cells by quantitative proteomics. J. Proteome. Res. *11*, 2078–2090. http://dx.doi.org/10.1021/pr201079y.

Xu, Q., Guikema, J.A., and Chitnis, P.R. (1994). Identification of surface-exposed domains on the reducing side of photosystem I. Plant Physiol. *106*, 617–624. http://dx.doi.org/10.1104/pp.106.2.617.

Xu, W., Tang, H., Wang, Y., and Chitnis, P.R. (2001). Proteins of the cyanobacterial photosystem I. Biochim. Biophys. Acta *1507*, 32–40. http://dx.doi.org/10.1016/S0005-2728(01)00208-0.

Yang, J., Meng, Q., and Liu, X.Q. (2004). Intein harbouring large tandem repeats in replicative DNA helicase of *Trichodesmium erythraeum*. Mol. Microbiol. *51*, 1185–1192. http://dx.doi.org/10.1046/j.1365-2958.2003.03902.x.

Yang, M.K., Qiao, Z.X., Zhang, W.Y., Xiong, Q., Zhang, J., Li, T., Ge, F., and Zhao, J.D. (2013). Global phosphoproteomic analysis reveals diverse functions of serine/threonine/tyrosine phosphorylation in the model cyanobacterium *Synechococcus* sp. strain PCC 7002. J. Proteome Res. *12*, 1909–1923. http://dx.doi.org/10.1021/pr4000043.

Yang, M.K., Yang, Y.H., Chen, Z., Zhang, J., Lin, Y., Wang, Y., Xiong, Q., Li, T., Ge, F., Bryant, D.A., *et al.* (2014).

Proteogenomic analysis and global discovery of posttranslational modifications in prokaryotes. Proc. Natl. Acad. Sci. U.S.A. *111*, E5633–42. http://dx.doi.org/10.1073/pnas.1412722111.

Yang, X.J. (2004). Lysine acetylation and the bromodomain: a new partnership for signaling. Bioessays *26*, 1076–1087. http://dx.doi.org/10.1002/bies.20104.

Yu, B.J., Kim, J.A., Moon, J.H., Ryu, S.E., and Pan, J.G. (2008). The diversity of lysine-acetylated proteins in *Escherichia coli*. J. Microbiol. Biotechnol. *18*, 1529–1536.

Zaffagnini, M., Bedhomme, M., Marchand, C.H., Morisse, S., Trost, P., and Lemaire, S.D. (2012). Redox regulation in photosynthetic organisms: focus on glutathionylation. Antioxid. Redox Signal. *16*, 567–586. http://dx.doi.org/10.1089/ars.2011.4255.

Zhang, C.C., Jang, J., Sakr, S., and Wang, L. (2005). Protein phosphorylation on Ser, Thr and Tyr residues in cyanobacteria. J. Mol. Microbiol. Biotechnol. *9*, 154–166. http://dx.doi.org/10.1159/000089644.

Zhang, H., and Forman, H.J. (2012). Glutathione synthesis and its role in redox signaling. Semin. Cell. Dev. Biol. *23*, 722–728. http://dx.doi.org/10.1016/j.semcdb.2012.03.017.

Zhang, J., Sprung, R., Pei, J., Tan, X., Kim, S., Zhu, H., Liu, C.F., Grishin, N.V., and Zhao, Y. (2009). Lysine acetylation is a highly abundant and evolutionarily conserved modification in *Escherichiacoli*. Mol. Cell. Proteomics *8*, 215–225. http://dx.doi.org/10.1074/mcp.M800187-MCP200.

Zhang, L., Selão, T.T., Pisareva, T., Qian, J., Sze, S.K., Carlberg, I., and Norling, B. (2013). Deletion of *Synechocystis* sp. PCC 6803 leader peptidase LepB1 affects photosynthetic complexes and respiration. Mol. Cell. Proteomics *12*, 1192–1203. http://dx.doi.org/10.1074/mcp.M112.022145.

Zheng, P., Xiong, Q., Wu, Y., Chen, Y., Chen, Z., Fleming, J., Gao, D., Bi, L., and Ge, F. (2015). Quantitative proteomics analysis reveals novel insights into mechanisms of action of long noncoding RNA hox transcript antisense intergenic RNA (HOTAIR) in HeLa cells. Mol. Cell. Proteomics *14*, 1447–1463. http://dx.doi.org/10.1074/mcp.M114.043984.

Zorina, A., Stepanchenko, N., Novikova, G.V., Sinetova, M., Panichkin, V.B., Moshkov, I.E., Zinchenko, V.V., Shestakov, S.V., Suzuki, I., Murata, N., *et al.* (2011). Eukaryotic-like Ser/Thr protein kinases SpkC/F/K are involved in phosphorylation of GroES in the cyanobacterium *Synechocystis*. DNA. Res. *18*, 137–151. http://dx.doi.org/10.1093/dnares/dsr006.

Metabolic Engineering and Systems Biology for Free Fatty Acid Production in Cyanobacteria

8

Anne M. Ruffing

Sandia National Laboratories, Department of Bioenergy and Defense Technologies, Albuquerque, NM, USA.

Correspondence: aruffin@sandia.gov

Abstract

Free fatty acids (FFAs) are essential cellular components and also potential precursors for biofuel production. The modification of cyanobacteria for FFA production allows for the production of this high density energy molecule from CO_2 and sunlight as the main carbon and energy sources. Efforts to engineer cyanobacteria for FFA production have provided a proof-of-concept demonstration for this approach, yet FFA yields are too low to support industrial-scale production for biofuel applications. This chapter highlights previous successes in engineering cyanobacterial FFA production, possible targets for future metabolic engineering efforts, and the many challenges that must be overcome. Due to the essential nature of fatty acid biosynthesis, modification of this major metabolic pathway includes interactions with many other pathways of carbon metabolism and complex regulatory mechanisms. In addition, cyanobacterial FFA production has been shown to result in cellular stress responses affecting growth, membrane integrity and composition, and photosynthesis. These complex interactions necessitate the use of systems biology approaches, such as omics and computational modelling, to understand and effectively manipulate cyanobacteria for enhanced FFA production. While there are few examples of the application of systems biology to FFA production in cyanobacteria, a review of technologies and tools developed for cyanobacteria is presented to guide future efforts in this area. The proposed integration of metabolic engineering and systems biology approaches may advance our understanding of cyanobacterial fatty acid biosynthesis and overcome current barriers in cyanobacterial FFA production.

Introduction

Fatty acids are essential components for the synthesis of cellular lipids including phospholipids, sphingolipids, and sterols (Janßen and Steinbüchel, 2014). As such, the fatty acid biosynthesis pathway is found in all eukaryotes and prokaryotes. Fatty acids are particularly important in cyanobacteria as constituents of glycoglycerolipids which form the thylakoid membranes, where photosynthesis occurs (Wada and Murata, 1998). Cyanobacterial fatty acids have predominantly been studied in terms of their contributions to these complex lipid structures. The change in saturation and composition of membrane fatty acids in response to temperature changes, desiccation, osmotic stress, and high light stress has also been studied in many cyanobacteria (Singh *et al.*, 2002; Wada and Murata, 1998). Moreover, the role of membrane lipids in cyanobacterial photosynthesis has been explored (Wada and Murata, 2007). As lipids constitute 15–20% of total cellular weight in cyanobacteria (Wahlen *et al.*, 2011), cyanobacteria represent a rich source for fatty acids.

Recent metabolic engineering efforts have focused on the potential of cyanobacterial fatty acids for fuel production. Cyanobacterial fatty acids are a renewable energy source, as their production requires only carbon dioxide, sunlight, and nutrients. Unlike eukaryotic algae, cyanobacteria do not accumulate high levels of triacylglycerol

(TAG) as a carbon storage compound, and therefore, cyanobacteria were initially dismissed by the Aquatic Species Program as not useful for biofuel production (Sheehan *et al.*, 1998). However, the advancement of recombinant DNA technology and the relative ease of genetic manipulation of cyanobacteria in comparison to eukaryotic algae have advanced cyanobacteria as candidates for fatty acid based biofuel production. Free fatty acids (FFAs) are candidate biofuel precursors, as chemical based transesterification of FFAs with either methanol or ethanol yields biodiesel in the form of fatty acid methyl or ethyl esters (FAMEs or FAEEs) (Marchetti and Errazu, 2008). FAEEs may also be produced biologically in cyanobacteria, as recently demonstrated in *E. coli* (Steen *et al.*, 2010). Biological pathways for the production of other fatty acid derived fuels such as fatty alcohols (Tan *et al.*, 2011) and alkanes (Schirmer *et al.*, 2010) may also be introduced in cyanobacteria (see Chapter 11). This chapter will focus solely on cyanobacterial manipulation and omics for FFA production; however, approaches resulting in high FFA production are also likely to be effective for improving yields of these other fatty acid-derived fuels.

Elucidation and characterization of the fatty acid biosynthesis (FAB) pathway in Gram-negative bacteria has primarily been conducted in *Escherichia coli* (Janßen and Steinbüchel, 2014; Lennen and Pfleger, 2012). Using these fundamental studies in *E. coli* and genomic comparisons of sequenced cyanobacteria, a putative cyanobacterial FAB pathway is proposed in Fig. 8.1. Similar to other Gram-negative bacteria, the acetyl-CoA node

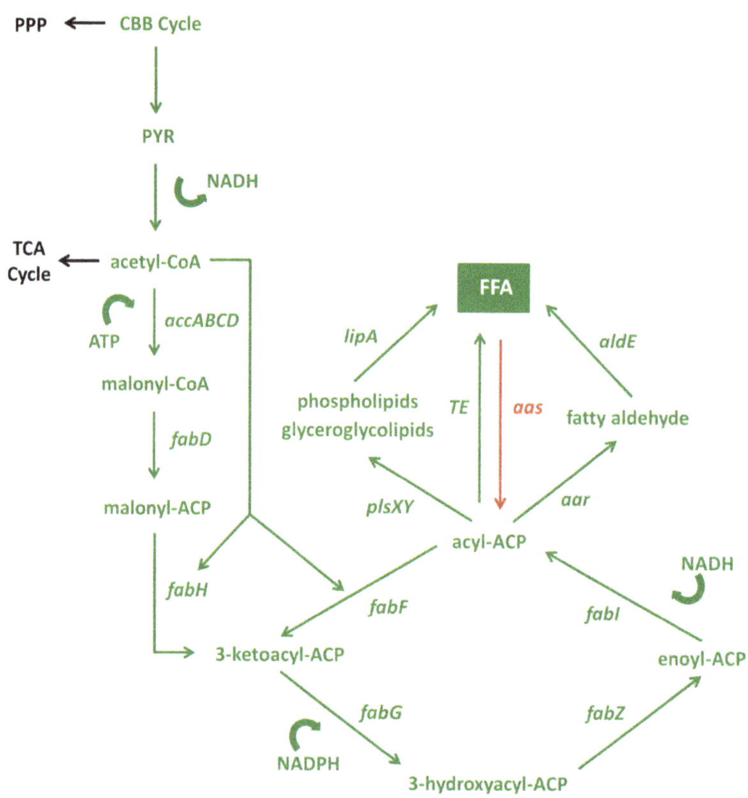

Figure. 8.1 Cyanobacterial metabolism for fatty acid biosynthesis. Pathways contributing to FFA production are highlighted in green; FFA degradation pathways are shown in red; and other metabolic pathways in black. Abbreviations: *aar*: acyl-ACP reductase, *aas*: acyl-ACP synthetase, *accACBD*: acetyl-CoA carboxylase, ACP: acyl carrier protein, *aldE*: aldehyde dehydrogenase, ATP: adenosine triphosphate, CBB: Calvin Benson Bassham, CoA: coenzyme A, *fabD*: malonyl-CoA-ACP transacylase, *fabH*: β-ketoacyl-ACP synthase III, *fabF*: β-ketoacyl-ACP synthase II, *fabG*: β-ketoacyl-ACP reductase, *fabI*: enoyl-ACP reductase, *fabZ*: β-hydroxyacyl-ACP dehydratase, FFA: free fatty acid, *lipA*: lipase, NADH: nicotinamide adenine dinucleotide reduced, NADPH: nicotinamide adenine dinucleotide phosphate reduced, *plsXY*: acyltransferases, PPP: pentose phosphate pathway, PYR: pyruvate, TCA: tricarboxylic acid, TE: thioesterase.

serves as the interface between FAB and primary metabolism. The multi-subunit acetyl-CoA carboxylase (AccABCD) converts acetyl-CoA into malonyl-CoA, and malonyl is transferred to the acyl-carrier-protein (ACP) by malonyl CoA-ACP transacylase (FabD). A β-ketoacyl-ACP synthase (FabH) converts malonyl-ACP into 3-ketoacyl-ACP, the first step in fatty acid chain elongation. An *in vitro* kinetic analysis of FAB enzymes from the cyanobacterium *Synechococcus* sp. PCC 7002 revealed FabH to be the sole rate-limiting step of this pathway (Kuo and Khosla, 2014). The elongation cycle consists of four enzymatic steps (FabF, FabG, FabZ, and FabI) for the addition of two carbons from acetyl-CoA, along with the consumption of two reducing equivalents (NADPH/NADH). Interestingly, the FabI of *Synechococcus* sp. PCC 7002 was shown to have a preference for NADH (Kuo and Khosla, 2014), despite the fact that cyanobacteria generate NADPH via photosynthesis. This contrasts with an early study of FAB in *Anabaena variabilis*, which determined higher FAB with NADPH as opposed to NADH (Lem and Stumpf, 1984). The *A. variabilis* study measured FAB of the entire pathway, including FabG which requires NADPH in *E. coli* (Heath and Rock, 1996b), while the *Synechococcus* sp. PCC 7002 study analysed FabI specifically. While this may explain the discrepancy between these two studies, the NADH versus NADPH requirement of cyanobacterial FAB remains to be definitively resolved across species. Another unconfirmed step of FAB in cyanobacteria is the contribution of FabB. In *E. coli*, FabB contributes to the generation of unsaturated acyl-ACPs (Janßen and Steinbüchel, 2014), yet fatty acid unsaturation in cyanobacteria occurs in lipid membranes rather than during FAB. Despite this difference, a putative FabB has been annotated in cyanobacterial genomes such as *Synechococcus* sp. PCC 7002. *In vitro* kinetic analysis of this putative FabB indicated that this enzyme is not essential for fatty acid chain elongation (Kuo and Khosla, 2014). This, along with the low homology of this enzyme to *E. coli* FabB, suggests that it serves a different function in *Synechococcus* sp. PCC 7002 and is likely mis-annotated.

The fatty acid chain from acyl-ACP is transferred to glycerol substrates for the formation of phospholipids and glycoglycerolipids. Studies in *Synechocystis* sp. PCC 6803 have identified and characterized acyltransferases involved in glycerolipid metabolism (Okazaki *et al.*, 2006; Weier *et al.*, 2005), yet these genes remain uncharacterized in many other sequenced cyanobacteria. These studies have shown that cyanobacterial acyltransferases prefer acyl-ACP as substrate as opposed to acyl-CoA. Once the fatty acid is converted into a lipid, cyanobacterial desaturases act upon these membrane lipids to form unsaturated fatty acid side chains; this differs from *E. coli* where unsaturated fatty acids are formed during fatty acid chain elongation of the acyl-ACP via FabA and FabB (Lennen and Pfleger, 2012). In wild type cyanobacteria, FFAs released by degradation of cell and thylakoid membranes are recycled into the FAB pathway via an acyl-ACP synthetase (Aas), previously annotated as a long chain fatty acid CoA ligase (Kaczmarzyk and Fulda, 2010). Again, this process differs from *E. coli* which possesses a β-oxidation pathway for fatty acid degradation (FAD) (Janßen and Steinbüchel, 2014). Some cyanobacterial genomes also include an aldehyde dehydrogenase, which can convert FFAs into aldehydes. Thus, in cyanobacteria, FFAs may be either recycled directly for lipid biosynthesis via Aas or converted into aldehydes.

Metabolic engineering of FFA production in cyanobacteria

Previous accomplishments

While cyanobacteria naturally produce FFAs during membrane degradation, these FFAs are typically recycled within the cell for lipid biosynthesis. Therefore, metabolic engineering is necessary to convert cyanobacteria into FFA producers. To date, three cyanobacterial hosts have been modified for FFA production: two freshwater species (*Synechocystis* sp. PCC 6803 and *Synechococcus elongatus* PCC 7942) and one marine species (*Synechococcus* sp. PCC 7002). Table 8.1 includes a summary of the genetic modifications of these hosts for FFA production. The highest FFA titres reported for each host are 197 mg/l for *Synechocystis* sp. PCC 6803, 55.6 mg/l for *S. elongatus* PCC 7942, and 131 mg/l for *Synechococcus* sp. PCC 7002. While high titres are important for batch production, FFA productivity of the modified cyanobacterium will ultimately determine feasibility for production. FFA productivities in Table 8.1 were estimated from the

Table 8.1 Genetic modifications of cyanobacterial hosts for FFA production

Genetic modifications	FFA productivity (mg/l/h)	Final FFA titre (mg/l)	Cultivation conditions	FFA measurement	Reference
***Synechocystis* sp. PCC 6803**					
SD216 – ΔaasPpsbA2-′tesA	0.209	83.6	BG-11, 30°C, continuous illumination (140 μmol/m²s), bubbled with 1% CO_2-enriched air	Secreted FFAs: acidification and hexane extraction followed by GC analysis	Liu *et al.* (2011b)
SD225 – Δslr1993–1994::Pcpc-accBC-Prbc-accDA, Δaas-PpsbA2-′tesA	0.209	83.6			
SD232 – Δsll1951-PpsbA2-UcfatB1, Δslr1993–1994::Pcpc-accBC-Prbc-accDA, Δaas-PpsbA2-′tesA	0.22625	90.5			
SD243 – Δslr2001–2002::PpsbA2-ChfatB2, Δsll1951-PpsbA2-UcfatB1, Δslr1993–1994::Pcpc-accBC-Prbc-accDA, ΔaasPpsbA2-′tesA	0.23225	92.9			
SD249 – Δslr1710::PpsbA2-CcfatB1, Δslr2001–2002::PpsbA2-ChfatB2, Δsll1951-PpsbA2-UcfatB1, Δslr1993–1994::Pcpc-accBC-Prbc-accDA, Δaas-PpsbA2-′tesA	0.365	146			
SD277 – Δslr2132::Ptrc-tesA137, Δslr1710, Δslr2001–2002::PpsbA2-ChfatB2, Δsll1951-PpsbA2-UcfatB1, Δslr1993–1994::Pcpc-accBC-Prbc-accDA, Δaas-PpsbA2-′tesA	0.4925	197			
Δaas	0.367/0.177	44/17	BG-11 + 6x phosphate, 30°C, continuous/diurnal (12:12) illumination (140 μmol/m²s), 50x addition of 1 M sodium bicarbonate every 12 h with shaking	Total FFAs: Bligh and Dyer extraction (chloroform-methanol) with acidification, dried under nitrogen, resuspended in hexane and analysed by GC	Cheah *et al.* (2015)
GG1.1 – Δaas::PpsbAII-′tesA^{CO}	0.383/0.219	46/21			
GG1.2 – Δaas::PpsbAII(Δhex-HLR)-′tesA^{CO}	0.49/0.198	47/19			
GG1.3 – Δaas::PpsbAII(Δhex)-′tesA^{CO}	0.4/0.188	48/18			
GG2.3 NF0168- Δaas::PpsbAII(Δhex)-′tesA^{CO}, Δ0168::PpsbAII(Δhex)-′tesA^{CO}	0.458/0.167	44/16			
GQ8 – Δaas	0.0708	17	BG-11, 30°C, continuous illumination (30 μmol/m²s), with shaking	Total FFAs: Cell lysis by sonication, chloroform:methanol extraction, dried under nitrogen, resuspended in hexane and analysed by GC-MS	Gao *et al.* (2012)
Δaas::UcTE	0.0268	13.5	BG-11 containing 5 mM TES-KOH, 30°C, continuous illumination (40 μmol/m²s), with air bubbling	Secreted FFAs: Acidification and hexane extraction followed by GC analysis	Kawahara *et al.* (2016)
ΔcyabrB2, Δaas::UcTE	0.0466	23.5			

Synechococcus elongatus **PCC 7942**					
SE01 – Δaas	0.0834	55.6	BG-11, 30°C, continuous illumination (60 μmol/m^{2}s), bubbled with 1% CO_2-enriched air with shaking	Secreted FFAs: Culture supernatant analysed using Free Fatty Acid Quantification Kit (Biovision) – enzymatic method for fluorescence-based detection with standard calibration curve	Ruffing and Jones (2012)
SE02 – Δaas::Ptrc-'tesA	0.0638	39.6			
SE03 – Δaas::Ptrc-Crfat1	0.0607	32	BG-11, 30°C, continuous illumination (60 μmol/m^{2}s), bubbled with 1% CO_2-enriched air with shaking		Ruffing (2013a)
SE04 – Δaas::Ptrc-Crfat1-rbcLS	0.0701	33			
SE05 – Δaas::Ptrc-Crfat1-rbcLS, NSII::PlLacO1-accBCDA	0.0580	25			
SE06 – Δaas::Ptrc-Crfat1-PpsbAI-rbcLS	0.0638	33.6			
SE07 – Δaas::Ptrc-Crfat1-PpsbAI-rbcLS, NSII::Prbc-accBC-Pcpc-accDA	0.0479	23			
Aar – NS2::Ptrc-aar	1.42 (total lipids)	170 (total lipids)	BG-11, 30°C, continuous illumination (50 μmol/m^{2}s), with shaking	Total lipid FAs: Modified Folch extraction (methanol and sulfuric acid) for transesterification followed by hexane extraction and FAME analysis by GC	Kraiser *et al.* (2013)
Aar (2x) – NS2::Ptrc-aar	0.854 (total lipids)	82 (total lipids)			
Synechococcus **sp. PCC 7002**					
S07 – Δaas::Ptrc-'tesA-PpsbAI-rbcLS	0.264	131	A+ medium, 34°C, continuous illumination (60 μmol/m^{2}s), bubbled with 1% CO_2-enriched air with shaking	Secreted FFAs: Culture supernatant analysed using Free Fatty Acid Quantification Kit (Biovision)	Ruffing (2014)
S06 – Δaas::Ptrc-'tesA-rbcLS	0.0125	5.98			
S05 – Δaas::Ptrc-Crtfat1	0.0584	30.7			
S03 – Δaas::Ptrc-Crfat1	0.0171	9			
S02 – Δaas::Ptrc-'tesA	0.0966	50.8			
S01 – Δaas	0.0097	5.12			
SA01 – Δaas::PcpcBA-fatB1-aadA	0.133	6.4	A+ medium, 34°C, continuous illumination (160 μmol/m^{2}s), shaking with 1% CO_2 atmosphere	Secreted FFAs/lipids: Methanol, hydrochloric acid for transesterification followed by hexane extraction and FAME analysis by GC	Work *et al.* (2015)
SA13 – ΔglgC, Δaas::PcpcBA-fatB1	0.125	6			

reported small-scale, batch experiments by dividing the highest FFA titre by the time required to reach that concentration. These estimates may underpredict the actual FFA productivities, but they provide a useful metric for comparison. It is also important to note that this measurement does not account for differences in cell density, which will also impact FFA productivity. *Synechocystis* sp. PCC 6803 achieved productivities as high as 0.49 mg FFA/l/h; *S. elongatus* PCC 7942 reached 0.083 mg FFA/l/h; and *Synechococcus* sp. PCC 7002 demonstrated productivities up to 0.26 mg FFA/l/h. Based on this preliminary comparison, *Synechocystis* sp. PCC 6803 has demonstrated the highest FFA titres and productivities; however, this strain was also the most heavily modified (6 gene knockouts and 8 genes expressed) (Liu *et al.*, 2011b).

A direct comparison of FFA production between different references is challenging due to the variability in cultivation conditions and FFA measurement techniques. Light conditions, including both the light intensity and spectral quality of the light source, can vary considerably between experimental set-ups. In Table 8.1, reported light intensities vary between 30 μmol photons/m^2/s and 160 μmol photons/m^2/s for characterization of the FFA-producing cyanobacterial strains, and the spectral quality of the light sources is often not reported. As cyanobacteria are photoautotrophs, the supply of inorganic carbon to the culture can significantly impact the rate of carbon fixation and subsequently the rates of both growth and FFA production. Inorganic carbon supplied to the FFA-producing cyanobacterial strains varies from simple aeration by shaking to bubbling air supplemented with 1% CO_2 to the addition of 20 mM of sodium bicarbonate every 12 h (Table 8.1). Furthermore, the mass flow rates of air or CO_2-enriched air are often not reported for laboratory studies due to the difficulties associated with both controlling and measuring these rates. In addition to variability in environmental conditions between studies, the methods of FFA measurement differ considerably. FFAs are traditionally measured by solvent extraction and gas chromatography (GC), either directly or after derivatization to FAMEs. Two different types of solvents have been utilized for cyanobacterial FFA extraction: hexane or chloroform/methanol. The efficiency of FFA extraction between these two solvents has not been directly compared for these cyanobacterial extractions. The derivatization of FFAs to FAMEs may also lead to erroneous FFA measurements due to the fact that all fatty acids, including side chains of glycoglycerolipids and phospholipids, are converted to FAMEs in this reaction. An alternative method of measuring FFAs utilizes an enzymatic reaction that leads to a fluorescent signal and is provided as a FFA quantification kit by BioVision. In this assay, fatty acids are enzymatically converted to their CoA derivatives, which are subsequently oxidized to produce a colorimetric and fluorescent signal. This method can be used for quantifying FFAs with chain lengths of more than eight carbons at concentrations down to 2 μM. Since this method does not extract FFAs from a cell culture, it has only been used in quantifying excreted FFAs in culture supernatant (Ruffing, 2013a, 2014; Ruffing and Jones, 2012). While each FFA quantification method has been previously developed and validated, there has been no direct comparison of measurement sensitivities for quantifying cyanobacterial FFAs. Differences in environmental conditions and FFA quantification likely explain the discrepancies between measured FFA titres and productivities for identical strains in different studies, e.g. Δ*aas* strains of *Synechocystis* sp. PCC 6803 (Cheah *et al.*, 2015; Gao *et al.*, 2012).

The genetic modification of cyanobacteria for FFA production follows some basic metabolic engineering strategies. First, the enzyme responsible for recycling FFAs, the acyl-ACP synthetase (*aas*), is knocked out to allow for FFA accumulation. In *Synechocystis* sp. PCC 6803 and *S. elongatus* PCC 7942, *aas* knockout led to significant FFA accumulation late in the growth cycle, up to 44 mg/l (Cheah *et al.*, 2015) and 55.6 mg/l (Ruffing and Jones, 2012), respectively. In *Synechococcus* sp. PCC 7002; however, only 5.1 mg/l of excreted FFA was detected with *aas* knockout (Ruffing, 2014). These results suggest that *Synechocystis* sp. PCC 6803 and *S. elongatus* PCC 7942 have more active processes for membrane degradation and recycling as compared to *Synechococcus* sp. PCC 7002. Next, a thioesterase is often expressed to release the fatty acid chain directly from the ACP, rather than relying solely on membrane degradation. In *E. coli*, thioesterase expression was also shown to alleviate feedback inhibition of enzymes involved in FAB, which are inhibited by long chain acyl-ACPs (Davis and Cronan, 2001; Heath and

Rock, 1996a). The thioesterase can also determine the chain length and degree of unsaturation of the FFA product. For example, the *E. coli* thioesterase ('TesA) shows a preference for releasing C_{16} fatty acids while plant thioesterases such as the thioesterase from *Umbellularia californica* (UcFat1) has a preference for shorter C_{12} fatty acids (Voelker and Davies, 1994). Chain length is a key factor in determining fuel properties, with C_4–C_{12} preferred for gasoline substitutes, C_8–C_{16} used for jet fuels, and C_9–C_{23} applied for diesel fuels (Lee *et al.*, 2008). The degree of fatty acid saturation also affects fuel properties. Unsaturated FFAs can reduce the stability of the fuel due to the increased reactivity of the unsaturated bond, yet unsaturated FFAs can also reduce the viscosity of fuels, which is desirable for flow in engines. In cyanobacteria, fatty acid unsaturation is only known to occur at the level of membrane lipids through the action of desaturases (Los and Murata, 1998; Murata *et al.*, 1992). Therefore, thioesterase activity should increase the amount of saturated FFAs while the expression of lipases could be used to control the release of unsaturated FFAs. A number of different thioesterases have been expressed in cyanobacteria, including bacterial 'TesA (Liu *et al.*, 2011b; Ruffing, 2014; Ruffing and Jones, 2012), plant UcfatB1, ChfatB2, and CcfatB1 (Kawahara *et al.*, 2016; Liu *et al.*, 2011b), and algal Crfat1/Crtfat1 (Ruffing, 2013a). Bacterial and algal thioesterases have led to predominantly C_{16} and C_{18} FFAs, whereas plant thioesterases increased C_{12} and C_{14} FFAs. While the expression of multiple thioesterases has been shown to improve FFA production in *Synechocystis* sp. PCC 6803 (Liu *et al.*, 2011b), another study of *Synechocystis* sp. PCC 6803 found no thioesterase activity from expression of the truncated *E. coli* thioesterase 'TesA (Cheah *et al.*, 2015). Additionally, 'TesA expression in *S. elongatus* PCC 7942 was shown to improve FFA excretion during growth but ultimately resulted in lower FFA titres later in the growth phase (Ruffing and Jones, 2012). *Synechococcus* sp. PCC 7002 demonstrated the greatest improvement in FFA production with thioesterase expression, a nearly 10-fold increase in final FFA titre (Ruffing, 2014).

In addition to the basic modifications of *aas* knockout and thioesterase expression, most other metabolic engineering efforts have focused on improving carbon flux through the FAB pathway. A major rate limiting step in *E. coli* FAB is the conversion of acetyl-CoA to malonyl-CoA, catalysed by acetyl-CoA carboxylase (*accABCD*). Because *accABCD* overexpression in *E. coli* led to a 6-fold improvement in FFA production (Davis *et al.*, 2000), this enzymatic step has been a target of several cyanobacterial engineering efforts. Overexpression of *accABCD* in *Synechocystis* sp. PCC 6803 or *S. elongatus* PCC 7942, along with *aas* knockout and thioesterase expression, resulted in no improvement in FFA production for either host (Liu *et al.*, 2011b; Ruffing, 2013a). These results agree with the recent *in vitro* kinetic study of the FAB pathway from *Synechococcus* sp. PCC 7002, which concluded that FabH is the sole rate-limiting step of FAB (Kuo and Khosla, 2014). The rate of carbon fixation, catalysed by ribulose-1,5-bisphosphate carboxylase/oxygenase (RuBisCO), is also predicted to be a rate-limiting step in photoautotrophic organisms due to its relatively low carboxylase activity (Sukenik *et al.*, 1987). As such, RuBisCO has been targeted for improvement in many photosynthetic hosts but has shown very limited success. The overexpression of native RuBisCO (*rbcLS*) in FFA-producing *S. elongatus* PCC 7942 yielded no change in either growth or FFA production; however, the expression of a heterogeneous *rbcLS* in FFA-producing *Synechococcus* sp. PCC 7002 resulted in more than a 2.5-fold increase in FFA production (Ruffing, 2013a, 2014). This result is similar to the expression of a heterogeneous *rbcLS* in *S. elongatus* PCC 7942 engineered for isobutyraldehyde production, which led to a 2-fold increase in isobutyraldehyde (Atsumi *et al.*, 2009). It is important to note that in both cases, showing either improved FFA or isobutyraldehyde production with foreign *rbcLS* expression, the total rate of CO_2 fixation does not increase, but rather, the CO_2 flux is re-routed to either FFA or isobutyraldehyde synthesis. Both the heterologous and native RuBisCO proteins are form B1 and are naturally contained in β-carboxysomes (Rae *et al.*, 2013). Furthermore, RbcL and RbcS from *S. elongatus* PCC 7942 share high homology to RbcL and RbcS of the FFA-producing host *Synechococcus* sp. PCC 7002 with 86.2% and 68.8% identities, respectively, and the RbcL and RbcS from *S. elongatus* PCC 6301 have 100% identity to the native RbcL and RbcS in the isobutyraldehyde host *S. elongatus* PCC 7942. Thus, while the underlying cause of carbon flux redistribution with heterogeneous *rbcLS* overexpression remains to be investigated, there is no apparent mechanism for this redistribution based on protein sequence alone. Additionally, the subcellular localization of heterogeneous RbcL and RbcS (i.e. whether these are incorporated into native carboxysomes within the cyanobacterial host) must also be analysed. Other metabolic engineering efforts have focused on deleting competing carbon pathways to improve FAB flux. In *Synechococcus* sp. PCC 7002, a glycogen deficient strain (Δ*glgC*) was modified for FFA production with *aas* knockout and thioesterase expression (Work *et al.*, 2015). The underlying hypothesis was that the excess carbon from blocking glycogen biosynthesis would be diverted to FFA production as a carbon sink. However, no improvement in FFA biosynthesis was observed, presumably due to changes in photosynthetic electron transport which resulted from *glgC* disruption. In *Synechocystis* sp. PCC 6803, three competing carbon pathways were disrupted: acetate production (Δ*pta*), cyanophycin biosynthesis (Δslr2001–2002), and polyhydroxybutyrate (PHB) synthesis (Δslr1993–1994) (Liu *et al.*, 2011b). The results of these gene knockouts on FFA production are difficult to interpret due to multiple parallel modifications (i.e. introduction of a thioesterase along with gene knockout) and variability in the genotype of the parent strain, as the modifications were additive. Despite these issues, elimination of acetate, cyanophycin, and PHB did not appear to significantly effect the production of FFAs in *Synechocystis* sp. PCC 6803. The widespread lack of success at improving carbon flux through FAB in cyanobacteria suggests that this pathway may be highly regulated and highlights our fundamental lack of understanding of cyanobacterial FAB.

While traditional metabolic engineering efforts for enhancing carbon flux through FAB have met with limited success, alternative targets, such as regulatory factors, cell membrane integrity, and alternative FFA pathways, have shown promise for improving FFA production. The transcriptional regulator CyAbrB2 coordinates nitrogen and carbon metabolism in *Synechocystis* sp. PCC 6803, and deletion of *cyAbrB2* leads to the accumulation of glycogen granules. As glycogen can be used as a carbon source for FAB, *cyAbrB2* was deleted in a *Synechocystis* sp. PCC 6803 strain modified for FFA production (Kawahara *et al.*, 2016). The *cyAbrB2* deletion strain produced 2-fold higher FFA concentrations on a per cell basis and approximately a 70% increase in final FFA titre compared to the parent strain. Since there was no observable degradation of the glycogen granules accumulated by *cyAbrB2* deletion in the FFA-producing strain, the mechanism for this improvement in FFA production is unclear. If the glycogen granules could also be utilized for FFA biosynthesis, further improvement may be feasible. While homologues of *cyAbrB2* exist in other cyanobacteria, including *S. elongatus* PCC 7942 and *Synechococcus* sp. PCC 7002, this target has only been explored for FFA production in *Synechocystis* sp. PCC 6803. Intracellular accumulation of FFAs may also limit FFA production, either through equilibrium dynamics, feedback inhibition, or toxic effects of FFAs (Wu *et al.*, 2006). To improve FFA release from *Synechocystis* sp. PCC 6803, Liu and colleagues deleted two genes involved in assembly of hydrophilic membrane components: the hemolysin-like surface layer (S-layer) protein and the penicillin binding protein involved in peptidoglycan layer assembly (Liu *et al.*, 2011b). Deletion of the S-layer contributed to an 8% improvement in FFA titre while knockout of the penicillin binding protein was associated with a 57% improvement in FFA titre. As discussed previously, these gene deletions were conducted with other parallel modifications and used parent strains of different genotypes, and as such, the results of these knockout mutants should be viewed with this caveat. Lastly, an alternative pathway for releasing FFAs, one that does not involve a thioesterase, has been demonstrated in *S. elongatus* PCC 7942. This pathway is native to *S. elongatus* PCC 7942 and involves the conversion of acyl-ACP to an aldehyde by Aar (acyl-ACP reductase), followed by conversion of the aldehyde to FFA by AldE (aldehyde dehydrogenase). The overexpression of native *aar* in *S. elongatus* PCC 7942 resulted in the accumulation of excreted FFAs, demonstrating feasibility of this pathway (Kaiser *et al.*, 2013). Interestingly, *Synechocystis* sp. PCC 6803 contains homologues for both *aar* and *aldE*, yet *Synechococcus* sp. PCC 7002 lacks homologues for these genes. The role of this alternative FFA production pathway in cyanobacteria remains unclear, and its potential for FFA production remains to be fully explored. The successes achieved by targeting a transcriptional regulator, membrane structural components, and alterative FFA pathways illustrate the benefit of considering non-traditional metabolic engineering targets and approaching strain development from a systems-level perspective.

Potential targets for improvement

The literature on FAB and engineering FFA production in *E. coli* is far more extensive than the studies in cyanobacteria, and thus, successful strategies for improving FFA production in *E. coli* may also be applicable in cyanobacteria. Many *E. coli* enzymes involved in FAB and FAD have been targeted for overexpression or deletion. While the overexpression of the transcriptional regulator FadR improved FFA production by 7.4-fold in *E. coli* (Zhang *et al.*, 2012), FadR homologues have not been identified in cyanobacterial hosts (Table 8.2). Overexpression of other FAB genes, such as *fabG*, *fabZ*, and *fabI* or *fabA* and *fabB*, have led to modest improvements in *E. coli* FFA production (1.1- to 2.3-fold) (Janßen and Steinbüchel, 2014). FabA and FabB are paralogues of FabZ and FabF, yet overexpression of *fabZ* or *fabF* in *E. coli* produced conflicting results in FFA production, with studies reporting either 3-fold improvement in FFA yield or no improvement (Janßen and Steinbüchel, 2014). These FAB genes have yet to be targeted in any cyanobacterial host, and while FabA and FabB homologues have not been identified in cyanobacteria, expression of these *E. coli* genes in a cyanobacterial host may improve FFA production. As cyanobacterial and *E. coli* FAB pathways appear to have different mechanisms of regulation, expression of *E. coli* FAB genes in cyanobacterial hosts may also avoid native regulatory mechanisms that limit fatty acid production. Lastly, a reversal of the *E. coli* β-oxidation pathway, normally used for FFA degradation, has generated FFAs at high titres, 85% of the theoretical yield (Dellomonaco *et al.*, 2011; Lennen and Pfleger, 2012). This was achieved by overexpression of *fadR* and *atoC*, regulators of the β-oxidation pathway; overexpression of *fadBA*, a 3-ketoacyl-CoA thiolase, and *fadM*, a thioesterase; replacement of the catabolite repressor protein *crp* with a cAMP-independent mutant *crp**; and deletion of a β-oxidation repressor (*arcA*), other fermentative pathways (*adhE*, *pta*, and *frdA*), lactaldehyde reductase (*fucO*), alcohol dehydrogenase (*yqhD*), and fatty acyl-CoA synthetase (*fadD*) in *E. coli* (Dellomonaco *et al.*, 2011). As the β-oxidation pathway is not present in cyanobacteria, introduction of this *E. coli* pathway with knockout of *aas* may enable FFA production via this efficient pathway.

Additional metabolic engineering targets for FFA production in cyanobacteria have also been directly identified in these hosts. An *in vitro* kinetic analysis of FAB enzymes from *Synechococcus* sp. PCC 7002 investigated the rate limiting concentrations of FabB, FabD, FabF, FabG, FabH, FabI, FabZ, and ACP (Kuo and Khosla, 2014). It should be noted that FabB from *Synechococcus* sp. PCC 7002 was identified based on its annotation in GenBank rather than its sequence homology to *E. coli* FabB, as *Synechococcus* sp. PCC 7002 FabF is a closer homologue to *E. coli* FabB (35% identity for FabF vs. 29% identity for FabB). Moreover, *Synechococcus* sp. PCC 7002 FabF was found to be essential for pathway activity, while FabB was non-essential. Therefore, the role of annotated FabB in *Synechococcus* sp. PCC 7002 remains unclear. Despite this ambiguity, the kinetic study revealed that FabH is the rate-limiting step in *Synechococcus* sp. PCC 7002 FAB. Overexpression of *fabH* is therefore a viable strategy for improving cyanobacterial FFA production. Additionally, this study revealed that a high level of ACP is required for optimal FAB activity in *Synechococcus* sp. PCC 7002 (15 μM compared to 1 μM of the enzymatic components) (Kuo and Khosla, 2014). In *E. coli*, ACP is encoded by *acpP*, yet in its active form, a phosphopantethein group is attached to a serine residue of apo-ACP by ACP synthase (*acpS*). An ACP phosphodiesterase (AcpH) cleaves off the phosphopantethein residue to inactivate ACP (Janßen and Steinbüchel, 2014). In cyanobacteria, however, there are no homologues of AcpS nor AcpH (Table 8.2), suggesting that ACP activation may not occur. The availability of active ACP in *E. coli* is also linked to other metabolic processes such as the inhibition of glycerol-3-phsophate acyltransferase and interaction with SpoT, an enzyme involved in regulating levels of the global regulator guanosine tetraphosphate (ppGpp). Whether these processes play a role in regulating ACP within cyanobacteria has yet to be determined, but they are potential targets for engineering FFA production. Co-factor availability is another metabolic engineering target in cyanobacterial FAB. Two steps in FAB require reducing equivalents: FabG and FabI. In *E. coli*, FabG requires NADPH as the reducing equivalent,

Table 8.2 Homology-based comparisons of fatty acid metabolism proteins from *E. coli* and three model cyanobacteria

E. coli K12 MG1655		*Synechocystis* sp. PCC 6803				*S. elongatus* PCC 7942				*Synechococcus* sp. PCC 7002			
Protein query	Product	Hit ID	Product	e-value	% identity	Hit ID	Product	e-value	% identity	Hit ID	Product	e-value	% identity
Fatty acid biosynthesis													
AccA	Acetyl-CoA carboxylase carboxyltransferase, alpha subunit	WP_010874235.1	accA – acetyl-CoA carboxylase carboxyltransferase, alpha subunit	2.00E-95	54.31	WP_011242365.1	Acetyl-CoA carboxylase carboxyltransferase, alpha subunit	1.00E-93	54.31	WP_012308039.1	Acetyl-CoA carboxylase carboxyltransferase, alpha subunit	1.00E-96	53.35
AccB	Acetyl-CoA carboxylase, biotin carboxyl carrier protein	WP_010873478.1	Acetyl-CoA carboxylase, biotin carboxyl carrier protein	4.00E-25	49.44	WP_011243856.1	Acetyl-CoA carboxylase, biotin carboxyl carrier protein	2.00E-24	52.86	WP_012305693.1	Acetyl-CoA carboxylase, biotin carboxyl carrier protein	6.00E-23	53.42
AccC	Acetyl-CoA carboxylase biotin carboxylase subunit	WP_010873525.1	Acetyl-CoA carboxylase biotin carboxylase subunit	3.00E-170	57.34	WP_011242491.1	Acetyl-CoA carboxylase biotin carboxylase subunit	1.00E-168	56.43	WP_012307729.1	Acetyl-CoA carboxylase biotin carboxylase subunit	2.00E-164	55.06
AccD	Acetyl-CoA carboxylase carboxyltransfease, beta subunit	WP_010873320.1	Acetyl-CoA carboxylase carboxyltransfease, beta subunit	6.00E-85	49.81	WP_011244449.1	Acetyl-CoA carboxylase carboxyltransfease, beta subunit	7.00E-86	51.67	WP_012305699.1	Acetyl-CoA carboxylase carboxyltransfease, beta subunit	9.00E-86	49.63
FabD	malonyl CoA-ACP transacylase	WP_010871898.1	FabD – malonyl CoA-ACP transacylase	2.00E-47	39.65	WP_011242414.1	Malonyl CoA-ACP transacylase	1.00E-44	41.87	WP_012308154.1	Malonyl CoA-ACP transacylase	1.00E-47	39.07
FabB	β-Ketoacyl-ACP synthase I	WP_010871941.1	β-Ketoacyl-ACP synthase II	2.00E-58	35.11	WP_011243296.1	β-Ketoacyl-ACP synthase II	4.00E-61	36.17	WP_012306650.1	β-Ketoacyl-ACP synthase II	2.00E-66	35.92
FabF	β-Ketoacyl-ACP synthase II	WP_010871941.1	β-Ketoacyl-ACP synthase II	1.00E-133	50	WP_011243296.1	β-Ketoacyl-ACP synthase II	2.00E-137	52.07	WP_012306650.1	β-Ketoacyl-ACP synthase II	2.00E-141	51.58
FabG	β-Ketoacyl-ACP reductase	WP_010872244.1	FabG – 3-ketoacyl-ACP reductase	3.00E-73	51.44	WP_011243157.1	β-Ketoacyl-ACP reductase	3.00E-73	50.4	WP_012307963.1	β-Ketoacyl-ACP reductase	5.00E-80	53.66
FabH	β-Ketoacyl-ACP synthase III	WP_010872643.1	β-Ketoacyl-ACP synthase III	1.00E-86	40.56	WP_011242415.1	β-Ketoacyl-ACP synthase III	4.00E-85	42.41	WP_012306156.1	β-Ketoacyl-ACP synthase III	6.00E-82	42.59
FabA	β-Hydroxyacyl-ACP dehydratase	WP_010872534.1	FabZ – β-hydroxyacyl-ACP dehydratase	0.001	31.87	WP_011377793.1	β-Hydroxyacyl-ACP dehydratase	9.00E-05	30.34	WP_012305711.1	β-Hydroxyacyl-ACP dehydratase	6.00E-06	32.65
FabZ	β-Hydroxyacyl-ACP dehydratase	WP_010872534.1	FabZ – β-hydroxyacyl-ACP dehydratase	8.00E-44	50	WP_011377793.1	β-Hydroxyacyl-ACP dehydratase	4.00E-41	46.81	WP_012305711.1	β-Hydroxyacyl-ACP dehydratase	1.00E-40	47.62
FabI	Enoyl-ACP reductase, NADH-dependent	WP_014407072.1	Enoyl-ACP reductase	2.00E-78	45.53	WP_011377454.1	Enoyl-ACP reductase	7.00E-87	49.61	WP_012307288.1	Enoyl-ACP reductase	4.00E-85	48.64
FabR	Transcriptional repressor of fabA and fabB	No homologue				No homologue				No homologue			

TE													
TesA	Acyl-CoA thioesterase I	No homologue				No homologue				No homologue			
TesB	Acyl-CoA thioesterase 2	No homologue				No homologue				No homologue			
β-Oxidation pathway													
FadD	Acyl-CoA synthetase	WP_010873726.1	Acetyl-CoA synthetase	2.00E-25	25.53	WP_011377787.1	Long-chain fatty acid CoA ligase	6.00E-31	25.16	WP_041443562.1	AMP-binding protein, acyl-CoA synthetase	4.00E-38	26.7
FadK	Short chain acyl-CoA synthetase, anaerobic	WP_014407070.1	Long-chain fatty acid CoA ligase	2.00E-21	28.27	WP_011242517.1	Acetyl-CoA synthetase	1.00E-24	24.18	WP_041443561.1	AMP-binding protein, long-chain fatty acid CoA ligase	3.00E-31	25.77
FadE	Acyl-CoA dehydrogenase	No homologue				No homologue				No homologue			
Ydio	Putative acyl-CoA dehydrogenase	No homologue				No homologue				No homologue			
FadB	3-Hydroxybutyryl-CoA epimerase/Δ(3)-cis-Δ(2)-trans-enoyl-CoA isomerase/enoyl-CoA hydratase/β-hydroxyacyl-CoA dehydrogenase	No homologue				No homologue				No homologue			
FadJ	Enoyl-CoA hydratase/epimerase and isomerase/3-hydroxyacyl-CoA dehydrogenase	No homologue				No homologue				No homologue			
FadA	β-Ketoacyl-CoA thiolase	WP_041425968.1	Acetyl-CoA acetyltransferase	6.00E-61	38.57	No homologue				No homologue			
FadI	β-Ketoacyl-CoA thiolase	WP_041425968.1	Acetyl-CoA acetyltransferase	2.00E-37	26.39	No homologue				No homologue			
FadR	Fatty acid metabolism regulon transcriptional regulator	No homologue				No homologue				No homologue			

Table 8.2 Continued

E. coli K12 MG1655		*Synechocystis* sp. PCC 6803				*S. elongatus* PCC 7942				*Synechococcus* sp. PCC 7002			
Protein query	Product	Hit ID	Product	e-value	% identity	Hit ID	Product	e-value	% identity	Hit ID	Product	e-value	% identity
Unsaturated FA synthesis													
No homologue	N/A	WP_010872792.1	Delta(12)-fatty-acid desaturase	0	100	WP_011243859.1	Acyl-CoA desaturase	3.00E-118	62	WP_012308348.1	Fatty acid desaturase	N/A	N/A
No homologue	N/A	WP_010872924.1	Fatty acid desaturase	0	99					WP_012305797.1	Fatty acid desaturase	N/A	N/A
No homologue	N/A	WP_020862061.1	Acyl-CoA desaturase	0	100					WP_041443588.1	Acyl-CoA desaturase	N/A	N/A
No homologue	N/A	WP_010873125.1	Linoleoyl-CoA desaturase	0	100					WP_012307598.1	Delta-9 desaturase	N/A	N/A
No homologue	N/A									WP_012308420.1	Delta 9 acyl-lipid fatty acid desaturase	N/A	N/A
Membrane biosynthesis													
plsX	Putative phosphate acyltransferase	WP_014411807.1	Phosphate acyltransferase	3.00E-50	38	WP_011242416.1	Phosphate acyltransferase	3.00E-51	38	WP_049761547.1	Phosphate acyltransferase	4.00E-46	36
plsY	Putative glycerol-3-phosphate acyltransferase	WP_010872624.1	Glycerol-3-phosphate acyltransferase	5.00E-08	41	WP_011244346.1	Glycerol-3-phosphate acyltransferase	2.00E-10	52	WP_012306637.1	Acyl-phosphate glycerol 3-phosphate acyltransferase	1.00E-13	42
plsB	Glycerol-3-phosphate O-acyltransferase	No homologue				No homologue				No homologue			
ACP synthesis													
acpP	Acyl carrier protein	WP_010871942.1	Acyl carrier protein	1.00E-15	67	WP_011243297.1	Acyl carrier protein	4.00E-15	64	WP_012306651.1	Acyl carrier protein	2.00E-11	57
acpS	Holo-ACP synthase 1	No homologue				No homologue				No homologue			
acpH	ACP phosphodiesterase	No homologue				No homologue				No homologue			

as opposed to NADH, while FabI utilizes either NADPH or NADH (Bergler *et al.*, 1996; Heath and Rock, 1996b). Interestingly, the FabI from *Synechococcus* sp. PCC 7002 was shown to prefer NADH over NADPH despite the fact that cyanobacteria produce reducing equivalents in the form of NADPH via photosynthesis (Kuo and Khosla, 2014). The availability of NADH may therefore limit FabI in vivo, and overexpression of a transhydrogenase may enhance the interconversion of NADPH and NADH.

Lastly, most engineering efforts for producing FFAs in cyanobacteria have relied on thioesterases for fatty acid release from the ACP. The demonstration of the Aar/AldE pathway for generating FFAs via aldehydes offers an alternative route (Kaiser *et al.*, 2013). The introduction of this pathway into non-native cyanobacterial hosts such as *Synechococcus* sp. PCC 7002 may improve FFA production, and the combination of this pathway along with thioesterase expression remains to be explored. In addition to thioesterases and the Aar/AldE pathway, lipases may also release FFAs, using membrane lipids as substrate rather than acyl-ACP. While the expression of lipases in cyanobacteria has been investigated as a mechanism for releasing FFAs concomitant with cell death (Liu *et al.*, 2011a), low-level lipase expression for FFA release while maintaining cell viability may also be feasible. Cyanobacteria naturally express lipases for membrane recycling, leading to FFA accumulation with *aas* knockout (Kaczmarzyk and Fulda, 2010), and modification of this natural process may either enhance FFA production or allow for control of the unsaturated fatty acid content.

Challenges

Host selection is a challenge for cyanobacterial FFA production. While decades of heterotrophic FFA production have focused primarily on a single production host (*E. coli*), cyanobacterial FFA production has involved three cyanobacterial hosts: *Synechocystis* sp. PCC 6803, *S. elongatus* PCC 7942, and *Synechococcus* sp. PCC 7002 (Table 8.1). Protocols for genetic manipulation and genetic tools (promoters, ribosome binding sites, transcription terminators, and RNAi) must be developed and optimized in each host. Most metabolic engineering efforts have relied on a limited set of genetic tools for each cyanobacterial host; however, recent efforts have focused on developing more versatile and characterized genetic tools for *Synechocystis* sp. PCC 6803 and *Synechococcus* sp. PCC 7002 (Huang *et al.*, 2010; Liu and Curtiss, 2009; Markley *et al.*, 2014; Zess *et al.*, 2016). Each cyanobacterial host also has different physiological properties. *Synechococcus* sp. PCC 7002 is a marine host, allowing for growth on non-freshwater sources. This may be an advantage for low-cost, large-scale production systems, such as those required for biofuel production, compared to *Synechocystis* sp. PCC 6803 and *S. elongatus* PCC 7942, which are freshwater hosts. In addition, *Synechococcus* sp. PCC 7002 has high light and temperature tolerances, which are advantageous for systems using natural sunlight and photobioreactors, where water temperatures can reach upwards of 40°C (Ong *et al.*, 2010). In a direct comparison of *Synechococcus* sp. PCC 7002 and *S. elongatus* PCC 7942 engineered for FFA production, *Synechococcus* sp. PCC 7002 produced greater FFA titres and showed improved tolerance to FFAs compared to *S. elongatus* PCC 7942. These two hosts also had different responses to similar genetic modifications. For example, *aas* knockout in *S. elongatus* PCC 7942 resulted in extracellular FFA titres up to 55.6 mg/l, while the same modification in *Synechococcus* sp. PCC 7002 produced approximately 5 mg/l of excreted FFAs (Ruffing, 2014). This suggests that each cyanobacterial host may have different rates of membrane degradation or different rates of FFA excretion. No other studies have directly compared cyanobacterial hosts for FFA production using the same experimental conditions and measurement techniques. Further characterization of FFA production and genetic tool development in cyanobacterial hosts is needed to optimize host selection for the final process conditions.

FFAs have demonstrated anti-microbial properties (Desbois and Smith, 2010). Engineering the production of a toxic metabolite is a challenge for any metabolic engineering effort and any host. FFAs may have detrimental effects on the microbial host by intercalating into the cellular membranes, disrupting membrane integrity and functions. Furthermore, unsaturated FFAs are highly susceptible to peroxidation by reactive oxygen species, generating toxic degradation products (Desbois and Smith, 2010). *E. coli* strains engineered for FFA production have shown signs of membrane stress, with an 85% reduction in viable cell counts, loss of inner membrane integrity, and up-regulation of phage shock proteins (Lennen *et al.*, 2011). Similarly, cyanobacteria engineered for FFA production have shown reduced cell growth, decreased photosynthetic yields, degradation and change in localization of photosynthetic pigments, change in fatty acid membrane composition, loss of cell membrane integrity, and up-regulation of stress response genes (Ruffing, 2013b; Ruffing and Jones, 2012).

FFA toxicity may be more pronounced in cyanobacteria than heterotrophs such as *E. coli* due to the disruption of photosynthesis via effects on thylakoid membrane integrity. In many cases, additional metabolic engineering for increased FFA production led to higher FFA production on a per cell basis but lower overall FFA yields due to reduced cell growth and viability of the cyanobacterial host (Kaiser *et al.*, 2013; Ruffing, 2013b). Thus, FFA toxicity must be taken into account in the development of metabolic engineering strategies for FFA production. Some strains, such as *Synechococcus* sp. PCC 7002, have shown improved physiological tolerance to FFA production, yet this tolerance was shown to be temperature dependent, with higher cultivation temperatures (38°C) resulting in toxic effects compared to lower temperatures (30°C) (Ruffing, 2014). Temperature dependent FFA toxicity may be associated with a thermodynamic increase in FFA production or increased solubility of FFAs. It is important to note that exogenous FFA toxicity differs considerably from toxicity associated with intracellular FFA synthesis. *S. elongatus* PCC 7942 was shown to be resistant to 5 μM of exogenous polyunsaturated fatty acid (α-linolenic acid) while *Synechococcus* sp. PCC 7002 was only resistant to exogenous polyunsaturated fatty acid concentrations less than 1 μM (Ruffing and Trahan, 2014). The opposite physiological response was observed with intracellular FFA production, with engineered *Synechococcus* sp. PCC 7002 producing higher FFA titres and showing less toxic effects compared to modified *S. elongatus* PCC 7942 (Ruffing, 2014). A similar phenomenon was observed in FFA-producing *E. coli*, where intracellular FFA production resulted in greater toxicity compared to exogenous FFAs (Lennen *et al.*, 2011). FFA cytotoxicity is a fundamental challenge for microbial FFA production, and both genetic as well as process engineering solutions are being considered (Lennen *et al.*, 2013; Liu *et al.*, 2012).

While much of our understanding of microbial FAB is derived from studies of *E. coli*, fundamental genetic, metabolic, and regulatory differences make the direct application of this knowledge to cyanobacterial hosts a flawed approach. A summary of key genetic differences between *E. coli* and three model cyanobacterial hosts is included in Table 8.2. Cyanobacteria utilize desaturases that act upon lipids for production of unsaturated fatty acids, while *E. coli* fatty acid desaturation occurs during FAB via FabA and FabB acting upon the acyl-ACP (Janßen and Steinbüchel, 2014). *E. coli* contains thioesterases (TesA and TesB) which can cleave fatty acids from either ACP or CoA, while cyanobacteria only contain lipases for fatty acid release. *E. coli* utilizes the β-oxidation pathway for fatty acid degradation (FAD), while this pathway is absent in cyanobacteria. The FabR and FadR regulatory proteins, which regulate both FAB and FAD in *E. coli*, are also lacking in cyanobacteria. In addition to these genomic differences, FFA-producing cyanobacteria have shown different responses than similarly engineered *E. coli* strains. For example, the overexpression of acetyl-CoA carboxylase (*accABCD*) in *E. coli* has improved FFA production in *E. coli*, while no improvement was observed with *accABCD* expression in cyanobacterial hosts (Janßen and Steinbüchel, 2014; Liu *et al.*, 2011b; Ruffing, 2013a). *In vitro* analysis of FAB enzymes from *Synechococcus* sp. PCC 7002 and *E. coli* revealed key kinetic differences among these microbial hosts, with FabH being the rate-limiting enzyme in *Synechococcus* sp. PCC 7002 while FabZ and FabI are rate-limiting in *E. coli* (Kuo and Khosla, 2014). Furthermore, this study highlighted variability in co-factor preference. The FabI of *E. coli* can utilize either NADH or NADPH, yet FabI from *Synechococcus* sp. PCC 7002 can only utilize NADH. Regulatory mechanisms associated with heterotrophy in *E. coli* also play a role in regulating fatty acid pathways, while a homologous catabolite repression system is absent in cyanobacteria. Genetic stability is also variable between *E. coli* and cyanobacteria. Many of the 'model' cyanobacterial hosts are naturally competent, allowing for facile DNA uptake for targeted genetic modification but also potentially enhancing rates of horizontal gene transfer. In cyanobacteria, genome integration is typically used for introducing genetic modifications, as opposed to primarily plasmid-based gene expression in *E. coli*. While genome integration is generally considered to be more stable than plasmid-based modification, the stability of cyanobacterial genome integration has recently been questioned (Jones, 2014). In my laboratory's own experience, we isolated spontaneous mutants of FFA-producing *S. elongatus* PCC 7942, modified with *aas* knockout and *'tesA* expression, with improved growth rates, yet when these mutants were analysed, the genome integrated *'tesA* construct was either found to be missing or to be mutated, making *'tesA* non-functional. Owing to the cytotoxicity of FFAs, genetic mutations that reduce FFA biosynthesis may be naturally selected. Cyanobacteria are also reported to have high genome copy numbers per cell, up to 218 genomes per cell for *Synechocystis* sp. PCC 6803 (Griese *et al.*, 2011). This high genome copy number may allow for increased rates of mutation. Although this topic requires further investigation, there is some evidence to suggest that cyanobacteria may have reduced genetic stability compared to *E. coli*.

Cyanobacterial FFA production presents new challenges with regard to production and scale-up. Most reported studies on cyanobacterial FFA biosynthesis use artificial, low-intensity, and continuous illumination conditions, although the final process conditions will use natural sunlight for renewable production. Sunlight varies considerably with respect to spectral quality when compared with artificial sources, particularly fluorescent lights. Additionally, sunlight can reach peak intensities greater than 2500 μmol photons/m^2/s, yet laboratory efforts have used intensities ranging from only 30 to 160 μmol photons/m^2/s (Table 8.1), more than an order of magnitude lower. The diurnal nature of sunlight also has a significant effect on gene expression and regulation in cyanobacteria (Liu *et al.*, 1995). The only study to compare diurnal and continuous light conditions for FFA-producing cyanobacteria found that diurnal conditions led to a significant reduction in FFA titres, with very little increase in FFA production for *aas* knockout strains of *Synechocystis* sp. PCC 6803 compared with the wild type (Cheah *et al.*, 2015). While this diurnal study used other non-traditional conditions such as bicarbonate feeding which may affect FFA production, it highlights a need for testing under realistic process conditions. Most large-scale FFA production systems will also be subjected to seasonal and daily temperature fluctuations along with the diurnal light cycle. FFA productivities with realistic temperature cycles have yet to be tested with modified cyanobacteria. Along with changing environmental conditions, contamination issues are likely to increase with production scale-up. The engineered cyanobacteria are producing and excreting FFAs into the culture medium, offering a rich carbon source for potential contaminants. Many environmental bacterial species such as staphylococci and lactic acid bacteria readily consume FFAs as a carbon source (Talon *et al.*, 2000). Thus, FFA excretion may make product separation less complicated and reduce the requirement for biomass harvesting and re-growth, but contamination will likely be a major issue with any scale-up effort.

At the current stage of cyanobacterial FFA production, proof-of-concept production has been demonstrated at the laboratory scale, yet many challenges remain to be resolved. These challenges should not be viewed as insurmountable, however. The first effort of cyanobacterial FFA production was reported in 2011 (Liu *et al.*, 2011b). Hence, in just nearly 5 years, our understanding of cyanobacterial FAB and our ability to manipulate these hosts has greatly expanded. The remaining challenges simply emphasize the need for additional fundamental research and metabolic engineering efforts before industrial-scale production can be realized.

Systems biology for FFA production in cyanobacteria

As an essential pathway for membrane biosynthesis, the FAB pathway interfaces with many other carbon flux pathways, requires cellular cofactors (ATP, NADH, and NADPH), and is likely to be strictly controlled by global, growth-associated regulatory networks. Metabolic engineering of FAB for FFA production introduces additional complexity by requiring a balance between cellular growth and FFA production and may also stimulate stress responses in the cyanobacterial host. Due to these multi-scale and cell-wide interactions, systems biology will play a key role in understanding and engineering cyanobacterial FFA production. In this chapter, systems biology is defined as the biomolecular characterization of cellular components including experimental omics analyses, computational modelling, and integrated experimental and computational approaches.

Experimental omics analyses generally include four levels of cellular characterization: genomics, transcriptomics, proteomics, and metabolomics. Genomics investigates cyanobacteria at the DNA level, determining which genes and pathways are present in a specific host. The inclusion of specific pathways or genes in the genome, such as additional stress response genes, or the exclusion of genes, such as FFA consuming pathways, may be used to select cyanobacterial hosts for FFA production. Transcriptomics analyses the level of RNA present in a culture at a given time and under specific conditions. Transcriptomics studies can be used to determine which pathways are active and how gene expression changes in response to environmental conditions or genetic modification, such as those for FFA production. While RNA levels are a critical factor in the production of functional proteins, experimental studies have shown relatively poor correlation between RNA and protein (Vogel and Marcotte, 2012). Therefore, proteomics, the cell-wide analysis of proteins within a culture, is useful for determining protein levels of specific pathways, such as those necessary for FFA production as well as competing pathways. Lastly, metabolomics measures the level of molecular metabolites within a culture. This quantification of metabolites provides an accurate view of potential rate limiting steps in cyanobacterial FFA production. Taken together, these four omics approaches provide a holistic view of cellular components, from DNA to metabolites, and are powerful tools for assessing FFA-producing cyanobacteria and identifying additional targets for metabolic engineering.

Computational modelling is often a critical element for analysing omics data, and it can also be used as a tool for assessing metabolic engineering strategies. While computational modelling is limited by our fundamental understanding of cyanobacterial biology and our ability to accurately identify cellular components, recent advances in systems-level and flux-based experimental analyses, as well as computational approaches, have enabled more accurate simulation and predictive capabilities in cyanobacterial metabolic modelling. These computational tools will likely improve future efforts to engineer FFA production in cyanobacteria.

Omics analyses

As organisms with relatively small genomes 2.5–13.2 Mb (Herdman *et al.*, 1979), cyanobacterial genomics is relatively straightforward and inexpensive given the advancements in next-gen DNA sequencing. Approximately 126 cyanobacterial genomes are available (Shih *et al.*, 2013). This provides a rich dataset for cyanobacterial comparative genomics. As a preliminary assessment, FAB and FAD genes from *E. coli* are compared to the genomes of three model cyanobacteria in Table 8.2. This assessment, along with additional analysis of carbon fixation pathways and stress response mechanisms in cyanobacterial genomes, may provide valuable insight for selecting a cyanobacterial host for FFA production and understanding the possible host responses to FFA biosynthesis. While genomics is now a common and useful tool for cyanobacterial characterization, there are several caveats when using sequenced genomes to guide your metabolic engineering strategy. First, mis-annotations have a high occurrence. As a first pass, most newly sequenced cyanobacterial genomes are annotated using homology-based comparison programs. The stringency of the selection criteria used for this comparison will influence the accuracy of the annotation. Furthermore, many of the original annotations are based on biochemical studies in common laboratory organisms like *E. coli*. As discussed in the previous section, cyanobacterial metabolisms differ significantly from *E. coli*, leading to higher frequencies of mis-annotation. For example, the cyanobacterial acyl-ACP synthetase (*aas*) was originally annotated as a long-chain acyl-CoA ligase, as this is the mechanism for FFA recycling in *E. coli*. Recent biochemical characterizations in cyanobacteria revealed that acyl-ACP, rather than acyl-CoA, is the means for FFA recycling in cyanobacteria (Kaczmarzyk and Fulda, 2010). This mis-annotation effects cofactor balance (ACP vs. CoA) and downstream pathway flux predictions in a computational model of cyanobacterial metabolism. While the lower cost of DNA sequencing now allows for higher genome coverage, sequencing errors still occur, and this is a second caveat. A sequencing error in the genome of *Synechococcus* sp. PCC 7002 was recently discovered using VESPA, a program that facilitates genome annotation through the integration of transcriptomic and proteomic data (Peterson *et al.*, 2012). The predicted open reading frame SYNPCC7002_A2842 was annotated as a pseudogene for glycerol kinase due to a frameshift detected through DNA sequencing, yet both transcript and protein matching to this region of the genome was identified with VESPA. Resequencing of the predicted pseudogene revealed that the frameshift was in fact a sequencing error. While metabolic engineers using genome sequences must be aware of the pitfalls of mis-annotation and sequencing errors, genomics remains a fundamental tool for metabolic analysis, and advanced tools for correcting these errors, such as VESPA, will help to improve the accuracy of cyanobacterial genomes.

Advances in DNA sequencing have also advanced transcriptomics analyses, with RNA-seq largely replacing conventional microarray analysis. While RNA-seq is widely applied to many prokaryotes, the application of RNA-seq for cyanobacterial analysis has several key differences. Conventional methods for RNA extraction, such as commercial RNA isolation kits, often yield poor results both in terms of yield and quality for cyanobacterial RNA isolation. Methods for cyanobacterial RNA isolation vary in the literature and may be misleading. In a comparison of RNA isolation methods for *S. elongatus* PCC 7942 in my laboratory, I found that RNA isolation methods containing guanidine thiocyanate resulted in RNA degradation. This result is surprising given that guanidine thiocyanate typically prevents RNase activity. Moreover, published methods for RNA isolation in cyanobacteria often include guanidine thiocyanate (Pinto *et al.*, 2009); these methods show high yields of cyanobacterial RNA but do not confirm RNA quality. For *S. elongatus* PCC 7942, the most reliable method for RNA isolation in my laboratory was hot acid phenol extraction followed by RNA purification using a commercial kit, initially described by O'Shea's laboratory (Vijayan *et al.*, 2009). Regardless of the method used, both RNA yield and quality should be confirmed prior to conducting RNA-seq or other quantitative RNA analyses. Cyanobacterial RNA-seq analysis is also complicated by the addition of polyA sequences to transcripts in cyanobacteria as a signal for RNA degradation (Rott *et al.*, 2003). The *in vivo* transcription rate of these polyA-tagged transcripts has not been quantified in cyanobacteria. Should these polyA-tagged transcripts be discarded in RNA-seq analysis, or should the polyA tags be removed for transcript quantitation? This remains an open question for cyanobacterial transcriptomics.

RNA-seq has been applied for transcriptomic analysis of FFA-producing cyanobacteria with *S. elongatus* PCC 7942 as host (Ruffing, 2013b). The transcriptome of the wild type was compared to that of two FFA-producing strains, one with knockout of *aas* and one with both *aas* knockout and thioesterase expression. As typical of many transcriptomics experiments, a high number of differentially regulated transcripts (>2-fold change) were identified in the study. In general, oxidative stress response, nitrogen metabolism, and photosystem II (PSII) genes were up-regulated while chlorophyll biosynthesis, PSI photosynthesis, carbon metabolism, and hydrogen metabolism genes were down-regulated. Many of these transcriptomic responses were found to be in agreement with physiological measurements. For example, elevated reactive oxygen species (ROS) were detected along with the upregulation of ROS-degrading genes (Ruffing, 2013b). Reduced chlorophyll-*a* levels were also detected in FFA-producing *S. elongatus* PCC 7942 (Ruffing and Jones, 2012); this is likely triggered by the upregulation of nitrogen metabolism genes, as chlorophyll-*a* degradation is a conserved mechanism of nitrogen scavenging (Hendry *et al.*, 1987). Additionally, imaging of the phycobiliproteins in FFA-producing *S. elongatus* PCC 7942 indicated detachment from the thylakoid membranes; this would lead to the reduced photosynthetic yield measurements and upregulation of PSII genes, as detected through RNA-seq. Transcriptomic analysis of FFA production in *S. elongatus* PCC 7942 helped to provide a link between physiological measurements and genetic response and also revealed insight into global regulation mechanisms associated with FFA production. Based on the RNA-seq results, 15 genetic mutants were constructed in the FFA-producing *S. elongatus* PCC 7942 (Ruffing, 2013b). Interestingly, many of the mutants displayed phenotypes that were contrary to what was expected based on their changes in expression level. Six hypothetical proteins with reduced expression during FFA production were targeted for overexpression with the expectation that the overexpression mutants would have reduced fitness or FFA production. However, four of the six overexpression mutants showed improved cell growth, increased extracellular FFA concentrations, and enhanced photosynthetic yields. Two porin proteins, found to be upregulated by transcriptomics and therefore putative FFA exporters, were targeted for gene knockout. The knockout mutants had increased extracellular FFA concentrations and improved photosynthetic yields, suggesting that they may allow for FFA uptake rather than export. These unexpected phenotypes highlight the importance of experimental investigation in conjunction with transcriptomics. This study also constructed mutants that were found to be in agreement with the transcriptomics analyses. The overexpression of four ROS-degrading enzymes was found to improve cell growth, FFA production, and photosynthetic yields in FFA-producing *S. elongatus* PCC 7942. A summary of the genetic targets with improved FFA production is shown in Table 8.3. Comparative transcriptomics of the RNA-seq results from FFA production in *S. elongatus* PCC 7942 and RNA-seq results from stress conditions in *Synechococcus* sp. PCC 7002 revealed conserved stress responses in cyanobacteria. In addition to the expected upregulation of stress response genes, genes involved in nitrogen metabolism had increased expression, indicating a conserved link between stress response and nitrogen metabolism in cyanobacteria. The down-regulation of carbon and hydrogen metabolism is also a possible conserved cyanobacterial stress response. While this remains the only transcriptomic analysis of FFA production in cyanobacteria, the wide-range of gene targets and pathways affected by FFA production in *S. elongatus* PCC 7942 illustrates the power and need for a systems-level approach.

In addition to traditional transcriptomics studies of changes in gene expression, RNA-seq also provides an opportunity to study cellular non-coding RNAs (ncRNAs). Non-coding, regulatory RNAs are important regulatory components in cyanobacteria. A method known as differential RNA-seq (dRNA-seq) has been developed to assess these small regulatory RNAs, antisense RNAs, and other non-coding RNAs (Borries *et al.*, 2012). dRNA-seq uses a differential exonuclease treatment that enriches primary transcripts by degrading processed RNAs. The application of dRNA-seq to cyanobacteria revealed that only 25–33% of all transcription start sites are associated with annotated genes, indicating a high level of possible regulatory RNAs (Kopf and Hess, 2015). The role of ncRNAs in regulating fatty acid biosynthesis in cyanobacteria has yet to be investigated and remains a potential target for engineering cyanobacterial FFA production.

Proteomic changes during cyanobacterial FFA production also have yet to be studied. However, previous methods developed for cyanobacterial

Table 8.3 Genetic modifications in FFA-producing *S. elongatus* PCC 7942 leading to enhanced extracellular FFA production (Ruffing, 2013b)

Gene target	Gene product	Modification	Increase in extracellular FFA concentration (% relative to wild type)
Synpcc7942_1655	Hypothetical protein	Overexpression	1.94 ± 0.59
Synpcc7942_0122	Hypothetical protein	Overexpression	1.94 ± 0.36
Synpcc7942_0900	Hypothetical protein	Overexpression	1.76 ± 0.49
Synpcc7942_1845	Hypothetical protein	Overexpression	1.92 ± 0.17
Synpcc7942_1214	Glutathione peroxidase	Overexpression	1.53 ± 0.43
Synpcc7942_0437	Glutathione peroxidase	Overexpression	1.59 ± 0.48
Synpcc7942_0801	Superoxide dismutase	Overexpression	1.91 ± 0.63
Synpcc7942_1656	Catalase/peroxidase	Overexpression	1.84 ± 0.47
Synpcc7942_1464	Porin	Knockout	1.50 ± 0.30
Synpcc7942_1607	Porin; major outer membrane protein	knockout	1.55 ± 0.43

proteomics can be readily applied (Ow and Wright, 2009). Cyanobacterial proteomics differs from other bacterial applications in that effective protein extraction methods are more similar to those used in plant proteomics. Furthermore, the high abundance of phycobiliproteins in cyanobacterial protein extracts may suppress the identification of other low abundance proteins (Ow and Wright, 2009). Proteomic studies of cyanobacteria exposed to exogenous biofuels, such as ethanol, butanol, and hexane, have shown the up-regulation of transporters, photosynthesis, carbon fixation, and ROS-degrading proteins (Pei *et al.*, 2014). The transcriptional changes in photosynthesis and ROS-degrading proteins resulting from FFA production in *S. elongatus* PCC 7942 must still be investigated at the protein level, yet these may be conserved cyanobacterial stress responses to lipophilic biofuels.

Metabolomics measures intracellular metabolites through the coupling of high-throughput separation techniques, such as capillary electrophoresis (CE) and liquid chromatography (LC), and small molecule identification techniques, such as mass spectrometry (MS) and nuclear magnetic resonance (NMR). Cyanobacterial metabolomics can be challenging due to complications with sample preparation and metabolite identification. Solvent extraction methods for metabolomics must be optimized for each host, and the extraction solvent for cyanobacteria such as *Synechococcus* sp. PCC 7002 have been found to differ significantly from those used for typical Gram-negative bacteria like *E. coli* (Bennette *et al.*, 2011). Additionally, cyanobacteria respond rapidly to changes in illumination, leading to metabolite changes during centrifugation due to dark conditions. Fast filtration with illumination and temperature control can be used to prevent these metabolite changes (Schwarz *et al.*, 2013). The identification of unknown metabolites in cyanobacteria can also be challenging. For example, only 50–102 metabolites with identified chemical structures are reported to be available for *Synechocystis* sp. PCC 6803 using conventional metabolomics (Narainsamy *et al.*, 2013; Schwarz *et al.*, 2013). In optimizing metabolomics techniques for *Synechocystis* sp. PCC 6803, a pentafluorophenylpropyl-(PFPP)-LC/MS method was found to be more effective than the widely used C_{18}-ultra high-performance liquid chromatography (UHPLC)/MS system (Narainsamy *et al.*, 2013). In addition to these traditional metabolomics techniques, several high-throughput, single cell techniques have been applied to analyse metabolites in cyanobacteria, including synchrotron radiation Fourier transform infrared (SR-FTIR) spectromicroscopy and Raman spectroscopy (Hu *et al.*, 2013; Wu *et al.*, 2011). These high-throughput, single cell techniques do not provide the molecular-level resolution of metabolite identification afforded by more traditional methods, but they can be used for rapid screening of mutants with variations in fatty acid unsaturation and chain length (Wu *et al.*, 2011). Metabolite analysis for the fatty acid biosynthesis pathway is particularly challenging due to the large number of intermediate chain lengths and the fact that the acyl chain is bound to a protein (ACP) (Lennen and Pfleger, 2012). Traditional analysis of fatty acid composition involves the conversion of fatty acids, including membrane fatty acids, into FAMEs or FAEEs, followed by GC/MS analysis. Recently, a GC/MS method for simultaneous detection of FFAs, FAMEs, FAEEs, fatty alcohols, and alkanes was developed for *Synechocystis* sp. PCC 6803 (Guan *et al.*, 2011). The advancement of lipidomics may offer further capabilities for detecting and quantifying lipid species. Natural lipid diversity is estimated to include between 10,000 to 100,000 different lipid structures, and many challenges remain to be resolved, including high purity extraction methods, detection of low abundance lipids, double bond identification, and elucidation of complex lipid structures (Wenk, 2010). Fortunately, standard GC/MS analysis is typically sufficient for analysing the FFAs produced by cyanobacteria (Cheah *et al.*, 2015; Liu *et al.*, 2011b; Ruffing and Jones, 2012), yet other methods are necessary to quantify cellular intermediates of FAB and the complex cellular lipids which may be degraded to produce FFAs. Dynamic changes in FFAs and lipids resulting from diurnal light conditions also remain to be investigated.

Metabolic modelling and data integration

Systems-level computational models are useful tools for analysing metabolic fluxes and regulatory networks to validate and characterize metabolic pathways and to select metabolic engineering targets for improving FFA production. To date, the cyanobacterium most extensively studied via computational modelling is *Synechocystis* sp. PCC 6803 (Knoop *et al.*, 2013; Steuer *et al.*, 2012). Due to the limited availability of kinetic and regulatory data for cyanobacteria, stoichiometric metabolic models are often the most informative type of systems-level computational modelling. Cyanobacterial metabolic models are subject to the same challenges as models for other organisms, namely errors in genome annotation, inconsistency in naming of metabolic intermediates and enzymes, enzymes with unclear specificity, metabolites that can exist in different protonation states, and enzymes with unknown cofactor requirements (Steuer *et al.*, 2012). Cyanobacterial metabolic networks are further complicated by the influence of light harvesting and photosynthesis, the circadian clock, and carbon concentrating mechanisms. Despite these challenges, cyanobacterial metabolic models have been instrumental in elucidating metabolic pathways. For example, the existence of a cyanobacterial glyoxylate shunt was analysed via metabolic modelling. The glyoxylate shunt consists of isocitrate lyase and malate synthase, converting isocitrate into malate. The existence of the glyoxylate shunt in cyanobacteria has been debated due to the lack of genetic homologues in cyanobacterial genomes along with conflicting reports of measured enzymatic activities. By using a metabolic reconstruction network for *Synechocystis* sp. PCC 6803 along with experimental measurements, the glyoxylate shunt was shown to be absent in *Synechocystis* sp. PCC 6803 (Knoop *et al.*, 2013). Cyanobacterial metabolic models have also been applied to identify metabolic engineering targets for the production of metabolites, including biofuels such as ethanol and butanol (Nogales *et al.*, 2013). This effort demonstrated that redirecting carbon flux in *Synechocystis* sp. PCC 6803 is more difficult under autotrophic growth as compared to either mixotrophic or heterotrophic growth. Experimental verification of this result must still be tested, however. Additionally, it is unclear whether this constraint would hold true for obligate photoautotrophs. While metabolic modelling has yet to be applied for engineering cyanobacterial FFA production, metabolic models for *Synechocystis* sp. PCC 6803 have been developed with both FAB and photosynthetic lipid biosynthesis pathways (Nogales *et al.*, 2012).

Metabolic models are often used in conjunction with experimental measurement of carbon fluxes, also known as flux balance analysis (FBA). Traditional FBA assumes that internal metabolite pools are stationary, making it difficult to apply FBA to organisms with diurnal metabolic changes. A diurnal metabolic model was recently constructed for *Synechocystis* sp. PCC 6803 by defining cellular objectives as a function of time and light availability (Knoop *et al.*, 2013). This diurnal model includes elevated expression of transport processes and TCA cycle activity with glycogen metabolism to drive respiration and ATP production at night. To match experimental data, a factor for photosynthetic pigments increases 2 h before sunrise and is reduced after noon, while a factor for glycogen storage increases at noon. These constrains allow for simulation of diurnal flux conditions with FBA. The inclusion of isotopic labels in the carbon source for an organism can provide additional insight into internal flux distributions within the cell, as the labelling pattern of metabolites can be used to distinguish between carbon fluxes of alternative pathways for metabolite production. The application of isotopic labelling in cyanobacteria is challenging for autotrophic growth, as labelled bicarbonate or CO_2 results in a uniform labelling pattern. A new method, isotopic non-stationary 13C metabolic flux analysis (INST-MFA), was developed in *Synechocystis* sp. PCC 6803 to analyse autotrophic growth by measuring dynamic changes of labelling within the metabolite pools (Young *et al.*, 2011). Both FBA and INST-MFA are powerful tools for understanding metabolic flux distributions and may be used to identify new metabolic targets for improving cyanobacterial FFA production.

Metabolic network models and genomic models are important tools for the analysis and integration of omics data. As transcriptomic, proteomic, and metabolomic datasets continue to increase due to more efficient and lower cost omics technologies, additional insights can be gained through computational integration and analysis of these multi-scale datasets. Preliminary applications of this technology, such as VESPA, have used transcriptomic and proteomic data to identify sequencing and annotation errors in cyanobacterial genomes (Peterson *et al.*, 2012). More recently, combined metabolic and expression models have been developed for cyanobacteria which utilize experimental expression data along with metabolic reconstruction to expand the computational model to include cellular regulation (Saha *et al.*, 2016). In the future, more complex algorithms and programs may allow for the integration of all levels of cellular control: genetic, transcriptomic, proteomic, and metabolomics in a single computational model for a holistic prediction tool to guide strain development efforts.

Conclusions and future trends

Cyanobacteria are promising hosts for the biosynthesis for FFAs, as demonstrated by the previous accomplishments in this area. However, our understanding of cyanobacterial FAB and FAD limits our ability to modify these organisms for high levels of FFA production. Omics technologies provide powerful tools for analysing cyanobacterial FFA production at the systems level, yet to date, there is only a single transcriptomics study investigating FFA-producing cyanobacteria (Ruffing, 2013b). Future studies will likely include additional transcriptomics studies along with proteomics and metabolomics. In turn, computational tools for omics analysis and metabolic flux analysis will play an important role in analysing these rich datasets and applying them to (1) better understand cyanobacterial FAB and FAD pathways and their regulation within each cyanobacterial host and (2) identify gene candidates for additional biochemical studies and as targets for improving cyanobacterial FFA production. The true benefit of omics technologies will only be realized within the context of rigorous experimental design and analysis. Common pitfalls, such as a low number of biological replicates and suboptimal isolation methods, must be avoided to prevent erroneous interpretation of omics data. The application of lipidomics may help to understand the physiological effects of FFA production on the cyanobacterial host, as preliminary data suggests changes in membrane saturation (Ruffing, 2013b).

Future targets for improving cyanobacterial FFA production will likely include mechanisms for avoiding toxicity, such as the identification and overexpression of transporters or the engineering of membrane components. The regulatory mechanisms controlling FAB in cyanobacteria remain to be identified. Differential RNA-seq may be a useful tool for identifying small regulatory RNAs that contribute to cyanobacterial fatty acid metabolism. Once determined, these regulatory components are also potential targets for metabolic engineering.

More rigorous testing of FFA-producing cyanobacteria may also improve FFA yields. As illustrated in Table 8.1, a wide variety of methods have been used to both grow the cyanobacterial strains and measure the FFA product. As a result, strain comparisons between publications or between laboratories are challenging, calling for the development of community standards for both growth and FFA measurement. Additionally, most FFA-producing cyanobacteria are proposed for industrial applications that would rely on sunlight as the source of illumination, yet most studies are performed at low light intensities and under continuous light. A preliminary study in *Synechocystis* sp. PCC 6803 suggests that FFA accumulation is significantly diminished under diurnal light conditions (Cheah *et al.*, 2015); however, even this study used conditions that are not relevant for industrial applications: low light intensity and bicarbonate feeding rather than CO_2 bubbling. Equipment developed to mimic light, temperature, and CO_2 conditions that may be experienced under industrial-scale production will be useful for testing of FFA-producing cyanobacterial strains. Lastly, as FFA-producing cyanobacterial strains are scaled up, bacterial contamination may become a problem. The FFA-producing cyanobacteria are excreting a rich carbon source (i.e. FFAs) which naturally occurring bacteria can metabolize for growth. Therefore, methods for controlling contamination, either through genetic modification or chemical addition, may need to be developed.

Web resources

Table 8.4 provides a list of web resources for lipid analysis, metabolic engineering, omics, and computational tools and includes links for each resource.

Table 8.4 Online resources for lipid analysis, metabolic engineering, omics, and computational tools

Resource	Link	Description
Lipid resource		
LIPID MAPS	http://www.lipidmaps.org/	A lipidomics resource that includes lipid standards, experimental data, databases, pathways, tools, tutorials, and more
Metabolic engineering resources		
OptForce	http://maranas.che.psu.edu/submission/optforce.htm	A computational program for determining a minimum set of gene additions and deletions to optimize production of a biochemical target
OptKnock	http://maranas.che.psu.edu/submission/optknock_2.htm	A computational program for selecting gene deletions for production of a biochemical target
Omics resources		
Cyanobase	http://genome.microbedb.jp/cyanobase	Contains genomic information about sequenced cyanobacteria. At the time of this publication, however, only 39 of the 126 sequenced cyanobacterial genomes are included
CyanoEXpress	http://cyanoexpress.sysbiolab.eu/	A web database for exploration of transcriptomic data from public repositories for *Synechocystis* sp. PCC 6803
CyanOmics	http://lag.ihb.ac.cn/cyanomics	A database for omics studies of *Synechococcus* sp. PCC 7002
VESPA	http://omictools.com/vespa-tool	A desktop application for integration of proteomic, transcriptomic, and genomic information with tools for identifying regions of mis-annotation
Computational resources		
openCOBRA Project	https://opencobra.github.io/	Access to core COBRA toolbox and a repository for community contributed modules that build off of core COBRA features
KBase	https://kbase.us/	Software and data platform for systems biology to enable the prediction and design of biological function
INCA	http://mfa.vueinnovations.com/licensing/mfa-inca	Isotopomer Network Compartmental Analysis can simulate both steady-state and transient isotope labelling experiments
PathPred	http://www.genome.jp/tools/pathpred/	A web-based pathway prediction tool for determining feasible enzymatic routes for chemical degradation or biosynthesis

References

Atsumi, S., Higashide, W., and Liao, J.C. (2009). Direct photosynthetic recycling of carbon dioxide to isobutyraldehyde. Nature Biotechnol. *27*, 1177–1180. http://dx.doi.org/10.1038/nbt.1586.

Bennette, N.B., Eng, J.F., and Dismukes, G.C. (2011). An LC-MS-based chemical and analytical method for targeted metabolite quantification in the model cyanobacterium *Synechococcus* sp. PCC 7002. Anal. Chem. *83*, 3808–3816. http://dx.doi.org/10.1021/ac200108a.

Bergler, H., Fuchsbichler, S., Högenauer, G., and Turnowsky, F. (1996). The enoyl-[acyl-carrier-protein] reductase (FabI) of *Escherichia coli*, which catalyzes a key regulatory step in fatty acid biosynthesis, accepts NADH and NADPH as cofactors and is inhibited by palmitoyl-CoA. Eur. J. Biochem. *242*, 689–694. http://dx.doi.org/10.1111/j.1432-1033.1996.0689r.x.

Borries, A., Vogel, J., and Sharma, C.M. (2012). Differential RNA sequencing (dRNA-Seq): Deep-sequencing-based analysis of primary transcriptomes. In Tag-Based Next Generation Sequencing, Borries, A., Vogel, J., and Sharma, C.M. eds. (Weinheim, Berlin, Wiley-VCH Verlag GmbH & Co.), pp. 109–121. http://dx.doi.org/10.1002/9783527644582.ch7.

Cheah, Y.E., Zimont, A.J., Lunka, S.K., Albers, S.C., Park, S.J., Reardon, K.F., and Peebles, C.A. (2015). Diel light: dark cycles significantly reduce FFA accumulation in FFA producing mutants of *Synechocystis* sp. PCC 6803 compared to continuous light. Algal Res. *12*, 487–496. http://dx.doi.org/10.1016/j.algal.2015.10.014.

Davis, M.S., and Cronan, J.E. (2001). Inhibition of *Escherichia coli* acetyl coenzyme A carboxylase by acyl-acyl carrier protein. J. Bacteriol. *183*, 1499–1503. http://dx.doi.org/10.1128/JB.183.4.1499-1503.2001.

Davis, M.S., Solbiati, J., and Cronan, J.E. (2000). Overproduction of acetyl-CoA carboxylase activity increases the rate of fatty acid biosynthesis in *Escherichia coli*. J. Biol. Chem. *275*, 28593–28598. http://dx.doi.org/10.1074/jbc.M004756200.

Dellomonaco, C., Clomburg, J.M., Miller, E.N., and Gonzalez, R. (2011). Engineered reversal of the β-oxidation cycle for the synthesis of fuels and chemicals. Nature *476*, 355–359. http://dx.doi.org/10.1038/nature10333.

Desbois, A.P., and Smith, V.J. (2010). Antibacterial free fatty acids: activities, mechanisms of action and biotechnological potential. Appl. Microbiol. Biotechnol. *85*, 1629–1642. http://dx.doi.org/10.1007/s00253-009-2355-3.

Gao, Q., Wang, W., Zhao, H., and Lu, X. (2012). Effects of fatty acid activation on photosynthetic production of fatty acid-based biofuels in *Synechocystis* sp. PCC6803. Biotechnol. Biofuels 5, 17. http://dx.doi.org/10.1186/1754-6834-5–17.

Griese, M., Lange, C., and Soppa, J. (2011). Ploidy in cyanobacteria. FEMS Microbiol. Lett. *323*, 124–131. http://dx.doi.org/10.1111/j.1574-6968.2011.02368.x.

Guan, W., Zhao, H., Lu, X., Wang, C., Yang, M., and Bai, F. (2011). Quantitative analysis of fatty-acid-based biofuels produced by wild-type and genetically engineered cyanobacteria by gas chromatography-mass spectrometry. J. Chromatogr. A. *1218*, 8289–8293. http://dx.doi.org/10.1016/j.chroma.2011.09.043.

Heath, R.J., and Rock, C.O. (1996a). Regulation of fatty acid elongation and initiation by acyl-acyl carrier protein in *Escherichia coli*. J. Biol. Chem. *271*, 1833–1836. http://dx.doi.org/10.1074/jbc.271.4.1833.

Heath, R.J., and Rock, C.O. (1996b). Roles of the FabA and FabZ β-hydroxyacyl-acyl carrier protein dehydratases in *Escherichia coli* fatty acid biosynthesis. J. Biol. Chem. *271*, 27795–27801. http://dx.doi.org/10.1074/jbc.271.44.27795.

Hendry, G.A., Houghton, J.D., and Brown, S.B. (1987). The degradation of chlorophyll—a biological enigma. New Phytol. *107*, 255–302. http://dx.doi.org/10.1111/j.1469-8137.1987.tb00181.x.

Herdman, M., Janvier, M., Rippka, R., and Stanier, R.Y. (1979). Genome size of cyanobacteria. Microbiology UK *111*, 73–85. http://dx.doi.org/10.1099/00221287-111-1-73.

Hu, P., Borglin, S., Kamennaya, N.A., Chen, L., Park, H., Mahoney, L., Kijac, A., Shan, G., Chavarría, K.L., and Zhang, C. (2013). Metabolic phenotyping of the cyanobacterium *Synechocystis* 6803 engineered for production of alkanes and free fatty acids. Appl. Energy *102*, 850–859. http://dx.doi.org/10.1016/j.apenergy.2012.08.047.

Huang, H.H., Camsund, D., Lindblad, P., and Heidorn, T. (2010). Design and characterization of molecular tools for a synthetic biology approach towards developing cyanobacterial biotechnology. Nucleic Acids Res. *38*, 2577–2593. http://dx.doi.org/10.1093/nar/gkq164.

Janßen, H.J., and Steinbüchel, A. (2014). Fatty acid synthesis in *Escherichia coli* and its applications towards the production of fatty acid based biofuels. Biotechnol. Biofuels *7*, 1–26. http://dx.doi.org/10.1186/1754-6834-7-7.

Jones, P.R. (2014). Genetic instability in cyanobacteria– an elephant in the room? Front. Bioeng. Biotechnol. 2, 12. http://dx.doi.org/10.3389/fbioe.2014.00012.

Kaczmarzyk, D., and Fulda, M. (2010). Fatty acid activation in cyanobacteria mediated by acyl-acyl carrier protein synthetase enables fatty acid recycling. Plant Physiol. *152*, 1598–1610. http://dx.doi.org/10.1104/pp.109.148007.

Kaiser, B.K., Carleton, M., Hickman, J.W., Miller, C., Lawson, D., Budde, M., Warrener, P., Paredes, A., Mullapudi, S., and Navarro, P. (2013). Fatty aldehydes in cyanobacteria are a metabolically flexible precursor for a diversity of biofuel products. PLoS One *8*, e58307. http://dx.doi.org/10.1371/journal.pone.0058307.

Kawahara, A., Sato, Y., Saito, Y., Kaneko, Y., Takimura, Y., Hagihara, H., and Hihara, Y. (2016). Free fatty acid production in the cyanobacterium *Synechocystis* sp. PCC 6803 is enhanced by deletion of the *cyAbrB2* transcriptional regulator. J. Biotechnol. *220*, 1–11. http://dx.doi.org/10.1016/j.jbiotec.2015.12.035.

Knoop, H., Gründel, M., Zilliges, Y., Lehmann, R., Hoffmann, S., Lockau, W., and Steuer, R. (2013). Flux balance analysis of cyanobacterial metabolism: the metabolic network of *Synechocystis* sp. PCC 6803. PLoS Comput. Biol. *9*, e1003081. http://dx.doi.org/10.1371/journal.pcbi.1003081.

Kopf, M., and Hess, W.R. (2015). Regulatory RNAs in photosynthetic cyanobacteria. FEMS Microbiol. Rev. *39*, 301–315. http://dx.doi.org/10.1093/femsre/fuv017.

Kuo, J., and Khosla, C. (2014). The initiation ketosynthase (FabH) is the sole rate-limiting enzyme of the fatty acid synthase of *Synechococcus* sp. PCC 7002. Metab. Eng. *22*, 53–59. http://dx.doi.org/10.1016/j.ymben.2013.12.008.

Lee, S.K., Chou, H., Ham, T.S., Lee, T.S., and Keasling, J.D. (2008). Metabolic engineering of microorganisms for biofuels production: from bugs to synthetic biology to fuels. Curr. Opin. Biotechnol. *19*, 556–563. http://dx.doi.org/10.1016/j.copbio.2008.10.014.

Lem, N.W., and Stumpf, P.K. (1984). In vitro fatty acid synthesis and complex lipid metabolism in the cyanobacterium *Anabaena variabilis*: I. Some characteristics of fatty acid synthesis. Plant Physiol. *74*, 134–138. http://dx.doi.org/10.1104/pp.74.1.134.

Lennen, R.M., Kruziki, M.A., Kumar, K., Zinkel, R.A., Burnum, K.E., Lipton, M.S., Hoover, S.W., Ranatunga, D.R., Wittkopp, T.M., and Marner, W.D. (2011). Membrane stresses induced by overproduction of free fatty acids in *Escherichia coli*. Appl. Environ. Microbiol. 77, 8114–8128. http://dx.doi.org/10.1128/AEM.05421-11.

Lennen, R.M., and Pfleger, B.F. (2012). Engineering *Escherichia coli* to synthesize free fatty acids. Trends Biotechnol. *30*, 659–667. http://dx.doi.org/10.1016/j.tibtech.2012.09.006.

Lennen, R.M., Politz, M.G., Kruziki, M.A., and Pfleger, B.F. (2013). Identification of transport proteins involved in free fatty acid efflux in *Escherichia coli*. J. Bacteriol. *195*, 135–144. http://dx.doi.org/10.1128/JB.01477-12.

Liu, H., Yu, C., Feng, D., Cheng, T., Meng, X., Liu, W., Zou, H., and Xian, M. (2012). Production of extracellular fatty acid using engineered *Escherichia coli*. Microb. Cell Fact. *11*, 41. http://dx.doi.org/10.1186/1475-2859-11-41.

Liu, X., and Curtiss, R. (2009). Nickel-inducible lysis system in *Synechocystis* sp. PCC 6803. Proc. Natl. Acad. Sci. U.S.A. *106*, 21550–21554. http://dx.doi.org/10.1073/pnas.0911953106.

Liu, X., Fallon, S., Sheng, J., and Curtiss, R. (2011a). CO_2-limitation-inducible green recovery of fatty acids from cyanobacterial biomass. Proc. Natl. Acad. Sci. U.S.A. *108*, 6905–6908. http://dx.doi.org/10.1073/pnas.1103016108.

Liu, X., Sheng, J., and Curtiss III, R. (2011b). Fatty acid production in genetically modified cyanobacteria. Proc.

Natl. Acad. Sci. U.S.A. *108*, 6899–6904. http://dx.doi.org/10.1073/pnas.1103014108.

Liu, Y., Tsinoremas, N.F., Johnson, C.H., Lebedeva, N.V., Golden, S.S., Ishiura, M., and Kondo, T. (1995). Circadian orchestration of gene expression in cyanobacteria. Genes. Dev. *9*, 1469–1478.

Los, D.A., and Murata, N. (1998). Structure and expression of fatty acid desaturases. Biochim. Biophys. Acta *1394*, 3–15.

Marchetti, J., and Errazu, A. (2008). Esterification of free fatty acids using sulfuric acid as catalyst in the presence of triglycerides. Biomass Bioenerg. *32*, 892–895. http://dx.doi.org/10.1016/j.biombioe.2008.01.001.

Markley, A.L., Begemann, M.B., Clarke, R.E., Gordon, G.C., and Pfleger, B.F. (2015). Synthetic biology toolbox for controlling gene expression in the cyanobacterium *Synechococcus* sp. strain PCC 7002. ACS Synth. Biol. *4*, 595–603. http://dx.doi.org/10.1021/sb500260k.

Murata, N., Wada, H., and Gombos, Z. (1992). Modes of fatty-acid desaturation in cyanobacteria. Plant Cell Physiol. 33, 933–941.

Narainsamy, K., Cassier-Chauvat, C., Junot, C., and Chauvat, F. (2013). High performance analysis of the cyanobacterial metabolism via liquid chromatography coupled to a LTQ-Orbitrap mass spectrometer: evidence that glucose reprograms the whole carbon metabolism and triggers oxidative stress. Metabolomics *9*, 21–32. http://dx.doi.org/10.1007/s11306-011-0382-4.

Nogales, J., Gudmundsson, S., Knight, E.M., Palsson, B.O., and Thiele, I. (2012). Detailing the optimality of photosynthesis in cyanobacteria through systems biology analysis. Proc. Natl. Acad. Sci. U.S.A. *109*, 2678–2683. http://dx.doi.org/10.1073/pnas.1117907109.

Nogales, J., Gudmundsson, S., and Thiele, I. (2013). Toward systems metabolic engineering in cyanobacteria: opportunities and bottlenecks. Bioengineered *4*, 158–163. http://dx.doi.org/10.4161/bioe.22792.

Okazaki, K., Sato, N., Tsuji, N., Tsuzuki, M., and Nishida, I. (2006). The significance of C16 fatty acids in the *sn*-2 positions of glycerolipids in the photosynthetic growth of *Synechocystis* sp. PCC6803. Plant Physiol. *141*, 546–556. http://dx.doi.org/10.1104/pp.105.075796.

Ong, S.C., Kao, C.Y., Chiu, S.Y., Tsai, M.T., and Lin, C.S. (2010). Characterization of the thermal-tolerant mutants of *Chlorella* sp. with high growth rate and application in outdoor photobioreactor cultivation. Bioresour. Technol. *101*, 2880–2883. http://dx.doi.org/10.1016/j.biortech.2009.10.007.

Ow, S.Y., and Wright, P.C. (2009). Current trends in high throughput proteomics in cyanobacteria. FEBS Lett. *583*, 1744–1752. http://dx.doi.org/10.1016/j.febslet.2009.03.062.

Pei, G., Chen, L., Wang, J., Qiao, J., and Zhang, W. (2014). Protein network signatures associated with exogenous biofuels treatments in cyanobacterium *Synechocystis* sp. PCC 6803. Front. Bioeng. Biotechnol. 2. 48. http://dx.doi.org/10.3389/fbioe.2014.00048.

Peterson, E.S., McCue, L.A., Schrimpe-Rutledge, A.C., Jensen, J.L., Walker, H., Kobold, M.A., Webb, S.R., Payne, S.H., Ansong, C., and Adkins, J.N. (2012). VESPA: software to facilitate genomic annotation of prokaryotic organisms through integration of proteomic and transcriptomic data. BMC Genomics *13*, 131. http://dx.doi.org/10.1186/1471-2164-13-131.

Pinto, F.L., Thapper, A., Sontheim, W., and Lindblad, P. (2009). Analysis of current and alternative phenol based RNA extraction methodologies for cyanobacteria. BMC Mol. Biol. *10*, 79. http://dx.doi.org/10.1186/1471-2199-10-79.

Rae, B.D., Long, B.M., Badger, M.R., and Price, G.D. (2013). Functions, compositions, and evolution of the two types of carboxysomes: polyhedral microcompartments that facilitate CO_2 fixation in cyanobacteria and some proteobacteria. Microbiol. Mol. Biol. Rev. 77, 357–379. http://dx.doi.org/10.1128/MMBR.00061-12.

Rott, R., Zipor, G., Portnoy, V., Liveanu, V., and Schuster, G. (2003). RNA polyadenylation and degradation in cyanobacteria are similar to the chloroplast but different from *Escherichia coli*. J. Biol. Chem. *278*, 15771–15777. http://dx.doi.org/10.1074/jbc.M211571200.

Ruffing, A.M. (2013a). Borrowing genes from *Chlamydomonas reinhardtii* for free fatty acid production in engineered cyanobacteria. J. Appl. Phycol. 25, 1495–1507. http://dx.doi.org/10.1007/s10811-013-9993-7.

Ruffing, A.M. (2013b). RNA-Seq analysis and targeted mutagenesis for improved free fatty acid production in an engineered cyanobacterium. Biotechnol. Biofuels *6*, 113. http://dx.doi.org/10.1186/1754-6834-6-113.

Ruffing, A.M. (2014). Improved free fatty acid production in cyanobacteria with *Synechococcus* sp. PCC 7002 as host. Front. Bioeng. Biotechnol. *2*, 17. http://dx.doi.org/10.3389/fbioe.2014.00017.

Ruffing, A.M., and Jones, H.D. (2012). Physiological effects of free fatty acid production in genetically engineered *Synechococcus elongatus* PCC 7942. Biotechnol. Bioeng. *109*, 2190–2199. http://dx.doi.org/10.1002/bit.24509.

Ruffing, A.M., and Trahan, C.A. (2014). Biofuel toxicity and mechanisms of biofuel tolerance in three model cyanobacteria. Algal Res. *5*, 121–132. http://dx.doi.org/10.1016/j.algal.2014.07.006.

Saha, R., Liu, D., Hoynes-O'Connor, A., Liberton, M., Yu, J., Bhattacharyya-Pakrasi, M., Balassy, A., Zhang, F., Moon, T.S., Maranas, C.D., *et al.* (2016). Diurnal regulation of cellular processes in the cyanobacterium *Synechocystis* sp. strain PCC 6803: Insights from transcriptomic, fluxomic, and physiological analyses. mBio 7, e00464–16. http://dx.doi.org/10.1128/mBio.00464-16.

Schirmer, A., Rude, M.A., Li, X., Popova, E., and del Cardayre, S.B. (2010). Microbial biosynthesis of alkanes. Science *329*, 559–562. http://dx.doi.org/10.1126/science.1187936.

Schwarz, D., Orf, I., Kopka, J., and Hagemann, M. (2013). Recent applications of metabolomics toward cyanobacteria. Metabolites 3, 72–100. http://dx.doi.org/10.3390/metabo3010072.

Sheehan, J., Dunahay, T., Benemann, J., and Roessler, P. (1998). A Look Back at the US Department of Energy's Aquatic Species. Close-Out Report National Renewable Energy Laboratory, US Department of Energy1s Office of Fuels Development, Golden, Colorado, USA.

Shih, P.M., Wu, D., Latifi, A., Axen, S.D., Fewer, D.P., Talla, E., Calteau, A., Cai, F., Tandeau de Marsac, N., Rippka, R., *et al.* (2013). Improving the coverage of the cyanobacterial phylum using diversity-driven genome sequencing.

Proc. Natl. Acad. Sci. U.S.A. *110*, 1053–1058. http://dx.doi.org/10.1073/pnas.1217107110.

Singh, S.C., Sinha, R.P., and Hader, D.P. (2002). Role of lipids and fatty acids in stress tolerance in cyanobacteria. Acta Protozool. *41*, 297–308. http://dx.doi.org/

Steen, E.J., Kang, Y., Bokinsky, G., Hu, Z., Schirmer, A., McClure, A., Del Cardayre, S.B., and Keasling, J.D. (2010). Microbial production of fatty-acid-derived fuels and chemicals from plant biomass. Nature *463*, 559–562. http://dx.doi.org/10.1038/nature08721.

Steuer, R., Knoop, H., and Machné, R. (2012). Modelling cyanobacteria: from metabolism to integrative models of phototrophic growth. J. Exp. Bot. *63*, 2259–2274. http://dx.doi.org/10.1093/jxb/ers018.

Sukenik, A., Bennett, J., and Falkowski, P. (1987). Light-saturated photosynthesis – limitation by electron transport or carbon fixation? Biochim. Biophys. Acta *891*, 205–215. http://dx.doi.org/10.1016/0005-2728(87)90216-7.

Talon, R., Walter, D., and Montel, M.C. (2000). Growth and effect of staphylococci and lactic acid bacteria on unsaturated free fatty acids. Meat Sci. *54*, 41–47. http://dx.doi.org/10.1016/S0309-1740(99)00068-6.

Tan, X., Yao, L., Gao, Q., Wang, W., Qi, F., and Lu, X. (2011). Photosynthesis driven conversion of carbon dioxide to fatty alcohols and hydrocarbons in cyanobacteria. Metab. Eng. *13*, 169–176. http://dx.doi.org/10.1016/j.ymben.2011.01.001.

Vijayan, V., Zuzow, R., and O'Shea, E.K. (2009). Oscillations in supercoiling drive circadian gene expression in cyanobacteria. Proc. Natl. Acad. Sci. U.S.A. *106*, 22564–22568. http://dx.doi.org/10.1073/pnas.0912673106.

Voelker, T.A., and Davies, H.M. (1994). Alteration of the specificity and regulation of fatty acid synthesis of *Escherichia coli* by expression of a plant medium-chain acyl-acyl carrier protein thioesterase. J. Bacteriol. *176*, 7320–7327.

Vogel, C., and Marcotte, E.M. (2012). Insights into the regulation of protein abundance from proteomic and transcriptomic analyses. Nat. Rev. Genet. *13*, 227–232. http://dx.doi.org/10.1038/nrg3185.

Wada, H., and Murata, N. (1998). Membrane lipids in cyanobacteria. In Lipids in Photosynthesis: Structure, Function and Genetics, S. Paul-André, and N. Murata, eds. (Dordrecht: Springer Netherlands), pp. 65–81. http://dx.doi.org/10.1007/0-306-48087-5_4.

Wada, H., and Murata, N. (2007). The essential role of phosphatidylglycerol in photosynthesis. Photosynth. Res. *92*, 205–215. http://dx.doi.org/10.1007/s11120-007-9203-z.

Wahlen, B.D., Willis, R.M., and Seefeldt, L.C. (2011). Biodiesel production by simultaneous extraction and conversion of total lipids from microalgae, cyanobacteria, and wild mixed-cultures. Bioresour. Technol. *102*, 2724–2730. http://dx.doi.org/10.1016/j.biortech.2010.11.026.

Weier, D., Müller, C., Gaspers, C., and Frentzen, M. (2005). Characterisation of acyltransferases from *Synechocystis* sp. PCC6803. Biochem. Biophys. Res. Commun. *334*, 1127–1134.

Wenk, M.R. (2010). Lipidomics: new tools and applications. Cell *143*, 888–895. http://dx.doi.org/10.1016/j.cell.2010.11.033.

Work, V.H., Melnicki, M.R., Hill, E.A., Davies, F.K., Kucek, L.A., Beliaev, A.S., and Posewitz, M.C. (2015). Lauric acid production in a glycogen-less strain of *Synechococcus* sp. PCC 7002. Front. Bioeng. Biotechnol. 3, 48. http://dx.doi.org/10.3389/fbioe.2015.00048.

Wu, H., Volponi, J.V., Oliver, A.E., Parikh, A.N., Simmons, B.A., and Singh, S. (2011). In vivo lipidomics using single-cell Raman spectroscopy. Proc. Natl. Acad. Sci. U.S.A. *108*, 3809–3814. http://dx.doi.org/10.1073/pnas.1009043108.

Wu, J.T., Chiang, Y.R., Huang, W.Y., and Jane, W.N. (2006). Cytotoxic effects of free fatty acids on phytoplankton algae and cyanobacteria. Aquat. Toxicol. *80*, 338–345. http://dx.doi.org/10.1016/j.aquatox.2006.09.011.

Young, J.D., Shastri, A.A., Stephanopoulos, G., and Morgan, J.A. (2011). Mapping photoautotrophic metabolism with isotopically nonstationary (*13*)C flux analysis. Metab. Eng. *13*, 656–665. http://dx.doi.org/10.1016/j.ymben.2011.08.002.

Zess, E.K., Begemann, M.B., and Pfleger, B.F. (2016). Construction of new synthetic biology tools for the control of gene expression in the cyanobacterium *Synechococcus* sp. strain PCC 7002. Biotechnol. Bioeng. *113*, 424–432. http://dx.doi.org/10.1002/bit.25713.

Zhang, F., Ouellet, M., Batth, T.S., Adams, P.D., Petzold, C.J., Mukhopadhyay, A., and Keasling, J.D. (2012). Enhancing fatty acid production by the expression of the regulatory transcription factor FadR. Metab. Eng. *14*, 653–660. http://dx.doi.org/10.1016/j.ymben.2012.08.009.

Terpene Hydrocarbons Production in Cyanobacteria

9

Anastasios Melis

Department of Plant and Microbial Biology, University of California, Berkeley, CA, USA.

Correspondence: melis@berkeley.edu

Abstract

A description is given on the development and state of the art of technologies for the renewable generation of terpene hydrocarbons to serve as renewable fuel for transportation, and chemicals for the synthetic chemistry industry, with cyanobacteria as the primary photocatalyst and isoprene (C_5H_8) and monoterpene ($C_{10}H_{16}$) hydrocarbons as the paradigm outputs. This approach integrates pivotal aspects of the photosynthetic biofuels concept:

a efficient sunlight utilization by cyanobacterial mass cultures;

b effective delivery and assimilation of quantities of CO_2 to fast-growing cyanobacteria;

c paradigm of synthesis, spontaneous product separation from the biomass, and isoprene and -phellandrene (monoterpene) sequestration and harvesting;

d application of novel low-cost photobioreactors for the successful and cost-effective implementation of (a) through (c); and

e evaluation of isoprene (C_5H_8) and monoterpene ($C_{10}H_{16}$) hydrocarbons as fuel molecules, compared to alkanes and alcohols.

The effort advocates the development and application of designer cyanobacterial strains, which can grow in mass culture under bright sunlight conditions and demonstrate efficient sunlight conversion and utilization. Metabolic engineering approaches endow and enhance the yield of isoprene and monoterpene hydrocarbons generated in photosynthesis, relative to the biomass accumulated. Gaseous/aqueous two-phase photobioreactors are used for efficient conversion of sunlight and CO_2 into isoprene or monoterpenes, molecules that are potential biofuels and highly valuable industrial chemicals. Following isoprene or monoterpene sequestration, the residual cyanobacterial biomass is used as feedstock of bacterial fermentations to generate biogas (a mixture of CH_4 and CO_2), with the residual biomass serving as bio-fertilizer. In terms of Energy Density, terpene hydrocarbons are favourably compared to alkanes, and are substantially better than similar size alcohols. The promise of a cyanobacterial 'Photosynthesis-to-Fuels' approach is the ability to transform the primary products of photosynthesis directly, in a single cyanobacterial cell, into commodity products for human industrial and domestic consumption.

Introduction: state of the art

The concept of 'photosynthetic biofuels', as envisaged and pioneered by Lindberg *et al.* (2010) entails the direct application of photosynthesis for the generation of fuel and chemicals, in a process where a single organism acts both as catalyst and processor, synthesizing and secreting ready to use product. Research in this field has reduced this concept to practice with specific photosynthetic microorganisms (e.g. cyanobacteria) and terpene hydrocarbon products (e.g. C_5H_8 isoprene, $C_{10}H_{16}$ β-phellandrene, $C_{15}H_{24}$ bisabolene, and even squalene $C_{30}H_{50}$). Process concepts in this Photosynthetic Biofuels approach are based on foundation knowledge in molecular, cellular and metabolic biology, biochemistry, biophysics, environmental science, and the engineering disciplines. An authoritative review on terpenoids and their biosynthesis appeared, authored by Pattanaik and Lindberg (2015).

The following integrated and interdisciplinary objectives have been achieved:

1 Heterologous expression of an isoprene synthase gene (*ISPS*) in cyanobacteria enabled the renewable generation of volatile isoprene (C_5H_8) hydrocarbons, derived entirely from sunlight, carbon dioxide (CO_2) and water (H_2O) (Lindberg *et al.*, 2010; Bentley and Melis 2012; Bentley *et al.*, 2014; Gao *et al.*, 2016; Pade *et al.*, 2016).
2 Heterologous expression of a β-phellandrene synthase or limonene synthase gene (*PHLS* or *LMNS*, respectively) in cyanobacteria enabled the renewable generation of liquid monoterpene hydrocarbons, also derived from sunlight, carbon dioxide (CO_2) and water (H_2O) (Bentley *et al.*, 2013; Davies *et al.*, 2014; Formighieri and Melis 2014, 2015, 2016).
3 Heterologous expression of sesquiterpene or triterpene synthases in cyanobacteria enabled the renewable generation of 15-carbon β-caryophyllene (Reinsvold *et al.*, 2011), bisabolene (Davies *et al.*, 2014), and squalene (Englund *et al.*, 2014).
4 Molecular and genetic approaches have been applied by different investigators in the pursuit of terpene hydrocarbon production objectives, aiming to improve the achievable terpene-to-biomass carbon–partitioning ratio (Melis 2013). Preliminary work on these questions has pointed out blueprints of organism genetic and metabolic engineering, by which to improve the performance of photosynthesis and also to amplify cellular metabolic flux towards the desirable high-impact terpene hydrocarbons (Zurbriggen *et al.*, 2012; Bentley *et al.*, 2014; Formighieri and Melis 2016).
5 Importantly, cyanobacteria with a truncated light-harvesting antenna size (TLA-strains) have been developed and applied in mass culture under bright sunlight conditions to demonstrate efficient sunlight utilization and improved biomass productivity (Kirst *et al.*, 2014).
6 The principle of Bentley and Melis (2012) has been applied for the effective delivery of large quantities of CO_2 to fast-growth cyanobacteria, which are capable of consuming up to 130 g CO_2/m^2/day under mass culture and bright sunlight conditions.
7 Low-cost, modular, and easy-to-deploy photobioreactors can be used to demonstrate high rates of biomass accumulation and product generation. Photobioreactor design concepts and gaseous/aqueous two-phase conditions for feeding CO_2 and capturing isoprene or monoterpenes demonstrate the potential for future scalable deployment from both an economic and environmental perspective (Melis, 2012).

The integrated process of generating isoprene (C_5H_8) and monoterpene ($C_{10}H_{16}$) hydrocarbons via photosynthesis, and biogas from the accumulated biomass could serve as a case study in the development of technologies for the renewable generation of other fuels and useful chemicals.

Perspective

The quest for fuel and chemicals to serve human industrial and domestic consumption can be met by the emerging science that combines photosynthesis, metabolism, and engineering. Progress has been made, although some fundamental barriers are still encountered in the scale-up of sustainable photobiological processes for the production of fuels. As stated above, interdisciplinary research has brought together five different technologies: (i) innovative principles for efficient utilization of sunlight and CO_2 by photosynthetic cyanobacteria, (ii) innovative and scalable engineered systems (photobioreactors) with high volumetric productivity of phototrophic microorganisms, (iii) novel chemicals (e.g. isoprene and other terpene hydrocarbons) produced by metabolically engineered cyanobacteria, affording example of a direct conversion of CO_2 to fuel, eliminating steps associated with growth of the biomass, cell harvesting, dewatering, deconstructing and fermenting of the resultant sugars in order to arrive at a desirable product, (iv) implementation of a sustainable bioprocess, particularly with respect to water and nutrient use in cyanobacterial cultures, CO_2 mitigation, and residual microbial biomass utilization for manufacturing additional products, and (v) biogas generation.

A variety of products can be generated by cyanobacteria, including commercially useful terpene hydrocarbons (see Chapters 8 and 10). In this chapter, useful molecules to be emphasized include the hemiterpene isoprene, a C_5H_8 volatile hydrocarbon, and a variety of $C_{10}H_{16}$ liquid-at-room temperature monoterpenes. Chemical formulae of the exemplified hemi- and mono-terpene compounds are shown in the schematic of Fig. 9.1.

Whereas isoprene has a distinct and unique chemical formula, monoterpenes come is a variety of shapes and forms, including acyclic monoterpenes (e.g. ocimene, myrcene), monocyclic (e.g. limonene, phellandrene, terpinene, terpinolene), and bicyclic (e.g. carene, camphene, pinene, sabinene). Cyclization of the 10-carbon monoterpene molecule affects their stereochemistry and configuration, their physicochemical properties, and especially so their density in the liquid phase, as summarized in Table 9.1. Briefly, the cyclization process results in the condensation of the volume occupied by the molecule, thereby increasing their relative density in the liquid phase. Monoterpene synthases of different plant origin are responsible for the catalytic reactions converting a geranyl-diphosphate (GPP) molecule into an acyclic, monocyclic, or bicyclic monoterpene (Bohlmann *et al.*, 1999; Lange *et al.*, 2000; Lichtenthaler 2000; Keeling *et al.*, 2011; Keeling and Bohlmann 2012; Demissie *et al.*, 2012).

Isoprene (2-methyl-1,3-butadiene; C_5H_8) is the monomeric five-carbon building block of isoprenoids or terpenoids, comprising the largest and most diverse group of naturally occurring organic compounds on earth. This five-carbon backbone is biologically important as the building block of more than 50,000 different naturally occurring compounds, the vast majority of which are products of the so-called 'plant secondary metabolism'. Isoprene can be used as feedstock in the synthetic chemistry industry, where it is used to manufacture products ranging from rubber to adhesives and perfumes. The monomeric isoprene units can be developed as a renewable fuel useful in direct applications similar to that of other volatile hydrocarbons, e.g. butane and pentane (Table 9.2). Owing to its physicochemical properties, isoprene with a boiling point of +34°C, can be an additive to natural gas for heating or electricity generation purposes, or it can serve as a fuel in the transportation

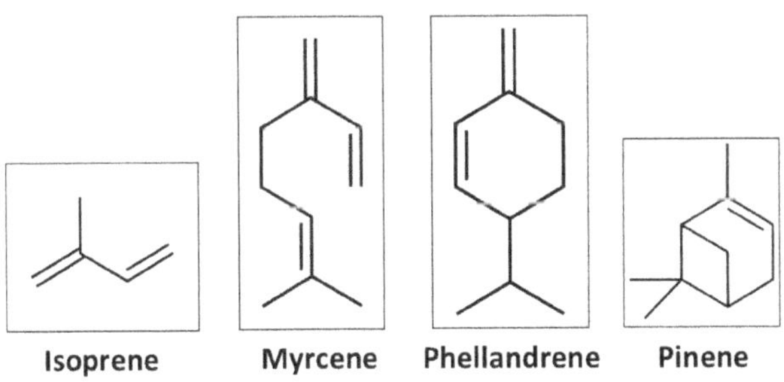

Figure 9.1 Chemical formulae of the hemiterpene isoprene (C_5H_8), the acyclic β-myrcene, the monocyclic β-phellandrene, and the bicyclic β-pinene ($C_{10}H_{16}$) monoterpene hydrocarbons.

Table 9.1 Density of monoterpenes (β-myrcene, β-phellandrene, and β-pinene) in liquid-phase measured at sea level and ambient temperature. Monoterpene cyclization lowers the space occupied by the molecule and correspondingly increases the density of the compound in the liquid phase

Monoterpene	Density (g/cm^3)
β-Myrcene (acyclic)	0.794
β-Phellandrene (monocyclic)	0.845
β-Pinene (bicyclic)	0.860

Table 9.2 Physicochemical properties of selected putative hydrocarbon fuels. Data were obtained from PubChem (https://pubchem.ncbi.nlm.nih.gov/compound/)

Hydrocarbon	Chemical formula	Molecular mass (g/mol)	Boiling point (°C)	H/C ratio (a:a)	Energy density (MJ/kg)
Methane	CH_4	16	–161	4:1	55.8
Ethane	C_2H_6	30	–128	3:1	52.0
Propane	C_3H_8	44	–42	2.67:1	50.3
Butane	C_4H_{10}	58	0.5	2.5:1	49.2
Isoprene	C_5H_8	68	34	1.6:1	43.8
Pentane	C_5H_{12}	72	36	2.41	48.6
Phellandrene (monoterpene)	$C_{10}H_{16}$	136	175	1.6:1	45.4
Decane	$C_{10}H_{22}$	142	174	2.2:1	47.7

industry upon blending with other volatile hydrocarbons such as propane and butane. Currently, the industrial source of isoprene is limited to fossil petroleum distillation. Sustained availability of fossil resources is questionable due to increasing global demand for energy and synthetic chemistry feedstock, and also because of environmental and climate concerns. Accordingly, there is a need to develop methods for creating alternative and renewable fuels and synthetic chemistry feedstock, such as isoprene.

Monoterpenes ($C_{10}H_{16}$) contain two isoprene units per molecule and have commercial value as key ingredients in synthetic chemistry, medical, cosmetic and cleaning products, and potentially as fuel (Bentley *et al.*, 2013; Formighieri and Melis, 2015). They are components of plant essential oils, naturally synthesized in plant trichomes from geranyl-diphosphate (GPP), or neryl-diphosphate, by nuclear-encoded and plastid localized monoterpene synthase enzymes. Biochemical modification of monoterpenes, such as hydroxylation or oxidation results in the generation of a variety of related monoterpenoids. In cyanobacteria, production of the monoterpenes β-phellandrene and limonene was achieved by genetically engineering *Synechocystis* PCC 6803 and *Synechococcus* PCC 7002, respectively, endowing upon these cells the heterologous synthesis, and spontaneous and quantitative separation of the monoterpene molecules from the biomass and their further separation from the extracellular aqueous phase (Bentley and Melis, 2013; Formighieri and Melis, 2014; Davies *et al.*, 2014). β-Phellandrene was shown to efficiently diffuse through the cell's plasma membrane and cell wall and, because of its hydrophobicity, it accumulated as a floater molecule on the surface of the liquid culture, from where it was harvested by siphoning (Bentley *et al.*, 2013; Formighieri and Melis 2014). Physicochemical properties of monoterpenes, including the relatively low –75°C melting point and the relatively high +175°C boiling point (Table 9.2) make these molecules excellent biogasoline substitutes and a good drop-in fuel.

Biogas can be generated as the byproduct of bacterial fermentations via the breakdown of bioorganic matter in the absence of oxygen (Zamalloa *et al.*, 2012; Bolzonella *et al.*, 2012; Hermann *et al.*, 2016; Passos *et al.*, 2016). Biogas generation takes advantage of the cyanobacterial biomass that would inevitably accumulate during the process of terpene hydrocarbons production. In this case, methanogenic bacteria perform the anaerobic digestion of the biodegradable cyanobacterial biomass to generate biogas. The latter comprises primarily methane (50–75%) and carbon dioxide (25–50%) but may also contain small amounts of hydrogen (H_2) and hydrogen sulfide (H_2S). Cyanobacterial biomass is superior to agricultural residue for biogas generation, as the former comprises compounds with a much higher hydrogen/carbon (H/C) and lower oxygen/carbon (O/C) ratios than the latter. Biogas applications include generation of electricity and fuel for heating (gas boiler) purposes, usage as fuel in internal combustion engines after compression, and bio-methane conversion (a natural gas supplement) following enrichment in CH_4 and removal of CO_2 from the primary biogas products.

The process

Cyanobacteria have emerged among the most promising biological organisms for renewable product generation applications because they combine the efficient sunlight-harvesting and CO_2-assimilating ability of photosynthesis with the high yields of controlled microbial cultivation. These properties make cyanobacteria a potentially valuable resource for environmentally and economically sustainable industrial-scale production processes. Cyanobacteria have the potential to generate considerably greater amounts of biomass, including protein, carbohydrate, and lipids per hectare than terrestrial plants, and can be cultivated on marginal lands, avoiding competition with food or industrial crops. They also possess and operate the methylerythritol-phosphate pathway for the generation of a variety of terpenoids naturally required for cell growth and development.

The prokaryotic bacterial origin methylerythritol-4-phosphate (MEP) biosynthetic pathway that operates in cyanobacteria (Rohmer *et al.*, 1993, 1996; Lange *et al.*, 2000; Lichtenthaler 2000; Lombard and Moreira 2011) accounts for the synthesis of all natural terpenoids in these cells. The MEP pathway is closely linked to cellular photosynthetic activity because it uses pyruvate and glyceraldehyde 3-phosphate (GA 3-P) as the primary substrates, and further utilizes reducing equivalents (reduced ferredoxin, NADPH) and cellular energy (CTP, ATP) for the synthesis of the cellular terpenoids. All of these reactants and cellular energy are derived from photosynthesis (Sharkey and Yeh 2001; Sharkey *et al.*, 2008). Isoprene and monoterpenes are also products of the MEP pathway that are naturally made in the trichomes of the leaf and stems of many terrestrial plants, but can be heterologously synthesized in cyanobacteria. Isoprene is derived directly from dimethylallyl-diphosphate upon expression of an isoprene synthase (*ISPS*) transgene. Monoterpenes, e.g. β-phellandrene, can be derived directly from geranyl-diphosphate (GPP) upon expression of a β-phellandrene synthase (*PHLS*) transgene in cyanobacteria (Fig. 9.2).

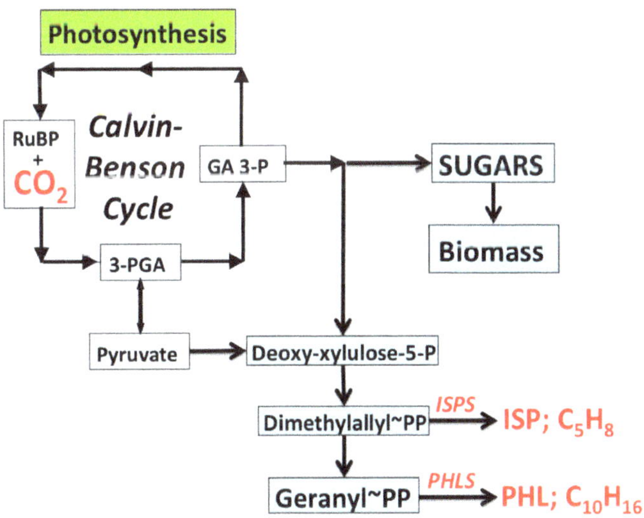

Figure 9.2 Photosynthetic carbon flux and partitioning in the terpenoid biosynthetic pathway of transformant cyanobacteria. Assimilated CO_2 yields 3-phosphoglyceric acid (3-PGA), which is converted into glyceraldehyde 3-phosphate (GA 3-P) or pyruvate. These are the primary reactants of the methylerythritol phosphate (MEP) biosynthetic pathway for the synthesis of deoxy-xylulose-5-P, leading to isopentenyl-diphosphate (IPP, not shown) and its isomer, dimethylallyl diphosphate (DMAPP) from which isoprene is derived. Covalent linkage of IPP to DMAPP yields geranyl-diphosphate (GPP), a 10-carbon terpenoid intermediate metabolite, which serves as the reactant for the heterologous synthesis of monoterpenes. Geranyl-diphosphate is naturally used for the synthesis of phytol, carotenoids, quinone prenyl tails, among others, needed for cell growth and development. Heterologous expression of the *ISPS* and *PHLS* genes in cyanobacteria drains a portion of the DMAPP and GPP pool towards the synthesis of isoprene and β-phellandrene, respectively. These products spontaneously diffuse out of the cyanobacterial cell and escape from the liquid culture. (Adapted from Lindberg *et al.*, 2010.)

Solar-to-product energy conversion efficiency and photosynthetic productivity

Yield of the targeted bio-compounds is a most significant parameter affecting commercialization of a cyanobacterial production process. The challenge of endogenous carbon partitioning towards the desirable product (isoprene or monoterpenes) was recently addressed (Melis 2013). In addition, the photosynthetic solar energy conversion efficiency of the cyanobacterial mass culture under bright sunlight conditions is of import (Melis 2009; Liberton *et al.*, 2013; Ort *et al.*, 2013). Cyanobacteria are endowed with a large phycobilisome–chlorophyll antenna structure, which functions in the thylakoid membrane, containing up to 850 phycocyanin, allophycocyanin and chlorophyll molecules per photosynthetic electron transport chain. This large antenna size has evolved to absorb sunlight efficiently in the natural aquatic ecotype, where cyanobacteria live, and where light intensities are typically very low, thus limiting growth and productivity. Cyanobacteria, microalgae, and all plants need to compete with each other for the absorption of sunlight, and this is achieved by having evolved a large light-harvesting chlorophyll antenna size for enhanced light absorption. On the other hand, in commercial exploitation situations, where a monoculture is cultivated under direct and bright sunlight, a large antenna size can cause excessive-absorption of irradiance. Under these conditions, excess absorbed photons are dissipated (lost) as fluorescence or heat. This phenomenon is very pronounced in high-density cultures, where most of the irradiance is intercepted by the top layer of cells in the reactor, leading to excessive heat dissipation, and leaving cells deeper in the culture literally in the dark (Fig. 9.3A). Such optical pitfall is counter to the requirement of efficiently utilizing all absorbed irradiance, and it causes losses in sunlight utilization efficiency and culture productivity (Melis, 2009). Under such conditions, a cyanobacteria culture can dissipate more than 80% of absorbed photons (Fig. 9.3A), thereby lowering the potential productivity of the culture to unacceptably low levels. Diminishing the pigment content of the phycobilisome by

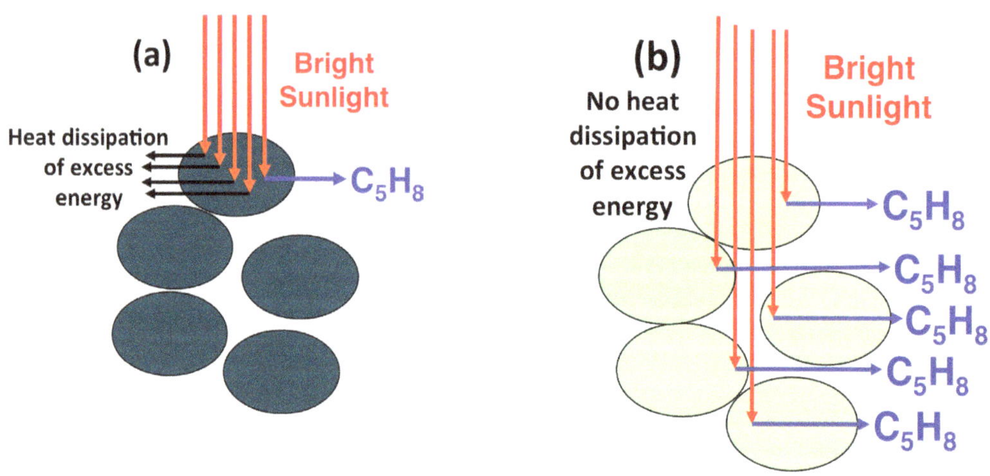

Figure 9.3 Schematic depiction of the pathway for incident sunlight absorption and processing in the photosynthetic apparatus of cyanobacteria. Following absorption of incoming sunlight, excitation energy is either converted and stored into chemical energy via photochemical utilization and product generation (C_5H_8), or dissipated as heat, when absorbed sunlight energy is in excess of what can be utilized in photosynthesis. (a) Fully pigmented (dark blue-green) cyanobacterial cells at the surface of the high-density culture over-absorb incoming sunlight (more than can be utilized by photosynthesis), and wastefully dissipate most of it in the form of heat, thereby compromising productivity. (b) Truncated phycobilisome antenna cyanobacteria (greenish TLA stains result from the loss of the dominant blue phycocyanin) in a high-density mass culture. Individual TLA cells have a diminished capacity of absorbing sunlight, thereby permitting greater penetration and a more uniform distribution and utilization of irradiance through the depth of the culture. Such optical properties prevent over-absorption of sunlight and the ensuing heat dissipation and result in enhancements in photosynthetic productivity and terpene (C_5H_8) production by the culture as a whole. (Adapted from Melis, 2009.)

deletion of the phycocyanin encoding genes (Kirst *et al.*, 2014) was shown to be a successful approach by which to alleviate the over-absorption and ensuing wasteful dissipation of energy, enabling deeper penetration of sunlight, enabling photosynthesis by cells deeper in the culture, and increasing the overall productivity of a cyanobacterial high-density culture (Fig. 9.3B) (Melis, 2009; Kirst *et al.*, 2014).

Principles of CO_2 delivery and terpene hydrocarbons sequestration

A novel gaseous–aqueous two-phase photobioreactor was designed, specifically to enable high rates of photosynthesis, delivering sufficient amounts of CO_2 so as not to impede the rate of photosynthesis, while at the same time enabling effective sequestration of the accumulating terpene hydrocarbons (Bentley and Melis, 2012; see also Fig. 9.4). The experimental approach was to inoculate the photobioreactor with the transformant cyanobacteria in the liquid phase, occupying about half of the reactor volume, followed by filling of the headspace (gaseous phase) with a stream of 100% CO_2. In the course of photosynthesis, CO_2 is spontaneously drawn from the headspace (via the power of diffusion) to the liquid phase, as needed, where it is assimilated in support of the growth and productivity of the cyanobacteria. High rates of photosynthesis and doubling of the cyanobacterial biomass every 24 h were recorded with such an approach (Bentley and Melis, 2012).

As discussed by Bentley and Melis (2012) this novel method avoids continuous bubbling of the culture (expensive and wasteful) and relies only on periodic product removal from the headspace and refilling of the gaseous phase with fresh CO_2. The headspace of the photobioreactor is otherwise maintained in a sealed and resting state, and the process relies on a diffusion-driven uptake of CO_2 from the gaseous phase and its exchange with O_2

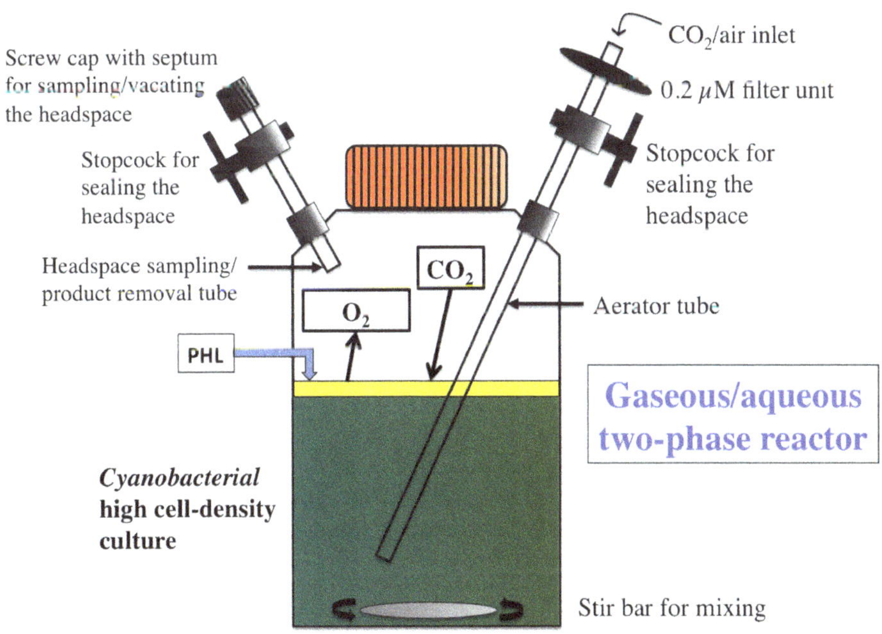

Figure 9.4 Cross-sectional view of a 1-litre photobioreactor bottle, showing the gaseous/aqueous two-phase configuration and function. The schematic depicts a 40:60 partition between the gaseous and aqueous phases. The aqueous phase with the growth medium is inoculated with the transformant cyanobacteria, whereas the gaseous phase is slowly filled with a stream of 100% CO_2. In a sealed reactor, this configuration permits spontaneous diffusion-driven CO_2 uptake from the gaseous phase and C-assimilation by the cells in the liquid phase. Consumed CO_2 is replaced, in nearly stoichiometric ratios, by O_2 emanating from photosynthesis. Whereas isoprene is a volatile product that accumulates in the gaseous phase, monoterpenes (e.g. β-phellandrene, PHL) are liquid at room temperature (Table 9.2) and accumulate as floater molecules on the surface of the liquid culture (Adapted from Bentley and Melis, 2012).

and the corresponding terpene hydrocarbon. A continuous process of biomass accumulation and terpene hydrocarbon generation was thus achieved. Periodic removal (harvesting) of the biomass and terpene hydrocarbons, followed by replacement with fresh growth media and supplemental CO_2 enabled such a continuous photosynthesis and production process to occur.

Efficacy of terpene hydrocarbons as biofuels

A convenient and reliable way to assess the efficacy of organic compounds as potential fuels is from the energy density of the molecule. The energy density of pure hydrocarbons (Table 9.2) can be described as a linear function of the atomic hydrogen-to-carbon (H/C) ratio of the molecule (Fig. 9.5). It is seen that methane (CH_4) with the highest H/C ratio of 4:1 has the highest energy density among all hydrocarbons, equal to about 56 MJ/kg. Longer chain alkanes (ethane, propane, butane, pentane and decane) have a gradually declining H/C ratio which, in decane, is equal to 2.2 : 1 (Table 9.2), translating into energy density of 47.7 MJ/kg (Fig. 9.5). Terpene hydrocarbons with unsaturated C=C bonds have a lower H content and lower H/C ratio, which is equal to 1.6:1 for both isoprene and monoterpenes (Table 9.2). Accordingly, the energy density of these hydrocarbons is lower than that of the equivalent alkanes, with an average of about 44.5 MJ/kg (Fig. 9.5). It is of interest that a linear relationship exists between the energy density of these pure hydrocarbons and the atomic H : C ratio of the respective molecules (Fig. 9.5).

Efficacy of alcohol molecules as biofuels

The same H/C ratio principle can be applied to alcohol molecules, as a way by which to assess their potential to be used as renewable biofuels. Physicochemical properties of selected alcohols, ranging from methanol to decanol are shown in Table 9.3.

A cursory look at the properties of these alcohols (Table 9.3), compared with those of equal carbon length hydrocarbons (Table 9.2), shows the relatively elevated boiling point and the substantially diminished energy density of the alcohol molecules. A plot of the energy density of these alcohols as a function of their atomic H/C ratio is shown in Fig. 9.6. Opposite to the trend in the results of Fig. 9.5, the energy content of alcohols declined as a function of the H/C atomic ratio, from a relative

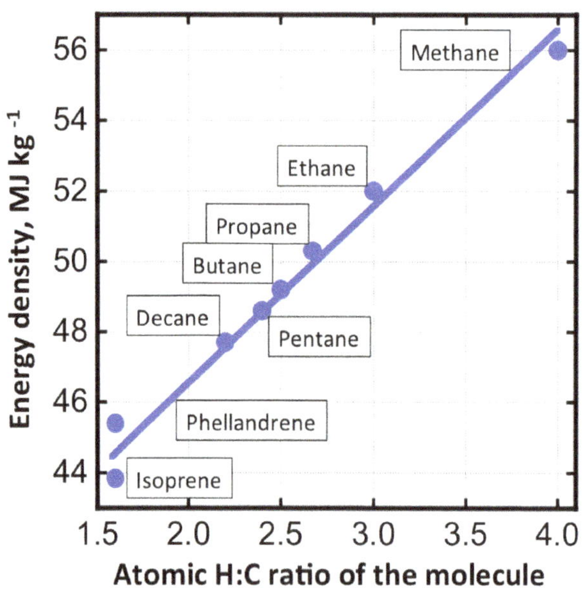

Figure 9.5 Energy density of selected alkanes and terpene hydrocarbons as a function of the hydrogen-to-carbon (H/C) atomic ratio of the respective molecules. These plotted results show that energy density is a linear function of the H/C atomic ratio of the hydrocarbon molecules.

Table 9.3 Physicochemical properties of the alcohol molecules listed were obtained from PubChem (https://pubchem.ncbi.nlm.nih.gov/compound/)

Alcohol	Chemical formula	Molecular mass (g/mol)	Boiling point (°C)	H/O (a:a ratio)	C/O (a a ratio)	H/C (a:a ratio)	Energy density (MJ/kg)
Methanol	CH_3OH	32	65	4	1	4	19.7
Ethanol	CH_3CH_2OH	46	78	6	2	3	26.7
Propanol	$CH_3(CH_2)_2OH$	60	97	8	3	2.67	33.6
Butanol	$CH_3(CH_2)_3OH$	74	118	10	4	2.5	36.1
Pentanol	$CH_3(CH_2)_4OH$	88.2	138	12	5	2.4	37.8
Heptanol	$CH_3(CH_2)_6OH$	116.2	175.8	16	7	2.29	39.9
Decanol	$CH_3(CH_2)_9OH$	158.3	233	22	10	2.2	41.8
Glucose	$C_6H_{12}O_6$	180	N/A	2	1		17.0

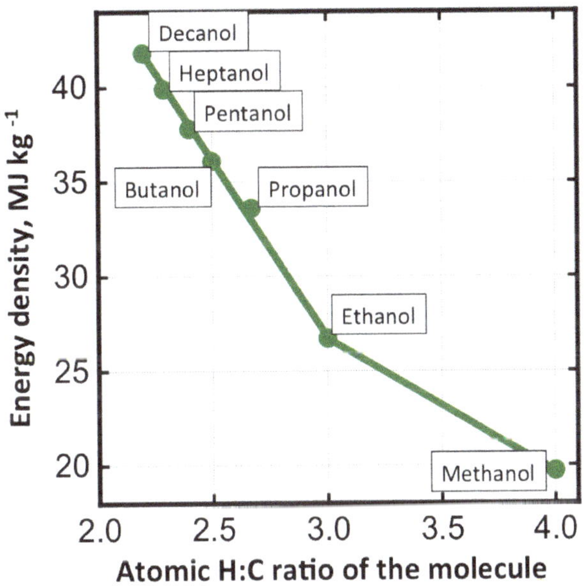

Figure 9.6 Energy density of selected alcohols as a function of the hydrogen-to-carbon (H/C) atomic ratio of the respective molecules. These plotted results show that energy density is a declining function of the H/C atomic ratio of the alcohol molecule.

high of 41.8 MJ/kg for decanol, to 19.7 MJ/kg for methanol. In all cases, the energy density of C_1 to C_{10} alcohols is considerably lower than the energy density of C_1 to C_{10} alkanes or terpenes, substantially attenuated because of the presence of oxygen in the alcohol molecule, which constitutes a partial oxidation, thereby lowering the standard enthalpy of combustion of the alcohol compound.

Oxygen-related losses in the standard enthalpy of combustion of alcohols containing a single hydroxyl (-OH) group are minimized with the length in the hydrocarbon chain of the compound. This is because longer carbon chain alcohol molecules have a higher H/O and C/O ratio (Table 9.3), endowing the compound with a greater standard enthalpy of combustion. This is graphically shown in the plot of Fig. 9.7, where the energy content of alcohols is plotted as a function of the atomic H/O ratio of the molecule. As intuitively known, longer chain alcohols have a higher energy density than smaller sized ones. Among single hydroxyl group (-OH) alcohols, methanol (CH_4OH) has the lowest energy density of 19.7 MJ/kg, which is in sharp contrast to methane (CH_4), the latter having the highest energy density among alkanes (55.8 MJ/kg). A similar result was obtained upon

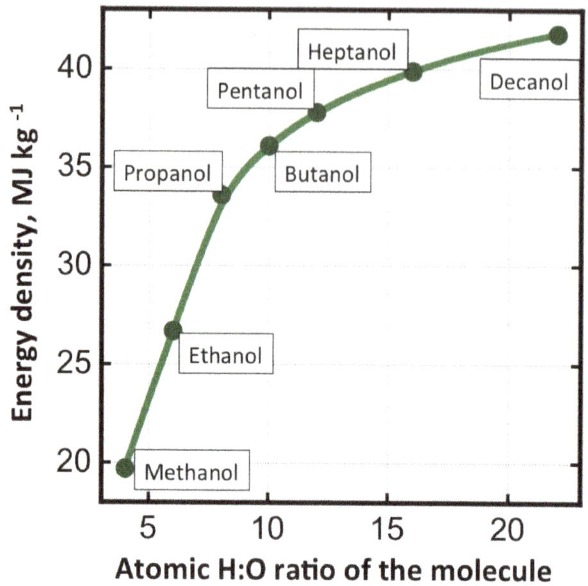

Figure 9.7 Energy density of selected alcohols as a function of the hydrogen-to-oxygen (H/O) atomic ratio of the respective molecules. These plotted results show that energy density increases as a function of the H/O atomic ratio of the alcohol molecule. A similar function was derived upon plotting the energy density of alcohols as a function of the carbon-to-oxygen (C/O) atomic ratio of the respective molecules (not shown).

plotting the energy density of alcohols as a function of the carbon-to-oxygen (C/O) atomic ratio of the respective molecules (results not shown). Interestingly, glucose ($C_6H_{12}O_6$), which is proposed as a feedstock for fermentative ethanol and other biofuels production, has a mere 17 MJ/kg, a property that is often overlooked when advocating sugar-based fermentative biofuels production.

Conclusion

In summary, this chapter showed the different facets of cyanobacterial terpene hydrocarbons production, put in the context of the overall cyanobacterial photosynthetic productivity under ambient sunlight conditions and also in view of the efficacy of these terpenes to serve as fuel molecules, when compared to alkanes and alcohols. The broadly based background of this study, with references to recent and classical original publications, would be useful both for instructional purposes and in the effort to achieve meaningful rates and yields of terpene product formation, leading to a commercially successful application of cyanobacteria as platforms for the generation of commodity products, such as terpene hydrocarbons. A concerted effort is still needed towards cost-efficient modes of mass cultivation and terpene product generation. The further development of genetically modified strains of cyanobacteria with suitably altered metabolism is recommended in this respect. Such strains should be able to accumulate larger amounts of the target compounds, at the expense of biomass accumulation, an achievement that would improve prospects of downstream processing and commercial exploitation.

References

Bentley, F.K., and Melis, A. (2012). Diffusion-based process for carbon dioxide uptake and isoprene emission in gaseous/aqueous two-phase photobioreactors by photosynthetic microorganisms. Biotech. Bioeng. *109*,100–109. http://dx.doi.org/10.1002/bit.23298.

Bentley, F.K., García-Cerdán, J.G., Chen, H.-C., and Melis, A. (2013). Paradigm of monoterpene (β-phellandrene) hydrocarbons production via photosynthesis in cyanobacteria. BioEnergy Res. *6*, 917–929. http://dx.doi.org/10.1007/s12155-013-9325-4.

Bentley, F.K., Zurbriggen, A., and Melis, A. (2014). Heterologous expression of the mevalonic acid pathway in cyanobacteria enhances endogenous carbon partitioning to isoprene. Mol. Plant. *7*, 71–86. http://dx.doi.org/10.1093/mp/sst134.

Bohlmann, J., Phillips, M., Ramachandiran, V., Katoh, S., and Croteau, R. (1999). cDNA cloning, characterization, and functional expression of four new monoterpene synthase members of the Tpsd gene family from grand fir (*Abies grandis*). Arch. Biochem. Biophys. *368*, 232–243. http://dx.doi.org/10.1006/abbi.1999.1332.

Bolzonella, D., Cavinato, C., Fatone, F., Pavan, P., and Cecchi, F. (2012). High rate mesophilic, thermophilic, and temperature phased anaerobic digestion of waste activated sludge: a pilot scale study. Waste Manage. (Oxford) *32*, 1196–1201. http://dx.doi.org/10.1016/j.wasman.2012.01.006.

Davies, F.K., Work, V.H., Beliaev, A.S., and Posewitz, M.C. (2014). Engineering limonene and bisabolene production in wild type and a glycogen-deficient mutant of *Synechococcus* sp. PCC 7002. Front. Bioeng. Biotechnol. *2*, 21. http://dx.doi.org/10.3389/fbioe.2014.00021.

Demissie, Z.A., Sarker, L.S., and Mahmoud, S.S. (2011). Cloning and functional characterization of β-phellandrene synthase from *Lavandula angustifolia*. Planta *233*, 685–696. http://dx.doi.org/10.1007/s00425-010-1332-5.

Englund, E., Pattanaik, B., Ubhayasekera, S.J., Stensjö, K., Bergquist, J., and Lindberg, P. (2014). Production of squalene in *Synechocystis* sp. PCC 6803. PLoS One *9*, e90270. http://dx.doi.org/10.1371/journal.pone.0090270.

Formighieri, C., and Melis, A. (2014). Regulation of β-phellandrene synthase gene expression, recombinant protein accumulation, and monoterpene hydrocarbons production in *Synechocystis* transformants. Planta *240*, 309–324. http://dx.doi.org/10.1007/s00425-014-2080-8.

Formighieri, C., and Melis, A. (2015). A phycocyanin·phellandrene synthase fusion enhances recombinant protein expression and β-phellandrene (monoterpene) hydrocarbons production in *Synechocystis* (cyanobacteria). Metab. Eng. *32*,116–124. http://dx.doi.org/10.1016/j.ymben.2015.09.010.

Formighieri, C., and Melis, A. (2016). Sustainable heterologous production of terpene hydrocarbons in cyanobacteria. Photosynth. Res. – in press. http://dx.doi.org/10.1007/s11120-016-0233-2.

Gao, X., Gao, F., Liu, D., Zhang, H., Nie, X., and Yang, C. (2016). Engineering the methylerythritol phosphate pathway in cyanobacteria for photosynthetic isoprene production from CO_2. Energy Environ. Sci. *9*, 1400–1411. http://dx.doi.org/10.1039/c5ee03102h.

Herrmann, C., Kalita, N., Wall, D., Xia, A., and Murphy, J.D. (2016). Optimised biogas production from microalgae through co-digestion with carbon-rich co-substrates. Bioresour. Technol. *214*, 328–337. http://dx.doi.org/10.1016/j.biortech.2016.04.119.

Keeling, C.I., Weisshaar, S., Ralph, S.G., Jancsik, S., Hamberger, B., Dullat, H.K., and Bohlmann, J. (2011). Transcriptome mining, functional characterization, and phylogeny of a large terpene synthase gene family in spruce (*Picea* spp.). BMC Plant Biol. *11*, 43. http://dx.doi.org/10.1186/1471-2229-11-43.

Keeling, C.I., and Bohlmann, J. (2012). Plant terpenoids. In: N Civjan (ed) Natural Products in Chemical Biology, 1st edn. Wiley, New York. pp. 121–142. http://dx.doi.org/10.1002/9781118391815.fmatter.

Kirst, H., Formighieri, C., and Melis, A. (2014). Maximizing photosynthetic efficiency and culture productivity in cyanobacteria upon minimizing the phycobilisome light-harvesting antenna size. Biochim. Biophys. Acta *1837*, 1653–1664. http://dx.doi.org/10.1016/j.bbabio.2014.07.009.

Lange, B.M., Rujan, T., Martin, W., and Croteau, R. (2000). Isoprenoid biosynthesis: the evolution of two ancient and distinct pathways across genomes. Proc. Natl. Acad. Sci. U.S.A. *97*, 13172–13177. http://dx.doi.org/10.1073/pnas.240454797

Liberton, M., Collins, A.M., Page, L.E., O'Dell, W.B., O'Neill, H., Urban, V.S., Timlin, J.A., and Pakrasi, H.B. (2013). Probing the consequences of antenna modification in cyanobacteria. Photosynth. Res. *118*, 17–24. http://dx.doi.org/10.1007/s11120-013-9940-0.

Lichtenthaler, H.K. (2000). Non-mevalonate isoprenoid biosynthesis: enzymes, genes and inhibitors. Biochem. Soc. Trans. *28*, 785–789. http://dx.doi.org/10.1042/0300-5127:0280785.

Lindberg, P., Park, S., and Melis, A. (2010). Engineering a platform for photosynthetic isoprene production in cyanobacteria, using *Synechocystis* as the model organism. Metab. Eng. *12*, 70–79. http://dx.doi.org/10.1016/j.ymben.2009.10.001.

Lombard, J., and Moreira, D. (2011). Origins and early evolution of the mevalonate pathway of isoprenoid biosynthesis in the three domains of life. Mol. Biol. Evol. *28*, 87–99. http://dx.doi.org/10.1093/molbev/msq177.

Melis, A. (2009). Solar energy conversion efficiencies in photosynthesis: minimizing the chlorophyll antennae to maximize efficiency. Plant Sci. *177*, 272–280. http://dx.doi.org/10.1016/j.plantsci.2009.06.005.

Melis, A. (2012). Photosynthesis-to-Fuels: From sunlight to hydrogen, isoprene, and botryococcene production. Energy Environ. Sci. *5*, 5531–5539. http://dx.doi.org/10.1039/c1ee02514g.

Melis, A. (2013). Carbon partitioning in photosynthesis. Curr. Opin. Chem. Biol. *17*, 453–456. http://dx.doi.org/10.1016/j.cbpa.2013.03.010.

Ort, D.R., Zhu, X.G., and Melis, A. (2011). Optimizing antenna size to maximize photosynthetic efficiency. Plant Physiol. *155*, 79–85. http://dx.doi.org/10.1104/pp.110.165886.

Pade, N., Erdmann, S., Enke, H., Dethloff, F., Dühring, U., Georg, J., Wambutt, J., Kopka, J., Hess, W.R., Zimmermann, R., *et al.* (2016). Insights into isoprene production using the cyanobacterium *Synechocystis* sp. PCC 6803. Biotechnol. Biofuels *9*, 89. http://dx.doi.org/10.1186/s13068-016-0503-4.

Passos, F., Felix, L., Rocha, H., de Oliveira Pereira, J., and de Aquino, S. (2016). Reuse of microalgae grown in full-scale wastewater treatment ponds: Thermochemical pretreatment and biogas production. Bioresour. Technol. *209*, 305–312. http://dx.doi.org/10.1016/j.biortech.2016.03.006.

Pattanaik, B., and Lindberg, P. (2015). Terpenoids and their biosynthesis in cyanobacteria. Life 5, 269–293. http://dx.doi.org/10.3390/life5010269.

Reinsvold, R.E., Jinkerson, R.E., Radakovits, R., Posewitz, M.C., and Basu, C. (2011). The production of the sesquiterpene β-caryophyllene in a transgenic strain of the cyanobacterium *Synechocystis*. J. Plant Physiol. *168*, 848–852. http://dx.doi.org/10.1016/j.jplph.2010.11.006.

Rohmer, M., Knani, M., Simonin, P., Sutter, B., and Sahm, H. (1993). Isoprenoid biosynthesis in bacteria: a novel pathway for the early steps leading to isopentenyl diphosphate. Biochem. J. *295*, 517–524.

Rohmer, M., Seemann, M., Horbach, S., Bringer-Meyer, S., and Sahm, H. (1996). Glyceraldehyde 3-phosphate and pyruvate as precursors of isoprenic units in an alternative non-mevalonate pathway for terpenoid biosynthesis. J. Am. Chem. Soc. *118*, 2564–2566. http://dx.doi.org/10.1021/ja9538344.

Sharkey, T.D., and Yeh, S. (2001). Isoprene emission from plants. Annu. Rev. Plant. Physiol. Plant. Mol. Biol. *52*, 407–436. http://dx.doi.org/10.1146/annurev.arplant.52.1.407.

Sharkey, T.D., Wiberley, A.E., and Donohue, A.R. (2008). Isoprene emission from plants: why and how. Ann. Bot. *101*, 5–18. http://dx.doi.org/10.1093/aob/mcm240.

Zamalloa, C., Boon, N., and Verstraete, W. (2012). Anaerobic digestibility of *Scenedesmus obliquus* and *Phaeodactylum tricornutum* under mesophilic and thermophilic conditions. Appl. Energy *92*, 733–738. http://dx.doi.org/10.1016/j.apenergy.2011.08.017.

Zurbriggen, A., Kirst, H., and Melis, A. (2012). Isoprene production via the mevalonic acid pathway in *Escherichia coli* (Bacteria). BioEnergy Res. *5*, 814–828. http://dx.doi.org/10.1007/s12155-012-9192-4.

Ethanol Production in Cyanobacteria: Impact of Omics of the Model Organism *Synechocystis* on Yield Enhancement

10

J. Tony Pembroke[1,2]*, Lorraine Quinn[1], Helen O'Riordan[1], Con Sheahan[3] and Patricia Armshaw[1]

[1]Department of Chemical and Environmental Sciences, University of Limerick, Limerick, Ireland.
[2]The Materials and Surface Science Institute, University of Limerick, Limerick, Ireland.
[3]Department of Design and Manufacturing Engineering, University of Limerick, Limerick, Ireland.

*Correspondence: tony.pembroke@ul.ie

Abstract

Owing to their phototrophic lifestyle, cyanobacteria have been investigated as a sustainable source of biofuel production. Model cyanobacteria, such as *Synechocystis*, have been metabolically engineered with heterologous genes encoding pyruvate decarboxylase and alcohol dehydrogenase to convert the metabolic intermediate pyruvate first to acetaldehyde and then to ethanol. The current state of the art suggests that there is some way to go before such a system can be exploited. Lessons from the genomics of the model organism and proteomic and transcriptomic analysis of the organism's response to ethanol production are being utilized to enhance productivity. These are informing a systems approach to rational modification for enhanced ethanol production, which will be a key to the biotechnological exploitation of such a system.

Introduction

The need to investigate replacements for fossil fuels, coupled to the need to find ways of decreasing atmospheric CO_2 and ocean acidification are major drivers in the search for new sources of sustainable energy production systems particularly for liquid transportation fuels. In the USA, ethanol accounts for some 90% of biofuels and can be blended with gasoline as an aid to reducing fossil fuel dependence. Much of this ethanol has traditionally been produced from corn via fermentation using various *Saccharomyces* or *Zymomonas* species with up to 20% ethanol substitution possible (Antoni *et al.*, 2007). However, such processes compete with agricultural crops and there has been a shift in thinking in the debate of food versus fuel. The economic implications of utilizing productive farmland for fuel purposes coupled to the burden this places on the environment, in terms of biological control agents, water usage and crop rotation are major concerns. The biosynthesis of ethanol as a biofuel from CO_2 and sunlight using metabolically engineered cyanobacterial species is thus of interest in terms of sustainable and renewable productivity. Model species such as *Synechocystis* and *Synechococcus* have been utilized widely not only for ethanol production but for a range of other biotech products (Dexter *et al.*, 2015) with *Synechocystis* being particularly useful as a model organism. It can grow both photoautotrophically and heterotrophically (Grigorieva and Shestakov, 1982; Morris *et al.*, 2014), it can be manipulated genetically, there is adequate genomic information available (Ikeuchi and Tabata, 2001), it undergoes natural transformation (Grigorieva and Shestakov, 1982) and grows rapidly relative to many cyanobacterial species.

Ethanol production in cyanobacteria

Although generally considered photoautotrophic, environmental conditions in cyanobacterial habitats are prone to change. Light and CO_2 can become limited, for instance, in cyanobacterial mats or during symbiotic relationships. Storage compounds, accumulated by cyanobacteria in the light, may be utilized in such situations and under anoxic conditions. Many cyanobacteria possess fermentative pathways, allowing rapid reaction to such changed environments (Stal and Moezelaar, 1997). Some environments may become permanently anoxic as in microbial mats, sulphide rich lakes or lake sediments. In such situations, fermentative pathways may be a major survival tool. Some cyanobacteria can naturally produce small amounts of ethanol via heterotrophic anaerobic metabolism. Heyer and Krumbein (1991) took an arbitrary 37 strains from the Oldenburg Collection of cyanobacterial strains to investigate dark metabolism associated with cyanobacterial mats. They found that *Oscillatoria* strain 08 and *Oscillatoria limosa* 23 produced ethanol at > 10 µmol per sample while other strains [*Oscillatoria terebriformis* 61, *Plectonema boryanum* PCC73110 and *Spirulina mimosa*] produced between 5–10 µmol per sample as excreted fermentation products in the dark. Other reports have indicated that *Microcystis aeruginosa* PCC7806, *Cyanothece* PCC7822, and *Microcoleus chthionoplastes* all produce ethanol, although levels were not determined (Stal and Moezelaar, 1997). In general, such levels are low and are not produced in the light but have not been further investigated from a biotechnological perspective.

The first cyanobacterial species to be genetically modified to produce ethanol in the light was *Synechococcus elongatus* PCC 7942 (formally called *Synechococcus* sp. PCC 7942) (Deng and Coleman, 1999). The strain was metabolically engineered by transforming the genes encoding pyruvate decarboxylase (*pdc*) (Deng and Coleman, 1999) and alcohol dehydrogenase II (*adh*) from *Zymomonas mobilis*, an obligatory fermentative prokaryote, into *Synechococcus*. These genes were expressed under the control of the cyanobacterial *rbcLS* operon promoter, alone and in combination with the *Escherichia coli lac* promoter. The reported yields of ethanol produced by the transformed strain reached 54 nmol/$OD_{730\,nm}$ unit/l/day (Deng and Coleman, 1999). This system was further modified in *Synechococcus* PCC 7942 (patent US6306639B1) to insert an inducible promoter system based on the *p*L promoter of the C1 repressor of the coliphage lambda. The same genes have also been expressed in the model organism *Synechocystis* sp. PCC 6803 under the control of a different promoter, the strong light driven *psbAII* promoter. This strain showed an increase in ethanol production reaching 5.2 mmol/$OD_{730\,nm}$ unit/l/day) (Dexter and Fu, 2009). The *pdc/adh* cassette utilized in *Synechocystis* was identical to that used to achieve ethanol production in *E. coli* (Ingram *et al.*, 1987) some two decades earlier.

Following the original metabolic engineering work (Deng and Coleman, 1999) (and patent US6306639B1), several US based companies entered the space, including Algenol Biofuels and Joule Unlimited. These companies began to further investigate ethanol productivity in these model organisms and a series of patents followed where promoter and construct modifications (patents US8163516B2 and WO2013098267A1) were generated which enhanced the productivity further. See Table 10.1 and Fig. 10.1 for a comparison of some notable ethanol-producing *Synechocystis* PCC6803 strains generated to date. Although there are significant issues with reporting of patent information, it is clear from the results presented that significant increases in productivity of ethanol were achieved. (Gao *et al.*, 2012) reported a double cassette where the genes for *adh* and *pdc* from *Zymomonas mobilis* were inserted via two separate cassettes into *Synechocystis* PCC6803, which further enhanced productivity. This indicated that gene dosage could play a significant role. However, the combination of strong promoters and gene dosage can presumably only work to a finite extent, although this upper limitation has yet to be reported. *Synechocystis* does have a number of native plasmids which could potentially be utilized to insert heterologous genes to achieve gene dosage (Berla and Pakrasi, 2012). These plasmids are stable and appear to increase in copy number during the cell cycle. Recently, the small native plasmid pCA2.4 was utilized to clone and overexpress the yellow fluorescent protein encoding gene (Armshaw *et al.*, 2015). This resulted in an average minimum 20 fold increase in fluorescence levels being achieved compared to integration of a single gene cassette into the chromosome. This work suggests that integration within

Table 10.1 Summary of known ethanol-producing *Synechocystis* PCC6803 strains and published productivity data. Note that reported productivities may have incorporated the use of 24 h light regimes which may not be representative of potential productivities under production conditions (light/dark)

No. of days	Production rate (g/l)	Rate (g/l/day)	Genetic construct	References
6	0.46	0.077	*Z. mobilis* PDC/ADHII *psbA2* promoter	Dexter and Fu (2009)
37	3.6	0.097	*Z. mobilis* PDC/*slr1192 petJ* promoter	Patent US8163516B2
30	7.1	0.236	*Z. mobilis* PDC/*slr1192 ziaA* promoter	Patent WO2013098267A1
26	5.5	0.212	*Z. mobilis* PDC/*slr1192 rbc* promoter	Gao *et al.* (2012)
18	4.7	0.261	*Z. mobilis* PDC/*slr1192 petJ* promoter	Dienst *et al.* (2014)
Varied	Varied	0.189 to 0.252	Varied	Patent EP2723872A1

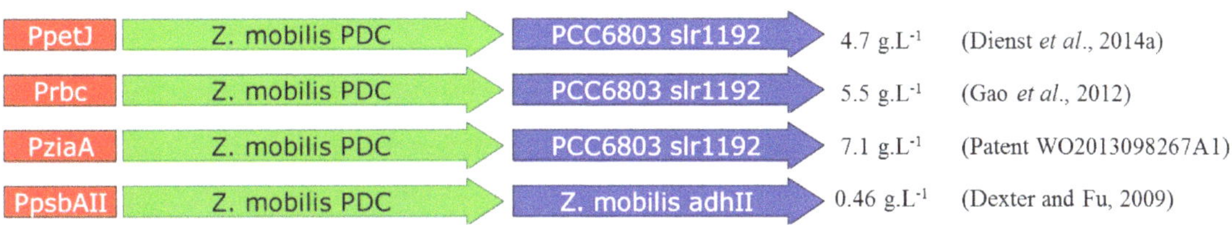

Figure 10.1 Critical characteristics of a selection of ethanol-producing *Synechocystis* PCC6803 strains generated to date. Promoters are shown in red, pyruvate decarboxylase in green and alcohol dehydrogenase in blue. Ethanol productivity is in g L^{-1}.

the native plasmids may be a useful tool for increasing gene expression in *Synechocystis*. As increasing gene dosage has led to an increase in ethanol productivity previously (Gao *et al.*, 2012), this may be a potential path to further increase ethanol titres (Armshaw *et al.*, 2015).

The construction of the *pdc/adh* cassette

The basis of metabolic engineering within the cyanobacteria relative to ethanol production is the ability to metabolically engineer and manipulate the pathways of central metabolism. Fig. 10.2 illustrates the key metabolic pathway steps associated with pyruvate production and the steps that need to be manipulated to convert it to ethanol. Carbon dioxide is fixed via the Calvin cycle intermediate ribulose-1,5-biphosphate to 3-phosphoglycerate which can cycle through the Calvin cycle or be converted via 2-phosphoglycerate to phosphoenolpyruvate (PEP) via enolase (Pepper *et al.*, 2010). PEP is then converted to the central intermediate pyruvate via pyruvate kinase. There are several possible metabolic modes of utilizing pyruvate including acetate production via acetyl-CoA, lactate, or isobutanol (Dexter *et al.*, 2015). Introduction of heterologous genes encoding pyruvate decarboxylase (*pdc*) and alcohol dehydrogenase (*adh*) can be utilized to convert pyruvate to ethanol and these activities have been central to early work on ethanol production via metabolic engineering (Deng and Coleman, 1999). However, given the central role of pyruvate, perturbations in the flux of pyruvate away from the tricarboxylic acid pathway (TCA cycle) can have major consequences for the supply of other metabolic intermediates such as amino acids, nucleotides and lipids, which will have consequence effects on biomass production. Thus production of ethanol through metabolic engineering is naturally going to have an effect on biomass yields.

The origin of the *pdc/adh* gene cassette

The obligate fermentative bacterium *Zymomonas mobilis* is one of few prokaryotes which generates ethanol as its predominant fermentative product (Montenecourt, 1986). The *Z. mobilis* pyruvate decarboxylase (PDC) and alcohol dehydrogenase (ADH) are abundant cell proteins with PDC

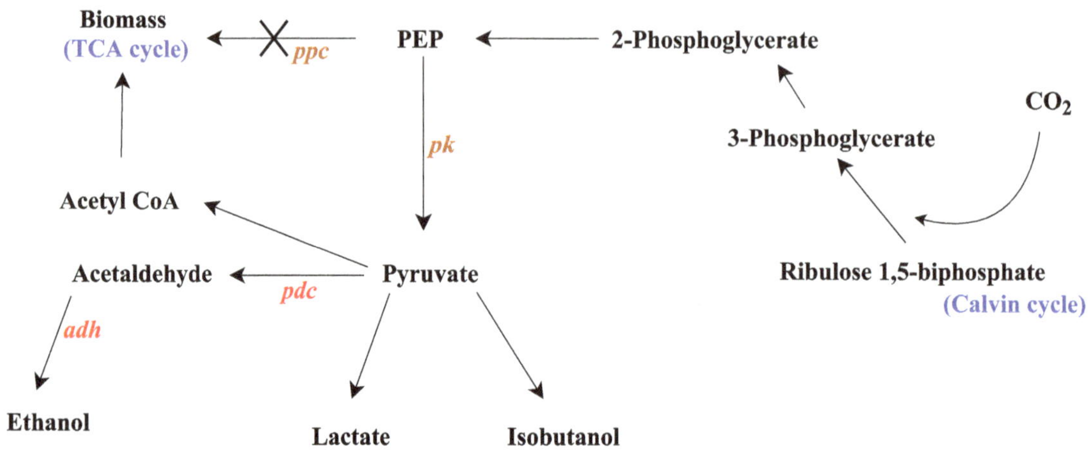

Figure 10.2 Summary of the key metabolic pathways involved in ethanol production in cyanobacteria. CO_2 is fixed via the Calvin Cycle to 2-Phosphoglycerate and diverted to PEP (Phospho Enol Pyruvate) and to the central metabolite Pyruvate. There are many sumps for Pyruvate but by introduction of heterologous *pdc* (Pyruvate decarboxylase and *adh* (Alcohol dehydrogenase), pyruvate can be converted to ethanol. Production of ethanol may come at the expense of TCA cycle intermediates and have consequent effects on biomass productivity. In addition, overexpression of pyruvate kinase (*pk*) or reduced expression of phosphoenolpyruvate carboxylase (*ppc*) can enhance pyruvate concentration leading to increased ethanol yields.

accounting for as much as 5% of the total soluble protein in the cells (Algar and Scopes, 1985). The *Z. mobilis* PDC is a tetramer composed of 4 identical subunits with a monomer molecular mass of 60 kDa, while there are two isozymes of ADH II (Hoppner and Doelle, 1983; Neale *et al.*, 1987). The more abundant isozyme is also a homotetramer, with monomeric molecular mass of 40 kDa. Cloning and molecular characterization of the genes encoding *Z. mobilis* PDC (Bräu and Sahm, 1986; Conway *et al.*, 1987a; Neale *et al.*, 1987) and ADH II (Conway *et al.*, 1987a) have been reported previously. These have formed the basis of metabolic engineering in cyanobacteria during most of the academic or industrial manipulations reported thus far.

Algenol patent (WO2013098267A1) has attempted to titrate the PDC in both *Synechococcus* PCC 7002 and *Synechocystis* PCC 6803 to optimize yields. They utilized a number of promoter constructs with Ni^{2+}-, Cu^{2+}- and Zn^{2+}-inducible promoters, used an expression vector containing the expression cassette, juxta-positioned the relative positions of the *pdc* and *adh* genes, and introduced potential gene terminators. These modifications resulted in a reported 7.1 g/l ethanol after a 30-day period, one of the highest yields ever reported (patent WO2013098267A1). These data indicate that the relative proportions of PDC and ADH may play a pivotal role in driving the flux from pyruvate to ethanol and that optimization may be an essential requirement and something that would evolve naturally in nature but which must be manipulated specifically in a metabolic engineering context.

The *Zymomonas mobilis* (Zm) pyruvate decarboxylase

The ZmPDC (pyruvate decarboxylase, EC 4.1.1.1) is a cytosolic enzyme which plays a key role in metabolic engineering of *Synechocystis* for ethanol. It carries out the decarboxylation of pyruvate to acetaldehyde in alcohol fermentation and requires thiamine diphosphate/pyrophosphate (ThDP) and the divalent cation Mg^{2+} as cofactors. Several other enzymes in various metabolic pathways also require these cofactors to function and it is believed that each of them use a similar mechanism of action. Pyruvate decarboxylases can be found in fungi, plants and yeast but are rather rare in bacteria and are not present in humans (Schenk *et al.*, 1997). Only a small number of bacterial species possess *pdc* genes some of which have been cloned and characterized: *Zymomonas mobilis*, (Candy and Duggleby, 1998), *Zymobacter palmae*, (Raj *et al.*, 2002) *Acetobacter pasteurianus* (Raj *et al.*, 2001),

Gluconacetobacter diazotrophicus (Van Zyl *et al.*, 2014a), *Thermococcus quaymacensis* (Eram *et al.*, 2014), *Geobacillus thermoglucosidasius* (Van Zyl *et al.*, 2014b) and *Sarcina ventricular* (Talarico *et al.*, 2001). The most commonly utilized of these in *Synechocystis* PCC 6803 has been the *Z. mobilis* PDC. Work on this *pdc* and others has shown that only bacterial *pdc* genes can be expressed highly in *E. coli* to give recombinant products that are active. Therefore only bacterial *pdc* genes are of use in metabolic engineering applications utilizing cyanobacterial systems (Raj *et al.*, 2002). As stated, ThDP is utilized as a co factor for PDC (Van Zyl *et al.*, 2014b). Generally enzymes that utilize ThDP as a cofactor are oligomeric and are arranged as either homodimers, homotetramers or heterotetramers (Pei *et al.*, 2010). An issue of concern in metabolic engineering of *Synechocystis* PCC 6803 systems is thus the cofactor availability for the PDC enzyme in a heterologous host. In particular, it is hypothesized that ThDP, may not be present in sufficient quantity in *Synechocystis* PCC 6803 when PDC is overexpressed. This is due to the lack of a thiamine transport system in *Synechocystis* PCC 6803. Thus steps may need to be taken to increase the cellular ThDP concentration. Steps to overcome this may involve overexpression of the *Synechocystis* thiamine synthesis system. Proteome analysis suggests that such overexpression is a feature of the cellular response to ethanol tolerance (Borirak *et al.*, 2015). Other potential methods of supplying thiamine exogenously may be by ultrasonic intensification to allow diffusion of added thiamine into the cell (Balakrishnan *et al.*, 2015).

The ZmPDC utilized in metabolic engineering in the *Synechocystis* system (Deng and Coleman, 1999) has been extensively studied and its crystal structure determined at a resolution of 1.9 Å (PDB entry 1zpd). The ZmPDC is present as four identical monomeric subunits, each consisting of 568 amino acids with a combined molecular mass of ~240 kDa. The optimum pH for activity is 6.0 but optimal activity is observed in the range of pH 4.5–8.5, which is important given the slightly alkaline growth conditions of *Synechocystis*. There may be issues with this relatively high pH with respect to the ZmPDC, particularly relating to the ThDP/PDC interaction and function at this pH (Van Zyl *et al.*, 2014b) with further increases in alkalinity potentially causing the cofactors separation leading to loss of enzymatic activity (Dobritzsch *et al.*, 1998).

To improve ethanol productivity in *Synechocystis* PCC 6803, bioinformatic analysis comparing *Z. mobilis* PDC to other less characterized PDCs may be a way forward. Aligning the amino acid sequences of the *Zymomonas mobilis*, *Zymobacter palmae* (Zp), *Acetobacter pasteurianus* (Ap) and *Sarcina ventriculi* (Sv) PDCs (Raj *et al.*, 2002) has shown that ZpPDC shares 72% identity with the ApPDC but only 62% to ZmPDC while these are more closely related to the PDCs of plants. The SvPDC showed higher similarity to PDCs from fungi and yeast and only shared 30–31% with the other bacterial PDCs. All of the enzymes had a similar amino acids length ranging from 552–568, with masses in the region of 59.8–61.8 kDa, while with respect to structure, the regions involved in cofactor and substrate binding were maintained in each of the bacterial PDCs. An important factor to characterize with PDCs of different origins is whether they possess different kinetic profiles. In metabolic engineering applications of *Synechocystis*, a key issue is to have a Km that will remove as much pyruvate as possible from the central metabolic pathway to channel this to ethanol. Comparison of kinetic parameters (Table 10.2) indicates that the ZpPDC and the GoPDCs may have some potential in this respect to enhance ethanol productivity over the ZmPDC, based on their lower K_m values.

The native ADH of *Synechocystis*

Synechocystis unusually contains an *adh* gene of its own. The native *Synechocystis* PCC 6803 gene, *adh*A (*slr1192*), encodes a member of the medium chain alcohol dehydrogenase/reductase superfamily. This superfamily catalyses the reversible oxidation of alcohols to aldehydes or ketones in a NAD(P)-dependent manner. The *Synechocystis* homologue resembles the yeast Y-ADH family (Nordling *et al.*, 2002a). This family is associated with ethanol or short chain alcohol metabolism with the group demonstrating broad substrate specificity (Vidal *et al.*, 2009). Why *Synechocystis* would contain this activity without possessing a native *pdc* gene is unusual and suggests that the *slr1192* gene (encoding the native *adh*A) product may play another as yet unknown metabolic role in *Synechocystis*. The

Table 10.2 Characteristics of various PDC enzymes. Data from Raj *et al.* (2002), Van Zyl *et al.* (2014a)

Name of organism and enzyme	PDB number	K_m (mM) pyruvate	Optimum pH	Optimum temperature (°C)
Zymomonas mobilis ZmPdc	1ZPD	0.43 [pH 6.0] 0.94 [pH 7.0]	6.0–6.5	60
Zymobacter palmae ZpPdc		0.24 [pH 6.0] 0.71 [pH 7.0]	7.0	55
Acetobacter pasteurianus ApPdc	2VBI	0.39 [pH 5.0] 5.1 [pH 7.0]	3.5–6.5	65
Sarcina ventriculi SvPdc		5.7 [pH 6.5] 4.0 [pH 7.0]	6.3–6.7	N/A
Gluconobacter oxydans GoPdc		0.12 [pH 5.0] 2.8 [pH 7.0]	4.5–5.0	53
Gluconacetobacter diazotrophicus GdPdc	4COK	0.06 [pH 5.0] 0.6 [pH 6.0] 1.2 [pH 7.0]	5.0–5.5	45–50

slr1192 gene can be induced by osmotic or salt stress which is often regulated by two component regulatory systems. In this case, the histidine kinase (Hik34) and a response regulator (Rre1) can induce or repress *adh*A (*slr1192*) expression in response to environmental stress (Vidal *et al.*, 2009). Hik34 senses heat, osmotic or salt stress (Marin *et al.*, 2003; Vidal *et al.*, 2009; Chapter 5) and once activated, it auto-phosphorylates a conserved histidine residue and then transfers the phosphate to an aspartate residue in the response regulator, Rre1. The Rre1 response regulator has been shown to then bind specifically to the promoter region of the *adh*A (*slr1192*) gene with two repeated palindromic GTTG sequences being the suspected binding site (Vidal *et al.*, 2009), resulting in gene activation.

The *adh*A (*slr1192*) gene product is a tetrameric 140 kDa zinc-dependent enzyme which uses the co-factor NADP(H) and which exhibits broad alcohol dehydrogenase activity on aromatic and aliphatic alcohols and aldehydes. It shows no activity on secondary alcohols or ketones but demonstrates efficient catalytic activity towards aldehyde reduction as opposed to alcohol oxidation (Vidal *et al.*, 2009). This co factor requirement differs from that of the *adh* of *Zymomonas mobilis* which requires NAD(H) (Cooley and Vermaas, 2001). Thus, in manipulating the *adh* genes, it is important that depending on which one is expressed the co-factor requirements will differ and indeed there may be merit in expressing both.

As mentioned, at first glance it may appear unusual that *Synechocystis*, which cannot natively synthesize ethanol, would possess a native alcohol dehydrogenase with high aldehyde reduction activity. In cyanobacteria, medium-chain dehydrogenases/reductases are considered the main scavengers of lipid-derived reactive carbonyls. Reactive carbonyls are widespread species in living organisms and are mainly known for their damaging effects, with the most abundant reactive carbonyl species derived from oxidation of carbohydrates, lipids, and amino acids. Reactive carbonyls such as methylglyoxal and acrolein have a toxic effect on the photosynthetic activity of cyanobacteria (Shimakawa *et al.*, 2014). These toxic effects are unique to photosynthetic organisms indicating that scavenging systems are thus essential for cyanobacterial survival. The aldo-keto reductase and the glyoxalase systems mainly scavenge sugar-derived reactive carbonyls. The 2-alkenal reductases and alkenal/alkenone reductase catalyse the reduction of lipid-derived reactive carbonyls while medium-chain dehydrogenases/reductases are the main scavengers of lipid-derived reactive carbonyls (Shimakawa *et al.*, 2014). This suggests one possible role for *adh*A (*slr1192*) in *Synechocystis*. In another study, it was demonstrated that AdhA enhanced odd-chain fatty alcohols production which may point to yet another function. Such fatty alcohols are produced for purposes of buoyancy, as a source of metabolic water and for energy purposes (Cao *et al.*, 2015).

That AdhA functions as an alcohol dehydrogenase has been shown directly (Vidal *et al.*, 2009) and by substituting it for the *Zymomonas mobilis* derived ADH and in such systems it functions to dehydrogenate acetaldehyde directly to ethanol (P. Armshaw, University of Limerick, personal communication).

The *Zymomonas mobilis* (Zm) ADH

Two ADH isozymes are known to be present within the genome of *Zymomonas mobilis*, Adh I and Adh II (EC 1.1.1.1, PDB entry 30WO/30X4). Both enzymes appear to take part in the fermentation process (Conway *et al.*, 1987b) and possess high affinities for their NADH and acetaldehyde substrates Hence, both appear to be suitable for the production of ethanol. However, at increased ethanol levels, Adh I undergoes substrate inhibition while Adh II, with its sensitivity to oxygen, may be affected in cyanobacteria due to the large amounts of oxygen they produce. The suitability of these enzymes for ethanol production in cyanobacteria has been examined and compared to the native *Synechocystis* PCC 6803 AdhA (*slr1192*). It was found that all three ADHs were able to convert acetaldehyde to ethanol in *Synechocystis* PCC 6803. However, deleterious growth effects were evident with the expression of either of the *Z. mobilis* ADHs. In addition, there was evidence of a back reaction at higher ethanol concentrations, converting ethanol to acetaldehyde (patent WO2013098267A1). Overall, it has been determined that expression of the *Z. mobilis* PDC and native *Synechocystis* PCC 6803 AdhA is optimal for ethanol production in *Synechocystis* PCC 6803. In fact, native expression levels of *slr1192* may be sufficient for ethanol production when one copy of the PDC enzyme of choice is present in the cell under the control of a mid-strength promoter (patent WO2013098267A1).

The effect of the ADH co-factor

It is important to evaluate the co-factor preference of an enzyme to ensure maximum productivity. Gao *et al.*, 2012 examined the co-factor preference of the exogenous Adh II from *Z. mobilis* and the native AdhA (*slr1192*) of *Synechocystis* PCC 6803 (Gao *et al.*, 2012). Adh II requires NAD(H) for activity and the native AdhA of *Synechocystis* PCC 6803 requires NADP(H). Cyanobacteria naturally have a high concentration of NADP(H) due to the release of electrons from water in the membranes of cells. Hence, cellular concentration of NADP(H) are approximately 10-fold higher than that of NAD(H). With this in mind, Gao *et al.* (2012) confirmed that utilization of the native *Synechocystis* PCC 6803 AdhA enzyme likely leads to higher ethanol productivity due to its co-factor preference for NADP(H).

In total, nine *adh* genes from three different cyanobacteria were investigated in conjunction with the *pdc* from *Zymomonas mobilis*. All nine strains produced ethanol but co-expression of the ZmPDC with AdhA (*slr1192*) was determined to be the most productive (Gao *et al.*, 2012). Further investigation by purification and characterization showed that the activity of AdhA (*slr1192*) was much higher than that of others: up to ~74,000 fold difference when acetaldehyde and NADPH were used as substrates with a K_m of 1.56 mM. However, *Z. mobilis* Adh II activity was 94-fold higher than AdhA (*slr1192*) when acetaldehyde and NADH were used as substrates which a K_m of 2.73 mM which reinforces cofactor importance.

Overall, AdhA (*slr1192*) only gave ~ 50% higher ethanol production than Adh II but yet its activity is ~74,000 fold higher than the latter under ideal conditions. This indicates that the pyruvate decarboxylation may be the rate limiting step in the ethanol pathway (Gao *et al.*, 2012).

Some key issues with metabolic engineering of *Synechocystis*

High level productivity yields of ethanol are dependent on the biomass yields which are dependent on the growth rate of the producing cell factory. *Synechocystis* PCC 6803 can grow photoautotrophically with sunlight and CO_2, or mixo-trophically and heterotrophically on substrates such as glucose. When growing photoautotrophically, depending on temperature (which is usually 30°C), and light intensity, doubling times can be 10–15 h. Optimum growth is observed at light intensities between 40–70 $\mu E/m^2/s$ (Vermaas, 1996). In ideal photobioreactors with optimal growth conditions set, *Synechocystis* PCC 6803 can achieve specific growth rates of 1.7–2.5/day (Kim *et al.*, 2011). However, because of the carbon flux to ethanol in engineered strains, there will be an effect on biomass yields as mentioned previously. From the perspective of genetic manipulation *Synechocystis* is spontaneously transformable (Barten and Lill, 1995) with highest efficiencies being achieved during exponential growth phase. Certain host genes may be specifically required for this with transformation frequencies varying from 10^{-3} to 10^{-5} (Vermaas, 1996). Yoshihara *et al.* (2001) have shown that the gene product of *slr0197* is required for DNA uptake and transformation competency and enhanced rates being achieved upon deletion of *sll1354* (encoding a RecJ exonuclease) (Yu *et al.*, 2013). In general, insertion of DNA into the host genome occurs by double recombination events with a requirement for approximately 200 bp of sequence identity at both sites of recombination locus. There are several resistance markers (Vermaas, 1996) that can be used in *Synechocystis* which aids in insertion and selection of cloned genes.

Synechocystis PCC 6803 has been shown to be polyploid, with the 'Kasusa' strain reported to contain 12 copies of the chromosome per cell (Labarre *et al.*, 1989). A recent study using real-time PCR has demonstrated that there may in fact be several hundred copies in exponential phase reducing to 58 copies in stationary phase (Griese *et al.*, 2011). This polyploid behaviour has implications for selection of inserted genes into *Synechocystis* with several rounds of selection needed to isolate stable insertions into chromosomal sites. Thus selection inevitably requires use of a selective marker such as an antibiotic resistant determinant, several plating's and reselection and utilization of PCR probes spanning the insert region to ensure full segregation within a single strain. However once selected, because of the polyploid nature of the selected strain, the insert is usually quite stable. For optimal expression of metabolically engineered genes there are also a number of transcription control tools that have been developed for *Synechocystis* PCC 6803. These include strong native promoters such as P_{rbcLS} and P_{psbA2} (the latter being light driven), promoters such as P_{rbpP} and P_{clpP}, which show a circadian patterns of expression, while metal controllable promoters such as the nickel inducible promoter, $P_{nrsBACD}$, are also available (Yu *et al.*, 2013). *E. coli* promoters have also been examined but appear to behave differently in many cases in *Synechocystis* (Huang *et al.*, 2010).

Gene dosage is also a technique that has been utilized in *Synechocystis* either by insertion to native multi-copy plasmids (Armshaw *et al.*, 2015), via cloning into a multi-copy vector (Angermayr *et al.*, 2014) or by multiple gene insertions (Gao *et al.*, 2012). In the case of ethanol, it has been shown that insertion of two copies of the *Z. mobilis pdc* and *adh* genes leads to enhanced ethanol yields (Gao *et al.*, 2012).

Lessons can also be learned from experience of pathway manipulation in other systems. One of the key objectives in optimizing and enhancing product yield is to redirect carbon from biomass into a specific product. In a study to examine overproduction of lactate via over expressing exogenous *ldh* (lactate dehydrogenase) from *Lactococcus lactis* in cyanobacteria there are lessons that can be learned (Angermayr and Hellingwerf, 2013). Increasing the expression of *ldh* resulted in more conversion of pyruvate to lactate, a lesson that may be useful in utilizing over expression of other enzymatic activities, such as the *pdc/adh* cassette. However, this may not be a universal truth as observations with single enzyme expression may not hold with multiple enzyme expression as for ethanol. Indeed matching the expression levels or optimizing one relative to another in a metabolic pathway may be appropriate. Co-expressing the pyruvate kinase (PK) enzyme which transfers a phosphate group from PEP to ADP (Pepper *et al.*, 2010) in order to increase pyruvate levels has been shown to enhance flux in the lactate system (Angermayr *et al.*, 2014) and may also have applicability in the ethanol system. Knockdown/knockout of a competing pathway for pyruvate; for example, the conversion of PEP to OAA (oxaloacetate, an intermediate of the TCA cycle) by phosphoenolpyruvate carboxylase (PPC) which depletes the flux towards pyruvate could potentially lead to more product. In the *ldh* study the use of various *ldh* enzymes with improved enzymatic activity (lower K_m and higher k_{cat} values for affinity and turnover rate respectively) also offer lessons in strategy to improve flux. Knockout of storage pathways can remove the carbon sink for competing products such as PHB (polyhydroxylbutyrate) or glycogen storage can also manifest in higher product yields (van der Woude *et al.*, 2014).

Developing high potential targeted strategies to significantly increase ethanol yields is beginning to happen using synthetic biology (Berla *et al.*, 2013), random mutagenesis and rapid screening techniques (Hammar *et al.*, 2015). When used combinatorially these techniques may lead to the development of novel high ethanol producing strains with significantly increased yields. Techniques that enable the production of large libraries of gene knockout/overexpression mutants of *Synechocystis* PCC 6803 and the incorporation of complementing rapid screening technologies such as single-cell high throughput microdroplet analysis have recently been reported (Abalde-Cela *et al.*, 2015) that may lead to novel high performing biofuel producers.

Strains and substrains and genomics of the cell factory *Synechocystis*

On the basis of their GC contents, strains of *Synechocystis* have been classified as the marine group, the low GC group and the high GC group (Holt *et al.*, 1994) with the type strain *Synechocystis* sp. strain PCC 6803 belonging to the high CG group. The original model strain of *Synechocystis* was isolated in Berkley (Stanier *et al.*, 1971) and subsequently deposited in the Pasteur Culture Collection (PCC6803 strain) and the American Type Culture Collection (ATCC 27184 strain). A glucose tolerant strain was isolated from the ATCC strain (Williams, 1988) and distributed to the Kazusa Research Institute in Japan which led to the first sequenced genome of *Synechocystis* (Kaneko *et al.*, 1996). However over the years many sub-strains have emerged and have been variously reported in the literature, the glucose tolerant strain (Nordling *et al.*, 2002b; Williams, 1988), the Kazusa strain, ATCC strain, a PCC strain (Ikeuchi and Tabata, 2001) and 'Otago' strain (Morris *et al.*, 2014), an 'Amsterdam' strain (Angermayr *et al.*, 2014), a 'China' strain (Dexter and Fu, 2009), and a Vermaas strain (Vermaas, 1998) to mention a few. Many of these have differing culture behaviour, transformability, and production variability traits which may prove important in using *Synechocystis* as a production cell factory.

Synechocystis sp. strain PCC 6803 was the first cyanobacterial species to be sequenced (Kaneko *et al.*, 1996). The genome length is 3,573,480 bp and several self-replicating plasmids were also identified. On the basis of the original sequencing approximately 45% of the genes could be assigned function. The genome contained several features of note, as seen in Table 10.3. The repetitive sequences discovered were mainly insertion sequences with disrupted elements, suggesting frequent

Table 10.3 The key genomic features present in *Synechocystis* PCC6803

Key genomic features	No. present	Key genomic features	No. present
Repetitive sequences	~77	DNA metabolism genes	49
RNA genes	2 clusters	Transcription genes	24
tRNA genes	42	Translation genes	144
Predicted protein encoded genes	3,168	Transport genes	158
Amino acid biosynthesis genes	84	ATP synthesis genes	9
Co-factor genes	108	CO_2 fixation genes	24
Cell envelope associated genes	64	Cytochromes and complex genes	15
Cellular process genes	68	NADH dehydrogenase genes	17
Energy metabolism genes	86	Photosystem I genes	12
Central metabolism genes	31	Photosystem II genes	26
Lipid synthesis genes	35	Phycobilosome genes	17
Photosynthesis genes	138	Soluble electron carrier genes	8
Nucleotide synthesis genes	39	Sensory kinase genes	26
Regulatory genes	135	Response regulator genes	38
		Hybrid sensory kinase genes	16

rearrangements and HIP1 (an octameric highly iterated palindromic sequence) reported at 2810 copies at an average spacing of 1268 bp. Generally HIP1 was located at the borders of gene deletions and it was speculated that it may participate in recombination events. The two RNA gene clusters were similar in number to those found in plant plastids and algae and the tRNA genes were scattered throughout the genome (Kotani and Tabata, 1998).

The genome analysis allowed interrogation of the genome information in terms of the metabolic activity of *Synechocystis* which is available at the Kyoto Encyclopedia of Genes and Genomes (KEGG) database (Kanehisa and Goto, 2000). This has allowed rational prediction of the pathways to and from pyruvate and aided in developing strategies to manipulate the system. In addition, it has revealed that there are no transporters for ThDP, the PDC co-factor, indicating that it may not be supplied exogenously. The genome survey has also allowed correlation of proteomic and transcriptomic data upon ethanol production or examining the response to added ethanol. In addition, there are lessons from the genome data that may be pertinent to other cyanobacterial species. There appears to be significant evolution of laboratory strains within the general group termed *Synechocystis* which may also have implications for other cyanobacterial cell factories. Genome analysis of the 'Kazusa' strain and comparison with GT variants identified 21 SNPs (single nucleotide polymorphisms) within the sequenced genomes (Tajima *et al.*, 2011). As an example a mutation was identified within the *pil*C gene which is believed to be the reason for the 'Kazusa' strain being non-motile (Bhaya *et al.*, 2000). Equally the location of a number of IS (insertion sequences) elements varied and small deletions were identified indicating quite a bit of evolutionary deviation from the original *Synechocystis* PCC 6803 strain. The GT-'Kazusa' strain was shown to contain three large inserted IS elements into different genes, *sll1780* (ISY203b), *sll1635* (ISY203e) and *sll1474* (ISY203g) relative to other strains and a deletion in the gene *slr1084* (Tajima *et al.*, 2011; Trautmann *et al.*, 2012). Two 'Otago' strains, GT01 and GT02, originating via Arizona (Vermaas, 1998) and derived from the GT strain (Williams, 1988), were compared by deep sequencing (Morris *et al.*, 2014) and 10 *de novo* mutations were discovered which were both SNPs and insertions and deletions relative to the original sequenced genome. In sequencing a 'Moscow' strain of *Synechocystis* PCC 6803 (Trautmann *et al.*, 2012), a number of both chromosomal and plasmid mutations were found and it was proposed that there was significant microevolution occurring within the sub-strains. This highlighted the need to know the genomic background of the parent strain when constructing mutants or using metabolic engineering in *Synechocystis* PCC 6803 strains (Morris *et al.*, 2014), a lesson that may have more widespread implications within the cyanobacteria.

Proteomic analysis in response to ethanol

Synechocystis cells that produce metabolically engineered products are termed cell factories for that particular substrate (Angermayr *et al.*, 2012). High level production may impose several stresses on the production strain depending on the product. Key stress issues that have been considered include the diversion of carbon flux from central metabolism to product, which can effect growth rate and biomass formation and the changes in physio-chemical environment caused by the product (Borirak *et al.*, 2015). High level production of ethanol in such cyanobacterial model systems will require strains capable of tolerating the production levels required to make such processes economic. If competing with industrial yeast fermentation from starch derived glucose then the competition is 20% (v/v) ethanol (Antoni *et al.*, 2007). Ethanol itself can be toxic to production strains (Dunlop, 2011) and this has necessitated the isolation of tolerant mutants. As knowledge on the systems have evolved, this has allowed a more targeted approach to making knowledge based changes to metabolic pathways to enhance tolerance (Park *et al.*, 2012). Thus in attempting to develop production systems in cyanobacterial species, the tolerance of the production strain to ethanol is important. Proteomic analysis of the responses of model cyanobacterial species to endogenous and engineered production of ethanol is thus important not just in terms of examining the response to tolerance but also in examining an organism's response to productivity. Such analyses may point the way to enhancement by overexpression of upregulated genes or knock out of down-regulated genes identified in such a screen in a production strain. Qiao *et al.*, have carried out a pioneering study utilizing quantitative iTRAQ LC-MS proteomics analysis examining the metabolic responses to ethanol in *Synechocystis* PCC6803 using the ATCC strain (Qiao *et al.*, 2012). In this case it is important to stress that ethanol was added to the strain at the beginning of cultivation and samples taken every 12 h subsequently. While giving an indication of the response to ethanol stress it is not exactly mimicking the production as would occur in a metabolically modified strain producing metabolically engineered ethanol intercellularly at the expense of biomass. In this analysis (Qiao *et al.*, 2012), it was discovered that following 24 h incubation at 1.5% (v/v) ethanol, there was a 50% reduction in growth with visible aggregation occurring as demonstrated by microscopy and flow cytometry (Qiao *et al.*, 2012; Wang *et al.*, 2012) Proteomics was carried out on 1.5% (v/v) treated cells with sampling time points at mid exponential and at the exponential-stationary phase transition. In this study a total of 24,887 unique spectra were matched to 1509 proteomes at good peptide coverage corresponding to 42% of the *Synechocystis* proteins predicted from the genome sequence. Interestingly after 24 h incubation, 32 unique proteins were up regulated while 42 were down-regulated, 54 were up-regulated and 17 down regulated in common at 24 and 48 h, while at 48 h treatment, 135 were uniquely up-regulated and 87 down-regulated (Qiao *et al.*, 2012). The analysis identified proteins involved in a common stress response differentially expressed relative to untreated controls. Oxidative stress responses were induced (Table 10.4), as were multiple transporters that play a role in resistance, proteins involved in modifications to the cell membrane and envelope and proteins related to photosystems. Measurement of chlorophyll *a* in ethanol treated cells relative to controls indicated that the chlorophyll *a* levels were increased consistent with proteomic observations of up-regulation of photosynthesis related proteins (Qiao *et al.*, 2012), suggesting that enhanced photosynthesis may play a role in tolerance to ethanol. The systems reported to be up regulated are similar to those reported for *Zymomonas mobilis* although the exact nature of the proteins is different in many cases (Widiastuti *et al.*, 2011).

Borirak *et al.* (2015) analysed the proteomic response of ethanol production by an engineered *Synechocystis* strain rather in response to added ethanol as with Qiao *et al.*, (2012) which is more applicable to the situation when a carbon sink for ethanol is engineered (Borirak *et al.*, 2015; Qiao *et al.*, 2012). In this study a glucose tolerant derivative of *Synechocystis* sp. PCC6803, SAA012, was examined upon ethanol production with between 60% and 70% of fixed carbon being diverted to product with the growth rate being reduced by 71% compared to wild type (Borirak *et al.*, 2015). The ethanol production observed in SAA012 was 9.88 ± 0.16 mM over a 19-day period, giving a maximum production rate of 1.33 ± 0.12 mmol/l/day in this particular engineered strain. Under these conditions, this did not result in a significant stress response. Instead, production of ethanol appeared to involve a realignment of existing systems. In general it was noted that there was up-regulation of carbon concentrating mechanisms, carbon fixation enzyme RuBisCo, and other enzymes within the Calvin cycle to accommodate the extra flux needed to produce product (Borirak *et al.*, 2015). Some 168 proteins showed alteration with carbon fixation (carbonic anhydrase) and proteins involved in carbon concentration (CcmM and CcmK1) as well as enzymes associated with the Calvin cycle (Pgk, Fda Zwf) and glutathione pathways upregulated. Photosynthetic antennae proteins (CpcA-G2, Haem PbsA1) and ribosomal proteins showed down regulation possibly as a result of growth retardation associated with the added cellular pressure of being a 'cell factory'. The enzyme phosphomethylpyrimidine synthase (ThiC), involved in ThDP biosynthesis, was also up-regulated (Borirak *et al.*, 2015). Significantly, ThDP is a co-factor for pyruvate decarboxylase (PDC), one of the engineered enzymes necessary to synthesize ethanol. It was also noted that the circadian clock may also be altered by ethanol production as evidenced by the up-regulation of *Kai*A and C (Borirak *et al.*, 2015) and *Kai*B (Qiao *et al.*, 2012).

In contrast to the study of (Qiao *et al.*, 2012), when ethanol was added exogenously as a stressor,

Table 10.4 Metabolic response to ethanol stress in *Synechocystis* PCC6803

Activity/role	Gene	Regulation
Glutathione peroxidase (oxidative protection)	*slr1992*	Up
Bacterioferritin proteins (detoxification of peroxides)	*slr0242*, *sll0221*	Up
Glutaredoxin (remodelling cellular defences)	*slr1846*	Up
Methionine sulfoxide reductase (amino acid damage repair)	*sll1394*, *sll1680*	Up
Flavodoxin (antioxidant protein)	*sll0248*	Up
Rehydrin (antioxidant protein)	*sll1198*	Up
Rubredoxin (peroxidase activity)	*slr0600*	Down
Thioredoxin reductase (redox regulation)	*slr0600*	Down
Cytochrome 553 (and other multiple cytochromes)	*sll1796*	Up
Bicarbonate transporter (enhanced carbon flux)	*slr0040*, *slr0043*	Up
Transcriptional regulator (most up regulated protein)	*sll1512*	Up
Na^+/H^+ antiporter (Ion transport)	*sll0689*	Up
Iron transporter (protection of PSII)	*slr1295*	Up
Squalene hopene cyclase (increase membrane order)	*slr2089*	Up
Peptidyl prolyl cis trans isomerases (increase folding rate)	*sll0408*, *slr1251*	Up
Conductance channel (transduce physical stresses)	*slr0875*	Up
cheY proteins (signal proteins in chemotaxis)	*sll0039*, *sll1293*	Up
Response regulators (regulation of salt stress)	*slr1783*	Up
PSI proteins (PSI assembly)	*slr0172*, *slr0823*	Up
PSII proteins and six ferredoxins	*sll1398*, *sll1194*	Up
Thylakoid membrane proteins (increased photosynthesis)	*slr0729*, *slr1034*	Up
Ribosomal proteins		Down

Quantitative iTRAQ LC-MS proteomics analysis was carried out on samples of mid-exponential *Synechocystis* cells incubated in 1.5% (v/v) ethanol. Ethanol was added to the strain at the beginning of cultivation and samples taken every 12 h subsequently (Qiao *et al.*, 2012).

internal production of ethanol in SAA012, at least at the levels reported, did not induce a universal stress response, although moderate up-regulation of glutathione synthetase involved in oxidative stress was observed (Borirak *et al.*, 2015).

Transcriptomic analysis in response to ethanol

Although one might expect the data from proteomics to match transcriptomic data, there are many instances where these do not correlate (Nie *et al.*, 2007). A comprehensive transcriptomic analysis using an RNA-seq library has been reported for growth of *Synechocystis* PCC6803 cultures supplemented with 1.25, 1.5% and 2% ethanol with sampling at various time periods from 24 h to 72 h post treatment (Wang *et al.*, 2012). This study confirmed that 1.5% ethanol caused a 50% reduction in growth rate and that visible aggregation could be observed, indicating that cells were stressed at this level of ethanol. In the transcriptomics study, some 112 million sequencing reads were completed with data filtering and a mapping ration of >60% and reproducibility at 0.98–0.99 with a cut-off of 1.5 times up or down regulation to be significant with all reads matching the 3189 genes of the model organism (Wang *et al.*, 2012). Many of the genes up-regulated (some 274) were associated with energy metabolism, particularly those involved in photosynthesis and ATP synthesis. Others included ribosomal subunits enzymes of the Calvin cycle which matched previous proteomic data (Qiao *et al.*, 2012). The transcriptomic data also showed up a large number of proteins of unknown function, e.g. *slr0144*, *slr1470*, *slr0373* and *ssl0483*, suggesting an unknown role in response to ethanol (Wang *et al.*, 2012).

Although there were some differences between the transcriptomic data and published proteomic data the general nature of activities induced were similar. The activities up- and down-regulated can be categorized into specific cellular activities that appear to be key to responding to the ethanol stress.

1 Genes associated with photosynthetic activity (e.g. *ssl0563* photosystem1 subunit V, *sll1051* phycocyanobilin lyase, *sll1471* phycobilisome core polypeptide, *slr1185* cytochrome b_6f) were upregulated. This general up-regulation of photosynthesis has been proposed to increase reactive oxygen species and trigger the oxidative stress (Wang *et al.*, 2012) which was a key observation in previous proteomic studies (Qiao *et al.*, 2012).
2 Transporter synthesis may be an important mechanism of reacting to ethanol. Using transcriptomics 12 transporters with a wide range of functions including ABC transporters (*sll0759*), a RND multi drug transporter (*slr2131*), a sulphate transporter (*sll1041*), ABC-Nat Permease (*slr0949*), a cation transporter (*sll0671*), Na-dependent transporter (*sll1428*) and a urea transporter (*sll0374*) (Wang *et al.*, 2012), were induced upon ethanol treatment while 5 putative promoters were detected during proteomics (Qiao *et al.*, 2012).
3 Equally proteins associated with membrane modifications are also an important mechanism of ethanol tolerance with genes associated with cis to trans unsaturated fatty acids (Dunlop, 2011), e.g. acyl lipid desaturase (desA), being up-regulated (Wang *et al.*, 2012). Squalene hopene cyclase, associated with membrane order (*slr2089*), was also shown to be up-regulated (Qiao *et al.*, 2012; Wang *et al.*, 2012), suggesting that strengthening the membrane may be a major tolerance mechanism.
4 A number of envelope proteins were also up-regulated (*slr0819, sll1370, slr1910*), homologues of which have been shown to function in stress adaptation in other organisms to solvent or salt by restructuring or modifying the cell envelope to strengthen it (Wang *et al.*, 2012).
5 Up-regulation of genes (*slr1994 and slr1993*) encoding enzymes associated with storage of polyhydroxyalkanoate as a storage compound was also observed, indicative of a storage stress response.
6 As another adaptation, the gene for lactonylglutathione lyase was also upregulated, which has been reported to aid detoxification of reactive dicarbonyls and can react with core proteins and DNA causing cellular damage (Pepper *et al.*, 2010) indicative that it may play a role in ethanol stress.
7 Ten signal transduction proteins were also shown to be induced by ethanol histidine kinases (*sll1473* and *slr1805*), response regulators (*slr0947* and *sll1330*), ser/thr kinase (*slr1225*) and several transcriptional regulators. These may be associated in several stress response mechanisms and possible resetting the circadian clock in response to ethanol stress (Wang *et al.*, 2012).

In addition, many genes were down-regulated (some 1874) with some 68% representing unknown or hypothetical function while the rest suggested a down-regulation of central metabolism consistent with the slower growth rate observed on ethanol treatment (Wang *et al.*, 2012).

Dienst *et al.*, carried out a transcriptomic wide analysis of the response of *Synechocystis* to continuous production of ethanol using high resolution microarrays (Dienst *et al.*, 2014). A 40% reduction in biomass was observed similar to growth rate reductions previously reported (Qiao *et al.*, 2012). Given that the levels produced were significantly lower than adding exogenous ethanol (0.03% v/v per day) one might not expect a stress response as with exogenous addition to levels of 1.5% v/v. The genes *slr1192* encoding the native AdhA were upregulated as were some ribosomal protein genes and an ABC transporter (*slr1897*). However the level of induction was rather low. Several genes were down-regulated including unusually two transporters, a biopolymer transporter (*sll0477*) and an iron transporter (*slr1295*). Yet there was little evidence of a general stress response at this low level of production rather somewhat specific response involving 31 genes (Dienst *et al.*, 2014).

These data suggests that with continued development of the *Synechocystis* cell factory for ethanol production, when levels of 1.5% (v/v) are reached, then tolerance mechanisms may become a major issue. Current production levels however remain much lower than this but the transcriptomic and proteomic studies point to ways that tolerance and indeed production stress in engineered strains might be tackled.

Modelling of metabolic flux in *Synechocystis* to enhance ethanol production

Although the genomes of many *Synechocystis* variants and sub-strains have been completed and whole genome predictions of metabolic pathways are available at the KEGG database, there is still not a complete picture available for *Synechocystis* as a metabolic system. With respect to the tricarboxylic acid cycle, there have been reports that it is not complete (Steinhauser *et al.*, 2012) but may contain different enzymatic activities to make it different when compared to other traditional bacterial cycles. As with many organisms for which biochemical studies have lagged behind genomic data there have been few kinetic studies and much of the reaction stoichiometry necessary for a complete metabolic picture is missing. For these reasons model based analysis has been coupled to genome scale data in an attempt to analyse flux within *Synechocystis*. When using *Synechocystis* as a cell factory it is obvious that the more product one obtains, the more carbon that will be diverted from biomass. It is proposed that flux analysis may identify pathways that might be diverting carbon and hence identify targets for mutagenesis or deletion. There are multiple modelling systems that might aid in developing overproducing strains (Table 10.5).

Based on this methodology, a number of models have been developed and applied to the *Synechocystis* cell factory for modelling diverse product formation. They fall into genome scale models which include all reactions annotated from genome data, or incorporate all substrates and products of metabolism or central carbon metabolic models which examine carbon based reactions of biomass without all building blocks (Montagud *et al.*, 2011). The first such model was developed by Yang *et al.*, 2002 but modelled only a limited number of reactions (Yang *et al.*, 2002). This was built upon by Shastri and Morgan, as a central model with many assumptions and optimization strategies and tripled the number of reactions modelled (Shastri and Morgan, 2005). Kun *et al* (2008) then developed a genome scale model of the annotated genome with over 900 reactions modelled. Other models (Navarro *et al.*, 2009) soon followed. However, given that a limited picture of the kinetics of key enzymes in key pathways is known, a complete picture from modelling is still some way off.

Theoretical estimates of ethanol yields

Tajima *et al.* (2011) calculated theoretical yields of ethanol under autotrophic conditions as 2 mmol/g DW/h, while in another study Montagud *et al.* (2011) estimated a maximum ethanol productivity of 0.85 mmol/g DW/h under autotrophic conditions and 1.7 mmol/g DW/h under mixotrophic conditions. These are quite at odds with the reported high producers (from patent US7785861B2) at 2.171×10^{-2} mmol/g DW/h. These data based on metabolic flux analysis and theoretical considerations suggest that there may

Table 10.5 Methods of Flux analysis utilized in *Synechocystis* PCC6803

Method	Basis of method	Reference
Flux balance analysis	Quantitative characterization of metabolic properties of genome scale networks steady state model based on mass balancing	Shastri and Morgan (2005)
^{13}C metabolic flux analysis	Labelling of functional pathway intermediates, relative to knowledge on a genome scale	Wiechert (2001)
Flux coupling	Examines flux connectivity at a genome scale and plots the optimal effects of deletions, identifying bottlenecks	Montagud *et al.* (2011)

be some way to go with metabolic engineering to reach theoretical yields.

Towards industrial production of ethanol from cyanobacteria

Based on the potential of model organisms to produce ethanol photoautotrophically two US companies, Algenol Biofuels and Joule Unlimited, have developed significant technology in the production of ethanol as a biofuel. In 1998, Algenol patented (US6306639B1) work relating to *Synechococcus* PCC 7942 (Deng and Coleman, 1999) and later work related to (Dexter and Fu, 2009) manipulation of *Synechocystis* PCC6803 for ethanol production (US8372613B2). This was followed with patents showing that expression of the native *Synechocystis* PCC6803 AdhA (*slr1192*) with the *Zymomonas mobilis* PDC under control of the copper inducible *petJ* promoter resulted in enhanced ethanol yields (US8163516B2). Although it is often difficult to quantify what exactly is produced because of differences in the way yields are expressed in the patent literature, it appears that this resulted in concentrations of up to 3.5 g/l after 37 days. Other patents concentrated on re-routing carbon flux by generating knockout mutants of competing pathways. For example, generating acetate kinase knockouts (Δ*slr1299*) (US20100297736A1) and phosphoacetyltransacetylase knockouts (Δ*slr2132*) resulted in a reported doubling of yield. The same patent demonstrated that overexpressing of RuBisCo resulted in a 10% increase (US20100297736A1). Mutants in storage of glycogen were also reported which aid in diverting flux towards increased ethanol yields (Carrieri *et al.*, 2012), indicating that some diversion of carbon flux may be an important aid in increasing ethanol yield. These companies have pioneered the push towards industrial production systems and added considerably to our knowledge of cyanobacterial cell factories producing ethanol.

Conclusion and future trends

The availability of genome data coupled to more precise systems modelling will in the future generate accurate metabolic flux data which will allow determination of carbon flux and partitioning within model species. This will allow more targeted deletion or up-regulation to divert carbon flux to products such as ethanol in cyanobacterial cell factories. Use of microfluidics to screen randomized transposon insertion libraries will also play a role in identifying higher producers. Optimization of gene copy number, ratio of converting enzymes and selection of enzymes with optimum kinetics will also play a role. Increasing the carbon flux within the model organism through manipulating CO_2 and bicarbonate flux will also be necessary to improve both product and biomass yield, particularly when production strains are used. At a fundamental level understanding the nature of polyploidy and potentially manipulating it downwards could result in large energy savings which could be diverted towards product. Examining transport systems and their manipulation could lead to higher cofactor generation particularly in the case of cassette cofactor, such as nicotinamide and thiamine. The availability of these cofactors could potentially be a limiting factor particularly at high gene copy number or enzyme concentration. There are also lessons to be learned from proteomic and transcriptomic studies which could lead to cell factory adaptation and higher product yields by enhancing or repressing production of up or down regulated genes respectively. Generation of ethanol tolerant strains may also improve the productivity of ethanol although at current levels tolerance does not appear to be an issue. From the production perspective, while model organisms can play a major role in developing systems and testing hypothesis, more robust, more competitive and faster-growing cyanobacteria may prove more useful in transitioning to large scale production. Thus although there has been a large interest in utilizing model organisms such as *Synechocystis* as cell factories for photoautotrophic production of ethanol and other products we are still at the beginning of understanding the capabilities, limitations and avenues towards higher yields.

Acknowledgement

This work was supported by funding from the EU Seventh Framework Programme under grant agreement no. 309086.

Web resources

The web resources are listed in Table 10.6.

Table 10.6 A list of appropriate web resources

Site title	Content	Web address
Cyanosite	Images, media recipes links	http://www-cyanosite.bio.purdue.edu/index.html
KEGG Database	Genome and function integration with cyanobacteria	http://www.genome.jp/kegg/kegg1.html
Joule Unlimited	CO_2 to fuel platform	http://www.jouleunlimited.com/
Algenol	Ethanol production commercially	http://www.algenol.com/
Cyanobase	Cyanobacterial genomes	http://genome.microbedb.jp/cyanobase/
Cyorf	Cyanobacterial gene annotation	http://cyano.genome.ad.jp/
FAME	Synechocystis metabolic models	http://f-a-m-e.org/synechocytis
JGI	Synechocystis genomes	https://gold.jgi.doe.gov/
BioCyc	*Synechocystis* information	http://www.biocyc.org/SSP1148/NEW-IMAGE?type=ORGANISM&object=TAX-1148
NCBI	Access to biomedical, genomic and genetic informational source	http://www.ncbi.nlm.nih.gov/

References

Abalde-Cela, S., Gould, A., Liu, X., Kazamia, E., Smith, A.G., and Abell, C. (2015). High-throughput detection of ethanol-producing cyanobacteria in a microdroplet platform. J. R. Soc. Interface *12*, 20150216. http://dx.doi.org/10.1098/rsif.2015.0216.

Algar, E.M., and Scopes, R.K. (1985). Studies on cell-free metabolism: Ethanol production by extracts of *Zymomonas mobilis*. J. Biotechnol. *2*, 275–287. http://dx.doi.org/10.1016/0168-1656(85)90030-6.

Angermayr, S.A., and Hellingwerf, K.J. (2013). On the use of metabolic control analysis in the optimization of cyanobacterial biosolar cell factories. J. Phys. Chem. B. *117*, 11169–11175. http://dx.doi.org/10.1021/jp4013152.

Angermayr, S.A., Paszota, M., and Hellingwerf, K.J. (2012). Engineering a cyanobacterial cell factory for production of lactic acid. Appl. Environ. Microbiol. *78*, 7098–7106. http://dx.doi.org/10.1128/AEM.01587-12.

Angermayr, S.A., van der Woude, A.D., Correddu, D., Vreugdenhil, A., Verrone, V., and Hellingwerf, K.J. (2014). Exploring metabolic engineering design principles for the photosynthetic production of lactic acid by *Synechocystis* sp. PCC6803. Biotechnol. Biofuels 7, 99. http://dx.doi.org/10.1186/1754-6834-7-99.

Antoni, D., Zverlov, V.V., and Schwarz, W.H. (2007). Biofuels from microbes. Appl. Microbiol. Biotechnol. 77, 23–35. http://dx.doi.org/10.1007/s00253-007-1163-x.

Armshaw, P., Carey, D., Sheahan, C., and Pembroke, J.T. (2015). Utilising the native plasmid, pCA2.4, from the cyanobacterium *Synechocystis* sp. strain PCC6803 as a cloning site for enhanced product production. Biotechnol. Biofuels 8, 201. http://dx.doi.org/10.1186/s13068-015-0385-x.

Balakrishnan, N., Armshaw, P., and Pembroke, T.J. (2015). Ultrasonic intensification as a tool for enhanced microbial biofuel yields. Biotechnol. Biofuels *8*, 140. http://dx.doi.org/10.1186/s13068-015-0321-0.

Barten, R., and Lill, H. (1995). DNA-uptake in the naturally competent cyanobacterium, *Synechocystis* sp. PCC 6803. FEMS Microbiol. Lett. *129*, 83–88. http://dx.doi.org/10.1111/j.1574-6968.1995.tb07561.x

Berla, B.M., and Pakrasi, H.B. (2012). Upregulation of plasmid genes during stationary phase in Synechocystis sp. strain PCC 6803, a cyanobacterium. Appl. Environ. Microbiol. *78*, 5448–5451. http://dx.doi.org/10.1128/AEM.01174-12.

Berla, B.M., Saha, R., Immethun, C.M., Maranas, C.D., Moon, T.S., and Pakrasi, H.B. (2013). Synthetic biology of cyanobacteria: unique challenges and opportunities. Front. Microbiol. *4*, 246. http://dx.doi.org/10.3389/fmicb.2013.00246.

Bhaya, D., Bianco, N.R., Bryant, D., and Grossman, A. (2000). Type IV pilus biogenesis and motility in the cyanobacterium *Synechocystis* sp. PCC6803. Mol. Microbiol. *37*, 941–951. http://dx.doi.org/10.1046/j.1365-2958.2000.02068.x.

Borirak, O., de Koning, L.J., van der Woude, A.D., Hoefsloot, H.C., Dekker, H.L., Roseboom, W., de Koster, C.G., and Hellingwerf, K.J. (2015). Quantitative proteomics analysis of an ethanol- and a lactate-producing mutant strain of *Synechocystis* sp. PCC6803. Biotechnol. Biofuels *8*, 111. http://dx.doi.org/10.1186/s13068-015-0294-z.

Bräu, B., and Sahm, H. (1986). Cloning and expression of the structural gene for pyruvate decarboxylase of *Zymomonas mobilis* in *Escherichia coli*. Arch Microbiol *144*, 296–301. http://dx.doi.org/10.1007/BF00410966.

Candy, J.M., and Duggleby, R.G. (1998). Structure and properties of pyruvate decarboxylase and site-directed mutagenesis of the *Zymomonas mobilis* enzyme. Biochim. Biophys. Acta *1385*, 323–338. http://dx.doi.org/10.1016/S0167-4838(98)00077-6.

Cao, Y.X., Xiao, W.H., Liu, D., Zhang, J.L., Ding, M.Z., and Yuan, Y.J. (2015). Biosynthesis of odd-chain fatty alcohols in *Escherichia coli*. Metab. Eng. *29*, 113–123. http://dx.doi.org/10.1016/j.ymben.2015.03.005.

Carrieri, D., Paddock, T., Maness, P.-C., Seibert, M., and Yu, J. (2012). Photo-catalytic conversion of carbon dioxide to organic acids by a recombinant cyanobacterium incapable of glycogen storage. Energy Environ. Sci. *5*, 9457–9461. http://dx.doi.org/10.1039/c2ee23181f.

Conway, T., Osman, Y.A., Konnan, J.I., Hoffmann, E.M., and Ingram, L.O. (1987a). Promoter and nucleotide sequences of the *Zymomonas mobilis* pyruvate decarboxylase. J. Bacteriol. *169*, 949–954.

Conway, T., Sewell, G.W., Osman, Y.A., and Ingram, L.O. (1987b). Cloning and sequencing of the alcohol dehydrogenase II gene from *Zymomonas mobilis*. J. Bacteriol. *169*, 2591–2597. http://dx.doi.org/10.1016/0734-9750(93)90082-x.

Cooley, J.W., and Vermaas, W.F. (2001). Succinate dehydrogenase and other respiratory pathways in thylakoid membranes of *Synechocystis* sp. strain PCC 6803: capacity comparisons and physiological function. J. Bacteriol. *183*, 4251–4258. http://dx.doi.org/10.1128/JB.183.14.4251-4258.2001.

Deng, M.D., and Coleman, J.R. (1999). Ethanol synthesis by genetic engineering in cyanobacteria. Appl. Environ. Microbiol. *65*, 523–528.

Dexter, J., Armshaw, P., Sheahan, C., and Pembroke, J.T. (2015). The state of autotrophic ethanol production in Cyanobacteria. J. Appl. Microbiol. *119*, 11–24. http://dx.doi.org/10.1111/jam.12821.

Dexter, J., and Fu, P. (2009). Metabolic engineering of cyanobacteria for ethanol production. Energy Environ. Sci 2, 857–864. http://dx.doi.org/10.1039/b811937f.

Dienst, D., Georg, J., Abts, T., Jakorew, L., Kuchmina, E., Börner, T., Wilde, A., Dühring, U., Enke, H., and Hess, W.R. (2014). Transcriptomic response to prolonged ethanol production in the cyanobacterium *Synechocystis* sp. PCC6803. Biotechnol. Biofuels 7, 21. http://dx.doi.org/10.1186/1754-6834-7-21.

Dobritzsch, D., König, S., Schneider, G., and Lu, G. (1998). High Resolution Crystal structure of pyruvate decarboxylase from *Zymomonas mobilis*. Implications for substrate activation in pyruvate decarboxylases. J. Biol. Chem. *273*, 20196–20204. http://dx.doi.org/10.1074/jbc.273.32.20196.

Dunlop, M.J. (2011). Engineering microbes for tolerance to next generation biofuels. Biotechnol. Biofuels *4*, 32. http://dx.doi.org/10.1186/1754-6834-4-32.

Eram, M.S., Oduaran, E., and Ma, K. (2014). The bifunctional pyruvate decarboxylase/pyruvate ferredoxin oxidoreductase from *Thermococcus guaymasensis*. Archaea *2014*, 349379. http://dx.doi.org/10.1155/2014/349379.

Gao, Z., Zhao, H., Li, Z., Tan, X., and Lu, X. (2012). Photosynthetic production of ethanol from carbon dioxide in genetically engineered cyanobacteria. Energy Environ. Sci. 5, 9857–9865. http://dx.doi.org/10.1039/C2EE22675H.

Griese, M., Lange, C., and Soppa, J. (2011). Ploidy in cyanobacteria. FEMS Microbiol. Lett. *323*, 124–131. http://dx.doi.org/10.1111/j.1574-6968.2011.02368.x.

Grigorieva, G., and Shestakov, S. (1982). Transformation in the cyanobacterium *Synechocystis* sp. 6803. FEMS Microbiol. Lett. *13*, 367–370. http://dx.doi.org/10.1111/j.1574-6968.1982.tb08289.x.

Hammar, P., Angermayr, S.A., Sjostrom, S.L., van der Meer, J., Hellingwerf, K.J., Hudson, E.P., and Joensson, H.N. (2015). Single-cell screening of photosynthetic growth and lactate production by cyanobacteria. Biotechnol. Biofuels *8*, 193. http://dx.doi.org/10.1186/s13068-015-0380-2.

Heyer, H., and Krumbein, W. (1991). Excretion of fermentation products in dark and anaerobically incubated cyanobacteria. Arch. Microbiol. *155*, 284–287. http://dx.doi.org/10.1007/BF00252213.

Holt, J., Krieg, N., Sneath, P., and Staley, J. (1994). Bergey's manual of determinative bacteriology, Lippincott Williams & Wilkins. Philadelphia, PA. http://dx.doi.org/10.5962/bhl.title.10728.

Hoppner, T., and Doelle, H. (1983). Purification and kinetic characteristics of pyruvate decarboxylase and ethanol dehydrogenase from *Zymomonas mobilis* in relation to ethanol production. Eur. J. Appl. Microbiol. Biotechnol. *17*, 152–157. http://dx.doi.org/10.1007/BF00505880.

Huang, H.-H., Camsund, D., Lindblad, P., and Heidorn, T. (2010). Design and characterization of molecular tools for a synthetic biology approach towards developing cyanobacterial biotechnology. Nucleic Acids Res. *38*, 2577–2593. http://dx.doi.org/10.1093/nar/gkq164.

Ikeuchi, M., and Tabata, S. (2001). *Synechocystis* sp. PCC 6803– a useful tool in the study of the genetics of cyanobacteria. Photosynth. Res. *70*, 73–83. http://dx.doi.org/10.1023%2FA%3A1013887908680.

Ingram, L.O., Conway, T., Clark, D.P., Sewell, G.W., and Preston, J.F. (1987). Genetic engineering of ethanol production in *Escherichia coli*. Appl. Environ. Microbiol. *53*, 2420–2425.

Kanehisa, M., and Goto, S. (2000). KEGG: kyoto encyclopedia of genes and genomes. Nucleic Acids Res. *28*, 27–30. http://dx.doi.org/10.1093/nar/28.1.27.

Kaneko, T., Sato, S., Kotani, H., Tanaka, A., Asamizu, E., Nakamura, Y., Miyajima, N., Hirosawa, M., Sugiura, M., Sasamoto, S., *et al.* (1996). Sequence analysis of the genome of the unicellular cyanobacterium *Synechocystis* sp. strain PCC6803. II. Sequence determination of the entire genome and assignment of potential protein-coding regions. DNA Res. *3*, 109–136. http://dx.doi.org/10.1007/978-1-4615-6369-3_36.

Kim, H.W., Vannela, R., Zhou, C., and Rittmann, B.E. (2011). Nutrient acquisition and limitation for the photoautotrophic growth of *Synechocystis* sp. PCC6803 as a renewable biomass source. Biotechnol. Bioeng. *108*, 277–285. http://dx.doi.org/10.1002/bit.22928.

Kotani, H., and Tabata, S. (1998). Lessons from sequencing of the genome of a unicellular cyanobacterium, *Synechocystis* sp. PCC6803. Annu. Rev. Plant. Physiol. Plant. Mol. Biol. *49*, 151–171. http://dx.doi.org/10.1146/annurev.arplant.49.1.151.

Labarre, J., Chauvat, F., and Thuriaux, P. (1989). Insertional mutagenesis by random cloning of antibiotic resistance genes into the genome of the cyanobacterium *Synechocystis* strain PCC 6803. J. Bacteriol. *171*, 3449–3457.

Marin, K., Suzuki, I., Yamaguchi, K., Ribbeck, K., Yamamoto, H., Kanesaki, Y., Hagemann, M., and Murata, N. (2003). Identification of histidine kinases that act as sensors in the perception of salt stress in *Synechocystis* sp. PCC 6803. Proc. Natl. Acad. Sci. U.S.A. *100*, 9061–9066. http://dx.doi.org/10.1073/pnas.1532302100.

Montagud, A., Zelezniak, A., Navarro, E., de Córdoba, P.F., Urchueguía, J.F., and Patil, K.R. (2011). Flux coupling and transcriptional regulation within the metabolic network of the photosynthetic bacterium *Synechocystis* sp. PCC6803. Biotechnol. J. *6*, 330–342. http://dx.doi.org/10.1002/biot.201000109.

Montenecourt, B. (1985). *Zymomonas*, a unique genus of bacteria. In: Biology of industrial microorganisms. Demain A.L., and Solomon, N.A., eds. (Menlo Park, CA: Benjamin-Cummings Publishing Co. Inc.) pp. 261–289.

Morris, J.N., Crawford, T.S., Jeffs, A., Stockwell, P.A., Eaton-Rye, J.J., and Summerfield, T.C. (2014). Whole genome re-sequencing of two 'wild-type' strains of the model cyanobacterium *Synechocystis* sp. PCC 6803. New Zeal. J. Bot. *52*, 36–47. http://dx.doi.org/10.1080/0028825X.2013.846267.

Navarro, E., Montagud, A., De Córdoba, P.F., and Urchueguía, J. (2009). Metabolic flux analysis of the hydrogen production potential in *Synechocystis* sp. PCC6803. Int. J. Hyd. Energy *34*, 8828–8838. http://dx.doi.org/10.1016/j.ijhydene.2009.08.036.

Neale, A.D., Scopes, R.K., Wettenhall, R.E., and Hoogenraad, N.J. (1987). Nucleotide sequence of the pyruvate decarboxylase gene from *Zymomonas mobilis*. Nucleic Acids Res. *15*, 1753–1761. http://dx.doi.org/10.1093/nar/15.4.1753.

Nie, L., Wu, G., Culley, D.E., Scholten, J.C., and Zhang, W. (2007). Integrative analysis of transcriptomic and proteomic data: challenges, solutions and applications. Crit. Rev. Biotechnol. *27*, 63–75. http://dx.doi.org/10.1080/07388550701334212.

Nordling, E., Jörnvall, H., and Persson, B. (2002). Medium-chain dehydrogenases/reductases (MDR). Family characterizations including genome comparisons and active site modeling. Eur. J. Biochem. *269*, 4267–4276. http://dx.doi.org/10.1046/j.1432-1033.2002.03114.x.

Park, A.H., Sugiyama, M., Harashima, S., and Kim, Y.H. (2012). Creation of an ethanol-tolerant yeast strain by genome reconstruction based on chromosome splitting technology. J. Microbiol. Biotechnol. *22*, 184–189. http://dx.doi.org/10.4014/jmb.1109.09046.

Pei, X.Y., Erixon, K.M., Luisi, B.F., and Leeper, F.J. (2010). Structural insights into the prereaction state of pyruvate decarboxylase from *Zymomonas mobilis*. Biochemistry *49*, 1727–1736. http://dx.doi.org/10.1021/bi901864j.

Pepper, E.D., Farrell, M.J., Nord, G., and Finkel, S.E. (2010). Antiglycation effects of carnosine and other compounds on the long-term survival of *Escherichia coli*. Appl. Environ. Microbiol. *76*, 7925–7930. http://dx.doi.org/10.1128/AEM.01369-10.

Qiao, J., Wang, J., Chen, L., Tian, X., Huang, S., Ren, X., and Zhang, W. (2012). Quantitative iTRAQ LC-MS/MS proteomics reveals metabolic responses to biofuel ethanol in cyanobacterial *Synechocystis* sp. PCC 6803. J. Proteome Res. *11*, 5286–5300. http://dx.doi.org/10.1021/pr300504w.

Raj, K.C., Ingram, L.O., and Maupin-Furlow, J.A. (2001). Pyruvate decarboxylase: a key enzyme for the oxidative metabolism of lactic acid by *Acetobacter pasteurianus*. Arch. Microbiol. *176*, 443–451. http://dx.doi.org/10.1007/s002030100348.

Raj, K.C., Talarico, L.A., Ingram, L.O., and Maupin-Furlow, J.A. (2002). Cloning and characterization of the *Zymobacter palmae* pyruvate decarboxylase gene (pdc) and comparison to bacterial homologues. Appl. Environ. Microbiol. *68*, 2869–2876. http://dx.doi.org/10.1128/AEM.68.6.2869-2876.2002.

Schenk, G., Leeper, F.J., England, R., Nixon, P.F., and Duggleby, R.G. (1997). The role of His113 and His114 in pyruvate decarboxylase from *Zymomonas mobilis*. Eur. J. Biochem. *248*, 63–71. http://dx.doi.org/10.1111/j.1432-1033.1997.t01-1-00063.x.

Shimakawa, G., Suzuki, M., Yamamoto, E., Saito, R., Iwamoto, T., Nishi, A., and Miyake, C. (2014). Why don't plants have diabetes? Systems for scavenging reactive carbonyls in photosynthetic organisms. Biochem. Soc. Trans. *42*, 543–547. http://dx.doi.org/10.1042/BST20130273.

Stal, L.J., and Moezelaar, R. (1997). Fermentation in cyanobacteria. FEMS Microbiol. Rev. *21*, 179–211. http://dx.doi.org/10.1111/j.1574-6976.1997.tb00350.x.

Stanier, R.Y., Kunisawa, R., Mandel, M., and Cohen-Bazire, G. (1971). Purification and properties of unicellular blue-green algae (order Chroococcales). Bacteriol. Rev. *35*, 171–205.

Steinhauser, D., Fernie, A.R., and Araújo, W.L. (2012). Unusual cyanobacterial TCA cycles: not broken just different. Trends Plant Sci. *17*, 503–509. http://dx.doi.org/10.1016/j.tplants.2012.05.005.

Tajima, N., Sato, S., Maruyama, F., Kaneko, T., Sasaki, N.V., Kurokawa, K., Ohta, H., Kanesaki, Y., Yoshikawa, H., Tabata, S., *et al.* (2011). Genomic structure of the cyanobacterium *Synechocystis* sp. PCC 6803 strain GT-S. DNA. Res. *18*, 393–399. http://dx.doi.org/10.1093/dnares/dsr026.

Talarico, L.A., Ingram, L.O., and Maupin-Furlow, J.A. (2001). Production of the Gram-positive Sarcina ventriculi pyruvate decarboxylase in *Escherichia coli*. Microbiology *147*, 2425–2435. http://dx.doi.org/10.1099/00221287-147-9-2425.

Trautmann, D., Voss, B., Wilde, A., Al-Babili, S., and Hess, W.R. (2012). Microevolution in cyanobacteria: re-sequencing a motile substrain of *Synechocystis* sp. PCC 6803. DNA Res. *19*, 435–448. http://dx.doi.org/10.1093/dnares/dss024.

van der Woude, A.D., Angermayr, S.A., Puthan Veetil, V., Osnato, A., and Hellingwerf, K.J. (2014). Carbon sink removal: Increased photosynthetic production of lactic acid by *Synechocystis* sp. PCC6803 in a glycogen storage mutant. J. Biotechnol. *184*, 100–102. http://dx.doi.org/10.1016/j.jbiotec.2014.04.029.

van Zyl, L.J., Schubert, W.D., Tuffin, M.I., and Cowan, D.A. (2014a). Structure and functional characterization of pyruvate decarboxylase from *Gluconacetobacter diazotrophicus*. BMC Struct. Biol. *14*, 21. http://dx.doi.org/10.1186/s12900-014-0021-1.

van Zyl, L.J., Taylor, M., Eley, K., Tuffin, M., and Cowan, D.A. (2014b). Engineering pyruvate decarboxylase-mediated ethanol production in the thermophilic host *Geobacillus thermoglucosidasius*. Appl. Microbiol. Biotechnol. *98*, 1247–1259. http://dx.doi.org/10.1007/s00253-013-5380-1.

Vermaas, W. (1996). Molecular genetics of the cyanobacterium *Synechocystis* sp. PCC 6803: Principles and possible biotechnology applications. J. Appl. Phycol. *8*, 263–273. http://dx.doi.org/10.1007/BF02178569.

Vermaas, W.F. (1998). Gene modifications and mutation mapping to study the function of photosystem II. Methods Enzymol. *297*, 293–310. http://dx.doi.org/10.1016/S0076-6879(98)97022-7.

Vidal, R., López-Maury, L., Guerrero, M.G., and Florencio, F.J. (2009). Characterization of an alcohol dehydrogenase from the Cyanobacterium *Synechocystis* sp. strain PCC 6803 that responds to environmental stress conditions via the Hik34-Rre1 two-component system. J. Bacteriol. *191*, 4383–4391. http://dx.doi.org/10.1128/JB.00183-09.

Wang, J., Chen, L., Huang, S., Liu, J., Ren, X., Tian, X., Qiao, J., and Zhang, W. (2012). RNA-seq based identification and mutant validation of gene targets related to ethanol resistance in cyanobacterial *Synechocystis* sp. PCC 6803. Biotechnol. Biofuels *5*, 89. http://dx.doi.org/10.1186/1754-6834-5-89.

Widiastuti, H., Kim, J.Y., Selvarasu, S., Karimi, I.A., Kim, H., Seo, J.S., and Lee, D.Y. (2011). Genome-scale modeling and in silico analysis of ethanologenic bacteria *Zymomonas mobilis*. Biotechnol. Bioeng. *108*, 655–665. http://dx.doi.org/10.1002/bit.22965.

Williams, J.G.K. (1988). [85] Construction of specific mutations in photosystem II photosynthetic reaction center by genetic engineering methods in *Synechocystis* 6803. Methods Enzymol. *167*, 766–778. http://dx.doi.org/10.1016/0076-6879(88)67088-1.

Yu, Y., You, L., Liu, D., Hollinshead, W., Tang, Y.J., and Zhang, F. (2013). Development of *Synechocystis* sp. PCC 6803 as a phototrophic cell factory. Mar. Drugs *11*, 2894–2916. http://dx.doi.org/10.3390/md11082894.

Engineering of Alkane Production in Cyanobacteria

11

Xuefeng Lu[1,2*] and Weihua Wang[1,3*]

[1]Key Laboratory of Biofuels, Qingdao Institute of Bioenergy and Bioprocess Technology, Chinese Academy of Sciences, Qingdao, China.
[2]Shandong Provincial Key Laboratory of Synthetic Biology, Qingdao Institute of Bioenergy and Bioprocess Technology, Chinese Academy of Sciences, Qingdao, China.
[3]Shandong Provincial Key Laboratory of Energy Genetics, Qingdao Institute of Bioenergy and Bioprocess Technology, Chinese Academy of Sciences, Qingdao, China.

*Correspondence: lvxf@qibebt.ac.cn and wangwh@qibebt.ac.cn

Abstract

Alkanes with defined carbon chain lengths possess higher energy density, low hygroscopicity and volatility, and compatibility with existing liquid fuel infrastructure, which are the predominant constituents of gasoline, diesel and jet fuels. Alkane biosynthesis is ubiquitous and biosynthetic pathways have been identified in cyanobacteria, photosynthetic microbes, which opens a door to engineer alkane production with high efficiency in cyanobacteria. Firstly, redirecting the carbon flux to fatty acids or acyl-acyl carrier proteins can provide larger precursor pools for further conversion to alkanes. In combination with the overexpression of alkane biosynthesis genes, alkane production can be significantly improved in engineered strains. Protein engineering for key enzymes in alkane biosynthesis pathways will further enhance the yield of alkanes. Secondly, systems biology research on cyanobacteria will increase our knowledge about the metabolism in cyanobacteria and lead to significant improvements in strain modification for alkane production. Convenient and effective molecular tools for genetic engineering of cyanobacteria will expand the ability to engineer cyanobacteria for alkane production. It is significant and promising to directly utilize solar energy and convert carbon dioxide into alkanes, drop-in biofuels, in cyanobacteria.

Introduction

Cyanobacteria are becoming more and more attractive as biofuel-producing microbial hosts because they can utilize solar energy to directly convert carbon dioxide (CO_2) into biofuels and chemicals in one single biological system (Sarsekeyeva *et al.*, 2015). A wide range of metabolic pathways for biofuel and chemical production have been constructed or engineered in cyanobacteria including ethylene (Ungerer *et al.*, 2012; Zhu *et al.*, 2015), ethanol (Deng and Coleman, 1999; Dexter and Fu, 2009; Gao *et al.*, 2012b), butanol (Lan and Liao, 2011), fatty alcohol (Tan *et al.*, 2011; Yao *et al.*, 2014), sucrose (Du *et al.*, 2013; Ducat *et al.*, 2012), isoprene (Lindberg *et al.*, 2010), fatty acid (Liu *et al.*, 2011b) and alkane (Wang *et al.*, 2013). Compared to plants and eukaryotic microalgae, cyanobacteria are prokaryotic microbes with higher solar conversion efficiency and growth rate (Ducat *et al.*, 2011; Lu, 2010). *Synechocystis* sp. PCC 6803 (hereafter *Synechocystis* 6803) was the first photosynthetic microorganism for which the complete genome was determined in 1996 (Kaneko *et al.*, 1996). So far more than 180 genomic sequences of cyanobacteria strains are available (Klahn *et al.*, 2014). Well established genetic manipulation platform makes cyanobacteria better candidates to build highly efficient biosynthetic pathways for biofuel and chemical production (Huang *et al.*, 2010).

Alkanes with defined carbon chain lengths possess higher energy density, low hygroscopicity and volatility, and compatibility with current engines and transportation facilities, which are the predominant constituents of gasoline, diesel and jet fuels (Peralta-Yahya *et al.*, 2012). Alkanes can be produced by a variety of organisms such as bacteria, yeasts, fungi, plants and insects (Schirmer *et al.*, 2010). In the late 1960s, production of alkanes was reported in a diversity of cyanobacteria (Han *et al.*, 1968). In 2010, a two-step alkane biosynthetic pathway in cyanobacteria was identified (Fig. 11.1). Acyl-acyl carrier protein (ACP) can be reduced to aldehyde by an acyl-ACP reductase (AAR, EC 1.2.1.50), and then aldehyde can be oxidized to alkane by an aldehyde-deformylating oxygenase (ADO, EC 4.1.99.5) (Li *et al.*, 2012; Schirmer *et al.*, 2010). In 2011, an olefin synthase (OLS) pathway was found in a cyanobacterium strain, *Synechococcus* sp. PCC 7002 (Mendez-Perez *et al.*, 2011) (Fig. 11.1). This route of hydrocarbon biosynthesis employs an elongation-decarboxylation mechanism converting acyl-ACPs to medium-chain olefins.

To date, it has been found that only two above native biosynthesis pathways of alkanes in cyanobacteria, one is the AAR-ADO pathway, which contains two key enzymes, AAR and ADO; the other is the OLS pathway, which involves a polyketide synthase (PKS). Both pathways can

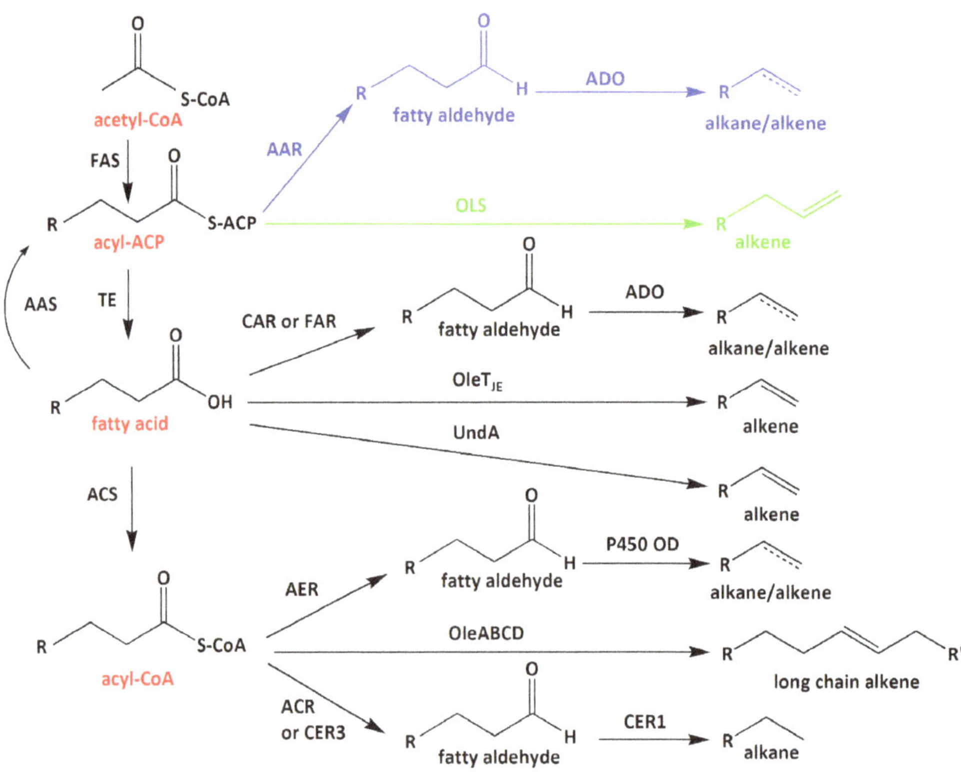

Figure 11.1 Fatty acid-based alkane biosynthesis pathway. Metabolic pathways for the biosynthesis of alkanes or alkenes have been established in various microorganisms. The AAR-ADO and OLS pathways for alkane/alkene biosynthesis in cyanobacteria are indicated in blue and green, respectively. FAS, Fatty acid synthase; ACP, acyl carrier protein; TE, thioesterase; AAS, acyl-ACP synthetase; ACS, acyl-CoA synthase; AAR, acyl-ACP reductase; ADO, aldehyde-deformylating oxygenase; OLS, a type I polyketide synthases for α-olefin biosynthesis; CAR, carboxylic acid reductase; FAR, fatty acid reductase; OleT$_{je}$, a cytochrome P450 enzyme from *Jeotgalicoccus* species that reduces fatty acids to alkenes; UndA, a nonheme iron oxidase from Pseudomonas that converts fatty acids into terminal olefins; AER, acyl-ester reductases; P450 OD, an insect P450 oxidative decarbonylase; OleABCD, a four protein families for long-chain olefin biosynthesis; ACR, acyl-CoA reductase; CER3, an acyl-CoA reductase from *Arabidopsis thaliana*; CER1, a fatty aldehyde decarbonylase from *Arabidopsis thaliana*.

catalyse the synthesis of alkanes from acyl-ACPs. The AAR-ADO pathway is consistent with the 'n-1' rule for alkanes, while the OLS pathway elongates the acyl chain with one carbon to produce olefins. Alkane biosynthesis pathways were identified in all genome-sequenced cyanobacteria over a broad phylogenetic range and none of the cyanobacteria appear to contain both the AAR-ADO and OLS pathways.

Alkane biosynthesis is ubiquitous in cyanobacteria. Heptadecane and heptadecene are the most abundant hydrocarbons in *Synechocystis* 6803. Two hydrocarbons, 1-nonadecene and 1,14-nonadecadiene, were identified in *Synechococcus* sp. PCC 7002 (hereafter – *Synechococcus* 7002). The hydrocarbon content is quite low (about 0.1% of the cell dry weight) in cyanobacteria (Coates *et al.*, 2014; Schirmer *et al.*, 2010). Biologically synthesized hydrocarbon molecules in cyanobacteria cells can be easily converted into alkanes with specific carbon chain length used as liquid biofuels by oil refining. The bottleneck is the relatively low biosynthetic efficiency of alkane in cyanobacteria compared with ethanol or butanol. So the key point is how to significantly improve the efficiency of alkane production in cyanobacteria. Two solutions can be proposed for this goal: (1) to genetically and metabolically modify alkane biosynthetic pathway in cyanobacteria strains; (2) to screen the alkane biosynthetic enzymes with higher catalytic efficiency though structure–function analysis-based rational design and high-throughput screening-based directed evolution strategies. Metabolic engineering of alkane biosynthetic pathways and protein engineering of alkane biosynthetic enzymes will be integrated, which will further enhance the yield of alkanes.

Alkane profiles in native cyanobacteria

Early research in the 1960s concerning hydrocarbon production in cyanobacteria was focused on their potential geochemical and taxonomic significance. The results showed that a variety of cyanobacterial strains (e.g. *Nostoc* sp., *Anacystis* sp., *Anacystis nidulans*, etc.) mainly produce hydrocarbons with carbon chain lengths varying from C15 to C18 with a predominance of n-C17 compounds (Han *et al.*, 1968; Winters *et al.*, 1969). The mixture of 7- and 8-methylheptadecanes, which is typical hydrocarbon of cyanobacteria, was shown to derive from *cis*-vaccenic acid (11-octadecenoic acid) by the addition of methyl group of methionine to a double bond with the subsequent reduction and decarboxylation of the fatty acids (Bird and Lynch, 1974; Fehler and Light, 1970). Two marine unicellular strains, *Synechococcus* sp. PCC 7002 and *Synechococcus* sp. PCC 7003 were indicated to synthesize C19 alkenes with one or two double bonds (Winters *et al.*, 1969). Morphological identification and alkane/alkene compositions of model cyanobacterial strains are shown in Fig. 11.2.

Recently, 19 common cyanobacterial strains distributed among 13 genera were autotrophically grown with continuously filtered air and then evaluated for alkane production (Liu *et al.*, 2013). The biomass concentration of different strains varied greatly, with the lowest value of 0.7 g/l dry cell weight (DW) in *Nostoc punctiforme* FACHB 252 and the highest value of 1.3 g/l DW in *Nostoc spongiaeforme* FACHB 130. All unicellular strains grew faster than filamentous strains before entering the exponential phase. The alkane yield per gram DW at the late stationary phase ranged from 0.2 to 1.8 mg. Among the different cyanobacterial strains distributed 13 genera, the alkane content of 9 strains exceeded 1 mg/g DW with the maximum content of 1.8 mg/g DW observed in *Nostoc spongiaeforme* FACHB 130 in a homogenously suspended mode (Liu *et al.*, 2013).

The 19 cyanobacteria strains were classified into five major types according to the hydrocarbon profiles, and heptadecanes were widespread in all five types. In addition to heptadecane, all six strains of type 1 contained 8-methylheptadecane in varied amounts. Two *Nostoc* strain in type 2 were first detected to produce 4-ethyltetradecane in cyanobacteria based on a preliminary gas chromatography–mass spectrometry (GC-MS) analysis. The type 3 strains, including two *Microcystis* species and other filamentous strains, only biosynthesized heptadecane. $C_{17:1}$ and $C_{15:0}$ hydrocarbons besides heptadecane, were found in those strains belonging to type 4 and type 5, respectively. The analysis of hydrocarbon profiles and the morphologies of those cyanobacterial strains showed a tendency for branched alkanes (ethyltetradecane and methylheptadecane) to be present in filamentous cyanobacteria.

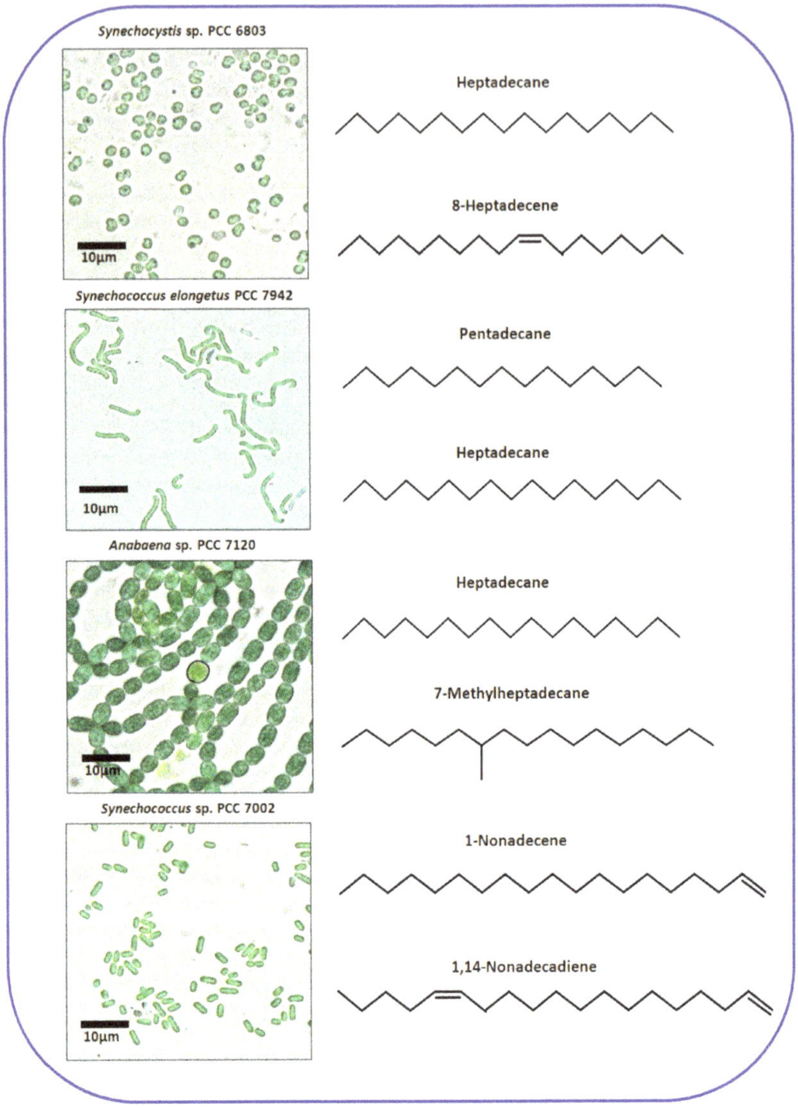

Figure 11.2 Morphological identification and alkane/alkene compositions of model cyanobacterial strains.

In another work Coates *et al.* (2014) reported the alkane compositions of 32 cyanobacterial strains were analysed and compared in the context of their respective biosynthetic pathways. Alkane yields as a per cent of DW ranged from 0.024% in *Cyanothece* sp. PCC 7425 to 0.262% in *Pleurocapsa* sp. PCC 7516, and averaged 0.11% across all strains. Strains possessing the OLS pathway appear to have significantly higher hydrocarbon yields (0.173%) than strains with the AAR-ADO pathway (0.070%). Alkanes such as heptadecane as well as all branched alkanes (e.g. 7-methylheptadecane), were only observed in strains that contained the AAR-ADO pathway while strains with the OLS pathway appeared to only produce alkenes. *Lyngbya* sp. PCC 8106 produced 5-heptadecene, an alkene with an internal double bond instead of a terminal olefin consistent with the OLS pathway. Therefore, 5-heptadecene in *Lyngbya* sp. PCC 8106 is likely derived from a monounsaturated fatty acid (Coates *et al.*, 2014).

Branched alkanes, most commonly as 7-methylheptadecane, were observed only in particular clades including heterocystous, ramified, and some filamentous cyanobacteria including *Planktothrix*. *Fischerella* sp. PCC 7414 was found to produce several less common 4-, 5- and 6-methylheptadecaneson the basis of fragmentation patterns. *Chlorogleopsis* sp. PCC 6912 also produces less common 3- and 4-methylheptadecane. However, except for 3- and 7-methylheptadecanes for which authentic standards were applied, the position of methylation in other methylheptadecanes remains tentative (Coates *et al.*, 2014).

Additional investigation revealed new double bond (2- and 3-heptadecene) and methyl group positions (3-, 4- and 5-methylheptadecane) for a variety of cyanobacteria. Cyanobacteria strains harbouring the OLS pathway formed a cohesive phylogenetic clade with the exception of three *Moorea* strains and *Leptolyngbya* sp. PCC 6406, which may have acquired the OLS pathway via horizontal gene transfer. Alkane pathways were identified in 142 strains of cyanobacteria over a broad phylogenetic range and there were no instances where both the AAR-ADO and the OLS pathways were found simultaneously in the same genome, which suggests that an unknown selective pressure maintains either AAR-ADO or OLS pathway (Coates *et al.*, 2014).

Molecular tools for genetic engineering of cyanobacteria

The bottleneck for alkane production in cyanobacteria is the relatively low efficiency of alkane biosynthesis. Genetic engineering of cyanobacteria will improve alkane production, and effective molecular tools will expand the ability to engineer cyanobacteria. In recent years, exciting results have been achieved for biological parts, tools and methods to engineer cyanobacteria, such as Clustered regularly interspaced short palindromic repeats interference (CRISPRi), riboswitch, shuttle vectors, markerless selection, and genetic parts for synthetic biology manipulations.

CRISPRi for gene repression was applied in *Synechocystis* 6803. The nuclease-deficient CRISPR-associated protein-9 nuclease (Cas9) from the type II CRISPR/Cas of *Streptrococcus pyogenes* was used to repress green fluorescent protein (GFP) to negligible levels. CRISPRi was also applied to repress production of carbon storage compounds polyhydroxybutryate (PHB) and glycogen during nitrogen starvation. The benefit of CRISPRi over traditional tools is the ability to repress multiple genes simultaneously.

A cyanobacterium strain with deficient in aldehyde reductases and dehydrogenases may provide more aldehyde precursors for alkane production. Four putative aldehyde reductases and dehydrogenases (slr0942, sll0990, slr1192, and slr0091) were simultaneously knocked down at 50–95% repression. These results also demonstrate that tightly repressed, low-strength promoters are most suitable for reversible repression in cyanobacteria (Yao *et al.*, 2016).

Theophylline-dependent riboswitch system was employed to control translational initiation of target genes in four diverse species of cyanobacteria, including *Synechococcus elongatus* PCC 7942 (hereafter *Synechococcus* 7942), *Leptolyngbya* sp. strain BL0902, *Anabaena* sp. strain PCC 7120, and *Synechocystis* sp. strain WHSyn. Reporter assays revealed that theophylline riboswitches performed as expected in the cyanobacteria. Riboswitches exhibited very low leaky expression, superior and dose-dependent on/off regulation of protein expression by theophylline. The wide-ranging induction level was responsive to a wide range of theophylline concentrations in the medium, facilitating the fine-tuning of protein expression. Expression of the circadian clock *kaiC* gene product and toxic protein SacB can be effectively regulated by the theophylline concentration. The results demonstrate the theophylline-dependent riboswitch system has potential for various applications as a useful genetic tool in diverse cyanobacteria (Ma *et al.*, 2014; Nakahira *et al.*, 2013).

Construction of markerless mutant by eliminating the selectable markers from the chromosomes will be necessary for the multi-site genetic modifications and the industrial application of cyanobacteria. Some site-specific recombinases can excise antibiotic markers flanked by the target sites through homologous recombination (Grindley *et al.*, 2006), which have emerged as powerful molecular tools for genetic engineering in both prokaryotes and eukaryotes (Askora *et al.*, 2012; Lyznik *et al.*, 2003; Schweizer, 2003; Siegal and Hartl, 2000). The Cre-loxP recombination system was firstly applied in the filamentous *Anabaena* sp.

PCC 7120 for generating the markerless mutant (Zhang *et al.*, 2007). The self-replicable shutter vectors harbouring the flipase (FLP) gene from *Saccharomyces cerevisiae* were introduced into two mutants of *Synechocystis* 6803 and *Synechococcus* 7942 whose genomes were inserted by a kanamycin resistance gene with flipase recombination target (FRT) flanking. Genotyping analysis indicated that FLP exactly excised and ligated the two FRT sites between which a kanamycin resistance gene was located. The shuttle vectors carrying the FLP gene were lost in these mutants afterwards by single colony separation in the absence of antibiotic (Tan *et al.*, 2013). In future, these powerful tools will facilitate the systematic genetic modification and biofuel production in cyanobacteria.

As vectors for introducing genetic information into the cell, plasmids are essential parts to consider when devising a strategy for engineering cyanobacteria. Because of the presence of several copies of the chromosome in cyanobacteria, it is important to make sure that an integrated construct is present in all copies, which necessitates an often time-consuming segregation procedure. Therefore, it is often easier to introduce and select for a replicative vector than to integrate a construct into the genome of a cyanobacterium.

As molecular cloning is generally performed in *Escherichia coli*, it is necessary to employ shuttle vectors, which can replicate both in *E. coli* and in the cyanobacterial strains. In cyanobacteria, two types of shuttle vectors have been applied: vectors harbouring a cyanobacterial specific replicon and an *E. coli* replicon, and vectors with a broad-host-range replicon. As a shuttle vector with two different replicons for *E. coli* and cyanobacteria, pSCR119 that contains the *E. coli* pMB1 replicon and the pDC1 replicon from *Nostoc* enables shuttling of DNA constructs between *E. coli* and *Nostoc* (Summers *et al.*, 1995). Such vectors may be more cyanobacterial host-specific than single origin broad-host-range plasmids. Examples of broad-host-range vectors are plasmids with a host-independent replicon derived from RSF1010 (Scholz *et al.*, 1989), exemplified by the BioBrick compatible pPMQAK1. Using broad-host-range plasmids, DNA constructs can be shuttled between the construction host and a wide range of final destination cyanobacteria, including *Synechocystis* 6803, *Synechococcus* 7942, *Pseudanabaena* sp. ABRG5-3, *Nostoc* sp. PCC 7120 and *Nostoc punctiforme* PCC 73102 (Asayama, 2012; Huang *et al.*, 2010). In most cases, integrative plasmids are suicide vectors that are unable to replicate in the cyanobacterium host, and therefore any plasmid that can be efficiently transferred into the recipient cyanobacterium may be used. Several integrative vectors, such as pFQ9R, pFQ20, pXT37a and pXT37b, have been developed for heterologous expression of alkane biosynthesis genes in cyanobacteria (Qi *et al.*, 2013; Tan *et al.*, 2011).

Synthetic biology encompasses an increasing number of tools to facilitate strain construction and optimization, which includes the use of standardized, well-characterized biological parts and modelling of the genetic and metabolic networks within cells (d'Espaux *et al.*, 2015). Preliminary analyses show that certain widely used promoters in *E. coli* behave very differently in model cyanobacteria, which reveals the necessity for part characterization in cyanobacteria and the need for characterization of new biological parts for cyanobacteria (Huang *et al.*, 2010).

Synthetic biology tools for cyanobacteria lag behind what has been developed for *E. coli* and yeast. Furthermore, genetic parts for *E. coli* or yeast often do not function as designed in cyanobacteria (Huang *et al.*, 2010). A diversity of constitutive or inducible promoters had been characterized in cyanobacteria (Huang *et al.*, 2010; Qi *et al.*, 2013). A nickel inducible promoter, P_{nrsB} was applied to construct a controllable lysis system in *Synechocystis* 6803 to facilitate extracting lipids for biofuel production (Liu and Curtiss, 2009). Lipolytic enzymes under the control of promoters induced by CO_2 limitation were used to convert membrane lipids into fatty acids (Liu *et al.*, 2011a). The genomic sequence of 560 bp upstream of the initiation codon of *cpcB* was verified as a super-strong promoter to drive gene expression in cyanobacteria (Zhou *et al.*, 2014).

Characterization of genetic parts depends on reporters to track the gene expression. Fluorescent proteins and bacterial luciferase are the most common reporters, because they can be detected without destruction of organisms or additional inputs. The lux operon is frequently used in cyanobacteria (Aoki *et al.*, 2002; Peca *et al.*, 2008) for real time reporting of gene expression due to the short half-lives of the relevant enzymes. A broad range of fluorescent protein genetic variants have been developed that feature fluorescence emission spectral profiles spanning almost the entire visible light spectrum. Although photosynthetic pigments in cyanobacteria may interfere with fluorescence emission of fluorescent proteins, Cerulean, GFPmut3B (a mutant of green fluorescent protein) and EYFP (enhanced yellow fluorescent protein) have been demonstrated to work as reporter proteins in cyanobacteria (Huang *et al.*, 2010; Landry *et al.*, 2013).

Systems biology research on cyanobacterial alkanes

Systems biology research on cyanobacteria can facilitate significant improvements in strain design for alkane production. Using innovative transcriptomic approaches such as differential RNA sequencing, the first genome-wide map of transcriptional start sites for *Synechocystis* 6803 was established (Mitschke *et al.*, 2011). Furthermore, a specific suite of microarrays for transcriptome analysis of *Synechocystis* 6803 has been established, including tiling microarrays (Georg *et al.*, 2009) and a comprehensive microarray platform, monitoring both protein-coding and non-coding transcripts in a compact and economic format. These data provided unprecedented insight into the complexity of the primary transcriptome of *Synechocystis* 6803, revealing the location of a total of 3527 transcriptional start sites (TSSs) on the chromosome and the four megaplasmids (Mitschke *et al.*, 2011).

A high number of 1013 start sites of antisense transcripts were identified in the *Synechocystis* 6803 genome with its altogether only 3172 annotated ORFs (NCBI annotation), which demonstrated a new level of complexity in the transcriptome of this model organism. Many of these non-coding RNAs are important regulators of gene expression in cyanobacteria (Georg *et al.*, 2014; Klahn *et al.*, 2015). In summary, the annotated primary transcriptome of *Synechocystis* 6803 will greatly facilitate the use of this organism as a simple photosynthetic model in systems biology research and for the biofuel production.

The analysis of 181 available cyanobacteria genomes showed that in 61% of them the AAR-ADO gene cluster together with genes encoding acetyl-CoA carboxyl transferase and a short-chain dehydrogenase, strengthen the link to fatty acid metabolism. Contrary to the expectations for an operon, specific promoters for the two genes, *sll0208* (*ado*) and *sll0209* (*aar*) were found in *Synechocystis* 6803, which gave rise to monocistronic transcripts. Moreover, on the upstream of *sll0208* gene, a ~160 nt noncoding RNA, SyR9, is transcribed. Thus, the transcriptional organization of alkane biosynthesis genes in *Synechocystis* 6803 is of substantial complexity. Independent function of three promoters was verified, and a similar promoter arrangement has been also found in more distant species – *Nodularia spumigena*, *Trichodesmium erythraeum*, *Anabaena* sp. PCC 7120, *Prochlorococcus* MIT9313, and MED4. The presence of separate regulatory elements and the dominance of monocistronic mRNAs indicate the possible independent regulation of *ado* and *aar* genes. The complicated transcriptional organization of the alkane biosynthesis genes has possible genetic and metabolic implications in cyanobacterial alkane production (Klahn *et al.*, 2014).

The responses of *Synechocystis* 6803 to hexane were investigated using a quantitative proteomics approach with isobaric tags for relative and absolute quantitation (iTRAQ)-liquid chromatography (LC)–MS/MS technologies. In total, 1492 proteins were identified, which represented about 42% of all predicted proteins in *Synechocystis* 6803. Among all proteins identified, 164 and 77 proteins were found up- and down-regulated, respectively. Functional annotation and KEGG pathway enrichment analyses showed that common stress responses were induced by hexane in *Synechocystis* 6803. Notably, a large number of transporters and membrane-bound proteins, proteins against oxidative stress, as well as proteins related to sulfur relay system and photosynthesis were induced, suggesting that they are possibly constitute the major protection mechanisms against hexane toxicity. The study provided a list of potential targets to engineer cyanobacteria against alkane stress (Liu *et al.*, 2012).

Fatty acid production in engineered cyanobacteria

Fatty acids can be used as precursors for further conversion to alkanes. Cyanobacteria were modified to produce and secrete free fatty acids (FFAs; Chapter 8 of this book). Starting with introducing an acyl–ACP thioesterase (TE) gene, six successive generations of engineered strains of

Synechocystis 6803 were constructed. The yield of secreted fatty acids was increased to about 200 mg/l in one improved strain by adding codon-optimized TE genes and weakening cell wall layers. The results suggest that fatty acid secretion in cyanobacteria is a promising approach for renewable biofuel production (Liu *et al.*, 2011b).

A recovery strategy was engineered to convert membrane lipids into fatty acids by utilising lipolytic enzymes under the control of promoters induced by CO_2 limitation. It was indicated that engineered *Synechocystis* 6803 strain underwent degradation of membrane lipids upon CO_2 limitation, leading to release of fatty acids into the culture medium. Recovered fatty acid yields of about $3.6. \times 10^{-11}$ mg/cell were measured in one of the engineered strains (Liu *et al.*, 2011a).

The thermostable lipases encoded by genes from thermophilic organisms have been transferred into the cyanobacterial genome. When synthesized in FFA-producing strains, the lipase Fnl from *Fervidobacterium nodosum* Rt17-B1 released about 43 mg/l FFAs at 47°C. Continuous semi-batch production experiment showed that daily harvested cultures released an average of about 44 mg/l fatty acids for at least 20 days without significant decline (Liu and Curtiss, 2012). Green recovery methods can be incorporated into previously constructed fatty acid -secretion strains, enabling fatty acid recovery from the remaining cyanobacterial biomass that will be generated during fatty acid production.

Alkane production in engineered cyanobacteria

A series of *Synechocystis* 6803 mutant strains have been constructed by redirecting the carbon flux to acyl-ACP and overexpressing alkane biosynthesis genes, which can provide larger precursor pools and increase the efficiency of alkane biosynthesis (Fig. 11.3). Alkane biosynthetic genes from *Synechocystis* 6803 (*sll0208* and *sll0209*), *Synechococcus* 7942 (Synpcc 7942_1593 and 1594) and *Nostoc punctiforme* PCC 73102 (*npun1710* and *npun1711*) were overexpressed in *Synechocystis* 6803. Alkane production can also be doubled in these mutants. Fatty acids as precursors of alkanes are biosynthesized by fatty acid synthase (FAS). Acetyl-CoA is converted to malonyl-CoA by a multi-subunit acetyl-CoA carboxylase (ACC) consisting of AccA, AccB, AccC and AccD, which is the rate-limiting step of fatty acid biosynthesis. In a *Synechocystis* 6803 mutant overexpressing *accBCDA* genes, a 56% increase of native alkane production was obtained (Tan *et al.*, 2011). With this mutant, the *sll0208* and *sll0209* were overexpressed to yield LX57 strain. The alkane production of LX57 mutant was enhanced by 3.6 times when compared to wild-type cells of *Synechocystis* 6803.

FFAs can be activated to acyl-ACPs by an acyl-ACP synthetase (AAS, EC 6.2.1.20). It is indicated that *slr1609* is the only AAS gene in *Synechocystis* 6803 (Kaczmarzyk and Fulda, 2010). The *slr1609*-knockout mutant was incapable of importing exogenous fatty acids and secreted fatty acids released from membrane lipids into the medium. This suggests a remarkable role for this cyanobacterial AAS in recycling released fatty acids (Kaczmarzyk and Fulda, 2010). The alkane production was significantly reduced by around 90% compared to the wild-type strain in *slr1609* deletion mutant of *Synechocystis* 6803. The latter indicates that AAS plays an essential role in alkane production. However, the alkane production was not enhanced by the overexpression of *slr1609* gene alone (Gao *et al.*, 2012a). However, a mutant overexpressing alkane biosynthetic genes together with *slr1609* showed enhanced productivity of alkane by 130% if compared to wild-type cells of *Synechocystis* 6803.

The lipolytic enzymes are capable of hydrolysing acyl chains from membrane lipids. FFAs released from membrane lipids can be activated to acyl-ACPs by an AAS. Alkane productivity of the *Synechocystis* mutant overexpressing *sll0208*, *sll0209*, and a gene for a putative lipolytic enzyme, *sll1969*, was enhanced by three times in comparison with wild-type strain (Wang *et al.*, 2013).

The β-ketothiolase encoded by *phaA* (*slr1993*) condenses two molecules of acetyl-CoA to acetoacetyl-CoA, which is the first step of poly-β-hydroxybutyrate (PHB) biosynthesis. Alkane productivity of the *Synechocystis* 6803 mutant lacking *slr1993* and overexpressing *sll0208 and sll0209* was enhanced by 150% compared with wild-type cells. Overexpression of *aar* and *ado* genes and deletion of PHB biosynthetic gene(s) simultaneously can divert acetyl-CoA and NADPH into production of alkanes (Wang *et al.*, 2013).

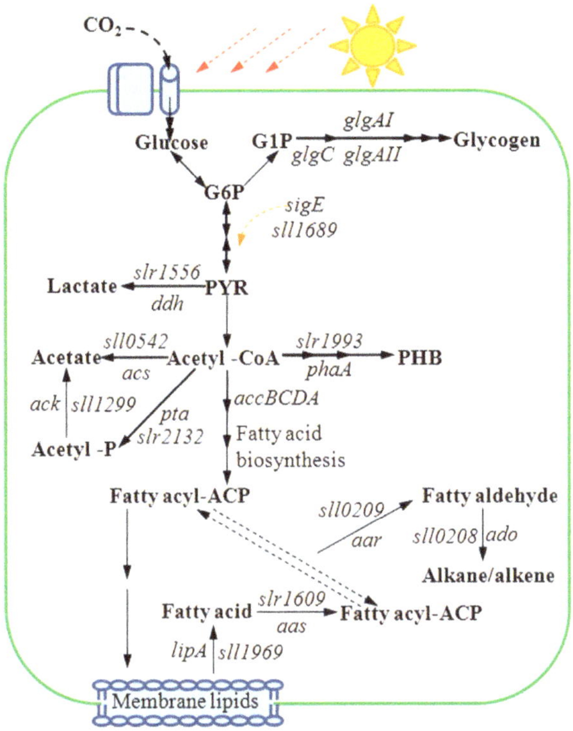

Figure 11.3 Schematic overview of fatty acid, alkane and main competing metabolic pathways in *Synechocystis* sp. PCC 6803. Key enzymes in those pathways are indicated. G6P, glucose 6-phosphate; G1P, glucose 1-phosphate; glgC, glucose-1-phosphate adenylyltransferase; glgA, glycogen synthase; sigE, a group 2 sigma factor; PYR, pyruvate; acs, acetyl-coenzyme A synthetase; ack, acetate kinase; pta, phosphotransacetylase; PHB, poly-β-hydroxybutyrate; acyl-ACP, acyl-acyl carrier protein; ddh, 2-hydroxyacid dehydrogenase; phaA, polyhydroxyalkanoates-specific beta-ketothiolase; accBCDA, multi-subunit acetyl-CoA carboxylase; lipA, lipolytic enzyme; aas, acyl-ACP synthetase; aar, acyl-ACP reductase; ado, aldehyde-deformylating oxygenase.

Alkane productivity of LX56 strain overexpressing *sll0208 and sll0209* in both *slr0168* and *slr1556* (2-hydroxyacid dehydrogenase gene, *ddh*) loci was constructed. Alkane production of LX56 strain was enhanced by 8.3 times, up to 1.3% of DW (6.5 mg/l). Small-scale photo-bioreactors were applied for precise characterization of wild-type and engineered strains to optimize the culture conditions and alkane production. The LX56 mutant and wild-type strain exhibited similar growth characteristics when cultivated in the bubble column photo-bioreactors. Both of the cultures in photo-bioreactors reached a much higher density (OD_{730} ~ 13) than cultures in shake flasks (OD_{730} ~ 3 to 4) at stationary phase. The alkane production of LX56 strain increased 8 times compared to wild-type cells; up to 26 mg/l, which was 1.1% of DW (Wang *et al.*, 2013).

Lipid droplets (LDs) were identified throughout the cell in the cyanobacterium *Nostoc punctiforme*. Analysis of isolated LDs revealed enrichment of hydrophobic energy-dense compounds containing triacylglycerol (TAG), α-tocopherol, and heptadecanes. The proportion of heptadecanes increased during stationary phase in comparison with exponentially grown cells (Peramuna and Summers, 2014). Heptadecane production in *N. punctiforme* was enhanced by introducing extrachromosomal copies of native *aar-ado* genes and a putative novel lipase (Npun_F5141) gene. A 16-fold higher accumulation of heptadecane (12.9% of DW) compared to the wild-type strain was obtained in the exponential phase under high light conditions by introducing extra copies of all three genes (Peramuna *et al.*, 2015).

Effects of nitrogen deficiency and salt stress on alkane biosynthesis were investigated in three kinds of cyanobacteria, including nitrogen-fixing cyanobacterium *Anabaena* sp. PCC 7120, non-diazotrophic cyanobacterium *Synechococcus* 7942, and halotolerant cyanobacterium *Aphanothece halophytica*. Intracellular alkane accumulation increased in *Anabaena* sp. PCC 7120, but decreased in *Synechococcus* 7942, and stayed constant in *A. halophytica* under nitrogen-deficient condition. It was shown that salt stress increased alkane accumulation in *Anabaena* sp. PCC 7120 and *A. halophytica*. The expression levels of *aar* and *ado* genes in *Anabaena* sp. PCC 7120 were not upregulated significantly under nitrogen deficiency or salt stress. The engineered *Anabaena* sp. PCC 7120 strain with introduced alkane biosynthesis gene from *A. halophytica* accumulated more intracellular alkanes compared to the control strain. These results provide a prospect to improve alkane production in cyanobacteria via abiotic stress treatments and genetic engineering (Kageyama *et al.*, 2015).

The 'OLS' pathway is present in a small number of cyanobacterial species including *Synechococcus* 7002 compared to the AAR-ADO pathway. Two hydrocarbons, 1-nonadecene and 1,14-nonadecadiene, were identified in *Synechococcus* 7002. The hydrocarbon titre was enhanced to approximately 4.2 µg/ml/OD_{730} by overexpressing the endogenous *ols* gene in *Synechococcus* 7002 (Mendez-Perez *et al.*, 2011). Alkane biosynthesis genes (*aar* and *ado*) from *Synechococcus* 7942 were heterologously expressed in *Synechococcus* 7002, and alkane production was claimed to reach 5% of DW. However, the technical route has limited practical utility due to high concentration of antibiotics (spectinomycin at 600 µg/ml) used to maintain the selective pressure and alkane content (Reppas *et al.*, US7794969-B1).

The marine cyanobacterium *Synechococcus* sp. NKBG15041c was shown to produce α-olefins similar to those of *Synechococcus* 7002 via the OLS pathway. Two different promoters, a heterologous P_{trc} promoter from *E. coli* or a homologous P_{cpc} promoter, were used for the expression of the *aar* and *ado* genes from *Synechococcus* 7942 in *Synechococcus* sp. NKBG15041c. The results showed that heptadecane was produced in both transformants, while 1-nonadecene and nonadecadiene were still the major hydrocarbons with decreased titres compared to that in the wild-type. The *aar* and *ado* genes under P_{cpc} were expressed at a higher level compared to P_{trc}. The yield of heptadecane was 4.2 ± 1.2 µg/g of DW in the transformant carrying the homologous P_{cpc} promoter, which is higher than the transformant carrying the heterologous P_{trc} promoter (Yoshino *et al.*, 2015).

Protein engineering for ADO

So far the alkane production in cyanobacteria has been modest, which is in part because of the low activity of ADO. Protein engineering would play an important role in enhancing the activity of ADO, and then improve cyanobacterial alkane production. It was reported that ADOs are capable of ~3–5 catalytic turnovers *in vitro*. Supplementing an exhausted reaction with additional ADO rather than extra NADPH, ferredoxin (Fd), or ferredoxin-$NADP^+$ reductase (FNR), or a combination of three partners result in additional alkane accumulation. The results showed that ADO becomes inactivated during incubation with the electron transfer system. ADO was inhibited by hydrogen peroxide (H_2O_2) with an apparent K_i of 16 ± 6 µM upon evaluating the peroxide shunt to drive ADO catalysis. Supplementing catalase (CAT) to exhausted assays could restore ADO activity, which demonstrated that the inhibition by H_2O_2 was reversible. During *in vitro* reactions, H_2O_2 originated from poor coupling of reductant consumption with alkane formation. A fusion protein comprising an N-terminal CAT coupled to a C-terminal ADO (CAT–ADO) was constructed to prevent H_2O_2 inhibition by converting it to the co-substrate O_2. *In vitro* assays showed that the number of turnovers per active site remained at ~3 for native ADO but reached 225 for the CAT–ADO bifunctional protein. CAT–ADO expression in *E. coli* catalysed approximately five-fold more catalytic turnovers per ADO active site relative to the native ADO (Andre *et al.*, 2013).

A reducing system is required for ADO activity. The commonly used chemical reducing system is including phenazine methosulfate (PMS) or 1-methyoxy-5-methylphenazinium methylsulfate (MeOPMS) and NADH (Li *et al.*, 2012). NADPH, surrogate Fd and FNR from spinach constitute the widely used biological reducing system (Li *et al.*, 2012; Schirmer *et al.*, 2010). Fd from *Zea mays* and FNR from *Anabaena* sp. PCC7120 have also been applied to support activity of ADO from *Prochlorococcus marinus* MIT9313 (Andre *et al.*, 2013)

An endogenous reducing system including Fd (SynPcc7942_1499) and FNR (SynPcc7942_0978) was identified to support

ADO (Synpcc7942_1593) activity in *Synechococcus* 7942. The genes for the above mentioned proteins were cloned, overexpressed, the corresponding proteins were purified, and characterized. The results showed that ADO activity was successfully supported with the endogenous electron transfer system, which worked more effectively than the heterologous and chemical ones. ADO was selective against Fd according to the results of the hybrid Fd/FNR reducing systems. Less H_2O_2 was produced when the cognate reducing system rather than heterologous was applied in ADO-catalysed reactions. Moreover, *kcat* value of ADO using the homologous Fd/FNR electron transfer system is 3.7-fold higher than that using the PMS. The identified Fd/FNR electron transfer system can improve catalytic efficiency of ADO and further enhance alkane production in cyanobacteria (Zhang *et al.*, 2013).

Seven self-sufficient ADOs from *Synechococcus* 7942 were constructed, including FusA, FusB, FusC (ADO from *Synechococcus* 7942 fused to the RhFRed from P450RhF of *Rhodococcus* sp. NCIMB 9784 through changing the linker and the order of components) and FusD, FusE, FusF, FusG (ADO from *Synechococcus* 7942 fused to the endogenous Fd/FNR system from *Synechococcus* 7942 through changing the order of the components). The results showed that the activity of ADO from *Synechococcus* 7942 (Synpcc7942_1593) has not been affected by fusion to the reducing systems and the flexible 10-amino acid linker between Synpcc7942_1593 and RhFRed in FusB is beneficial for the activity of Synpcc7942_1593. Besides, the electron transfer ability of the Fd/FNR fusion proteins is stronger than that of the RhFRed fusion proteins, whereas the fusion proteins containing the Fd/FNR system exhibited much greater apparent *kcat* values, indicating that the endogenous electron transfer system matched better with ADO and supported greater ADO activity. FusG showed the highest catalytic efficiency (*kcat/Km*), which is 2 to 3-fold greater than those of FusD, FusE, FusF, and native Synpcc7942_1593 in the presence of non-fused Fd/FNR, 11.3-fold greater than that of native Synpcc7942_1593 in the presence of the chemical reducing system, demonstrating the order of the ADO-Fd-FNR in FusG is the best in these fusion ADOs (Wang *et al.*, 2014).

Five crystal structures of ADO from the *Synechococcus* 7942 in both its iron-free and iron-bound forms were determined, which represent different states during its catalytic process. As one of the iron-coordinating residues, Glu144 is revealed by structural comparisons that switch its location around active centre, which indicates its vital role in the catalytic reaction. The speculation is further confirmed by enzymatic activity and iron content assays. Moreover, the H5 helix where Glu144 resides exhibits two distinct conformations that may contribute to the formation of oxygen access to reactive centre and correlate with the different binding states of the di-iron centre in ADO structures. The helix H5 is unwound in the middle, forming two short helices (H5A and H5B) that are connected by a loop L5 in the iron-free ADO (Fig. 11.4). The helix to loop transition leads to conformation changes for a number of amino acids (from residue 144 to 150), thus the two iron-coordinating residues (Glu144 and His147) move away from the di-iron site (Jia *et al.*, 2015). Similar results were found in structures of ADOs from *Prochlorococcus marinus* (Buer *et al.*, 2014). These results provide

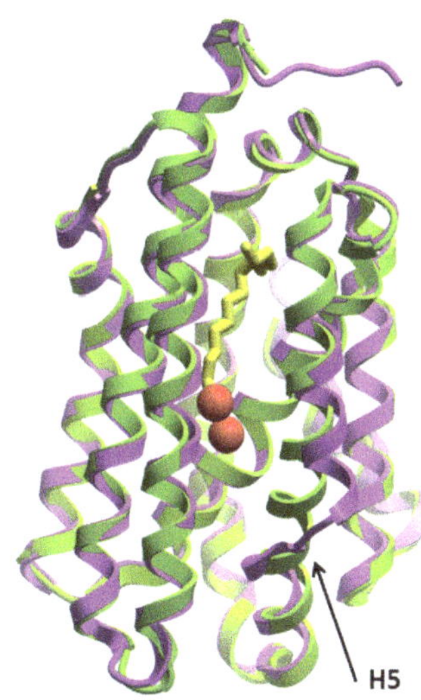

Figure 11.4 Overlay of the iron-bound and iron-free conformations of *Synechococcus* 7942 ADO. Iron bound form of ADO (molecule A of Y122F, PDB ID 4RC6) in green, iron-free form of ADO (WT0, PDB ID 4QUW) in purple. Helix 5 (H5) is partially unfolded between the residues 144 and 150. The iron atoms of the di-iron centre are shown in orange and the co-crystallized fatty alcohol in WT0 is shown in yellow.

a structural insight into the highly labile feature of ADO di-iron centre, which might be related to its low activity. On the basis of structural and biochemical data, a possible catalytic mechanism of ADO was proposed, which will contribute to improve ADO activity.

It is difficult to detect alkanes by a high-throughput method. So far, GC-MS is one of the limited methods available for quantification of alkanes, which are time-consuming and costly and thus not able to analyse large amount of samples. Although alkane response elements (AREs) like AlkR-PalkM or AlkS-PalkB have been successful in sensing alkanes in *Acinetobacter baylyi* or *E. coli*, none AREs have been reported to sense medium-chain alkanes *in vivo* in *E. coli*. For example, AlkS-PalkB could only respond to short-chain alkanes (i.e. C_5–C_{10}, C_8 preferred). It was reported that AlkS-PalkB failed to sense endogenous alkanes in the alkane-producing *E. coli*, in which medium-chain alkanes (C_{13}–C_{17}) were the main products due to the relatively abundant C_{14}–C_{18} fatty acids in the cells. AlkR-PalkM also failed to detect alkanes in *E. coli* probably due to the improper expression of AlkR in the heterologous host. A chimera ARE (cARE) was constructed by reassembled AlkR-PalkM and AlkS-PalkB, which consisting of PalkS, AlkR and PalkB. On the basis of the cARE and green fluorescent protein (GFP), a sensitive fluorescent biosensor for rapid and *in situ* monitoring of intracellular alkane titre was developed (Wu *et al.*, 2015).

To evaluate the correlation of fluorescence response with the alkane productivity of the cells, the fluorescence response of the cells as well as final alkane concentration of the cells in presence of different concentration of IPTG was evaluated. The fluorescent intensity was positively correlated with intracellular alkane concentration. The distributions of cells with different fluorescence levels in presence of varying IPTG concentration was analysed via flow cytometry. The cell counts approximately showed an normal distribution to the fluorescence intensity basing on nearly 50,000 cells, and the percentage of cells with stronger fluorescence response obviously increased with IPTG concentration (higher alkane productivity accordingly). The results suggests the feasibility of developing high-throughput strategies on the basis of the alkane biosensor in *E. coli*, and thus will greatly facilitate the application of directed evolution strategies to further improve the ADO activity (Wu *et al.*, 2015).

Conclusions

Cyanobacteria are capable of naturally and directly converting sunlight, CO_2 and H_2O into alkanes. Alkanes are less toxic to cyanobacteria than alcohols, acids and aldehydes (Kamarainen *et al.*, 2012). They also have high-energy density, low hygroscopicity and volatility, and compatibility with the existing infrastructure for fuel storage, transport and utilization. Therefore, alkanes are good candidates for biofuel production (Peralta-Yahya *et al.*, 2012).

Recently, there have been a number of studies on the development of cyanobacteria strains for the alkane production as biofuel, including *Synechocystis* 6803, *Nostoc punctiforme* PCC 73102, and *Synechococcus* 7002 (Table 11.1). However, most of the studies are only on its start-up, since the alkane titres are suboptimal for commercial production. One of the main reasons for the relatively low titres of alkane is low flux towards alkane biosynthesis and the low activities of the enzymes involved (Lee *et al.*, 2015). It was determined that 85% of photosynthetic carbon partitioning was directed towards sugar biosynthesis, while the fatty acid and terpenoid biosynthetic pathways were allocated only about 10% and 5% of photosynthetic carbon, respectively. Constitutive 10% carbon partitioning towards fatty acid biosynthetic pathway is primarily for membrane lipid formation, so little endogenous carbon flux can thus be directed towards fatty acid derived alkane accumulation (Lindberg *et al.*, 2010). It is evident that to improve titre of alkane production, metabolic engineering strategies would be needed to redirect photosynthetic carbon partitioning towards the fatty acid and alkane biosynthetic pathways.

Alkanes can be used as drop-in fuels similar to gasoline, diesel, and jet fuel. Cyanobacteria have emerged as a promising chassis for alkane production. A series of progresses in cyanobacterial alkane production has been made at the multi-scale study with genetic resources, systems biology research, metabolic and protein engineering. Further improvements are required to increase the efficiency of alkane biosynthesis in cyanobacteria.

Table 11.1 A summary of the alkane/alkene production in cyanobacteria

Alka(e)nes	Host Strains	Genetic Engineering design(s)	Cultivation conditions	Maximum production	Reference
Heptadecane and heptadecene	*Synechocystis* sp. PCC 6803	None	30°C, 30 µmol photons/m²/s	0.13% of DW	Wang *et al.* (2013)
Pentadecane and heptadecane	*Synechococcus elongatus* PCC 7942	None	30°C, 30–50 µmol photons/m²/s	0.05% of DW	Liu *et al.* (2013)
1-Nonadecene and 1,14-nonadecadiene	*Synechococcus* sp. PCC 7002	None	140 µmol photons/m²/s, 35°C	1.8 µg/ml/ OD_{730}	Mendez-Perez *et al.* (2011)
Heptadecane and methylheptadecane	*Anabaena* sp. PCC 7120	None	30°C, 30–50 µmol photons/m²/s	0.16% of DW	Liu *et al.* (2013)
Heptadecane and heptadecene	*Synechocystis* sp. PCC 6803	overexpressing two copies of native *aar/ado* in both slr0168 and slr1556 loci	30°C, 30 µmol photons/m²/s	1.3% of DW	Wang *et al.* (2013)
Heptadecane	*Nostoc punctiforme* PCC 73102	Co-overexpressing extra copies of *aar/ado* genes encoded by Npun._F1710/ Npun_F1711 and lipase encoded by Npun_F5141	25°C, 135–160 µmol photons/m²/s	12.9% of DW	Peramuna *et al.* (2015)
Heptadecane	*Anabaena* sp. PCC 7120	Overexpressing *aar* and *ado* from *Aphanothece halophytica*	40 µmol photons/m²/s, 30°C, 140 mM NaCl	Over 0.12% of DW	Kageyama *et al.* (2015)
1-Nonadecene and 1,14-nonadecadiene	*Synechococcus* sp. PCC 7002	Overexpressing endogenous *ols* gene	140 µmol photons/m²/s, 35°C	4.2 µg/ml/ OD_{730}	Mendez-Perez *et al.* (2011)
Pentadecane and heptadecane	*Synechococcus* sp. PCC 7002	Overexpressing aar/ado genes from *Synechococcus* PCC 7942	37°C, 100 µmol photons/m²/s, 600 µg/ml spectinomycin	5% of DW	Reppas *et al.*
Nonadecene, nonadecadiene and heptadecane	*Synechococcus* sp. NKBG 15041c	Overexpressing *aar/ado* genes from *Synechococcus* 7942	50 µmol photons/m²/s, 26°C	4.2 µg/g of DW heptadecane	Yoshino *et al.* (2015)

Future trends

Research on alkane production in cyanobacteria is only at the initial stage. Alkane biosynthesis is ubiquitous in cyanobacteria, while the regulatory mechanism and physiological roles of alkane metabolism are not understood (Coates *et al.*, 2014). Bypassing the native regulation and understanding the physiological roles of alkane metabolism will further pave a way to build a highly efficient alkane-producing cyanobacterial cell factory. New synthetic biology tools and techniques are needed for engineering cyanobacteria (Berla *et al.*, 2013). The use of computational and high-throughput approaches will allow development of better enzymes for alkane production. Considering the cost associated with harvesting and extraction, releasing alkane to the medium, either through cell lysis or by secretion, will be a good strategy for downstream processes (Doshi *et al.*, 2013). Based on further thorough understanding of regulation mechanism and physiological function of alkane metabolism in cyanobacteria, it will be crucial to construct highly efficient alkane-producing cyanobacterial cell factories and develop economical downstream cultivation and product separation process technologies.

Acknowledgements

We thank Drs Xufeng Liu, Lei Zhang and Tao Zhu in our lab for assist in image processing. This work was supported by the National Science Fund for Distinguished Young Scholars of China (31525002 to X. Lu), the Excellent Youth Award of the Shandong Natural Science Foundation (JQ201306 to X. Lu), the Shandong Taishan Scholarship (X. Lu), the Innovation Leading Talent of Qingdao (15-10-3-15-31-zch to X. Lu), SINO-GERMAN Research Project (GZ 877 to X. Lu), and the National Natural Science Foundation of China (31300036 to W. Wang).

Web resources: tools and databases

Biobike – powerful tools to manipulate and analyse biological data of cyanobacteria: http://biobike.csbc.vcu.edu:8003/biologin

Cyano2Dbase – a protein-gene linkage map of the unicellular cyanobacterium *Synechosystis* sp. strain PCC6803: http://wiki.annotation.jp/Cyano2Dbase

CopraRNA – a tool for sRNA target prediction: http://rna.informatik.uni-freiburg.de/CopraRNA/Input.jsp

Synergy – a web resource for exploring *Synechocystis* gene regulation, which integrates co-expression network analysis with motif analysis: http://synergy.plantgenie.org/

Cyanobacterial KnowledgeBase (CKB) – a free access database that contains the genomic and proteomic information of 74 fully sequenced cyanobacterial genomes belonging to seven orders: http://nfmc.res.in/ckb/index.html

References

Andre, C., Kim, S., Yu, X., and Shanklin, J. (2013). Fusing catalase to an alkane-producing enzyme maintains enzymatic activity by converting the inhibitory byproduct H_2O_2 to the cosubstrate O_2. Proc. Natl. Acad. Sci. U.S.A. *110*, 3191–3196. http://dx.doi.org/10.1073/pnas.1218769110

Aoki, S., Kondo, T., and Ishiura, M. (2002). A promoter-trap vector for clock-controlled genes in the cyanobacterium *Synechocystis* sp. PCC 6803. J. Microbiol. Methods *49*, 265–274.

Asayama, M. (2012). Overproduction and easy recovery of target gene products from cyanobacteria, photosynthesizing microorganisms. Appl. Microbiol. Biotechnol. *95*, 683–695. http://dx.doi.org/10.1007/s00253-012-3989-0.

Askora, A., Abdel-Haliem, M.E., and Yamada, T. (2012). Site-specific recombination systems in filamentous phages. Mol. Genet. Genomics.*287*, 525–530. http://dx.doi.org/10.1007/s00438-012-0700-1.

Berla, B.M., Saha, R., Immethun, C.M., Maranas, C.D., Moon, T.S., and Pakrasi, H.B. (2013). Synthetic biology of cyanobacteria: unique challenges and opportunities. Front. Microbiol. *4*, 246. http://dx.doi.org/10.3389/fmicb.2013.00246.

Bird, C.W., and Lynch, J.M. (1974). Formation of hydrocarbons by microorganisms. Chem. Soc. Rev. *3*, 309–328. http://dx.doi.org/10.1039/CS9740300309

Buer, B.C., Paul, B., Das, D., Stuckey, J.A., and Marsh, E.N. (2014). Insights into substrate and metal binding from the crystal structure of cyanobacterial aldehyde deformylating oxygenase with substrate bound. ACS Chem. Biol. *9*, 2584–2593. http://dx.doi.org/10.1021/cb500343j.

Coates, R.C., Podell, S., Korobeynikov, A., Lapidus, A., Pevzner, P., Sherman, D.H., Allen, E.E., Gerwick, L., and Gerwick, W.H. (2014). Characterization of cyanobacterial hydrocarbon composition and distribution of biosynthetic pathways. PLoS One *9*, e85140. http://dx.doi.org/10.1371/journal.pone.0085140.

d'Espaux, L., Mendez-Perez, D., Li, R., and Keasling, J.D. (2015). Synthetic biology for microbial production of lipid-based biofuels. Curr. Opin. Chem. Biol. *29*, 58–65. http://dx.doi.org/10.1016/j.cbpa.2015.09.009.

Deng, M.D., and Coleman, J.R. (1999). Ethanol synthesis by genetic engineering in cyanobacteria. Appl. Environ. Microbiol. *65*, 523–528.

Dexter, J., and Fu, P. (2009). Metabolic engineering of cyanobacteria for ethanol production. Energ. Environ. Sci. *2*, 857–864. http://dx.doi.org/10.1039/B811937F

Doshi, R., Nguyen, T., and Chang, G. (2013). Transporter-mediated biofuel secretion. Proc. Natl. Acad. Sci. U.S.A. *110*, 7642–7647. http://dx.doi.org/10.1073/pnas.1301358110.

Du, W., Liang, F., Duan, Y., Tan, X., and Lu, X. (2013). Exploring the photosynthetic production capacity of sucrose by cyanobacteria. Metab. Eng. *19*, 17–25. http://dx.doi.org/10.1016/j.ymben.2013.05.001.

Ducat, D.C., Avelar-Rivas, J.A., Way, J.C., and Silver, P.A. (2012). Rerouting carbon flux to enhance photosynthetic productivity. Appl. Environ. Microbiol. *78*, 2660–2668. http://dx.doi.org/10.1128/AEM.07901-11.

Ducat, D.C., Way, J.C., and Silver, P.A. (2011). Engineering cyanobacteria to generate high-value products. Trends Biotechnol. *29*, 95–103. http://dx.doi.org/10.1016/j.tibtech.2010.12.003.

Fehler, S.W., and Light, R.J. (1970). Biosynthesis of hydrocarbons in *Anabaenavariabilis*. Incorporation of [*methyl*-^{14}C]- and [*methyl*-$^{2}H_3$]methionine into 7- and 8-methylheptadecanes. Biochemistry *9*, 418–422.

Gao, Q., Wang, W., Zhao, H., and Lu, X. (2012a). Effects of fatty acid activation on photosynthetic production of fatty acid-based biofuels in *Synechocystis* sp. PCC6803. Biotechnol. Biofuels *5*, 17. http://dx.doi.org/10.1186/1754-6834-5-17.

Gao, Z., Zhao, H., Li, Z., Tan, X., and Lu, X. (2012b). Photosynthetic production of ethanol from carbon dioxide in genetically engineered cyanobacteria. Energ. Environ. Sci. *5*, 9857–9865. http://dx.doi.org/10.1039/C2EE22675H

Georg, J., Dienst, D., Schürgers, N., Wallner, T., Kopp, D., Stazic, D., Kuchmina, E., Klähn, S., Lokstein, H., Hess, W.R., *et al.* (2014). The small regulatory RNA SyR1/PsrR1 controls photosynthetic functions in cyanobacteria. Plant Cell *26*, 3661–3679. http://dx.doi.org/10.1105/tpc.114.129767.

Georg, J., Voss, B., Scholz, I., Mitschke, J., Wilde, A., and Hess, W.R. (2009). Evidence for a major role of antisense RNAs in cyanobacterial gene regulation. Mol. Syst. Biol. *5*, 305. http://dx.doi.org/10.1038/msb.2009.63.

Grindley, N.D., Whiteson, K.L., and Rice, P.A. (2006). Mechanisms of site-specific recombination. Annu. Rev. Biochem. *75*, 567–605. http://dx.doi.org/10.1146/annurev.biochem.73.011303.073908.

Han, J., McCarthy, E.D., Hoeven, W.V., Calvin, M., and Bradley, W.H. (1968). Organic geochemical studies,

II. A preliminary report on the distribution of aliphatic hydrocarbons in algae, in bacteria, and in a recent lake sediment. Proc. Natl. Acad. Sci. U.S.A. *59*, 29–33.

Huang, H.H., Camsund, D., Lindblad, P., and Heidorn, T. (2010). Design and characterization of molecular tools for a Synthetic Biology approach towards developing cyanobacterial biotechnology. Nucleic Acids Res. *38*, 2577–2593.

Jia, C., Li, M., Li, J., Zhang, J., Zhang, H., Cao, P., Pan, X., Lu, X., and Chang, W. (2015). Structural insights into the catalytic mechanism of aldehyde-deformylating oxygenases. Protein Cell *6*, 55–67. http://dx.doi.org/10.1007/s13238-014-0108-2.

Kaczmarzyk, D., and Fulda, M. (2010). Fatty acid activation in cyanobacteria mediated by acyl-acyl carrier protein synthetase enables fatty acid recycling. Plant Physiol. *152*, 1598–1610. http://dx.doi.org/10.1104/pp.109.148007.

Kageyama, H., Waditee-Sirisattha, R., Sirisattha, S., Tanaka, Y., Mahakhant, A., and Takabe, T. (2015). Improved alkane production in nitrogen-fixing and halotolerant cyanobacteria via abiotic stresses and genetic manipulation of alkane synthetic genes. Curr. Microbiol. *71*, 115–120.

Kämäräinen, J., Knoop, H., Stanford, N.J., Guerrero, F., Akhtar, M.K., Aro, E.M., Steuer, R., and Jones, P.R. (2012). Physiological tolerance and stoichiometric potential of cyanobacteria for hydrocarbon fuel production. J. Biotechnol. *162*, 67–74. http://dx.doi.org/10.1016/j.jbiotec.2012.07.193.

Kaneko, T., Sato, S., Kotani, H., Tanaka, A., Asamizu, E., Nakamura, Y., Miyajima, N., Hirosawa, M., Sugiura, M., Sasamoto, S., *et al.* (1996). Sequence analysis of the genome of the unicellular cyanobacterium *Synechocystis* sp. strain PCC6803. II. Sequence determination of the entire genome and assignment of potential protein-coding regions. DNA Res. *3*, 185–209.

Klähn, S., Baumgartner, D., Pfreundt, U., Voigt, K., Schön, V., Steglich, C., and Hess, W.R. (2014). Alkane biosynthesis genes in cyanobacteria and their transcriptional organization. Front. Bioeng. Biotechnol. *2*, 24. http://dx.doi.org/10.3389/fbioe.2014.00024

Klähn, S., Schaal, C., Georg, J., Baumgartner, D., Knippen, G., Hagemann, M., Muro-Pastor, A.M., and Hess, W.R. (2015). The sRNA NsiR4 is involved in nitrogen assimilation control in cyanobacteria by targeting glutamine synthetase inactivating factor IF7. Proc. Natl. Acad. Sci. U.S.A. *112*, E6243–E6252.

Lan, E.I., and Liao, J.C. (2011). Metabolic engineering of cyanobacteria for 1-butanol production from carbon dioxide. Metab. Eng. *13*, 353–363. http://dx.doi.org/10.1016/j.ymben.2011.04.004.

Landry, B.P., Stöckel, J., and Pakrasi, H.B. (2013). Use of degradation tags to control protein levels in the cyanobacterium *Synechocystis* sp. strain PCC 6803. Appl. Environ. Microbiol. *79*, 2833–2835. http://dx.doi.org/10.1128/AEM.03741-12.

Lee, S.Y., Kim, H.M., and Cheon, S. (2015). Metabolic engineering for the production of hydrocarbon fuels. Curr. Opin. Biotechnol. 33, 15–22. http://dx.doi.org/10.1016/j.copbio.2014.09.008.

Li, N., Chang, W.C., Warui, D.M., Booker, S.J., Krebs, C., and Bollinger, J.M. (2012). Evidence for only oxygenative cleavage of aldehydes to alk(a/e)nes and formate by cyanobacterial aldehyde decarbonylases. Biochemistry *51*, 7908–7916. http://dx.doi.org/10.1021/bi300912n.

Lindberg, P., Park, S., and Melis, A. (2010). Engineering a platform for photosynthetic isoprene production in cyanobacteria, using *Synechocystis* as the model organism. Metab. Eng. *12*, 70–79. http://dx.doi.org/10.1016/j.ymben.2009.10.001.

Liu, A., Zhu, T., Lu, X., and Song, L. (2013). Hydrocarbon profiles and phylogenetic analyses of diversified cyanobacterial species. Appl. Energ. *111*, 383–393. http://dx.doi.org/10.1016/j.apenergy.2013.05.008

Liu, J., Chen, L., Wang, J., Qiao, J., and Zhang, W. (2012). Proteomic analysis reveals resistance mechanism against biofuel hexane in *Synechocystis* sp. PCC 6803. Biotechnol. Biofuels 5, 68. http://dx.doi.org/10.1186/1754-6834-5-68.

Liu, X., and Curtiss, R. (2009). Nickel-inducible lysis system in *Synechocystis* sp. PCC 6803. Proc. Natl. Acad. Sci. U.S.A. *106*, 21550–21554. http://dx.doi.org/10.1073/pnas.0911953106.

Liu, X., and Curtiss, R. (2012). Thermorecovery of cyanobacterial fatty acids at elevated temperatures. J. Biotechnol. *161*, 445–449. http://dx.doi.org/10.1016/j.jbiotec.2012.08.013.

Liu, X., Fallon, S., Sheng, J., and Curtiss, R. (2011a). CO_2-limitation-inducible Green Recovery of fatty acids from cyanobacterial biomass. Proc. Natl. Acad. Sci. U.S.A. *108*, 6905–6908. http://dx.doi.org/10.1073/pnas.1103016108.

Liu, X., Sheng, J., and Curtiss, R. (2011b). Fatty acid production in genetically modified cyanobacteria. Proc. Natl. Acad. Sci. U.S.A. *108*, 6899–6904. http://dx.doi.org/10.1073/pnas.1103014108.

Lu, X. (2010). A perspective: Photosynthetic production of fatty acid-based biofuels in genetically engineered cyanobacteria. Biotechnol. Adv. *28*, 742–746. http://dx.doi.org/10.1016/j.biotechadv.2010.05.021.

Lyznik, L.A., Gordon-Kamm, W.J., and Tao, Y. (2003). Site-specific recombination for genetic engineering in plants. Plant. Cell. Rep. *21*, 925–932.

Ma, A.T., Schmidt, C.M., and Golden, J.W. (2014). Regulation of gene expression in diverse cyanobacterial species by using theophylline-responsive riboswitches. Appl. Environ. Microbiol. *80*, 6704–6713. http://dx.doi.org/10.1128/AEM.01697-14.

Mendez-Perez, D., Begemann, M.B., and Pfleger, B.F. (2011). Modular synthase-encoding gene involved in α-olefin biosynthesis in *Synechococcus* sp. strain PCC 7002. Appl. Environ. Microbiol. 77, 4264–4267. http://dx.doi.org/10.1128/AEM.00467-11.

Mitschke, J., Georg, J., Scholz, I., Sharma, C.M., Dienst, D., Bantscheff, J., Voss, B., Steglich, C., Wilde, A., Vogel, J., *et al.* (2011). An experimentally anchored map of transcriptional start sites in the model cyanobacterium *Synechocystis* sp. PCC6803. Proc. Natl. Acad. Sci. U.S.A. *108*, 2124–2129. http://dx.doi.org/10.1073/pnas.1015154108.

Nakahira, Y., Ogawa, A., Asano, H., Oyama, T., and Tozawa, Y. (2013). Theophylline-dependent riboswitch as a novel genetic tool for strict regulation of protein expression in cyanobacterium *Synechococcus elongatus* PCC 7942. Plant Cell Physiol. *54*, 1724–1735.

Peca, L., Kós, P.B., Máté, Z., Farsang, A., and Vass, I. (2008). Construction of bioluminescent cyanobacterial reporter strains for detection of nickel, cobalt and zinc. FEMS Microbiol. Lett. *289*, 258–264. http://dx.doi.org/10.1111/j.1574-6968.2008.01393.x.

Peralta-Yahya, P.P., Zhang, F., del Cardayre, S.B., and Keasling, J.D. (2012). Microbial engineering for the production of advanced biofuels. Nature *488*, 320–328. http://dx.doi.org/10.1038/nature11478.

Peramuna, A., Morton, R., and Summers, M.L. (2015). Enhancing alkane production in cyanobacterial lipid droplets: a model platform for industrially relevant compound production. Life *5*, 1111–1126. http://dx.doi.org/10.3390/life5021111.

Peramuna, A., and Summers, M.L. (2014). Composition and occurrence of lipid droplets in the cyanobacterium *Nostocpunctiforme*. Arch. Microbiol. *196*, 881–890. http://dx.doi.org/10.1007/s00203-014-1027-6.

Qi, F., Yao, L., Tan, X., and Lu, X. (2013). Construction, characterization and application of molecular tools for metabolic engineering of *Synechocystis* sp. Biotechnol. Lett*35*, 1655–1661. http://dx.doi.org/10.1007/s10529-013-1252-0.

Reppas, N.B., Ridley, C.P., Reppas, N., Ridley, C., and Rodley, C.P. Producing hydrocarbons comprises culturing engineered cyanobacterium in culture medium and exposing engineered cyanobacterium to light and carbon dioxide. US Patent 7794969-B1.

Sarsekeyeva, F., Zayadan, B.K., Usserbaeva, A., Bedbenov, V.S., Sinetova, M.A., and Los, D.A. (2015). Cyanofuels: biofuels from cyanobacteria. Reality and perspectives. Photosynth. Res. *125*, 329–340. http://dx.doi.org/10.1007/s11120-015-0103-3.

Schirmer, A., Rude, M.A., Li, X., Popova, E., and del Cardayre, S.B. (2010). Microbial biosynthesis of alkanes. Science *329*, 559–562. http://dx.doi.org/10.1126/science.1187936.

Scholz, P., Haring, V., Wittmann-Liebold, B., Ashman, K., Bagdasarian, M., and Scherzinger, E. (1989). Complete nucleotide sequence and gene organization of the broad-host-range plasmid RSF1010. Gene *75*, 271–288.

Schweizer, H.P. (2003). Applications of the *Saccharomyces cerevisiae* Flp-FRT system in bacterial genetics. J. Mol. Microbiol. Biotechnol. *5*, 67–77. http://dx.doi.org/10.1159/000069976.

Siegal, M.L., and Hartl, D.L. (2000). Application of *Cre/loxP* in *Drosophila*. Site-specific recombination and transgene coplacement. Methods Mol. Biol. *136*, 487–495. http://dx.doi.org/10.1385/1-59259-065-9:487.

Summers, M.L., Wallis, J.G., Campbell, E.L., and Meeks, J.C. (1995). Genetic evidence of a major role for glucose-6-phosphatedehydrogenase in nitrogen fixation and dark growth of the cyanobacterium *Nostoc* sp. strain ATCC 29133. J. Bacteriol. *177*, 6184–6194.

Tan, X., Liang, F., Cai, K., and Lu, X. (2013). Application of the FLP/FRT recombination system in cyanobacteria for construction of markerless mutants. Appl. Microbiol. Biotechnol. *97*, 6373–6382. http://dx.doi.org/10.1007/s00253-013-4837-6.

Tan, X., Yao, L., Gao, Q., Wang, W., Qi, F., and Lu, X. (2011). Photosynthesis driven conversion of carbon dioxide to fatty alcohols and hydrocarbons in cyanobacteria. Metab. Eng. *13*, 169–176. http://dx.doi.org/10.1016/j.ymben.2011.01.001.

Ungerer, J., Tao, L., Davis, M., Ghirardi, M., Maness, P.C., and Yu, J. (2012). Sustained photosynthetic conversion of CO_2 to ethylene in recombinant cyanobacterium *Synechocystis* 6803. Energ. Environ. Sci. *5*, 8998–9006. http://dx.doi.org/10.1039/C2EE22555G

Wang, Q., Huang, X., Zhang, J., Lu, X., Li, S., and Li, J. (2014). Engineering self-sufficient aldehyde deformylating oxygenases fused to alternative electron transfer systems for efficient conversion of aldehydes into alkanes. Chem. Commun. *50*, 4299–4301. http://dx.doi.org/10.1039/c4cc00591k

Wang, W., Liu, X., and Lu, X. (2013). Engineering cyanobacteria to improve photosynthetic production of alka(e)nes. Biotechnol. Biofuels *6*, 69. http://dx.doi.org/10.1186/1754-6834-6-69.

Winters, K., Parker, P.L., and Van Baalen, C. (1969). Hydrocarbons of blue-green algae: geochemical significance. Science *163*, 467–468. http://dx.doi.org/10.1126/science.163.3866.467.

Wu, W., Zhang, L., Yao, L., Tan, X., Liu, X., and Lu, X. (2015). Genetically assembled fluorescent biosensor for in situ detection of bio-synthesized alkanes. Sci. Rep. *5*, 10907. http://dx.doi.org/10.1038/srep10907.

Yao, L., Cengic, I., Anfelt, J., and Hudson, E.P. (2016). Multiple gene repression in cyanobacteria using CRISPRi. ACS Synth. Biol. *5*, 207–212. http://dx.doi.org/10.1021/acssynbio.5b00264.

Yao, L., Qi, F., Tan, X., and Lu, X. (2014). Improved production of fatty alcohols in cyanobacteria by metabolic engineering. Biotechnol. Biofuels *7*, 94. http://dx.doi.org/10.1186/1754-6834-7-94

Yoshino, T., Liang, Y., Arai, D., Maeda, Y., Honda, T., Muto, M., Kakunaka, N., and Tanaka, T. (2015). Alkane production by the marine cyanobacterium *Synechococcus* sp. NKBG15041c possessing the α-olefin biosynthesis pathway. Appl. Microbiol. Biotechnol. *99*, 1521–1529. http://dx.doi.org/10.1007/s00253-014-6286-2.

Zhang, J., Lu, X., and Li, J. (2013). Conversion of fatty aldehydes into alk(a/e)nes by *invitro* reconstituted cyanobacterial aldehyde-deformylating oxygenase with the cognate electron transfer system. Biotechnol. Biofuels *6*, 86–95. http://dx.doi.org/10.1186/1754-6834-6-86

Zhang, Y., Pu, H., Wang, Q., Cheng, S., Zhao, W., Zhang, Y., and Zhao, J. (2007). PII is important in regulation of nitrogen metabolism but not required for heterocyst formation in the cyanobacterium *Anabaena* sp. PCC 7120. J. Biol. Chem. *282*, 33641–33648.

Zhou, J., Zhang, H., Meng, H., Zhu, Y., Bao, G., Zhang, Y., Li, Y., and Ma, Y. (2014). Discovery of a super-strong promoter enables efficient production of heterologous proteins in cyanobacteria. Sci. Rep. *4*, 4500. http://dx.doi.org/10.1038/srep04500.

Zhu, T., Xie, X., Li, Z., Tan, X., and Lu, X. (2015). Enhancing photosynthetic production of ethylene in genetically engineered *Synechocystis* sp PCC 6803. Green Chem. *17*, 421–434. http://dx.doi.org/10.1039/C4GC01730G

Photoautotrophic Polyhydroxyalkanoate Production in Cyanobacteria 12

Ka-Kei Sam[1], Nyok-Sean Lau[1,2], Amirul Al-Ashraf Abdullah[1,3] and Minami Matsui[2]*

[1]Centre for Chemical Biology, Universiti Sains Malaysia, Bayan Lepas, Malaysia.
[2]Synthetic Genomics Research Group, RIKEN Centre for Sustainable Resource Science, Biomass Engineering Research Division, Yokohama, Japan.
[3]School of Biological Sciences, Universiti Sains Malaysia, Minden, Malaysia.

*Correspondence: minami@riken.jp

Abstract

Accumulation of recalcitrant plastics waste in the natural environment leading to environmental problems and still unresolved plastic litter issues are of significant environmental concern. Development and production of biobased and biodegradable plastics including polyhydroxyalkanoate (PHA) are rapidly expanding because these bioplastics can serve as alternatives for petrochemical based polymers. Industrial-scale PHA production currently depends on a fermentation process using heterotrophic bacteria. Economic competitiveness of this production process is limited by the cost of carbon feedstock. It is therefore desirable to develop production platforms based on carbon-neutral PHA producers with capability to fix atmospheric carbon dioxide by which carbon feedstock can be omitted, and at the same time reduce carbon dioxide emissions. Throughout this chapter, potentials of photosynthetic cyanobacteria as hosts for sustainable PHA production from renewable sunlight and atmospheric carbon dioxide are discussed. Some future trends for improving PHA productivity in photosynthetic cyanobacteria are also briefly described.

Introduction

Polyhydroxyalkanoate: bioplastics from bacteria

More than 100 million tons of plastics wastes is generated worldwide annually, and the amount is still increasing at an alarming rate. Petrochemical based plastics production is not sustainable, as these plastics are derived from non-renewable fossil resources that will be depleted in the near future (Churkunti *et al.*, 2016). After their useful life, some of these non-biodegradable plastics end up in marine ecosystems (Halden, 2010; Thompson *et al.*, 2009) and pose a threat to marine wildlife including invertebrates, turtles, fish, birds and mammals (Gregory, 2009). Rapid depletion of natural resources being used for synthetic plastics production along with environmental concerns associated with plastics waste management have led to efforts to develop biodegradable 'green' plastics.

Polyhydroxyalkanoate (PHA) is a family of structurally diverse intracellular biopolyesters that can serve as renewable plastic and is completely biosynthesized and biopolymerized in bacteria (Chen and Patel, 2012; Wang *et al.*, 2014). Poly(3-hydroxybutyrate), P(3HB) was firstly discovered in *Bacillus megaterium* in 1925 and is still the most widely studied PHA family member (Doi, 1990; Tsuge, 2002). Thus far, over 150 types of PHA monomers have been reported in different PHA-producing bacteria (Xiao *et al.*, 2015). Classification of PHA monomers is generally divided into two groups: short-chain-length PHA (scl-PHA) with three to five carbons and medium-chain-length PHA (mcl-PHA) with 6–14 carbons (Shrivastav *et al.*, 2013; Tsuge, 2002). The majority of PHA-producing organisms accumulate either scl-PHA or mcl-PHA. Nonetheless, some bacteria are capable of naturally synthesizing polymer with both scl and mcl monomer constituents (Chen *et al.*, 2015a; Steinbüchel and Hein, 2001). Different types of PHA have distinct physical and chemical properties: scl-PHA possesses thermoplastic material properties reminiscent of polypropylene; mcl-PHA is an elastomeric material similar to rubber and elastomer; whereas scl-mcl-PHA has properties very similar to those of low density polyethylene (Lee, 1996; Madison and Huisman, 1999; Park and Lee, 2004; Park *et al.*, 2012). To date, advanced techniques in molecular sciences have led to biosynthesis of various forms of PHA with different microstructures via native or synthetic pathways and mechanisms, namely homopolymers, random copolymers, block copolymers, functional polymers, graft polymers, etc. (Chen and Hajnal, 2015; Meng *et al.*, 2014). Such diverse platforms enable manipulation in terms of functional and material properties for tailor-made synthesis of PHA that can be used in different applications.

Exploring potentials of microbial PHA to be applied in various fields gained much attention due to PHA's desirable features such as non-toxicity, biodegradability, biocompatibility, and controllable thermal and mechanical properties (Chen and Wu, 2005; Sudesh *et al.*, 2000). Biodegradation of PHA is one of the most attractive properties with respect to ecology as PHA can be completely degraded to water and carbon dioxide through microbial mineralization under either aerobic or anaerobic environmental conditions (Mergaert *et al.*, 1992; Muhammadi *et al.*, 2015). Therefore, this environmentally friendly microbial PHA can be a suitable substitute as a replacement of non-degradable petrochemical-based plastics. More than 30 PHA depolymerases with experimentally validated PHA biodegradation activity have been reported (Knoll *et al.*, 2009). At present, microorganisms that harbour PHA depolymerase genes for intra- or extracellular PHA degradation have been reported among different taxa, including bacteria from both Gram-positive (*Bacillus*, *Rhodococcus*) and Gram-negative (*Alcaligenes*, *Comamonas*, *Illyobacter*, *Pseudomonas*) as well as fungi (*Ascomycetes*, *Basidiomycetes*, *Deuteromycetes*, *Mastigiomycetes* and *Myxomycetes*) (Kim *et al.*, 2007; Jendrossek and Handrick, 2002; Shimao, 2001; Tokiwa *et al.*, 2009; Volova *et al.*, 2015). Microbial PHA is suitable material to be used in packaging, coating, agricultural and chemical industries (Bonartsev *et al.*, 2007; Chee *et al.*, 2010; Verlinden *et al.*, 2007). On the other hand, PHA is also applicable in biomedical and biopharmaceutical fields as drug deliverers and implant materials (Brigham and Sinskey, 2012; Chen and Wu, 2005; Shrivastav *et al.*, 2013). Although many types of PHA are being generated, only P(3HB), poly 4-hydroxybutyrate P(4HB), copolymer of 3-hydroybutyrate and 3-hydroxyvalerate P(3HB-*co*-3HV), copolymer of 3-hydroybutyrate and 4-hydroxybutyrate P(3HB-*co*-4HB), copolymer of 3-hydroybutyrate and 3-hydroxyhexanoate P(3HB-*co*-3HHx) and poly-3-hydroxyoctanoate P(3HO) are being produced in sufficient quantity for commercial applications (Chen *et al.*, 2015b). Extensive efforts are being carried out to develop new PHA production systems with capability to produce a sufficient quantity of PHA to diversify PHA production to meet various applications in the near future.

At present, the commercial production of PHA mainly relies on a fermentative biotechnological process utilizing heterotrophic microorganisms to metabolize carbohydrate compounds such as glucose, fructose and methanol. However, the overall production cost of PHA is much higher than petroleum-derived synthetic plastic. Cost of carbon feedstock is considered one of the main drawbacks to this heterotrophic production system (Lee, 1996). Instead of focusing on use of heterotrophic bacteria for PHA production, PHA can also be biosynthesized using a photosynthetic biological system. Natural occurrences of PHA are not limited to heterotrophic bacteria; PHA production is also attainable in photosynthetic cyanobacteria. Photosynthetic cyanobacteria are suggested as alternative platforms for 'green' PHA production from carbon dioxide, at which point carbohydrate feedstock cost can be eliminated. Current advances in genetic modification techniques led to engineering of several cyanobacterial strains with encouraging PHA productivity. These positive outcomes drove researchers to further engineer and explore PHA production potentials in photosynthetic cyanobacteria and other photoautotrophic PHA producers. Beyond considering cost of carbon feedstock, application of photosynthetic cyanobacteria for carbon-neutral PHA production is highly desirable for alleviating environmental issues related to use of synthetic plastics. In this chapter, uses of photosynthetic cyanobacteria for environmentally friendly PHA production along with current developments are discussed.

Cyanobacteria as platforms for biomass production

Cyanobacteria (formerly known as blue-green algae) are unicellular Gram-negative bacteria endowed with a complex photosynthetic system that can perform plant-like, oxygen-evolving photosynthesis (Knoop *et al.*, 2010; Mulkidjanian *et al.*, 2006; Stanier and Bazine, 1977). They are the pioneer oxygenic phototrophs of planet Earth (Kulasooriya, 2011) because they are the evolutionarily oldest prokaryotes found in microfossils originating more than 3.5 billion years ago (Knoll and Barghoorn, 1977; Schopf, 1993). The presence of photosynthetic pigments in the thylakoid membrane enables absorption of free energy from sunlight to generate both ATP and NADPH intermediates that are essential for carbon fixation and other cellular processes (Vermaas, 2001). Cyanobacteria utilize the Calvin–Benson cycle to fix carbon dioxide, one of the main greenhouse gases during photosynthesis for generation of storage compounds for carbon and energy (Smith, 1983). Although cyanobacteria usually perform oxygenic photosynthesis and respiration, certain cyanobacteria can fix atmospheric nitrogen in anaerobic conditions (Vermaas, 2001). In nature, cyanobacteria exist in diverse unicellular, filamentous, planktonic, benthic and colonial (cocoid) morphologies (Burja *et al.*, 2001; Whitton and Potts, 2000). They are also metabolically flexible as they can switch rapidly from one metabolism mode to another (Stal, 1995). Metabolic flexibility enables cyanobacteria to colonize a wide range of ecological niches and be found to thrive in diverse marine, freshwater, and terrestrials environments (Paerl *et al.*, 2000; Whitton, 1992).

Cyanobacteria represent ideal platforms for sustainable production of PHA because they exhibit several properties, including that they are the only known photosynthetic prokaryotes with higher photosynthetic rates than higher plants, have minimal nutrient requirements, have superior genetic manipulation capabilities, are non-pathogenic, have low risk of contamination, have an inducible lytic system for product recovery and have databases of genomic information (Fig. 12.1). Solar power from sunlight is an inexpensive and renewable energy source. Utilization of solar energy for production of biomass has gained attention over the years. Photosynthetic cyanobacteria are desirable hosts for biomass production as they possess efficient photosynthetic and biomass production rates. Cyanobacteria can convert solar energy into biomass with a photosynthetic efficiency of approximately 3–9%, which is more efficient in comparison to ≤0.25–3% for higher plants (Dismukes *et al.*, 2008;

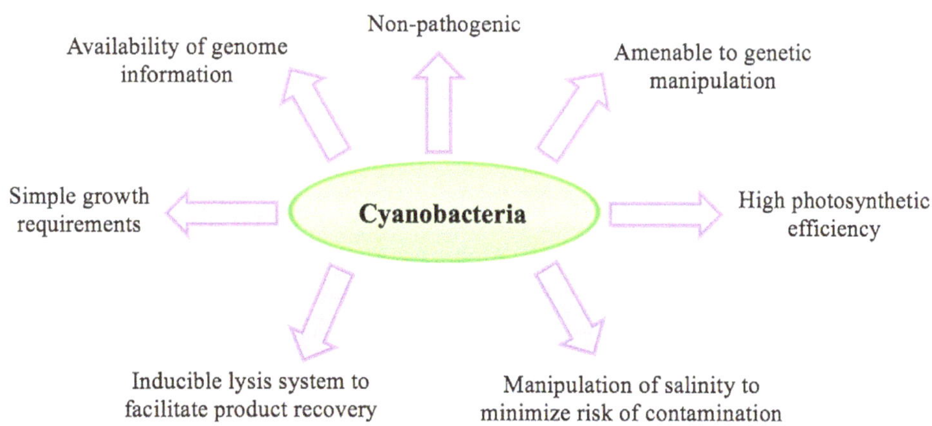

Figure 12.1 Desirable properties of cyanobacteria as platforms for sustainable biomass production.

Ducat *et al.*, 2011). In addition, cyanobacteria utilize atmospheric carbon dioxide as a source of carbon for organic compounds production (Jajesniak *et al.*, 2014). Global warming leading to climate change is one of the biggest issues facing the world today and carbon dioxide is the primary greenhouse gas contributing to global warming (Knutti and Rogelj, 2015; Solomon *et al.*, 2009; Tian *et al.*, 2016). Turning greenhouse gas into industrial products using photosynthetic cyanobacteria therefore offers a desirable production approach by which carbon dioxide emission can be reduced. At the same time, cyanobacteria's capability to fix carbon dioxide eliminates the cost of carbon feedstocks supplied for production of desired products (Ducat *et al.*, 2011). High photosynthetic efficiency and ability to fix carbon dioxide as the primary carbon source have led to utilization of cyanobacteria as cell factories for biosynthesis of various products such as ethanol (Gao *et al.*, 2012; see Chapter 10), isobutanol (Atsumi *et al.*, 2008), fatty acids (Liu *et al.*, 2011; Chapter 8) and isoprene (Bentley *et al.*, 2014; Chapter 9).

Compared with other bacteria, cultivation of cyanobacteria is relatively simple and inexpensive because they only require simple input of nutrients such as sunlight, carbon dioxide, water and mineral salts (Lau *et al.*, 2015; Quintana *et al.*, 2011). Cyanobacteria do not require arable land for growth and thus competition with terrestrial crops for land is not an issue (Dismukes *et al.*, 2008; Nogales *et al.*, 2013; Rittmann, 2008). It is noteworthy that many cyanobacterial strains have been used in numerous molecular studies that take advantage of their amenability to genetic manipulation. To date, strains of cyanobacteria successfully utilized for genetic modification in research studies include *Synechococcus elongatus* sp. PCC 7942 (*Synechococcus* PCC 7942) (Gao *et al.*, 2016a), *Synechocystis* sp. PCC 6803 (*Synechocystis* PCC 6803) (Wang *et al.*, 2016), *Synechococcus* sp. PCC 7002 (*Synechocystis* PCC 7002) (Zess *et al.*, 2016), *Leptolyngbya* sp. BL0902 (*Leptolyngbya* BL0902) (Taton *et al.*, 2012), *Anabaena variabilis* sp. PCC 7120 (*Anabaena* PCC 7120) (Narayan *et al.*, 2016), *Synechocystis* sp. WHSyn (*Synechocystis* WHSyn) (Ma *et al.*, 2014) and *Synechococcus elongatus* UTEX 2973 (*Synechococcus* UTEX 2973) (Yu *et al.*, 2015). Exogenous DNA can be introduced into the cyanobacteria genome either by conjugation or by natural transformation, depending on their transformation ability (Ma *et al.*, 2014). Model strains of cyanobacteria, namely *Synechocystis* 6803 (Barten and Lill, 1995; Kufryk *et al.*, 2002), *Synechococcus* 7002 (Xu *et al.*, 2011) and *Synechococcus* 7942 (Tsinoremas *et al.*, 1994) (except *Leptolyngbya* BL0902, *Anabaena* PCC 7120, *Synechocystis* WHSyn, and *Synechococcus* UTEX 2973), are known to be naturally transformable, in which exogenous DNA can be integrated spontaneously into the genome through homologous recombination (Ruffing, 2011; Vermaas, 1996). Although *Leptolyngbya* BL0902, *Anabaena* PCC 7120, *Synechocystis* WHSyn and *Synechococcus* UTEX 2973 are not naturally competent, they can be engineered through conjugation (Ma *et al.*, 2014; Olmedo-Verd *et al.*, 2005; Taton *et al.*, 2014; Yu *et al.*, 2015).

So far, widely used model strains of cyanobacteria for research are non-pathogenic, making them potential candidates for large-scale biotechnological application. Although these commonly used cyanobacterial strains are non-toxic, certain strains of cyanobacteria do produce toxins harmful to humans and are less likely to be used for large-scale application (Codd *et al.*, 2005). One challenge for culturing cyanobacteria or algae in open pond cultures is contamination from bacteria, fungi, viruses or zooplankton. Manipulation of salinity is a common approach to reduce contamination risk (von Alvensleben *et al.*, 2013). At present, examples of cyanobacterial strains with salt-tolerant capabilities in nature or by genetic engineering include *Synechocystis* PCC 6803, *Synechococcus* PCC 7942, *Leptolyngbya* BL0902, *Synechocystis* WHSyn and *Anabaena* PCC 7120. Although *Synechocystis* PCC 6803 was originally isolated from a freshwater lake, it is able to grow and proliferate in seawater (Iijima *et al.*, 2015) and under salt stress conditions (0.45 M NaCl) (Ferjani *et al.*, 2003). Similarly, cyanobacterium *Leptolyngbya* BL0902 isolated from an algal production pond can tolerate up to 0.5 M NaCl and specimens from the genus can be found in diverse environments ranging from hypersaline, marine and freshwater (Taton *et al.*, 2012). The relatively new experimental model strain *Synechocystis* WHSyn originated from an estuary, and was shown to grow in a wide range of salinities ranging from freshwater to seawater (Ma *et al.*, 2014; Taton *et al.*, 2014). On the other hand, engineered freshwater cyanobacterium *Synechococcus* PCC 7942 showed drastically enhanced salt tolerance after introduction of genes encoding for Na^+/H^+ antiporter, betaine, catalase and a chaperone, respectively (Deshnium *et al.*, 1995; Kaku *et al.*, 2000; Nomura *et al.*, 1995; Waditee *et al.*, 2002). In another attempt, freshwater *Anabaena* PCC 7120 showed improvement in salt tolerance after being transformed with genes encoding for glycine methylation from *Aphanothece halophytica* (Waditee-Sirisattha *et al.*, 2012). Salt-tolerant cyanobacteria discussed here are examples of attractive hosts for synthesis of industrial products by which risk of culture contamination can be minimized.

Efficiency of recovery of the product of interest is a main concern and also the crucial criteria that needs to be taken into consideration in selection of suitable host organisms for desired product production. Driven by advances in genetic engineering and synthetic biology approaches, several inducible lytic systems were tested in cyanobacteria. Feasibility of this idea was demonstrated by introducing lysis genes encoding holin and endolysin that are derived from T4 bacteriophage into *Synechocystis* PCC 6803 under regulation of green-light *cpcG2* promoter. Results from the study clearly showed that growth rate of cultured cells decreased and accumulation of cellular phycocyanin in the medium increased markedly under green light illumination, indirectly proving that cell lysis had occurred (Miyake *et al.*, 2014). In another example, the nickel-inducible lysis system was introduced into the genome of *Synechocystis* PCC 6803. Cell lysis was induced in recombinant cyanobacterium in presence of $NiSO_4$ (Liu and Curtiss, 2009). It is noteworthy that application of the inducible lysis system in cyanobacteria can eliminate use of mechanical forces or chemicals to induce cell break, reducing cost of product recovery. To date, more than 90 cyanobacterial genomes are sequenced (Gao *et al.*, 2016b; Hernández-Prieto *et al.*, 2014). Availability of the genome sequence provides invaluable information on the genes and metabolic pathways in cyanobacteria which facilitate the manipulation on the pathways for various industrial product syntheses. Considering aforementioned inherent merits of photosynthetic cyanobacteria, they can be considered ideal platforms for sustainable production of environmentally friendly PHA. Ongoing extensive research on cyanobacteria will further improve suitability of cyanobacteria as platforms for industrial product synthesis.

Cyanobacteria are native PHA producers

To date, more than 10 genera of cyanobacteria are known to accumulate PHA naturally (Table 12.1). They are the sole indigenous prokaryotic organisms that synthesize PHA through carbon-neutral photosynthesis. PHA granules are synthesized as monolayer-enclosed insoluble inclusions, usually globule in shape when observed under transmission electron microscope (Stal, 1992). PHA production in cyanobacteria was first reported in 1966 following extraction of P(3HB) from *Chlorogloea fristchii* (Carr, 1966). Further work performed by Jensen and Sicko confirmed presence of PHA in *C. fristchii* based on results from ultrastructural studies (Jensen and Sicko, 1971). Thereafter, PHA-synthesizing ability has been reported in other cyanobacteria including *Anabaena cylindrical* 10C (Lama *et al.*, 1996), *Anabaena* sp. BD47 (Ten *et al.*, 2015), *Aphanocapsa* PCC 6308 (Allen, 1984), *Aulosira fertilissima* (Samantaray and Mallick, 2012, 2014), *Chlorogloea fritschii* (Carr, 1966; Jensen and Sicko, 1971), *Gloeocapsa* sp. (Rippka *et al.*, 1971; Stal, 1992), *Nostoc muscorum* (Bhati and Mallick, 2012, 2015; Haase *et al.*, 2012; Mallick *et al.*, 2007; Sharma *et al.*, 2007; Sharma and Mallick, 2005), *Osillatoria limosa* (Stal, 1992), *Spirulina platensis* (Campbell *et al.*, 1982; Jau *et al.*, 2005), *Spirulina subsalsa* (Shrivastav *et al.*, 2010), *Spirulina maxima* (De Philippis *et al.*, 1992), *Synechococcus* sp. (Miyake *et al.*, 1996; Nishioka *et al.*, 2001), *Synechocystis* sp. (Hein *et al.*, 1998; Panda *et al.*, 2006; Panda and Mallick, 2007; Toh *et al.*, 2008), *Thermosynechococcus elongates* (Eberly and Ely, 2012) and *Trichodesmium thiebautii* (Siddiqui *et al.*, 1992).

Although diverse types of PHA are detected among PHA-producing bacteria (Chen and Hajnal, 2015), P(3HB) is the most widespread PHA found accumulated in different strains of PHA-producing cyanobacteria (Balaji *et al.*, 2013). With provision of a structurally related carbon source, P(3HB-*co*-3HV) synthesis has been reported in the genera *Anabaena*, *Aulosira* and *Nostoc*. Although PHA contents reported in some strains of cyanobacteria are generally low, manipulation of culture conditions and provided

Table 12.1 PHA production from wild-type cyanobacteria

Cyanobacterial species	Carbon sources	Types of PHA	PHA content (wt%)	References
Anabaena cylindrica 10C	Acetate	P(3HB)	2	Lama *et al*. (1996)
A. cylindrica 10C	Propionate/valerate	P(3HB-*co*-3HV)	0.2	Lama *et al*. (1996)
Anabaena sp. BD47	CO_2	P(3HB)	n.q.	Ten *et al*. (2015)
Aphanocapsa PCC 6308	Acetate	P(3HB)	n.q.	Allen (1984)
Aulosira fertilissima	Acetate + citrate	P(3HB)	85	Samantaray and Mallick (2012)
A. fertilissima CCC 444	Fructose + valerate	P(3HB-*co*-3HV)	77	Samantaray and Mallick (2014)
Chlorogloea fritschii	Acetate	P(3HB)	10	Carr (1966)
C. fritschii	CO_2	P(3HB)	n.q.	Jensen and Sicko (1971)
Gloeocapsa sp. PCC 6501	CO_2	P(3HB)	n.q.	Rippka *et al*. (1971)
Gloeothece sp. PCC 6909	CO_2	P(3HB)	6	Stal (1992)
Nostoc muscorum	Acetate	P(3HB)	35	Sharma and Mallick (2005)
N. muscorum	Acetate + glucose	P(3HB)	46	Sharma *et al*. (2007)
N. muscorum	Acetate + propionate	P(3HB-*co*-3HV)	31	Mallick *et al*. (2007)
N. muscorum Agardh	Acetate + valerate	P(3HB-*co*-3HV)	60	Bhati and Mallick (2012)
N. muscorum Agardh	Acetate + glucose + valerate	P(3HB-*co*-3HV)	69	Bhati and Mallick (2015)
Osillatoria limosa	Acetate	P(3HV)	6	Stal (1992)
Spirulina platensis	CO_2	P(3HB)	6	Campbell *et al*. (1982)
S. platensis	Acetate	P(3HB)	10	Jau *et al*. (2005)
Spirulina subsalsa	CO_2	P(3HB)	8	Shrivastav *et al*. (2010)
Spirulina maxima	CO_2	P(3HB)	1	De Philippis *et al*. (1992)
S. maxima	Acetate	P(3HB)	2	De Philippis *et al*. (1992)
Synechococcus sp. MA19	CO_2	P(3HB)	55	Nishioka *et al*. (2001)
Synechocystis sp. PCC6803	Acetate	P(3HB)	10	Hein *et al*. (1998)
Synechocystis sp. PCC6803	Acetate + glucose	P(3HB)	29	Panda *et al*. (2006)
Synechocystis sp. PCC6803	Acetate + fructose	P(3HB)	38	Panda and Mallick (2007)
Synechosystis sp. UNIWG	CO_2 + Acetate	P(3HB)	14	Toh *et al*. (2008)
Thermosynechococcus elongatus	CO_2	P(3HB)	15	Eberly and Ely (2012)
Trichodesmium thiebautii	CO_2	P(3HB)	n.q.	Siddiqui *et al*. (1992)

n.q., not quantified.

carbon feedstock can boost PHA accumulation in cyanobacteria to a commercially viable level. The thermophilic *Synechococcus* sp. MA19 was isolated from wet volcanic rock in Japan, and produced 27% dry cell weight (DCW) of P(3HB) in the dark and under nitrogen-limiting conditions (Miyake *et al*., 1996). Under phosphate starvation conditions, P(3HB) content in *Synechococcus* sp. MA19 increased markedly to 55% DCW (Nishioka *et al*., 2001). Another cyanobacterium, *Synechocystis* sp. PCC 6803 synthesized P(3HB) up to 38% under gas-exchange limitation, phosphate starvation and mixotrophic cultivation conditions when supplemented by fructose and acetate (Panda and Mallick, 2007). Manipulation of nutrient provided, for example, using 0.17% acetate, 0.16% glucose, and 5 mg/l K_2HPO_4 instead of 0.4% acetate, 0.4% glucose, and 40 mg/l K_2HPO_4 for culturing *Nostoc muscorum* increased P(3HB) content slightly from 41.6% to 45.6% (Sharma *et al*., 2007). A study from Bhati and Mallick (2012) reported accumulation of P(3HB-*co*-3HV) copolymer in *Nostoc muscorum* with provision of either propionate or valerate as a co-carbon source. Subsequently, the group demonstrated P(3HB-*co*-3HV) production in *N. muscorum* with a maximal content of 69% DCW by using multifactor optimization strategy (Bhati and Mallick, 2015). Notably, nitrogen-fixing *Aulosira fertilissima* was the first cyanobacterium able to accumulate P(3HB) as high as up to 85% DCW (Samantaray and Mallick, 2012). Interestingly, this cyanobacterium can also synthesize P(3HB-*co*-3HV) up to 76.9% with provision of fructose and valerate as carbon sources (Samantaray and Mallick, 2014).

The reactions involved in the metabolic pathway for P(3HB) synthesis were first discovered by Stanier and his co-workers in 1959. They studied biosynthesis of P(3HB) from acetic acid in *Rhodospirillum rubrum* (Doudoroff and Stanier, 1959). Since then, many researches on P(3HB) have been reported, but enzymes involved in the biosynthetic pathway were identified later in 1973 (Oeding and Schlegel, 1973; Senior and Dawes, 1973). Biosynthesis of P(3HB) occurs via the acetoacetyl-CoA pathway catalysed by three important enzymes, known as β-ketothiolase, acetoacetyl-CoA reductase and PHA synthase. These enzymes catalyse formation of PHA from acetyl-CoA via a series of enzymatic reactions: β-ketothiolase catalyses condensation of two acetyl-CoA molecules to form one acetoacetyl-CoA molecule, then acetoacetyl-CoA is converted to reduced form of D-3-hydroxybutyryl-CoA by acetoacetyl-CoA reductase and subsequently followed by formation of P(3HB) via polymerization of D-3-hydroxylbutyryl-CoA by PHA synthase (Anderson and Dawes, 1990; Steinbüchel and Schlegel, 1991). It has been reported that intracellular P(3HB) biosynthesis is regulated by the concentrations of acetyl-CoA, free coenzyme A and NAD(P)H/NAD(P). Previous study revealed that high level of NAD(P)H/NAD(P) can promote P(3HB) biosynthesis (Lee *et al.*, 1995). Similarly, high concentration of acetyl-CoA (low concentration free CoA) facilitate the drive of carbon and energy toward P(3HB) production (Amara and Moawad, 2011; Kessler and Witholt, 2001; Miyake *et al.*, 1997). Among PHA-producing cyanobacteria, *Synechocystis* sp. PCC 6803 (*Synechocystis* 6803) is one of the well-studied cyanobacteria for PHA biosynthesis. In *Synechocystis* 6803, genes encoding for key enzymes involved in P(3HB) biosynthesis are identified and characterized, including *phaA2* (β-ketothiolase), *phaB2* (acetoacetyl-CoA reductase), and *phaEC* (PHA synthase) (Fig. 12.2). Expressions of *phaA2* (*slr1993*) and *phaB2* (*slr1994*) are required for formation of

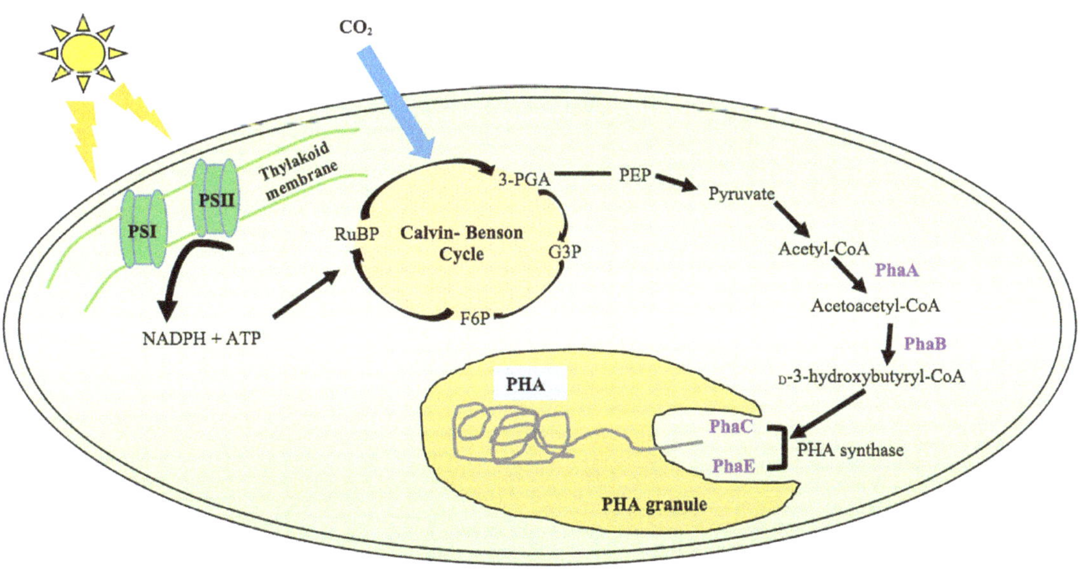

Figure 12.2 Schematic of native PHA biosynthetic pathway for the well-characterized cyanobacteria strain *Synechocystis* sp. PCC 6803. Abbreviations: F6P, fructose-6-phosphate; G3P, glyceraldehyde-3-phosphate; PEP, phosphoenolpyruvate; 3-PGA, 3-phosphoglycerate; RuBP, ribulose-1,5-bisphosphate.

scl-PHA building block, D-3-hydroxylbutyryl-CoA (Taroncher-Oldenburg *et al.*, 2000).

PHA synthases, the essential biocatalysts for polymerization of PHA, are essential for the last reaction of the PHA biosynthetic pathway (Rehm, 2003; Yang *et al.*, 2013). To date, there are four major groups of PHA synthases, including class I, II, III, and IV that are classified based on subunit compositions and substrate specificities of the enzymes. Class I and II PHA synthases are homodimer and have only one PhaC subunit. On the other hand, Class III and IV PHA synthases are heterodimer and composed of two different subunits, for example: PhaC and PhaE subunits in class III; PhaC and PhaR subunits in class IV (Park *et al.*, 2012; Rehm, 2003). Previous study by Hein and his colleagues revealed that the PHA synthase of *Synechocystis* PCC 6803 is a heterodimer and composed of subunits PhaE and PhaC that are encoded by genes *slr1829* and *slr1830*, respectively. The PHA synthase from *Synechocystis* PCC 6803 belongs to class III that is active towards short-chain-length hydroxyacyl-CoAs during PHA polymerization. The group also demonstrated expression of catalytically active *Synechocystis* PCC 6803 PHA synthase by co-expression of both the *phaE* and *phaC* genes (Hein *et al.*, 1998). Most identified PHA synthases in cyanobacteria are class III-like, except for those reported in *Cyanothece* sp. PCC 8955, *Gloeothece* sp. PCC 6501, and *Stanieria* sp. PCC 7437 (Hai *et al.*, 2001). Besides *Synechocystis* PCC 6803, PHA biosynthetic pathways in other PHA-producing cyanobacteria are not well-studied. Better understanding of the PHA biosynthetic mechanism of cyanobacteria is needed for development or manipulation of these photosynthetic bacteria for efficient PHA production.

Engineering cyanobacteria for PHA production

As discussed in the previous section, some cyanobacteria are natural PHA producers and equipped with capability to accumulate PHA via native PHA biosynthetic pathway. Recent advances in metabolic engineering have demonstrated potential to introduce artificial PHA biosynthetic pathway into non-PHA producing cyanobacteria and confer to them PHA-synthesizing ability. It is likely that genetic engineering approaches can be applied for development of PHA producing cyanobacteria to further enhance their PHA-synthesizing ability. Exploitation of current technologies to develop cyanobacteria for efficient PHA biosynthesis is achievable with available genomic information and synthetic biology tools. *Synechocystis* PCC 6803, *Synechococcus* PCC 7942 and *Synechococcus* PCC 7002 are the most widely used model cyanobacterial strains for biotechnological studies (Yu *et al.*, 2015) and these strains are also promising hosts of choice for engineered production of target compounds (PHA). Herein, trends and examples of engineering production of PHA in recombinant strains of cyanobacteria are briefly described.

One of the early efforts to improve *Synechocystis* PCC 6803's PHA production capability involved introduction of PHA biosynthetic genes from *Cupriavidus necator* into *Synechocystis* PCC 6803 (Table 12.2). PHA biosynthetic genes from the high-yield PHA producer, *C. necator*, were expressed in plasmid in *Synechocystis* PCC 6803. Enzymatic analysis performed in the work revealed that PHA synthase activity of the recombinant strain is 0.5 U/mg, which is slightly higher than that of the wild type (0.2 U/mg). Cultivation of both the wild-type and recombinant *Synechocystis* PCC 6803 under conditions favourable for PHA production indicated that only a marginal improvement on P(3HB) content was observed in the recombinant strain (11 wt%) in comparison with the wild-type (7 wt%). Findings from this research suggested that PHA synthase activity was not the main limiting factor that influenced the amount of P(3HB) content in *Synechocystis* PCC 6803 (Sudesh *et al.*, 2001, 2002). On the other hand, the knock-out of glycogen biosynthetic enzyme, ADP-glucose pyrophosphorylase, encoded by the *agp* gene of *Synechocystis* PCC 6803, was found to increase PHA accumulation up to 18.6% in the deletion mutant (Wu *et al.*, 2002). Another interesting study demonstrated PHA accumulation up to 41 wt% in an engineered *Synechocystis* PCC 6803. In the study, acetoacetyl-CoA synthase ($nphT7_{Ss}$) from *Streptomyces* sp. CL190, acetoacetyl-CoA reductase ($phaB_{Cn}$) from *C. necator* and PHA synthase ($phaC_{Cs}$) from *Chromobacterium* sp. USM2 were introduced into the genome of *Synechocystis* PCC 6803 by homologous recombination (Lau *et al.*, 2014). An attempt to increase PHA production by overexpressing RNA polymerase sigma factor, *sigE*

Table 12.2 PHA production from recombinant cyanobacteria

Cyanobacterial species	Genetic modification	Carbon sources	Types of PHA	PHA content (wt%)	References
Synechocystis sp. PCC6803	PHA synthase, β-ketothiolase and acetoacetyl-CoA reductase from *Cupriavidus necator*	Acetate	P(3HB)	11	Sudesh *et al.* (2001)
Synechocystis sp. PCC6803	Deletion of *agp* gene encoding for ADP-glucose pyrophosphorylase	Acetate	P(3HB)	19	Wu *et al.* (2002)
Synechocystis sp. PCC6803	Acetoacetyl-CoA synthase (*nphT7*$_{Ss}$) from *Streptomyces* sp. CL190, acetoacetyl-CoA reductase (*phaB*$_{Cn}$) from *C. necator* and PHA synthase (*phaC*$_{Cs}$) from *Chromobacterium* sp. USM2	Acetate	P(3HB)	41	Lau *et al.* (2014)
Synechocystis sp. PCC6803	PHA synthase, β-ketothiolase and acetoacetyl-CoA reductase from *Microcystis aeruginosa* NIES-843	CO_2	P(3HB)	7	Hondo *et al.* (2015)
Synechococcus sp. PCC7002	PHA synthase, β-ketothiolase and acetoacetyl-CoA reductase from *C. necator*	CO_2	P(3HB)	17	Miyasaka *et al.* (1998)
Synechocystis sp. PCC7002	PHA synthase, β-ketothiolase and acetoacetyl-CoA reductase from *C. necator* using *recA* complementation	CO_2	P(3HB)	52	Akiyama *et al.* (2011)
Synechococcus sp. PCC7002	PHA synthase, β-ketothiolase and acetoacetyl-CoA reductase from *Chlorogloeopsis fritschii* PCC 9212, deletion of *ccmR* gene encoding LysR-type transcription factor and construction of biosynthetic pathway for 4-hydroxybutyryl-CoA	CO_2	P(3HB-*co*-4HB)	5	Zhang *et al.* (2015)
Synechococcus sp. PCC7942	PHA synthase, β-ketothiolase and acetoacetyl-CoA reductase from *C. necator*	CO_2	P(3HB)	1	Suzuki *et al.* (1996)
Synechococcus sp. PCC7942	PHA synthase, β-ketothiolase and acetoacetyl-CoA reductase from *C. necator*	CO_2 + Acetate	P(3HB)	26	Takahasi *et al.* (1998)

in *Synechocystis* PCC 6803 was made. The altered metabolic pathway from glycogen to PHA during nitrogen starvation condition increased PHA content by 2.3 times in the *sigE*-overexpressed strain (Osanai *et al.*, 2013). Besides that, a transconjugant of *Synechocystis* PCC 6803 showed P(3HB) accumulation up to 7 wt%, which is 12-fold higher than the control strain following introduction of PHA genes from *Mycrocystis aeruginosa* NIES-843 (Hondo et al., 2015).

Marine cyanobacterium *Synechococcus* sp. PCC 7002 is not a natural producer of PHA and native PHA metabolism pathway has not been reported in this strain. Molecular engineering approaches can employ recombinant bacteria that are not native PHA producers to synthesize PHA. For example, an attempt to genetically engineer *Synechococcus* PCC 7002 was performed by introducing PHA biosynthetic genes from *C. necator* and the recombinant successfully produced PHA up to 17 wt% DCW (Miyasaka *et al.*, 1998). Considering the difficulty to maintain plasmid stability for PHA production, an antibiotics-free plasmid expression system was developed for recombinant *Synechococcus* PCC 7002. PHA biosynthetic genes from *C. necator* were cloned into the plasmid and transformed into *Synechococcus* PCC 7002. Instead of using antibiotic as selective pressure for plasmid stability, *recA* complementation was used as selection pressure and the recombinant accumulated PHA up to 52% DCW (Akiyama *et al.*, 2011). In another study, PHA accumulation up to 2.77% DCW was obtained with expression of PHA biosynthetic genes from *Chlorogloeopsis fritschii* PCC 9212 in *Synechococcus* PCC 7002. Further improvement in PHA accumulation to 3.7 wt% DCW was achieved following deletion of the *ccmR* gene. Besides 3HB monomer, 12% of 4HB monomer was incorporated into P(3HB-*co*-4HB) copolymer after construction of a 4-hydroxybutyrate-CoA pathway in *Synechococcus* PCC 7002 (Zhang *et al.*, 2015).

Synechococcus elongatus PCC 7942 is a non-PHA-producing cyanobacterium. However, a previous study successfully manipulated this strain by transforming it with plasmid carrying *C. necator* PHA biosynthetic genes and the transformant was proved to accumulate PHA up to 1 wt% DCW under nitrogen starved condition (Suzuki *et al.*, 1996). Further work on the recombinant strain by varying cultivation conditions increased PHA content to 25.6 wt% DCW with the provision of acetate and under nitrogen starvation (Takahashi *et al.*, 1998).

Future trends

Cyanobacteria offer promising greener paths for sustainable bioplastics production from sunlight and carbon dioxide through photosynthesis. Carbon-neutral bioplastics production using photosynthetic cyanobacteria not only eliminates the cost of carbon feedstock, but also addresses environmental issues associated with carbon dioxide emissions. Nevertheless, there are some challenges ahead for transforming cyanobacteria into competitive organisms for industrial-scale PHA production. Although there are many cyanobacteria that produce PHA naturally, the PHA content is usually low and this limitation is considered the major challenge to be overcome to boost PHA produced from cyanobacteria to a commercially viable level. An emerging trend to improve biomass production is to attempt to apply synthetic biology approaches to improve or redesign cellular function and behaviour to establish efficient cell factories for industrial products (Singh *et al.*, 2016; Wang *et al.*, 2012).

Transcriptome analysis of *Synechocystis* 6803 cells subjected to conditions favourable for photoautotrophic PHA biosynthesis suggested that the cells may utilize enhanced photosynthesis to cope with high PHA production (Lau *et al.*, 2014). Photosynthesis efficiency inevitably plays a pivotal role in determining light-driven biomass productivity. Light availability, light harvesting capacity and conversion efficiency of light energy to chemical energy greatly influence biomass productivity (Ort *et al.*, 2015; Zhu *et al.*, 2010). Light-harvesting complexes including photosystem I, photosystem II and phycobilisomes located around cyanobacteria thylakoid membranes enable absorption of sunlight for energy generation (Grossman *et al.*, 1993; Liu *et al.*, 2005; Ting *et al.*, 2002; Vermaas, 2001). As cyanobacteria can only capture solar radiation from the visible spectrum of 400 to 700 nm, approximately 50% of the solar energy is dissipated or lost during photosynthesis (Blankenship *et al.*, 2011; Ort *et al.*, 2015; Zhu *et al.*, 2010). Improving cyanobacteria light harvesting capacity is necessary to enhance cyanobacteria photosynthesis efficiency, which will eventually lead to enhanced biomass production. One possible approach to enhance photosynthesis efficiency is introduction of photosynthetic pigments (chlorophyll *d* and chlorophyll *f*), which absorb in the far red, into cyanobacteria to extend the utilizable photosynthetically active radiation (Blankenship and Chen, 2013). It is noteworthy that excess photon absorption causes 'photon burden' that leads to thermal dissipation and increases photodamage, which will in turn affect efficiency in conversion of light energy to biomass (Li *et al.*, 2009; Melis, 2009). Reduction of the light-harvesting antenna size was shown to enhance photosynthesis efficiency by minimizing oversaturation and photoinhibition and this concept has been demonstrated in engineered *Synechocystis* PCC 6803. A *Synechocystis* mutant strain having a knock-out of phycobilisome core-membrane linker gene was able to accumulate biomass 1.6 times higher than that of the wild-type under phototrophic condition (Joseph *et al.*, 2014). Although there are significant challenges in engineering cyanobacteria to improve their photosynthesis efficiency, extensive ongoing research efforts in this field will bring us one step closer to achieving this research direction.

In addition to improving photosynthesis, enhancing fixed carbon dioxide will ultimately boost biomass production. In cyanobacteria, carbon fixation takes place via the Calvin-Benson cycle catalysed by ribulose-1,5-bisphosphate carboxylase (Rubisco), the key enzyme involved in carbon dioxide utilization and the main target for genetic modification (Spreitzer and Salvucci, 2002; Wang *et al.*, 2012). A previous metabolic engineering effort using *Synechococcus* PCC 7942 demonstrated a two-fold improvement in isobutyraldehyde production with overexpression of Rubisco (Atsumi *et al.*, 2009). Further work using the same genetic modification strategy but with overexpression of Rubisco under *psbA1* promoter increased free fatty acid production to more than three-fold (Ruffing, 2014). Nevertheless, manipulation of Rubisco activity to improve carbon dioxide fixation for high yield biomass production is a challenging task and only limited success has been accomplished so far (Ruffing, 2011; Chapter 8). On the other hand, overexpression of *flv3* gene encoding flavodiiron protein in *Synechocystis* PCC 6803 was shown to enhance carbon dioxide incorporation, cell growth, and intracellular ATP level, providing a new idea in research aimed at improving carbon dioxide fixation (Hasunuma *et al.*, 2014). Another useful strategy to improve desired end-product synthesis is manipulation of intracellular carbon flux. For example, inactivation of the competing glycogen biosynthetic pathway in *Synechocystis* PCC 6803 directed the flow of carbon flux to P(3HB) synthesis, increasing P(3HB) accumulation in the mutated strain up to 18.6% DCW (Wu *et al.*, 2002). Pathway design by introducing foreign or additional native pathways to alter carbon partitioning between useful product and biomass productions is an interesting approach for engineered production of industrial products. Heterologous expression of isoprene synthase together with non-native mevalonate pathway genes in *Synechocystis* PCC 6803 resulted in a 2.5-fold higher isoprene synthesis in comparison with that of cyanobacteria transformed with isoperen synthase only (Bentley *et al.*, 2014). Although the examples of knocking out a competing pathway and directing carbon flux for desired product synthesis discussed above for enhancing carbon dioxide fixation were demonstrated for productions of isobutyraldehyde, fatty acid, isoprene and plant-type terpenoids (Chapters 8–11), it is likely these ideas can be applied for engineering PHA biosynthesis. Future work using either of the genetic engineering strategies discussed above or a combination would be promising towards realizing the potentials of cyanobacteria for PHA production through light-driven photosynthesis.

PHA biosynthesis driven by photosynthesis using renewable solar energy and atmospheric carbon dioxide as the carbon source is a possible direction for sustainable production in the near future. Owing to cyanobacteria's capability to convert solar energy into biomass, cyanobacteria hold great promise as suitable platforms for this carbon-neutral PHA biosynthesis concept. Thus far, PHA-producing ability has been described in more than 30 native producers and genetically engineered strains of cyanobacteria, indicating the potential of cyanobacteria as 'green' hosts for PHA synthesis. Nonetheless, there are still challenges and limitations for utilization of cyanobacteria as cell factories for PHA production at an industrial-scale. Genetic engineering strategies to improve photosynthesis efficiency, to enhance carbon fixation and to direct carbon flux towards desired product synthesis are promising to increase PHA productivity in cyanobacteria. Turning cyanobacteria into robust and efficient PHA production platforms is not an easy task, but the one that attracts a lot of research interest. Tremendous efforts to develop new strategies to address the challenges will make it possible to use cyanobacteria for commercial PHA production.

Acknowledgements

N.-S. Lau acknowledges the postdoctoral fellowship by Universiti Sains Malaysia. K.-K. Sam thanks USM Graduate Assistant Scheme for financial support. We also thank USM–RIKEN partnership for scientific researches.

References

Akiyama, H., Okuhata, H., Onizuka, T., Kanai, S., Hirano, M., Tanaka, S., Sasaki, K., and Miyasaka, H. (2011). Antibiotics-free stable polyhydroxyalkanoate (PHA) production from carbon dioxide by recombinant cyanobacteria. Bioresour. Technol. *102*, 11039–11042. http://dx.doi.org/10.1016/j.biortech.2011.09.058

Allen, M.M. (1984). Cyanobacterial cell inclusions. Annu. Rev. Microbiol. *38*, 1–25. http://dx.doi.org/10.1146/annurev.mi.38.100184.000245

Amara, A.A., and Moawad, H. (2011). PhaC synthases and PHA depolymerases: The enzymes that produce and degrade plastic. IIUM Engineering Journal *12*. 21–37.

Anderson, A.J., and Dawes, E.A. (1990). Occurrence, metabolism, metabolic role, and industrial uses of bacterial polyhydroxyalkanoates. Microbiol. Rev. *54*, 450–472.

Atsumi, S., Hanai, T., and Liao, J.C. (2008). Non-fermentative pathways for synthesis of branched-chain higher alcohols as biofuels. Nature *451*, 86–89. http://dx.doi.org/10.1038/nature06450

Atsumi, S., Higashide, W., and Liao, J.C. (2009). Direct photosynthetic recycling of carbon dioxide to isobutyraldehyde. Nat. Biotechnol. *27*, 1177–1180. http://dx.doi.org/10.1038/nbt.1586

Balaji, S., Gopi, K., and Muthuvelan, B. (2013). A review on production of poly β hydroxybutyrates from cyanobacteria for the production of bio plastics. Algal Res. *2*, 278–285. http://dx.doi.org/10.1016/j.algal.2013.03.002

Barten, R., and Lill, H. (1995). DNA-uptake in the naturally competent cyanobacterium, *Synechocystis* sp. PCC 6803. FEMS Microbiol. Lett. *129*, 83–88. http://dx.doi.org/10.1111/j.1574-6968.1995.tb07561.x

Bentley, F.K., Zurbriggen, A., and Melis, A. (2014). Heterologous expression of the mevalonic acid pathway in cyanobacteria enhances endogenous carbon partitioning to isoprene. Mol. Plant *7*, 71–86. http://dx.doi.org/10.1093/mp/sst134

Bhati, R., and Mallick, N. (2012). Production and characterization of poly(3-hydroxybutyrate-co-3-hydroxyvalerate) co-polymer by a N_2-fixing cyanobacterium, *Nostoc muscorum* Agardh. J. Chem. Technol. Biotechnol. *87*, 505–512. http://dx.doi.org/10.1002/jctb.2737

Bhati, R., and Mallick, N. (2015). Poly(3-hydroxybutyrate-co-3-hydroxyvalerate) copolymer production by the diazotrophic cyanobacterium *Nostoc muscorum* Agardh: process optimization and polymer characterization. Algal Res. *7*, 78–85. http://dx.doi.org/10.1016/j.algal.2014.12.003

Blankenship, R.E., and Chen, M. (2013). Spectral expansion and antenna reduction can enhance photosynthesis for energy production. Curr. Opin. Chem. Biol. *17*, 457–461. http://dx.doi.org/10.1016/j.cbpa.2013.03.031

Blankenship, R.E., Tiede, D.M., Barber, J., Brudvig, G.W., Fleming, G., Ghirardi, M., Gunner, M.R., Junge, W., Kramer, D.M., Melis, A., *et al.* (2011). Comparing photosynthetic and photovoltaic efficiencies and recognizing the potential for improvement. Science *332*, 805–809. http://dx.doi.org/10.1126/science.1200165

Bonartsev, A.P., Myshkina, V.L., Nikolaeva, D.A., Furina, E.K., Makhina, T.A., Livshits, V.A., Boskhomdzhiev, A.P., Ivanov, E.A., Iordanskii, A.L., and Bonartseva, G.A. (2007). Biosynthesis, biodegradation, and application of poly (3-hydroxybutyrate) and its copolymers-natural polyesters produced by diazotrophic bacteria. In Communicating Current Research and Educational Topics and Trends in Applied Microbiology, A. Mendez-Vilas, ed. (Badajoz, Spain: Formatex), pp. 295–307.

Brigham, C.J., and Sinskey, A.J. (2012). Applications of polyhydroxyalkanoates in the medical industry. Int. J. Biotechnol. Wellness Ind. *1*, 52–60. http://dx.doi.org/10.6000/1927-3037.2012.01.01.03

Burja, A.M., Banaigs, B., Abou-Mansour, E., Grant Burgess, J., and Wright, P.C. (2001). Marine cyanobacteria – a prolific source of natural products. Tetrahedron *57*, 9347–9377. http://dx.doi.org/10.1016/S0040-4020(01)00931-0

Campbell, J., Stevens, S.E., and Balkwill, D.L. (1982). Accumulation of poly-beta-hydroxybutyrate in *Spirulina platensis*. J. Bacteriol. *149*, 361–363.

Carr, N.G. (1966). The occurrence of poly-beta-hydroxybutyrate in the blue-green alga, *Chlorogloeafritschii*. Biochim. Biophys. Acta *120*, 308–310. http://dx.doi.org/10.1016/0926-6585(66)90353-0

Chee, J.Y., Yoga, S.S., Lau, N.S., Ling, S.C., Abed, R.M.M., and Sudesh, K. (2010). Bacterially produced polyhydroxyalkanoate (PHA): converting renewable resources into bioplastics. In Current Research, Technology and Education Topics in Applied Microbiology and Applied Biotechnology, A. Mendez-Vilas, ed. (Badajoz, Spain: Formatex), pp. 1395–1404.

Chen, G.Q., and Hajnal, I. (2015). The 'PHAome'. Trends Biotechnol. *33*, 559–564. http://dx.doi.org/10.1016/j.tibtech.2015.07.006

Chen, G.Q., Hajnal, I., Wu, H., Lv, L., and Ye, J. (2015a). Engineering biosynthesis mechanisms for diversifying polyhydroxyalkanoates. Trends Biotechnol. *33*, 565–574. http://dx.doi.org/10.1016/j.tibtech.2015.07.007

Chen, G.Q., and Patel, M.K. (2012). Plastics derived from biological sources: present and future: a technical and environmental review. Chem. Rev. *112*, 2082–2099. http://dx.doi.org/10.1021/cr200162d

Chen, G.Q., and Wu, Q. (2005). The application of polyhydroxyalkanoates as tissue engineering materials. Biomaterials *26*, 6565–6578. http://dx.doi.org/10.1016/j.biomaterials.2005.04.036

Chen, G.Q., Zhang, J., and Wang, Y. (2015b). White biotechnology for biopolymers: hydroxyalkanoates and polyhydroxyalkanoates: production and applications. In Industrial Biorefineries & White Biotechnology, A. Pandey, R. Höfer, M. Taherzadeh, K.M. Nampoothiri, and C. Larroche, eds. (Amsterdam, USA: Elsevier), pp. 555–574. http://dx.doi.org/10.1016/B978-0-444-63453-5.00018-5

Churkunti, P.R., Mattson, J., Depcik, C., and Devlin, G. (2016). Combustion analysis of pyrolysis end of life plastic fuel blended with ultra low sulfur diesel. Fuel Process. Technol. *142*, 212–218. http://dx.doi.org/10.1016/j.fuproc.2015.10.021

Codd, G.A., Morrison, L.F., and Metcalf, J.S. (2005). Cyanobacterial toxins: risk management for health protection. Toxicol. Appl. Pharmacol. *203*, 264–272. http://dx.doi.org/10.1016/j.taap.2004.02.016

De Philippis, R., Ena, A., Guastiini, M., Sili, C., and Vincenzini, M. (1992). Factors affecting poly-β-hydroxybutyrate accumulation in cyanobacteria and in purple non-sulfur bacteria. FEMS Microbiol. Lett. *103*, 187–194. http://dx.doi.org/10.1111/j.1574-6968.1992.tb05837.x

Deshnium, P., Los, D.A., Hayashi, H., Mustardy, L., and Murata, N. (1995). Transformation of *Synechococcus* with a gene for choline oxidase enhances tolerance to salt stress. Plant Mol. Biol. *29*, 897–907. http://dx.doi.org/10.1007/BF00014964

Dismukes, G.C., Carrieri, D., Bennette, N., Ananyev, G.M., and Posewitz, M.C. (2008). Aquatic phototrophs: efficient alternatives to land-based crops for biofuels. Curr. Opin. Biotechnol. *19*, 235–240. http://dx.doi.org/10.1016/j.copbio.2008.05.007

Doi, Y. (1990). Microbial polyesters (New York, UK: VCH). http://dx.doi.org/10.1002/pola.1991.080290916

Doudoroff, M., and Stanier, R.Y. (1959). Role of poly-beta-hydroxybutyric acid in the assimilation of organic carbon by bacteria. Nature *183*, 1440–1442.

Ducat, D.C., Way, J.C., and Silver, P.A. (2011). Engineering cyanobacteria to generate high-value products. Trends Biotechnol. *29*, 95–103. http://dx.doi.org/10.1016/j.tibtech.2010.12.003

Eberly, J.O., and Ely, R.L. (2012). Photosynthetic accumulation of carbon storage compounds under CO_2 enrichment by the thermophilic cyanobacterium *Thermosynechococcus elongatus*. J. Ind. Microbiol. Biotechnol. *39*, 843–850. http://dx.doi.org/10.1007/s10295-012-1092-2

Ferjani, A., Mustardy, L., Sulpice, R., Marin, K., Suzuki, I., Hagemann, M., and Murata, N. (2003). Glucosylglycerol, a compatible solute, sustains cell division under salt stress. Plant Physiol. *131*, 1628–1637. http://dx.doi.org/10.1104/pp.102.017277

Gao, X., Gao, F., Liu, D., Zhang, H., Nie, X., and Yang, C. (2016a). Engineering the methylerythritol phosphate pathway in cyanobacteria for photosynthetic isoprene production from CO_2. Energy Environ. Sci. *9*, 1400–1411. http://dx.doi.org/10.1039/C5EE03102H

Gao, X., Sun, T., Pei, G., Chen, L., and Zhang, W. (2016b). Cyanobacterial chassis engineering for enhancing production of biofuels and chemicals. Appl. Microbiol. Biotechnol. *100*, 3401–3413. http://dx.doi.org/10.1007/s00253-016-7374-2

Gao, Z., Zhao, H., Li, Z., Tan, X., and Lu, X. (2012). Photosynthetic production of ethanol from carbon dioxide in genetically engineered cyanobacteria. Energy Environ. Sci. *5*, 9857–9865. http://dx.doi.org/10.1039/C2EE22675H

Gregory, M.R. (2009). Environmental implications of plastic debris in marine settings—entanglement, ingestion, smothering, hangers-on, hitch-hiking and alien invasions. Phil. Trans. R. Soc. B. *364*, 2013–2025. http://dx.doi.org/10.1098/rstb.2008.0265

Grossman, A.R., Schaefer, M.R., Chiang, G.G., and Collier, J.L. (1993). The phycobilisome, a light-harvesting complex responsive to environmental conditions. Microbiol. Mol. Biol. Rev. *57*, 725–749.

Haase, S.M., Huchzermeyer, B., and Rath, T. (2012). PHB accumulation in *Nostoc muscorum* under different carbon stress situations. J. Appl. Phycol. *24*, 157–162. http://dx.doi.org/10.1007/s10811-011-9663-6

Hai, T., Hein, S., and Steinbüchel, A. (2001). Multiple evidence for widespread and general occurrence of type-III PHA synthases in cyanobacteria and molecular characterization of the PHA synthases from two thermophilic cyanobacteria: *Chlorogloeopsis fritschii* PCC 6912 and *Synechococcus* sp. strain MA19. Microbiology *147*, 3047–3060. http://dx.doi.org/10.1099/00221287-147-11-3047

Halden, R.U. (2010). Plastics and health risks. Annu. Rev. Public. Health *31*, 179–194. http://dx.doi.org/10.1146/annurev.publhealth.012809.103714

Hasunuma, T., Matsuda, M., Senga, Y., Aikawa, S., Toyoshima, M., Shimakawa, G., Miyake, C., and Kondo, A. (2014). Overexpression of *flv3* improves photosynthesis in the cyanobacterium *Synechocystis* sp. PCC6803 by enhancement of alternative electron flow. Biotechnol. Biofuels 7, 493. http://dx.doi.org/10.1186/s13068-014-0183-x

Hein, S., Tran, H., and Steinbüchel, A. (1998). *Synechocystis* sp. PCC6803 possesses a two-component polyhydroxyalkanoic acid synthase similar to that of anoxygenic purple sulfur bacteria. Arch. Microbiol. *170*, 162–170. http://dx.doi.org/10.1007/s002030050629

Hernández-Prieto, M.A., Semeniuk, T.A., and Futschik, M.E. (2014). Toward a systems-level understanding of gene regulatory, protein interaction, and metabolic networks in cyanobacteria. Front. Genet. *5*, 191. http://dx.doi.org/10.3389/fgene.2014.00191

Hondo, S., Takahashi, M., Osanai, T., Matsuda, M., Hasunuma, T., Tazuke, A., Nakahira, Y., Chohnan, S., Hasegawa, M., and Asayama, M. (2015). Genetic engineering and metabolite profiling for overproduction of polyhydroxybutyrate in cyanobacteria. J. Biosci. Bioeng. *120*, 510–517. http://dx.doi.org/10.1016/j.jbiosc.2015.03.004

Iijima, H., Nakaya, Y., Kuwahara, A., Hirai, M.Y., and Osanai, T. (2015). Seawater cultivation of freshwater cyanobacterium *Synechocystis* sp. PCC 6803 drastically alters amino acid composition and glycogen metabolism. Front. Microbiol. *6*, 326. http://dx.doi.org/10.3389/fmicb.2015.00326

Jajesniak, P., Ali, H.E.M.O., and Wong, T.S. (2014). Carbon dioxide capture and utilization using biological systems: opportunities and challenges. J. Bioprocess Biotech. *4*, 155. http://dx.doi.org/10.4172/2155-9821.1000155

Jau, M.H., Yew, S.P., Toh, P.S., Chong, A.S., Chu, W.L., Phang, S.M., Najimudin, N., and Sudesh, K. (2005). Biosynthesis and mobilization of poly(3-hydroxybutyrate) [P(3HB)] by *Spirulina platensis*. Int. J. Biol. Macromolec. *36*, 144–151. http://dx.doi.org/10.1016/j.ijbiomac.2005.05.002

Jendrossek, D., and Handrick, R. (2002). Microbial degradation of polyhydroxyalkanoates. Annu. Rev. Microbiol. *56*, 403–432. http://dx.doi.org/10.1146/annurev.micro.56.012302.160838

Jensen, T.E., and Sicko, L.M. (1971). Fine structure of poly-beta-hydroxybutyric acid granules in a blue-green alga, *Chlorogloea fritschii*. J. Bacteriol. *106*, 683–686.

Joseph, A., Aikawa, S., Sasaki, K., Matsuda, F., Hasunuma, T., and Kondo, A. (2014). Increased biomass production and glycogen accumulation in *apcE* gene deleted *Synechocystis* sp. PCC 6803. AMB Express *4*, 17. http://dx.doi.org/10.1186/s13568-014-0017-z

Kaku, N., Hibino, T., Tanaka, Y., Ishikawa, H., Araki, E., Takabe, T., and Takabe, T. (2000). Effects of overexpression of *Escherichia coli katE* and *bet* genes on the tolerance for salt stress in a freshwater cyanobacterium *Synechococcus* sp. PCC 7942. Plant Sci. *159*, 281–288. http://dx.doi.org/10.1016/S0168-9452(00)00353-8

Kessler, B., and Witholt, B. (2001). Factors involved in the regulatory network of polyhydroxyalkanoate metabolism. J. Biotechnol. *86*, 97–104. http://dx.doi.org/10.1016/S0168-1656(00)00404-1

Kim, D.Y., Kim, H.W., Chung, M.G., and Rhee, Y.H. (2007). Biosynthesis, modification, and biodegradation of bacterial medium-chain-length polyhydroxyalkanoates. J. Microbiol. *45*, 87–97

Knoll, A.H., and Barghoorn, E.S. (1977). Archean microfossils showing cell division from the swaziland system of South Africa. Science *198*, 396–398. http://dx.doi.org/10.1126/science.198.4315.396

Knoll, M., Hamm, T.M., Wagner, F., Martinez, V., and Pleiss, J. (2009). The PHA depolymerase engineering database: a systematic analysis tool for the diverse family of polyhydroxyalkanoate (PHA) depolymerases. BMC Bioinformatics *10*, 89. http://dx.doi.org/10.1186/1471-2105-10-89

Knoop, H., Zilliges, Y., Lockau, W., and Steuer, R. (2010). The metabolic network of *Synechocystis* sp. PCC 6803: systemic properties of autotrophic growth. Plant Physiol. *154*, 410–422. http://dx.doi.org/10.1104/pp.110.157198

Knutti, R., and Rogelj, J. (2015). The legacy of our CO_2 emissions: a clash of scientific facts, politics and ethics. Clim. Change *133*, 361–373. http://dx.doi.org/10.1007/s10584-015-1340-3

Kufryk, G.I., Sachet, M., Schmetterer, G., and Vermaas, W.F. (2002). Transformation of the cyanobacterium *Synechocystis* sp. PCC 6803 as a tool for genetic mapping: optimization of efficiency. FEMS Microbiol. Lett. *206*, 215–219. http://dx.doi.org/10.1111/j.1574-6968.2002.tb11012.x

Kulasooriya, S.A. (2011). Cyanobacteria: pioneers of planet earth. Ceylon J. Sci. Biol. Sci. *40*, 71–88. http://dx.doi.org/10.4038/cjsbs.v40i2.3925

Lama, L., Nicolaus, B., Calandrelli, V., Manca, M.C., Romano, I., and Gambacorta, A. (1996). Effect of growth conditions on endo- and exopolymer biosynthesis in *Anabaena cylindrica* 10 C. Phytochemistry *42*, 655–659. http://dx.doi.org/10.1016/0031-9422(95)00985-X

Lau, N.S., Foong, C.P., Kurihara, Y., Sudesh, K., and Matsui, M. (2014). RNA-seq analysis provides insights for understanding photoautotrophic polyhydroxyalkanoate production in recombinant *Synechocystis* sp. PLoS ONE *9*, e86368. http://dx.doi.org/10.1371/journal.pone.0086368

Lau, N.S., Matsui, M., and Amirul, A.A. (2015). Cyanobacteria: photoautotrophic microbial factories for the sustainable synthesis of industrial products. Biomed Res. Int. *2015*, 754934. http://dx.doi.org/10.1155/2015/754934

Lee, I.Y., Kim, M.K., Chang, H.N., and Park, Y.H. (1995). Regulation of poly-β-hydroxybutyrate biosynthesis by nicotinamide nucleotide in *Alcaligenes eutrophus*. FEMS Microbiol. Lett. *131*, 35–39. http://dx.doi.org/10.1111/j.1574-6968.1995.tb07750.x

Lee, S.Y. (1996). Bacterial polyhydroxyalkanoates. Biotechnol. Bioeng. *49*, 1–14. http://dx.doi.org/10.1002/(SICI)1097-0290(19960105)49:1<1::AID-BIT1>3.0.CO;2-P

Li, Z., Wakao, S., Fischer, B.B., and Niyogi, K.K. (2009). Sensing and responding to excess light. Annu. Rev. Plant Biol. *60*, 239–260. http://dx.doi.org/10.1146/annurev.arplant.58.032806.103844

Liu, L.N., Chen, X.L., Zhang, Y.Z., and Zhou, B.C. (2005). Characterization, structure and function of linker polypeptides in phycobilisomes of cyanobacteria and red algae: an overview. Biochim. Biophys. Acta *1708*, 133–142. http://dx.doi.org/10.1016/j.bbabio.2005.04.001

Liu, X., and Curtiss, R. (2009). Nickel-inducible lysis system in *Synechocystis* sp. PCC 6803. Proc. Natl. Acad. Sci. U.S.A. *106*, 21550–21554. http://dx.doi.org/10.1073/pnas.0911953106

Liu, X., Sheng, J., and Curtiss, R. (2011). Fatty acid production in genetically modified cyanobacteria. Proc. Natl. Acad. Sci. U.S.A. *108*, 6899–6904. http://dx.doi.org/10.1073/pnas.1103014108

Ma, A.T., Schmidt, C.M., and Golden, J.W. (2014). Regulation of gene expression in diverse cyanobacterial species by using theophylline-responsive riboswitches. Appl. Environ. Microbiol. *80*, 6704–6713. http://dx.doi.org/10.1128/AEM.01697-14

Madison, L.L., and Huisman, G.W. (1999). Metabolic engineering of poly(3-hydroxyalkanoates): from DNA to plastic. Microbiol. Mol. Biol. Rev. *63*, 21–53.

Mallick, N., Gupta, S., Panda, B., and Sen, R. (2007). Process optimization for poly(3-hydroxybutyrate-co-3-hydroxyvalerate) co-polymer production by *Nostoc muscorum*. Biochem. Eng. J. *37*, 125–130. http://dx.doi.org/10.1016/j.bej.2007.04.002

Melis, A. (2009). Solar energy conversion efficiencies in photosynthesis: minimizing the chlorophyll antennae to maximize efficiency. Plant Sci. *177*, 272–280. http://dx.doi.org/10.1016/j.plantsci.2009.06.005

Meng, D.C., Shen, R., Yao, H., Chen, J.C., Wu, Q., and Chen, G.Q. (2014). Engineering the diversity of polyesters. Curr. Opin. Biotechnol. *29*, 24–33. http://dx.doi.org/10.1016/j.copbio.2014.02.013

Mergaert, J., Anderson, C., Wouters, A., Swings, J., and Kersters, K. (1992). Biodegradation of polyhydroxyalkanoates. FEMS Microbiol. Lett. *103*, 317–321. http://dx.doi.org/10.1111/j.1574-6968.1992.tb05853.x

Miyake, K., Abe, K., Ferri, S., Nakajima, M., Nakamura, M., Yoshida, W., Kojima, K., Ikebukuro, K., and Sode, K. (2014). A green-light inducible lytic system for cyanobacterial cells. Biotechnol. Biofuels 7, 1–8. http://dx.doi.org/10.1186/1754-6834-7-56

Miyake, M., Erata, M., and Asada, Y. (1996). A thermophilic cyanobacterium, *Synechococcus* sp. MA19, capable of accumulating poly-β-hydroxybutyrate. J. Ferment. Bioeng. *82*, 512–514. http://dx.doi.org/10.1016/S0922-338X(97)86995-4

Miyake, M., Kataoka, K., Shirai, M., and Asada, Y. (1997). Control of poly-beta-hydroxybutyrate synthase mediated by acetyl phosphate in cyanobacteria. J. Bacteriol. *179*, 5009–5013.

Miyasaka, H., Nakano, H., Akiyama, H., Kanai, S., and Hirano, M. (1998). Production of PHA (polyhydroxyalkanoate) by genetically engineered marine cyanobacterium. Stud. Surf. Sci. Catal. *114*, 237–242.

Muhammadi, S., Afzal, M., and Hameed, S. (2015). Bacterial polyhydroxyalkanoates-eco-friendly next generation plastic: production, biocompatibility, biodegradation, physical properties and applications. Green Chem. Lett. Rev. *8*, 56–77. http://dx.doi.org/10.1080/17518253.2015.1109715

Mulkidjanian, A.Y., Koonin, E.V., Makarova, K.S., Mekhedov, S.L., Sorokin, A., Wolf, Y.I., Dufresne, A., Partensky, F., Burd, H., Kaznadzey, D., *et al.* (2006). The cyanobacterial genome core and the origin of photosynthesis. Proc. Natl. Acad. Sci. U.S.A. *103*, 13126–13131. http://dx.doi.org/10.1073/pnas.0605709103

Narayan, O.P., Kumari, N., Bhargava, P., Rajaram, H., and Rai, L.C. (2016). A single gene *all3940* (Dps) overexpression in *Anabaena* sp. PCC 7120 confers multiple abiotic stress tolerance via proteomic alterations. Funct. Integr. Genomics *16*, 67–78. http://dx.doi.org/10.1007/s10142-015-0467-7

Nishioka, M., Nakai, K., Miyake, M., Asada, Y., and Taya, M. (2001). Production of poly-β-hydroxybutyrate by thermophilic cyanobacterium, *Synechococcus* sp. MA19, under phosphate-limited conditions. Biotechnol. Lett. *23*, 1095–1099. http://dx.doi.org/10.1023/A:1010551614648

Nogales, J., Gudmundsson, S., and Thiele, I. (2013). Toward systems metabolic engineering in cyanobacteria: opportunities and bottlenecks. Bioengineered *4*, 158–163. http://dx.doi.org/10.4161/bioe.22792

Nomura, M., Ishitani, M., Takabe, T., Rai, A.K., and Takabe, T. (1995). *Synechococcus* sp. PCC7942 transformed with *Escherichia coli bet* genes produces glycine betaine from choline and acquires resistance to salt stress. Plant Physiol. *107*, 703–708.

Oeding, V., and Schlegel, H.G. (1973). β-Ketothiolase from *Hydrogenomonas eutropha* H16 and its significance in the regulation of poly-β-hydroxybutyrate metabolism.

Biochem. J. *134*, 239–248. http://dx.doi.org/10.1042/bj1340239

Olmedo-Verd, E., Flores, E., Herrero, A., and Muro-Pastor, A.M. (2005). HetR-dependent and -independent expression of heterocyst-related genes in an *Anabaena* strain overproducing the NtcA transcription factor. J. Bacteriol. *187*, 1985–1991. http://dx.doi.org/10.1128/JB.187.6.1985-1991.2005

Ort, D.R., Merchant, S.S., Alric, J., Barkan, A., Blankenship, R.E., Bock, R., Croce, R., Hanson, M.R., Hibberd, J.M., Long, S.P., *et al.* (2015). Redesigning photosynthesis to sustainably meet global food and bioenergy demand. Proc. Natl. Acad. Sci. U.S.A. *112*, 8529–8536. http://dx.doi.org/10.1073/pnas.1424031112

Osanai, T., Numata, K., Oikawa, A., Kuwahara, A., Iijima, H., Doi, Y., Tanaka, K., Saito, K., and Hirai, M.Y. (2013). Increased bioplastic production with an RNA polymerase sigma factor *SigE* during nitrogen starvation in *Synechocystis* sp. PCC 6803. DNA Res. *20*, 525–535. http://dx.doi.org/10.1093/dnares/dst028

Paerl, H.W., Pinckney, J.L., and Steppe, T.F. (2000). Cyanobacterial-bacterial mat consortia: examining the functional unit of microbial survival and growth in extreme environments. Environ. Microbiol. *2*, 11–26. http://dx.doi.org/10.1046/j.1462-2920.2000.00071.x

Panda, B., Jain, P., Sharma, L., and Mallick, N. (2006). Optimization of cultural and nutritional conditions for accumulation of poly-β-hydroxybutyrate in *Synechocystis* sp. PCC 6803. Bioresour. Technol. *97*, 1296–1301. http://dx.doi.org/10.1016/j.biortech.2005.05.013

Panda, B., and Mallick, N. (2007). Enhanced poly-β-hydroxybutyrate accumulation in a unicellular cyanobacterium, *Synechocystis* sp. PCC 6803. Lett. Appl. Microbiol. *44*, 194–198. http://dx.doi.org/10.1111/j.1472-765X.2006.02048.x

Park, S.J., and Lee, S.Y. (2004). Biosynthesis of poly (3-hydroxybutyrate-co-3-hydroxyalkanoates) by metabolically engineered *Escherichia coli* strains. Appl. Biochem. Biotechnol. *114*, 335–346. http://dx.doi.org/10.1385/ABAB:114:1-3:335

Park, S.J., Kim, T.W., Kim, M.K., Lee, S.Y., and Lim, S.C. (2012). Advanced bacterial polyhydroxyalkanoates: towards a versatile and sustainable platform for unnatural tailor-made polyesters. Biotechnol. Adv. *30*, 1196–1206. http://dx.doi.org/10.1016/j.biotechadv.2011.11.007

Quintana, N., Van der Kooy, F., Van de Rhee, M.D., Voshol, G.P., and Verpoorte, R. (2011). Renewable energy from cyanobacteria: energy production optimization by metabolic pathway engineering. Appl. Microbiol. Biotechnol. *91*, 471–490. http://dx.doi.org/10.1007/s00253-011-3394-0

Rehm, B.H. (2003). Polyester synthases: natural catalysts for plastics. Biochem. J. *376*, 15–33. http://dx.doi.org/10.1042/BJ20031254

Rippka, R., Neilson, A., Kunisawa, R., and Cohen-Bazire, G. (1971). Nitrogen fixation by unicellular blue-green algae. Arch. Mikrobiol. *76*, 341–348. http://dx.doi.org/10.1007/BF00408530

Rittmann, B.E. (2008). Opportunities for renewable bioenergy using microorganisms. Biotechnol. Bioeng. *100*, 203–212. http://dx.doi.org/10.1002/bit.21875

Ruffing, A.M. (2011). Engineered cyanobacteria: teaching an old bug new tricks. Bioeng. Bugs 2, 136–149.

Ruffing, A.M. (2014). Improved free fatty acid production in cyanobacteria with *Synechococcus* sp. PCC 7002 as host. Front. Bioeng. Biotechnol. *2*, 17. http://dx.doi.org/10.3389/fbioe.2014.00017

Samantaray, S., and Mallick, N. (2012). Production and characterization of poly-β-hydroxybutyrate (PHB) polymer from *Aulosira fertilissima*. J. Appl. Phycol. *24*, 803–814. http://dx.doi.org/10.1007/s10811-011-9699-7

Samantaray, S., and Mallick, N. (2014). Production of poly (3-hydroxybutyrate-co-3-hydroxyvalerate) co-polymer by the diazotrophic cyanobacterium *Aulosira fertilissima* CCC 444. J. Appl. Phycol. *26*, 237–245. http://dx.doi.org/10.1007/s10811-013-0073-9

Schopf, J.W. (1993). Microfossils of the early archean apex chert: new evidence of the antiquity of life. Science *260*, 640–646. http://dx.doi.org/10.1126/science.260.5108.640

Senior, P.J., and Dawes, E.A. (1973). The regulation of poly-beta-hydroxybutyrate metabolism in *Azotobacter beijerinckii*. Biochem. J. *134*, 225–238. http://dx.doi.org/10.1042/bj1340225

Sharma, L., Kumar Singh, A., Panda, B., and Mallick, N. (2007). Process optimization for poly-β-hydroxybutyrate production in a nitrogen fixing cyanobacterium, *Nostoc muscorum* using response surface methodology. Bioresour. Technol. *98*, 987–993. http://dx.doi.org/10.1016/j.biortech.2006.04.016

Sharma, L., and Mallick, N. (2005). Accumulation of poly-β-hydroxybutyrate in *Nostoc muscorum*: regulation by pH, light–dark cycles, N and P status and carbon sources. Bioresour. Technol. *96*, 1304–1310. http://dx.doi.org/10.1016/j.biortech.2004.10.009

Shimao, M. (2001). Biodegradation of plastics. Curr. Opin. Biotechnol. *12*, 242–247. http://dx.doi.org/10.1016/S0958-1669(00)00206-8

Shrivastav, A., Kim, H.Y., and Kim, Y.R. (2013). Advances in the applications of polyhydroxyalkanoate nanoparticles for novel drug delivery system. Biomed Res. Int. *2013*, 581684. http://dx.doi.org/10.1155/2013/581684

Shrivastav, A., Mishra, S.K., and Mishra, S. (2010). Polyhydroxyalkanoate (PHA) synthesis by *Spirulina subsalsa* from Gujarat coast of India. Int. J. Biol. Macromolec. *46*, 255–260. http://dx.doi.org/10.1016/j.ijbiomac.2010.01.001

Siddiqui, P.J., Bergman, B., Björkman, P.O., and Carpenter, E.J. (1992). Ultrastructural and chemical assessment of poly-β-hydroxybutyric acid in the marine cyanobacterium *Trichodesmium thiebautii*. FEMS Microbiol. Lett. *73*, 143–148. http://dx.doi.org/10.1111/j.1574-6968.1992.tb05304.x

Singh, V., Chaudhary, D.K., Mani, I., and Dhar, P.K. (2016). Recent advances and challenges of the use of cyanobacteria towards the production of biofuels. Renew. Sust. Energ. Rev. *60*, 1–10. http://dx.doi.org/10.1016/j.rser.2016.01.099

Smith, A.J. (1983). Modes of cyanobacterial carbon metabolism. Ann. Microbiol. *134B*, 93–113. http://dx.doi.org/10.1016/S0769-2609(83)80099-4

Solomon, S., Plattner, G.K., Knutti, R., and Friedlingstein, P. (2009). Irreversible climate change due to carbon dioxide emissions. Proc. Natl. Acad. Sci. U.S.A. *106*, 1704–1709. http://dx.doi.org/10.1073/pnas.0812721106

Spreitzer, R.J., and Salvucci, M.E. (2002). Rubisco: structure, regulatory interactions, and possibilities for a better enzyme. Annu. Rev. Plant Biol. *53*, 449–475. http://dx.doi.org/10.1146/annurev.arplant.53.100301.135233

Stal, L.J. (1992). Poly(hydroxyalkanoate) in cyanobacteria: an overview. FEMS Microbiol. Rev. *9*, 169–180. http://dx.doi.org/10.1111/j.1574-6968.1992.tb05835.x

Stal, L.J. (1995). Physiological ecology of cyanobacteria in microbial mats and other communities. New Phytol. *131*, 1–32. http://dx.doi.org/10.1111/j.1469-8137.1995.tb03051.x

Stanier, R.Y., and Cohen-Bazire, G. (1977). Phototrophic prokaryotes: the cyanobacteria. Annu. Rev. Microbiol. *31*, 225–274. http://dx.doi.org/10.1146/annurev.mi.31.100177.001301

Steinbüchel, A., and Hein, S. (2001). Biochemical and molecular basis of microbial synthesis of polyhydroxyalkanoates in microorganisms. Biotechnol. In Biopolyesters, W. Babel, and A. Steinbüchel, eds. (New York, US: Springer Berlin Heidelberg), pp. 81–123. http://dx.doi.org/10.1007/3-540-40021-4_3

Steinbüchel, A., and Schlegel, H.G. (1991). Physiology and molecular genetics of poly(β-hydroxyalkanoic acid) synthesis in *Alcaligenes eutrophus*. Mol. Microbiol. *5*, 535–542. http://dx.doi.org/10.1111/j.1365-2958.1991.tb00725.x

Sudesh, K., Abe, H., and Doi, Y. (2000). Synthesis, structure and properties of polyhydroxyalkanoates: biological polyesters. Prog. Polym. Sci. *25*, 1503–1555. http://dx.doi.org/10.1016/S0079-6700(00)00035-6

Sudesh, K., Taguchi, K., and Doi, Y. (2001). Can cyanobacteria be a potential PHA producer? Riken Review 75–76.

Sudesh, K., Taguchi, K., and Doi, Y. (2002). Effect of increased PHA synthase activity on polyhydroxyalkanoates biosynthesis in *Synechocystis* sp. PCC6803. Int. J. Biol. Macromolec. *30*, 97–104. http://dx.doi.org/10.1016/S0141-8130(02)00010-7

Suzuki, T., Miyake, M., Tokiwa, Y., Saegusa, H., Saito, T., and Asada, Y. (1996). A recombinant cyanobacterium that accumulates poly-(hydroxybutyrate). Biotechnol. Lett. *18*, 1047–1050. http://dx.doi.org/10.1007/BF00129729

Takahashi, H., Miyake, M., Tokiwa, Y., and Asada, Y. (1998). Improved accumulation of poly-3- hydroxybutyrate by a recombinant cyanobacterium. Biotechnol. Lett. *20*, 183–186. http://dx.doi.org/10.1023/A:1005392827791

Taroncher-Oldenburg, G., Nishina, K., and Stephanopoulos, G. (2000). Identification and analysis of the polyhydroxyalkanoate-specific β-ketothiolase and acetoacetyl Coenzyme A reductase genes in the cyanobacterium *Synechocystis* sp. strain PCC 6803. Appl. Environ. Microbiol. *66*, 4440–4448. http://dx.doi.org/10.1128/AEM.66.10.4440-4448.2000

Taton, A., Lis, E., Adin, D.M., Dong, G., Cookson, S., Kay, S.A., Golden, S.S., and Golden, J.W. (2012). Gene transfer in *Leptolyngbya* sp. strain BL0902, a cyanobacterium suitable for production of biomass and bioproducts. PLoS ONE *7*, e30901. http://dx.doi.org/10.1371/journal.pone.0030901

Taton, A., Unglaub, F., Wright, N.E., Zeng, W.Y., Paz-Yepes, J., Brahamsha, B., Palenik, B., Peterson, T.C., Haerizadeh, F., Golden, S.S., *et al.* (2014). Broad-host-range vector system for synthetic biology and biotechnology in cyanobacteria. Nucleic Acids Res. *42*, e136. http://dx.doi.org/10.1093/nar/gku673

Ten, L., Chae, S., and Yoo, S.A. (2015). Production of poly (3-hydroxybutyrate) by cyanobacterium *Anabaena* sp. BD47. Chem. Nat. Compd. *51*, 350–351. http://dx.doi.org/10.1007/s10600-015-1279-7

Thompson, R.C., Moore, C.J., vom Saal, F.S., and Swan, S.H. (2009). Plastics, the environment and human health: current consensus and future trends. Philos. Trans. R. Soc. Lond. B. Biol. Sci. *364*, 2153–2166. http://dx.doi.org/10.1098/rstb.2009.0053

Tian, D., Dong, W., Yan, X., Chou, J., Yang, S., Wei, T., Zhang, H., Guo, Y., Wen, X., and Yang, Z. (2016). Climate system responses to a common emission budget of carbon dioxide. J. Climate *29*, 2433–2442. http://dx.doi.org/10.1175/JCLI-D-15-0213.1

Ting, C.S., Rocap, G., King, J., and Chisholm, S.W. (2002). Cyanobacterial photosynthesis in the oceans: the origins and significance of divergent light-harvesting strategies. Trends Microbiol. *10*, 134–142. http://dx.doi.org/10.1016/S0966-842X(02)02319-3

Toh, P.S.Y., Jau, M.H., Yew, S.P., Abed, R.M.M., and Sudesh, K. (2008). Comparison of polyhydroxyalkanoates biosynthesis, mobilization and the effects on cellular morphology in *Spirulina platensis* and *Synechocystis* sp. UNIWG. J. Biosci. *19*, 21–38.

Tokiwa, Y., Calabia, B.P., Ugwu, C.U., and Aiba, S. (2009). Biodegradability of plastics. Int. J. Mol. Sci. *10*, 3722–3742. http://dx.doi.org/10.3390/ijms10093722

Tsinoremas, N.F., Kutach, A.K., Strayer, C.A., and Golden, S.S. (1994). Efficient gene transfer in *Synechococcus* sp. strains PCC 7942 and PCC 6301 by interspecies conjugation and chromosomal recombination. J. Bacteriol. *176*, 6764–6768.

Tsuge, T. (2002). Metabolic improvements and use of inexpensive carbon sources in microbial production of polyhydroxyalkanoates. J. Biosci. Bioeng. *94*, 579–584. http://dx.doi.org/10.1016/S1389-1723(02)80198-0

Verlinden, R.A.J., Hill, D.J., Kenward, M.A., Williams, C.D., and Radecka, I. (2007). Bacterial synthesis of biodegradable polyhydroxyalkanoates. J. Appl. Microbiol. *102*, 1437–1449. http://dx.doi.org/10.1111/j.1365-2672.2007.03335.x

Vermaas, W. (1996). Molecular genetics of the cyanobacterium *Synechocystis* sp. PCC 6803: principles and possible biotechnology applications. J. Appl. Phycol. *8*, 263–273. http://dx.doi.org/10.1007/BF02178569

Vermaas, W.F.J. (2001). Photosynthesis and respiration in cyanobacteria. Encyclopedia of Life Sciences, 1–7. http://dx.doi.org/10.1038/npg.els.0001670

Volova, T.G., Boyandin, A.N., and Prudnikova, S.V. (2015). Biodegradation of polyhydroxyalkanoates in natural soils. J. Siberian Fed. Univ. Biol. *8*, 152. http://dx.doi.org/10.17516/1997-1389-2015-8-2-152-167

von Alvensleben, N., Stookey, K., Magnusson, M., and Heimann, K. (2013). Salinity tolerance of *Picochlorum atomus* and the use of salinity for contamination control by the freshwater cyanobacterium *Pseudanabaena limnetica*. PLoS ONE *8*, e63569. http://dx.doi.org/10.1371/journal.pone.0063569

Waditee-Sirisattha, R., Singh, M., Kageyama, H., Sittipol, D., Rai, A.K., and Takabe, T. (2012). *Anabaena* sp.

PCC7120 transformed with glycine methylation genes from *Aphanothece halophytica* synthesized glycine betaine showing increased tolerance to salt. Arch. Microbiol. *194*, 909–914. http://dx.doi.org/10.1007/s00203-012-0824-z

Waditee, R., Hibino, T., Nakamura, T., Incharoensakdi, A., and Takabe, T. (2002). Overexpression of a Na+/H+ antiporter confers salt tolerance on a freshwater cyanobacterium, making it capable of growth in sea water. Proc. Natl. Acad. Sci. U.S.A. *99*, 4109–4114. http://dx.doi.org/10.1073/pnas.052576899

Wang, B., Wang, J., Zhang, W., and Meldrum, D.R. (2012). Application of synthetic biology in cyanobacteria and algae. Front. Microbiol. *3*, 344. http://dx.doi.org/10.3389/fmicb.2012.00344

Wang, Y., Sun, T., Gao, X., Shi, M., Wu, L., Chen, L., and Zhang, W. (2016). Biosynthesis of platform chemical 3-hydroxypropionic acid (3-HP) directly from CO_2 in cyanobacterium *Synechocystis* sp. PCC 6803. Metab. Eng. *34*, 60–70. http://dx.doi.org/10.1016/j.ymben.2015.10.008

Wang, Y., Yin, J., and Chen, G.Q. (2014). Polyhydroxyalkanoates, challenges and opportunities. Curr. Opin. Biotechnol. *30*, 59–65. http://dx.doi.org/10.1016/j.copbio.2014.06.001

Whitton, B.A. (1992). Diversity, ecology, and taxonomy of the cyanobacteria. In Photosynthetic Prokaryotes, N.H. Mann, and N.G. Carr, eds. (New York, UK: Plenum Press), pp. 1–51. http://dx.doi.org/10.1007/978-1-4757-1332-9_1

Whitton, B.A., and Potts, M. (2000). Introduction to the cyanobacteria. In Ecology of Cyanobacteria II: Their Diversity in Space and Time, A.B. Whitton, ed. (Dordrecht, The Netherlands: Kluwer Academic), pp. 1–13. http://dx.doi.org/10.1007/0-306-46855-7

Wu, G.F., Shen, Z.Y., and Wu, Q.Y. (2002). Modification of carbon partitioning to enhance PHB production in *Synechocystis* sp. PCC6803. Enzyme Microb. Technol. *30*, 710–715. http://dx.doi.org/10.1016/S0141-0229(02)00044-3

Xiao, Z., Zhang, Y., Xi, L., Huo, F., Zhao, J.Y., and Li, J. (2015). Thermophilic production of polyhydroxyalkanoates by a novel *Aneurinibacillus* strain isolated from Gudao oilfield, China. J. Basic Microbiol. *55*, 1125–1133. http://dx.doi.org/10.1002/jobm.201400843

Xu, Y., Alvey, R.M., Byrne, P.O., Graham, J.E., Shen, G., and Bryant, D.A. (2011). Expression of genes in cyanobacteria: adaptation of endogenous plasmids as platforms for high-level gene expression in *Synechococcus* sp. PCC 7002. Methods Mol. Biol. *684*, 273–293. http://dx.doi.org/10.1007/978-1-60761-925-3_21

Yang, C., Zhang, W., Liu, R., Zhang, C., Gong, T., Li, Q., Wang, S., and Song, C. (2013). Analysis of polyhydroxyalkanoate (PHA) synthase gene and PHA-producing bacteria in activated sludge that produces PHA containing 3-hydroxydodecanoate. FEMS Microbiol. Lett. *346*, 56–64. http://dx.doi.org/10.1111/1574-6968.12201

Yu, J., Liberton, M., Cliften, P.F., Head, R.D., Jacobs, J.M., Smith, R.D., Koppenaal, D.W., Brand, J.J., and Pakrasi, H.B. (2015). *Synechococcus elongatus* UTEX 2973, a fast growing cyanobacterial chassis for biosynthesis using light and CO_2. Sci. Rep. *5*, 8132. http://dx.doi.org/10.1038/srep08132

Zess, E.K., Begemann, M.B., and Pfleger, B.F. (2016). Construction of new synthetic biology tools for the control of gene expression in the cyanobacterium *Synechococcus* sp. strain PCC 7002. Biotechnol. Bioeng. *113*, 424–432. http://dx.doi.org/10.1002/bit.25713

Zhang, S., Liu, Y., and Bryant, D.A. (2015). Metabolic engineering of *Synechococcus* sp. PCC 7002 to produce poly-3-hydroxybutyrate and poly-3-hydroxybutyrate-co-4-hydroxybutyrate. Metab. Eng. *32*, 174–183. http://dx.doi.org/10.1016/j.ymben.2015.10.001

Zhu, X.G., Long, S.P., and Ort, D.R. (2010). Improving photosynthetic efficiency for greater yield. Annu. Rev. Plant Biol. *61*, 235–261. http://dx.doi.org/10.1146/annurev-arplant-042809-112206

Index

www.ingramcontent.com/pod-product-compliance
Ingram Content Group UK Ltd.
Pitfield, Milton Keynes, MK11 3LW, UK
UKHW061959041026
15591UKWH00016B/1092

9 781910 190555